Saas-Fee Advanced Course 23
Lecture Notes 1993

A.R. Sandage R.G. Kron M.S. Longair

The Deep Universe

Saas-Fee Advanced Course 23

Lecture Notes 1993

Swiss Society for Astrophysics and Astronomy

Edited by B. Binggeli and R. Buser

With 204 Figures

Springer

Professor A. R. Sandage
The Observatories, 813 Santa Barbara Street, Pasadena, CA 91101-1292, USA

Professor R. G. Kron
University of Chicago, Yerkes Observatory, Box 0258, Williams Bay, WI 53191, USA

Professor M. S. Longair
University of Cambridge, Cavendish Laboratories, Madingley Road,
Cambridge, CB3 0HE, U. K.

Volume Editors:

Dr. B. Binggeli
Professor R. Buser
Universität Basel, Astronomisches Institut, Venusstrasse 7, CH-4102 Binningen, Switzerland

This series is edited on behalf of the Swiss Society for Astrophysics and Astronomy:

Société Suisse d'Astrophysique et d'Astronomie
Observatoire de Genève, ch. des Maillettes 51, CH-1290 Sauverny, Switzerland

Cover photograph: Black-body spectrum of the cosmic microwave background radiation (after COBE measurement) superimposed on a deep image of distant galaxies taken by J. A. Tyson. With permission. Artwork by D. Cerrito.

ISBN 978-3-642-08209-2 e-ISBN 978-3-540-31635-0

CIP data applied for

Preface

Cosmology has not been the subject matter of a Saas-Fee Advanced Course since 1978. That occasion was the memorable course entitled *Observational Cosmology*, with lecturers J.E. Gunn, M.S. Longair, and M.J. Rees. A new course on cosmology seemed overdue, and we subsequently planned and organized, with the endorsement of the Swiss Society of Astrophysics and Astronomy, the present, 23rd Saas-Fee Advanced Course. Probing (drilling) the Universe to greatest depths, in terms of diffuse radiation and discrete objects, with all consequences for our understanding of cosmic evolution, was meant to be the unifying aspect of this course. Hence the title *The Deep Universe*.

We are very happy to have won such competent and brilliant scientists as lecturers. *Malcolm Longair* – Saas-Fee lecturer for the second time! – lectured on cosmological background radiation ("The Physics of Background Radiation"); *Richard Kron* on observations of extremely distant galaxies ("Evolution in the Galaxy Population"); while the Grand Master of Cosmology, *Allan Sandage*, provided the fundaments of it all, also in a historical sense ("Practical Cosmology: Inventing the Past").

The course took place from 29 March to 3 April, 1993, in Les Diablerets, with around 90 participants. The format was as usual, every lecturer giving 9 or 10 lectures of 45 minutes.

This volume contains the written versions of the lectures delivered. They capture much of the liveliness of the oral presentations, giving the reader who could not attend an idea of what he/she missed (but also, for the same reason, giving him/her some compensation). We believe these lecture notes will serve as an invaluable textbook on cosmology for years to come.

Our thanks are due, foremost, to the three lecturers for their commitment and great performance throughout this venture. The Saas-Fee Courses are financed in large part by the Swiss Academy of Natural Sciences, to which the Swiss Society of Astrophysics and Astronomy belongs. We also thank Martin Federspiel for his help with the word processing of the manuscripts, and Daniel Cerrito for preparing the cover photograph.

Basel
November 1994

Bruno Binggeli
Roland Buser

Contents

Practical Cosmology: Inventing the Past

By *Allan Sandage*

Evolution in the Galaxy Population

By *Richard G. Kron*

The Physics of Background Radiation

By *Malcolm S. Longair*

List of Previous Saas-Fee Advanced Courses

1994 Plasma Astrophysics
J.G. Kirk, D.B. Melrose, E.R. Priest

1993 The Deep Universe
A.R. Sandage, R.G. Kron, M.S. Longair

1992 Interacting Binaries
S.N. Shore, M. Livio, E.P.J. van den Heuvel

1991 The Galactic Interstellar Medium
W.B. Burton, B.G. Elmegreen, R. Genzel

1990 Active Galactic Nuclei
R. Blandford, H. Netzer, L. Woltjer

1989 The Milky Way as a Galaxy
G. Gilmore, I. King, P. van der Kruit

1988 Radiation in Moving Gaseous Media
H. Frisch, R.P. Kudritzki, H.W. Yorke

1987 Large Scale Structures in the Universe
A.C. Fabian, M. Geller, A. Szalay

1986 Nucleosynthesis and Chemical Evolution
J. Audouze, C. Chiosi, S.E. Woosley

1985 High Resolution in Astronomy
R.S. Booth, J.W. Brault, A. Labeyrie

1984 Planets, Their Origin, Interior and Atmosphere
D. Gautier, W.B. Hubbard, H. Reeves

1983 Astrophysical Processes in Upper Main Sequence Stars
A.N. Cox, S. Vauclair, J.P. Zahn

*1982 Morphology and Dynamics of Galaxies
J. Binney, J. Kormendy, S.D.M. White

1981 Activity and Outer Atmospheres of the Sun and Stars
F. Praderie, D.S. Spicer, G.L. Withbroe

*1980 Star Formation
J. Appenzeller, J. Lequeux, J. Silk

*1979 Extragalactic High Energy Astrophysics
F. Pacini, C. Ryter, P.A. Strittmatter

*1978 Observational Cosmology
J.E. Gunn, M.S. Longair, M.J. Rees

*1977 Advanced Stages in Stellar Evolution
I. Iben Jr., A. Renzini, D.N. Schramm

---

* Out of print

*1976 GALAXIES
K. Freeman, R.C. Larson, B. Tinsley
*1975 ATOMIC AND MOLECULAR PROCESSES IN ASTROPHYSICS
A. Dalgarno, F. Masnou-Seeuws, R.V.P. McWhirter
*1974 MAGNETOHYDRODYNAMICS
L. Mestel, N.O. Weiss
*1973 DYNAMICAL STRUCTURE AND EVOLUTION OF STELLAR SYSTEMS
G. Contopoulos, M. Hénon, D. Lynden-Bell
*1972 INTERSTELLAR MATTER
N.C. Wrickramasinghe, F.D. Kahn, P.G. Mezger
*1971 THEORY OF THE STELLAR ATMOSPHERES
D. Mihalas, P. Pagel, P. Souffrin

Books up to 1989 may be ordered from: SAAS-FEE COURSES
GENEVA OBSERVATORY
ch. des Maillettes 51
CH-1290 Sauverny, Switzerland

Books from 1990 on may be ordered from Springer-Verlag.

Practical Cosmology: Inventing the Past

Allan Sandage

The Observatories of the Carnegie Institution of Washington, 813 Santa Barbara Street, Pasadena, CA 91101, USA

Preface

The purpose of this series of 10 lectures is to provide a practical basis for calculating distribution functions important for observational cosmology. The lectures constitute a "methods" course, not a theoretical exposé, of which many excellent textbooks exist in the literature. The level is set for the senior year in undergraduate studies or the first year of graduate work as practiced in the current education scheme in the United States. No prior knowledge of cosmology is assumed.

The thesis of the course is that an understanding of the subject can be obtained most easily by both an ability to solve the problems and a knowledge of the history of the development.

The few problems suggested in the text generally consist of the calculation of various distribution functions related to space-time geometry and the expansion properties of Riemannian curved spaces. To this end, the work centers around (a) the geometrical properties of the line element of space-time curvature and the resulting volume elements as a function of the curvature in Lecture 1, (b) the resulting count-apparent magnitude relation for galaxies that have a distributed luminosity function in Lectures 2 and 3, (c) the angular diameter test of the standard model using "standard rods" seen at different distances in the various space-time manifolds in Lecture 4, (d) the classical Hubble diagram of apparent magnitude vs. redshift for various distance regimes and space curvatures in Lecture 5, (e) the Tolman surface brightness test for the reality of the expansion in Lecture 6, (f) the timing test, made by comparing the age of the universe found from stellar evolution (i.e. the ages of the stars) with the expansion age found from the calibration of the absolute extragalactic distance scale, depending thereby on the value of the Hubble constant in Lectures 7 and 8, (g) a description of the experiment through supernovae of type Ia as the most precise determination of the Hubble constant presently available in Lecture 9, and (h) the central problem of observational selection bias leading to false values of the extragalactic distance scale unless properly corrected for in Lecture 10. As much history as is necessary is given to whet the appetite of the student to persuade a reading of the older literature.

As an explanation for some of the footnotes, it was my opportunity to witness a part of the development of the subject that occurred in Southern California after 1948. I was a PhD student of Walter Baade from 1949 through 1952, worked for Edwin Hubble from 1950 to 1953, was trained at Palomar by Humason, took formal courses from H.P Robertson and Fred Hoyle at Caltech, was a post-doctoral fellow with Martin Schwarzschild at Princeton in 1952, had met and talked with Shapley, Lemaître, Heckmann, Gamow, and McCrea, and went fishing with Humason and Don Hendrix, genius optician of the Mount Wilson and Palomar observatories. History sprang from the walls of the office building, from the seminar room in the old "Government Building", and from the shops of the Mount Wilson Observatory in Pasadena where the optics and many of the mechanical parts had been made for the 60 and 100 inch Mount Wilson telescopes and later for the 48 inch Schmidt telescope at Palomar. Some of the footnotes are from these remembrances.

The lectures will, of course, be tedious, bearing as they do on detailed methods. But at the end I would expect the student to be able to predict such things as (1) the galaxy count-apparent magnitude relation for any space curvature, any luminosity function, and any change in galaxy luminosity in the look-back time due to stellar evolution, (2) the expected distribution of redshifts of a random sample of field galaxies between apparent magnitudes m_i and m_j, (3) the expected apparent magnitude-redshift distribution for any sample of galaxies (or quasars) (i.e. the Hubble diagram) with particular luminosity functions, space distributions, luminosity or density evolution, and observation selection bias, and (4) any number of similar functions in observational cosmology. The object of this course is simply to set out the tools to do this.

Students who do not develop a strong intuitive feeling of how to proceed in these and similar problems on the basis set out in outline here should better turn to the deep and modern aspects of the theory (for example, as in the lectures by Longair whose approach is classical physics set on a strong theoretical base). To work in theory requires clear direction, imagination, analytical brilliance, and courageous speculation rather than practicality. It may, of course, be that the student, after reading both approaches will decide that the best decision is to abandon the subject altogether.

The preparation of these lectures for publication was done in Pasadena at the Carnegie Observatories and in Basel at the Astronomisches Institut der Universität Basel. My gratitude goes to Bruno Binggeli and Roland Buser for all aspects of the development of this Saas Fee course and for the editing of these lectures, to Martin Federspiel for the exacting task of preparing the form of the lectures to press, and to G.A. Tammann not only for the hospitality of the institute where the writing was completed, but also for the scientific collaboration over the past 30 years leading to the work.

1 Introduction to the Standard Model

1.1 Scope of the Subject

These lectures pertain to the standard Robertson-Walker-Friedmann-Mattig cosmology. Most of the work will be concerned with the distribution functions containing the observables of redshifts, apparent magnitudes, and angular diameters of galaxies used as markers of space. Observational cosmology, often called the "old cosmology" by the new astronomers, was about these functions until high energy astrophysics expanded into the discipline with the discovery of the microwave background suggested by Gamow (Alpher, Bethe, & Gamow 1948), and predicted explicitly by the models of creation of the universe by Alpher and Herman (1948, 1950), and Alpher, Follin, & Herman (1953). In these and other papers of a series, these authors invented the physics of the early universe, tracing the consequences for all components of the universe of the cooling of the primeval matter (ylem) from the postulated high energy beginning at the "creation".

Their work was before its time. As always, such work was either unknown or was forgotten[1, 2] by the second wave of inventors that discovered the same

[1] Remembrance of the 1948 prediction of the cosmic microwave background (CMB) radiation, hinted at by Gamow and thereafter predicted in detail in a sweeping vision of "creation" by Alpher and Herman, was the occasion of a famous reply by Gamow to a question put by a session chairman at (probably) the fourth Texas Symposium on High Energy Astrophysics in 1965. The session was on the discovery by Penzias and Wilson (1965) of the CMB, which had just been announced. The chairman, after the observational data had been presented and discussed asked Gamow, in the audience of perhaps 1000 persons, if this was the background radiation which Alpher, Herman, Follin, and he had predicted. Gamow replied "Well, I lost 5 cents around here someplace and now someone has found a nickel nearby. It looks the same as mine. I would have thought that this is the exact same that I lost, but then all nickels look pretty much the same." A thousand heard and saw this lost-and-found exchange.

[2] The prediction of the radiation was so explicit by Alpher and Herman that many believe the CMB should be called the Alpher-Herman radiation. Critics of this view have stated that, although Alpher, Herman, and Follin did make the prediction, they could not have believed much in it because, it has been said, they mounted no plans to discover it experimentally, contrary, for example, to the plans explicitly made at Princeton, based on the later rediscovery of the physics of the hot universe at the creation. The criticism is not correct.

Sometime in the mid 1950's James Follin appeared in my office in Pasadena to talk about observational cosmology. Not knowing of his and his colleague's developments concerning genesis, I asked what his line of work was. He said that the George Washington University group was engaged in mounting efforts to observe the birthday of the universe directly. Their plan was to use rockets to observe the highly redshifted Lyman alpha line which they believed was related in some way to their predicted "5°K" relic radiation. This all sounded off the wall to me, not knowing of the "new cosmology", having come from a training as a simple classical (observational) cosmologist concerned solely with problems of galaxy distances and redshifts. However, Follin was dead serious, talking Sanskrit to my trail to the frontier even after that trail had been pioneered a generation before and where the wagon-wheel ruts could clearly be seen by those with a knowledge of the literature. (It is truly said that if you think you have discovered something fundamentally new, you must not read your predecessors. This is the theoretician's version of the observer's lament that "If you have discovered something fundamentally new at the telescope, wait a night and it will disappear.")

The subject of the early universe was reinvented and the trail rediscovered 20 years later by particle physicists that married into the original Ur cosmic family, not knowing of their previous parents.

We will not be concerned in these lectures with either the early universe or its sister field of the high energy astrophysics of fundamental particle physics at the energies extant in the free quark era. Here we shall consider only those internal parts of this now integrated field of physics and astronomy from the observational justification for the broad view of "scientific creation" and later evolution. Although "creation" and the early evolution thereafter is not the main menu, to set the stage we briefly outline the steps of this broad synthesis. It is a story-telling that starts with a hot creation event, which by subsequent mythical steps (who was there?) leads to the present-day astronomical structures. There is, of course, generally some truth in every myth.

The Grand Synthesis of the New Standard Model

The Grand Synthesis of the evolution of the universe is divided into the two phases of (1) the chemical era in the vast "cooling sequence", and (2) the "recycling sequence" which is where we are now.

In the chemical era, all the basic constituents of matter (quarks, gluons, baryons and atoms) were made. In the recycling era, the atoms form into galaxies whose stars make heavy elements that are recycled through the interstellar medium to make new stars, the new stars continuously increasing in the fraction of the heavy elements they contain.

The rawest of summaries of these postulated processes and the changes brought about thereby by recycling are:

The Cooling Sequence

(Synopsis)

The expansion of the universe cools all the undifferentiated matter in the manifold. Different temperature regimes cause different processes to occur in sequence at the various times after the extremely hot beginning.

---

naivety in the offices still dominated by Hubble's ghost.

More recently Alpher and Herman told that they had, in fact, begun contacting radar physicists in the late 1940's, discussing the idea of an experiment. In a letter of December 1989 they wrote "We also discussed this subject with some of the radar experts at the Naval Research Laboratories and the National Bureau of Standards during that same period when radio astronomy was in its infancy."

(Details)

1. Matter and antimatter annihilate to form gamma rays in the first 10^{-35} seconds after the big bang "event".
2. Matter-antimatter asymmetry causes an excess of matter over antimatter by a factor of 1 part in 5×10^9. This gives the current photon to baryon number ratio of that size, fixed for all time after the first 10^{-35} seconds.
3. Free quarks and free gluons exist from 10^{-35} to 10^{-5} seconds.
4. Quarks and gluons are confined to form protons and neutrons at 10^{-5} seconds, forming the baryons of matter (not antimatter which was totally annihilated in the first stage) .
5. As cooling continues, baryons combine to form ^{3}D, ^{4}He, and ^{7}Li nuclei at 1 minute.
6. Photons and baryons decouple, the universe becomes transparent, and atoms up to ^{7}Li form at 100,000 years.
7. Stars and galaxies form by gravitational collapse out of initial density fluctuations, completing their initial collapse phase in less than 10^9 years. The first stars, making the chemical elements heavier than He, Li, and Be, appear in as short a time as $10^{8.5}$ years from the "beginning".

The Recycling Sequence

8. First generation stars produce some of the heavy elements beyond He in their interiors.
9. Stars explode, and in the process complete the range of the chemical elements by a combination of (a) manufacturing processes making elements up to carbon in the post-main sequence stellar evolutionary path to the supernovae state, and (b) the SN explosion itself acting on the seed nuclei produced earlier in the stellar interior. Rapid neutron addition occurs during the events of the explosion (e.g. Burbidge *et al.* 1957; Clayton 1968, 1983) making a number of the heavy elements (the r-process nuclei). Other processes in other sites also operate to complete the periodic table, either in the interior of stars or in their SN explosion products (Wilson & Rood 1994 for a review).
10. The debris of the explosions, and/or stellar winds, enrich the interstellar medium (ISM) of the parent galaxy (Hoyle 1953, Fig. 1).
11. New stars of higher metal abundance form out of the enriched ISM.
12. The Galactic disk forms in our Galaxy by dissipative collapse from a larger volume (Eggen, Lynden-Bell, & Sandage 1962). Processes (8) through (12) occur at the same time (i.e. in parallel rather than in series) in the early galaxy.
13. The metal-rich sun is a late arrival 4.5 Gyr ago in the Galaxy which is 14×10^9 Gyr old.
14. The planets of the Solar System form.
15. Life: a property of the universe, forms on Earth 4 Gyr ago.

This is the Grand Synthesis of the new cosmology, where, at the "beginning" in item (1) the four known forces of nature are said to be unified before they decouple by various symmetry breaking events. Hence, the era is called "Grand Unification" at the hot beginning[3], one of the consequences of which is a globally flat space-time. Much will be said later concerning the required mass density (g/cm^3) of $3H_0^2/8\pi G$, as the requirement to "close the universe".

The lesson from this scheme is that the cosmology must produce the following necessities for any of the above to happen.

A. We need a hot creation event.
B. There must be enormous cooling.
C. The expansion must be real (i.e. the redshift must be due to a real time-dependent scale factor to the metric).
D. Space-time must be flat, i.e. the density must be the critical density just mentioned.
E. For all this to be true requires that the deceleration of the expansion with time must be such that the timing test, relating the actual age of the universe to the inverse Hubble constant, must give $T_U = 2/3H_0^{-1}$.

Items (C) through (E) have constituted the problems of classical cosmology since the early 1930's. This is what these Lectures are all about.

1.2 The Hall of Fame

In the present epoch when most astronomers are cosmologists, either in every day practice or in their fantasies, it is to be noted that perhaps only 20 primary pioneers partook in the development of the main subject from the 1920's into the early 1950's. Throughout these lectures nearly all of their names will appear, and it is useful to list the principal characters in the Hall of Fame.

The list is set out by dividing the terrain by those who contributed to the theory and those whose principal contributions were with the observations. Approximate dates are affixed for the time of their major work. The names are listed without comment at this stage; the work of each forms much of the body of the lectures.

Theory

Gauss	1820	(experimental geometry, space curvature)
Einstein	1916	(field equations)
de Sitter	1916	(special solutions of the field equations)
Friedmann	1922	(time dependent metric scale factor of the equations)
Lemaître	1927	(the whole ball of wax)
Robertson	1928	(line element, geometry, expansion)
McVittie	1929	(linked Lemaître & Eddington)
Eddington	1930	(expositor)
Tolman	1930+	(observations + models, surface brightness test)

[3] Pauli, speaking of Grand Unification as it was proposed in different circumstances in the 1950's admonished "What God has put asunder let no man put together".

Milne	1934	(Newtonian cosmology)
McCrea	1934	(Newtonian cosmology)
Heckmann	1942	(fundamental book, expositor)
Mattig	1958	(developed closed expressions for the equations)
		Observation
Wirtz	1920	(the European Hubble without a telescope)
Lundmark	1920+	(discovered that galaxies are galaxies)
Slipher	1925	(first observations of galaxy spectra)
Hubble	1922+	(galaxy counts, linear redshift law, rate)
Humason	1929+	(galaxy redshifts, expansion)
Mayall	1935+	(galaxy counts, redshifts)

1.3 Example of Simplicity: The Luminosity Density of Shining Baryons: A Study in Methods

As a demonstration of the simplicity of how to make estimates of parameters of cosmological significance without requiring the numbing detail demanded by the rigorous calculations (that are seldom actually germane for the accuracy needed), we show here how to estimate the mean global density of luminous matter in the universe. The estimate is based on (1) the observed galaxy counts per square degree brighter than apparent magnitude m, (2) a value for the average galaxy luminosity, and (3) the average mass-to-light ratio of shining baryons in stars and individual galaxies.

The average number of galaxies per square degree brighter than apparent B magnitude m, compensating for the over-density of the local Virgo complex, well known since the initial counts by Hardcastle, Seares, Hubble, and Shapley (see Sandage, Tammann, & Hardy 1972; Sandage 1975, 1988 for reviews of the early history) is

$$\log N(m) = 0.6m - 9.31, \tag{1.1}$$

where the constant is often called the "space density factor" (e.g. Shapley 1943).

This equation shows that the number of galaxies between, say, apparent magnitude 14.5 and 15.5 per square degree is 0.73. If the average absolute magnitude of such galaxies is $-20^{\rm m}$ (this is approximately M^* for a Schechter luminosity function on a distance scale of $H_0 = 50$ km s^{-1} Mpc^{-1}; Binggeli *et al.* 1988, Fig. 1), then the spatial volume enclosed within a shell of inner radius $r_1 = 79.4$ Mpc and outer radius $r_2 = 125.9$ Mpc (these are the distances corresponding to apparent magnitudes $14\overset{\rm m}{.}5$ and $15\overset{\rm m}{.}5$ for objects with $M = -20^{\rm m}$) in a cone 1° on a side is 1.52×10^2 (Mpc)^3. [Recall from your past that the number of square degrees in the complete sky is $4\pi(180/\pi)^2$]. Hence, the spatial density of such galaxies is 4.8×10^{-3} galaxies/Mpc^3.

The luminosity density due to galaxies follows directly. An absolute B magnitude of $M = -20^{\rm m}$ corresponds to a luminosity of 1.4×10^{10} solar luminosities, taking the absolute B magnitude of the sun to be $M_B = 5\overset{m}{.}34$. Hence, the luminosity density determined in this simplest of ways is 6.8×10^7 solar B luminosities per Mpc^3. (A rigorous calculation, properly integrating over the galaxian luminosity function, gives $LD = 8.8 \times 10^7$ solar B luminosities per Mpc^3; Yahil, Sandage, & Tammann 1979, 1980).

Of what use is this number?

Changing the luminosity density into mass units, using the mass of the sun as 2×10^{33} g, gives the mass density due to shining baryons as

$$\rho_{\mathrm{obs}} = 6.1 \times 10^{-33}(M/L) \ \mathrm{g} \ \ \mathrm{cm}^{-3}, \tag{1.2}$$

where the M/L ratio is in solar units. (Note that 1 $\mathrm{Mpc}^3 = 2.9 \times 10^{73}$ cm^3). The critical density needed to close the universe is

$$\rho_{\mathrm{crit}} = 3H_0^2/8\pi G = 4.7 \times 10^{-30}(H_0/50)^2 \ \mathrm{g\ cm}^{-3}. \tag{1.3}$$

Hence, the ratio of the luminous matter density to the critical density is

$$\Omega = \rho/\rho_{\mathrm{crit}} = 1.3 \times 10^{-3}(M/L)(H_0/50)^{-2}. \tag{1.4}$$

If we require $\Omega = 1$ from the new cosmology of Grand Unification, the required mass-to-light ratio (solar units) is

$$M/L = 770$$

if $H_0 = 50$.

If true, most of the matter in the universe must be dark because no luminous aggregate (individual stars, star clusters, or galaxies) has a mass-to-light ratio greater than, at most, 10.

The purpose of this section is simply to show how such a far-reaching conclusion can be derived from such a small amount of observational data and the most elementary analysis of it. Only the count density in equation (1.1) and knowledge of the average absolute magnitude of galaxies has been required.

1.4 Experimental Geometry: Introduction to Curved Spaces

A central theme of theoretical cosmology is *curvature*. It is the replacement by Einstein, preceded by Riemann (1826-1866) and by Clifford (1845-1879), of the Newtonian mystery of action at a distance by the equally profound mystery of dynamics determined by space-time curvature in the presence of matter.

What is curvature? The concept centers about the possibility of non-Euclidean geometries, invented by Saccheri (1667-1733) as an adumbration, brought full blown onto reality by Gauss (1777-1855), Lobachewsky (1793-1856), and Bolyai (1802-1860), and developed into a strong analytical discipline principally by Riemann (1826-1866), Christoffel (1829-1900), Ricci (1853-1925), and Levi-Cività (1873-1941).

Gauss built the stage in his historic paper "Disquisitiones Generales Circa Superficies Curvas" published in 1827 and reprinted in Vol. IV of the "Werke" (Göttingen, 1873). This is the paper from which much of the Gauss theory of surfaces is taken, as set out in all calculus textbooks in their sections on Differential Geometry. The theme is the premise upon which all observational cosmology concerning the measurement of space curvature rests.

There is a number (the curvature) that can be calculated for any arbitrary surface *from data obtained by measurements of the metric properties internal to that surface.* Said differently, surveyors confined to the surface can make practical measurements by standard surveying techniques from which calculations will reveal the Euclidean or the non-Euclidean nature of the surface. In particular the measurements can be made to yield a number which is the departure from flatness, called "curvature". It is calculated (in a complicated way to be sure) from an equation, due to Gauss, between the metric coefficients g_{11}, g_{12}, and g_{22} which define the distance interval $ds^2 = g_{ij}dx^i dx^j$ measured on the two-dimensional surface. Gauss considered his curvature-determining equation among his more satisfying discoveries, with which he was most pleased. He called the result the "theorema egregium" which means "the most beautiful theorem". The Gauss curvature-indicating equation can be seen in its delightful complication in the textbook, for example, by Berry (1976, equation B25).

For surfaces such as the surface of a sphere, the curvature is the reciprocal of the radius of the sphere. This is the simplest case of a two-space of constant curvature. The surface of an ellipsoid of revolution (an oblate spheroid) is a two-space where the curvature changes at places along the surface, but is nevertheless constant in certain particular directions (i.e. along latitude lines defined by small circles in planes perpendicular to the axis of revolution).

The central idea of Gauss's method is that metric measurement of properties *internal* to a space can determine a local vector perpendicular to the surface, or a tensor in the Riemannian generalization. The curvature vector lies *outside* the surface. It would be an abstract generalization, not intuitively obvious to a two-dimensional mathematician living on the surface. Nevertheless, survey data obtained by measurements on the surface could convince the surveyor that his space was not flat.

It is the difficulty of visualizing "curvature" of a 3-space that causes the intuition to fail in appreciation of the curvature of 3-space. If the 3-space could be "embedded" in a true four-dimensional space that we could step into, just as the 2-space of a surface is embedded in the three-dimensional space, we could "see" the curvature in the same way we "see" the curvature of a 2-space surface since, standing as we are in the "embedding" 3-space. The lack of an embedding 4-space is what causes non-Euclidean geometry to be so non-intuitive for beginning students, and often even long thereafter[4].

[4] As a beginning astronomer not yet 10 years out of the PhD, I once had the priviledge to be seated at lunch next to Georges Lemaître, a chief architect of

We can illustrate Gauss's proposition by considering how particular measurements by a surveyor, confined to the surface of a sphere, can be made to yield the radius of the sphere.

Problem 1.1. In one of the great experiments of the human intellect, Eratosthenes (276–195 BC) in the third century BC determined the radius of the earth. From that, using the method of Aristarchus of Samos (310–230 BC) that gives the ratio of the distance to the sun and moon, the linear values of such distances could be determined by means available to anyone with a measuring rod (to again perform the Eratosthenes measurement), a clock, and a device to measure angular distances between astronomical objects. Reconstruct the method and show how accurate the angular measurement of the angle between the moon and the sun (to be made at the relevant phase that is necessary for the experiment to succeed) must be to obtain the correct answer. Hint: use data from a central lunar eclipse and the clock for one of the needed data items.

1.4.1 Measurements in the 2-Space of the Surface of a Sphere

(a) Circumferences
Suppose the surveyor pounds a stake in the ground, ties a string to the stake, goes a distance d along the surface from the stake, and then walks around the stake making a small circle on the surface. If the surface were flat, the circumference of the small circle would be

$$l = 2\pi d. \tag{1.5}$$

However, direct calculation from the geometry shows that the true circumference is

$$l = 2\pi R \sin\theta \ , \tag{1.6}$$

where $\theta = d/R$ is the angle as seen from the center of the sphere of radius R between the direction of the stake and the end of the string of length d.

Expand eq. (1.6) in powers of d/R, giving

$$l = 2\pi d\left[1 - \frac{1}{6}\left(\frac{d}{R}\right)^2 + \mathrm{O}\left(\frac{d}{R}\right)^4\right], \tag{1.7}$$

the theory of the expanding universe; some say *the* father of practical cosmology. The occasion was the 1961 Santa Barbara IAU Symposium no. 15 on "Problems in Extragalactic research", ed. McVittie (1962). Lemaître asked me if I could visualize space curvature, its beauty, and its non-Euclidean character. I confessed that, much as I had tried, it still was not sensible to my imagination. Lemaître replied, "What a pity. The world is not understandable without it. If the intuition does not come around for you, it might be best for you to change fields".

which is smaller than eq. (1.5) by the factor $(1/6)(d/R)^2$. Because the surveyor measures both l and d, the only unknown in eq. (1.7) is R, thereby permitting its determination using the data of measurements *that are confined to the surface.*

The size of the departure from the Euclidean case is tiny for small d/R ratios, which is the practical case in daily experience. Even at the enormous distance along the surface of $d = R$, equation (1.6) is only a factor of 0.84 smaller than the $2\pi d$ Euclidean value. The more practical case of say $d = 0.1R$, or $d = 560$ km along the Earth's surface (a surveyor's possibility), equation (1.7) shows that the departure from Euclidean geometry is only 0.16%, suggesting the extreme difficulty of measurement of the radius of the Earth, R, in this way.

(b) Areas

A similar calculation of the *area* enclosed in the distance d along the surface gives

$$A = 2\pi R^2[1 - \cos\theta], \tag{1.8}$$

which expands in second order of d/R to

$$A = \pi d^2\left[1 - \left(\frac{1}{12}\right)\left(\frac{d}{R}\right)^2 + O\left(\frac{d}{R}\right)^4\right]. \tag{1.9}$$

Again, the surveyor measures d and the area by standard means (e.g. planting trees uniformly and counting the number of trees enclosed in distance d). The only unknown in eq. (1.9) again is R. As before, the departure from the Euclidean area is minute, being only 8% at the enormous distance of $d = R$. Again, at the more practical distance of $d = 0.1R$, the departure of the measurement from Euclidean is only 0.08%, yet this is the two-dimensional principle for the three-dimensional attempt by Hubble in his 1936 experiment (Lectures 2 and 3) to measure "space curvature" by galaxy counts.

(c) Angles

Other measurements on the surface can be made in principle to give R. The well known classical angle test, also invented by Gauss in his 1827 paper, is that the sum of the angles of a (geodesic) triangle drawn on the surface is larger than π rad by the area of the triangle divided by R_1R_2, where the two orthogonal radii, R_1 and R_2, are the two curvatures (reciprocal radii) of the osculating circles in the two principle sections of a spheroid (i.e. a surface of non-constant curvature). A modern derivation of this result is in Robertson and Noonan (1968, section 7.1).

Gauss actually made the experiment on a triangle he had measured in the Harz mountains between three peaks (area approximately 1000 km^2) where he obtained an angle excess of 14″ relative to 180°. (Weinberg [1972] is not correct, stating that Gauss obtained a null result in his actual experiment). A modern account of the geodesy by Gauss is by Heller *et al.* (1989).

A second type of surveying on a surface that gives the curvature of that surface is based on the concept of "parallel transport" of a vector at the surface point $P(x^i)$ to a surface point $P(x^i + dx^i)$, where x^i are the so-called geodesic coordinates on the surface. By parallel displacement is meant a displacement such that the angle between the vector and the tangent to the geodesic remains fixed during the displacement. It can be shown (cf. Robertson & Noonan 1968, section 8.12 and their Fig. 8.1) that the deviation in angle from the original direction upon a complete circuit around a geodesic triangle (a triangle made with three geodesics as sides) is given by the area of the curve divided by R^2, where R is the (constant) radius of curvature. In case of surfaces of non-constant curvature, the angle deviation upon a complete traversal of a closed curve in the surface is the surface integral of the (variable) Gaussian curvature over the area of the curve around which the parallel displacement has been performed.

This is the partial basis of the famous proposed gyroscope space experiment in a satellite circling the Earth to test the curvature of space-time caused by the mass of the earth, although the theory is considerably more complicated than the simple effect here (Will 1992, section 9.1).

1.5 The Four Standard Tests of Classical Cosmology

The curvature of 3-space as set out in section 1.6.1 later must ultimately be expressed by equations containing only those quantities that can be observed at the telescope. These are (a) fluxes (i.e. apparent magnitudes), (b) redshifts, denoted by z where $1 + z = \lambda/\lambda_0$, which is the spectral shift relative to the laboratory wavelength λ_0 of a particular spectral line (of course, the continuum also partakes in the redshift by the same ratio), (c) angular diameter, and (d) surface brightness. Each of these observables is affected differently by space curvature. The task of these lectures is to display these equations in operationally useful forms, i.e. containing observables only.

Four examples of this program are the four classical tests of the old standard model. The measure of curvature is the "deceleration parameter", which by definition is $q_0 = -\ddot{R}_0/R_0H_0^2$ (Hoyle & Sandage 1956), which we show later (section 1.6.2) to be related to the Gaussian curvature of 3-space by

$$kc^2/R_0^2 = H_0^2(2q_0 - 1). \tag{1.10}$$

Here, H_0 is the present-epoch value of the Hubble constant giving the rate of the expansion (both q_0 and H_0 vary with epoch as the expansion proceeds; see Sandage 1961b, Figure 1), and R_0 is the present value of the metric scale factor.

The equations needed for the tests involving observables only are (a) the galaxy count relation $N(m, q_0)$, where $N(m)$ is the number of galaxies per square degree brighter than apparent magnitude m, (b) the angular size-redshift relation as a function of redshift and q_0, (c) the Hubble diagram (redshift-apparent magnitude relation), $m(z, q_0)$, and (d) the time scale-q_0 relation relating the inverse Hubble constant with the age of the universe. Most of the remaining lectures are concerned with these relations.

1.6 The Classical Standard Model

1.6.1 The Robertson-Walker Metric and Proper Volumes

The Minkowski space-time interval of special relativity, with all its magic of being invariant under a Lorentz transformation, is valid as a metric for a homogeneous, isotropic space of constant 3-space curvature, kc^2/R^2, as

$$ds^2 = c^2 dt^2 - R(t)^2 du^2, \tag{1.11}$$

where the spatial part of the metric is non-Euclidean (i.e. there is curvature) with the form

$$du^2 = \frac{dr^2}{1 - kr^2} + r^2 d\theta^2 + r^2 \sin^2\theta d\phi^2, \tag{1.12}$$

as proved generally by Robertson (1929, 1935) and Walker (1936) independently. Here r is a co-moving *coordinate* "distance". It is *not* the distance measured by an astronomer. Confusion on this point was widespread during the 1930's. The co-moving coordinate r is constant in time for any given galaxy. It is the fixed mesh-point in the manifold whose scale factor can change with the change of $R(t)$ with time. The signature of the space curvature is, as usual, $k = 1$ for closed space, $k = 0$ for flat space, and $k = -1$ for open (hyperbolic) space.

The particular form of the metric in (1.12) can be changed by various coordinate transformations, such as using the distance that would be measured by an astronomer [equations (1.14) below]. These forms of (1.12) are well discussed in the lectures by Longair.

From (1.11) and (1.12) it is evident that the *interval* distance, i.e. the distance that can be measured by a surveyor as in section 1.4, but now in three dimensions, is

$$d = R(t)u = R(t) \int_0^r \frac{dr}{\sqrt{1 - kr^2}}. \tag{1.13}$$

This integrates to

$$\begin{aligned} d &= R(t) \arcsin r && \text{for } k = 1 \\ d &= R(t) r && \text{for } k = 0 \\ d &= R(t) \text{arsinh } r && \text{for } k = -1. \end{aligned} \tag{1.14}$$

Hence, the coordinate "distance" is either

$$r = \sin d/R; \quad d/R; \quad \text{or} \quad \sinh d/R. \tag{1.15}$$

Note the perfect analogy with the d/R factor in equations (1.7) and (1.9) for the 2-space analog.

The volume enclosed within coordinate "distance" r is evidently

$$\left.\begin{aligned} V(r) &= R^3 \int_0^r \frac{r^2 \mathrm{d}r}{\sqrt{1-kr^2}} \int_0^\pi \sin\theta \mathrm{d}\theta \int_0^{2\pi} \mathrm{d}\phi \\ &= 4\pi R^3 \int_0^r \frac{r^2 \mathrm{d}r}{\sqrt{1-kr^2}} \\ &= 2\pi (Rr)^3 \left[\frac{\arcsin r}{r^3} - \frac{\sqrt{1-r^2}}{r^2}\right], \quad \text{for } k=1. \end{aligned}\right\} \tag{1.16}$$

One half the complete volume is surveyed when $r = 1$. Hence the finite volume of the "total" space is

$$V(\infty) = 2\pi^2 R^3, \tag{1.17}$$

which is a famous result.

Similar expressions to (1.16) exist for the $k = -1$ case. Of course, the volume is Euclidean for $k = 0$.

To appreciate equation (1.16) in the same way as (1.7) and (1.9), expand (1.16) to give

$$V(r) = 4/3\pi (Rr)^3 \Big[1 + \frac{3}{10} kr^2 + \mathrm{O}(r^4)\Big]. \tag{1.18}$$

But, as above, Rr is not the interval distance measured by a surveyor (astronomer); rather $r = \sin d/R$, where d is the operationally defined measured quantity (i.e. perhaps a photometric distance or an angular size distance – see Lectures 2 to 5). In a sense, as in section 1.4, d is the "distance along the hyper- (i.e. 3-space) surface". Note in Longair's lecture on the metric, this is the "distance" coordinate he uses.

Expanding $r = \sin d/R$ and entering the terms in equation (1.18) gives

$$V(d) = \frac{4}{3}\pi d^3 \Big(1 - \frac{k}{5}\frac{d^2}{R^2} + \ldots\Big). \tag{1.19}$$

Note now that the positive curved space ($k = 1$) has a smaller volume out to interval distance d than its Euclidean cousin. To anticipate Lecture 2, this differs from equation (1.18), which Hubble mistakenly used assuming Rr is the operationally measured "photometric distance" (Lecture 5).

The philosophy of how galaxy counts are, in principle, used to determine the radius of curvature of non-Euclidean space is contained in equation (1.19) [but used in its exact, closed form in (1.16)]. However, to do this requires r, or d, to be expressed in the observable quantities of redshift and q_0, where q_0 is related to the curvature by equation (1.10), to be derived. Our next task is, then, to derive the relation between $R(t)r$ and the redshift z, given a q_0 value. To do this requires the theory of the time dependence of the metric scale factor r, i.e. we need to know the dynamics of the expansion. This is given by the solution of the fundamental Friedmann equation for the expansion.

1.6.2 The Friedmann Equation and Its Solution

The problem is to relate Rr with the redshift, z, and then to substitute the $Rr = f(z, q_0)$ and the $r = g(z, q_0)$ equations into equation (1.16) for the $k = 1$ case, and the equivalent equations for $k = 0$, and $k = -1$. We need the dynamics to find $Rr = f(z, q_0)$ which is the core problem of the subject.

The Friedmann equation is rigorously derived from the general relativity field equations. An appreciation of the equation is obtained from the analogy of Newtonian cosmology invented by McCrea and Milne (1934, see also McCrea 1953) which gives the correct equation (but with a problem of "the boundary conditions at infinity" swept under the carpet) as follows.

Consider a homogeneous distribution of matter of density ρ. From any arbitrary origin in the manifold (centered on an observer) consider the mass inside a radius R and neglect all the mass outside the R boundary[5]. The acceleration of a test particle at R due to the matter inside R is, evidently,

$$\ddot{R} = -\frac{GM(R)}{R^2}, \text{ or } R\ddot{R} = -\frac{GM(R)}{R}, \tag{1.20}$$

where the mass $M(R)$ inside R is constant during the expansion because the expansion is such that two shells at different R values do not cross.

Integrating gives the energy equation

$$\dot{R}^2 - \frac{2GM(R)}{R} = \text{const} = -kc^2, \tag{1.21}$$

where the constant is put as kc^2 (the curvature term) from a deeper knowledge of the problem from general relativity.

Putting (1.20) into (1.21) gives one form of the Friedmann equation as

$$\frac{\dot{R}^2}{R^2} + \frac{2\ddot{R}}{R} = -\frac{kc^2}{R^2}. \tag{1.22}$$

The other useful form is to use (1.21) with

$$M(R) = \frac{4}{3}\pi R^3 \rho \tag{1.23}$$

[5] This is the problem with the "boundary condition". In ordinary dynamics the force everywhere in the manifold is zero when the matter outside R is retained, all forces canceling by symmetry. But here we compute the force on a particle at the boundary of a sphere of radius R from the matter inside R as if all the outside matter in the universe were absent.

to give

$$\frac{\dot{R}^2}{R^2} - (8\pi/3)G\rho = -\frac{kc^2}{R^2}. \tag{1.24}$$

Note that $M(R)$ in (1.23) is constant with time, although both R and ρ are functions of time. This, of course, is because ρ varies as R^{-3}. This is why $M(R)$ in (1.20) must be treated as a constant.

We can now derive two central relations from the formalism. Define the Hubble constant at any epoch as

$$H \equiv \frac{\dot{R}}{R}, \tag{1.25}$$

and define the deceleration parameter, also at any given epoch as

$$q \equiv -\frac{\ddot{R}}{RH^2}. \tag{1.26}$$

Put these definitions into (1.22) to obtain

$$\frac{kc^2}{R^2} = H^2(2q-1), \tag{1.27}$$

which holds at every epoch. In particular it holds at the present epoch where the present-day values are denoted by R_0, H_0, and q_0, giving equation (1.10). There are two important points to be made from this equation. (A) The left hand term is the Gaussian curvature of the geometrical space. (B) H_0 and q_0 can be determined by measurements at the telescope. Hence, we are now on operationally solid ground (i.e. it is practical cosmology), and the result is, in principle, in the hands of the observers.

The second central relation is obtained by substituting (1.24) into (1.25), remembering that $H = \dot{R}/R$ by definition, to obtain

$$\rho_0 = \frac{3H_0^2 q_0}{4\pi G}, \tag{1.28}$$

showing that the deceleration parameter, q_0, is determined simply by the density and the expansion rate. Equation (1.28) with (1.27) expresses a fundamental theme of general relativity that the space curvature kc^2/R^2 is determined by the matter density of the manifold.

1.6.3 The Mattig Equation Relating the Coordinate Interval with z and q_0

We now require the relations between the coordinate interval r and the observables z, H_0, and q_0. This is given by the solution of equation (1.22), and, subsequently, using the definitions of q_0 and H_0 as functions of R.

The well known solutions of (1.22) are given by the parametric equations

$$R = a(1 - \cos\theta) \tag{1.29}$$

$$t = \frac{a}{c}(\theta - \sin\theta) \tag{1.30}$$

for $k = +1$, and

$$R = a(\cosh\theta - 1) \tag{1.31}$$

$$t = \frac{a}{c}(\sinh\theta - \theta) \tag{1.32}$$

for $k = -1$, where θ is the so called development angle (the angle which the generating circle of the cycloid and the hypercycloid of these equations rolls through). The constant a, which is the radius of the generating circle, can be shown to be

$$a = \frac{4\pi G\rho R^3}{3c^2}. \tag{1.33}$$

Many of the equations of practical cosmology can be most easily derived using this development angle formalism (see for example Sandage 1961b, 1962).

The other required relation is between redshift and the metric scale factor $R(t)$ given by the Lemaître equation, which is general no matter what the form of the $R(t)$ function may be. Let R_1 be the scale factor at the time light left a galaxy with redshift z. Let R_0 be the factor at the time of photon reception (now). The redshift is a stretching of the wave between these two epochs, derived rigorously by Lemaître (1927, 1931), and hence called the Lemaître equation, to be

$$1 + z = \frac{R_0}{R_1}. \tag{1.34}$$

The way to relate r and R to z is via equations (1.11) and (1.12) which, without loss of generality by orienting the coordinate system so that $\mathrm{d}\theta = \mathrm{d}\phi = 0$, can be written with $\mathrm{d}s = 0$ for the photon path as

$$c\int_{t_1}^{t_0} \frac{\mathrm{d}t}{R} = \int_0^r \frac{\mathrm{d}r}{\sqrt{1 - kr^2}} = \begin{cases} \arcsin r, & k = +1 \\ r, & k = 0 \\ \mathrm{arsinh}\ r, & k = -1. \end{cases} \tag{1.35}$$

By putting (1.35) into either (1.29), (1.30) or (1.31), (1.32), and then using (1.34) and the definitions of H_0 and q_0 [eqs. (1.25), (1.26)] we obtain, after considerable reduction, the famous Mattig (1958) equation

$$R_0 r = \frac{c}{H_0 q_0^2 (1 + z)}\left\{q_0 z + (q_0 - 1)\left[-1 + \sqrt{1 + 2q_0 z}\right]\right\}, \tag{1.36}$$

valid for all three values of the curvature signature[6]. A detailed derivation of this equation in a way different from Mattig, using the development angle formalism is in the literature (Sandage 1962).

The equation for the co-moving coordinate distance r is found from (1.36) using (1.10), giving

$$r = \frac{\sqrt{2q_0 - 1}}{q_0^2(1+z)}\left\{q_0 z + (q_0 - 1)\left[-1 + \sqrt{1 + 2q_0 z}\right]\right\}, \tag{1.37}$$

Problem 1.2. Using the equations derived so far, prove that $q_0 = 1/2$ exactly for $k = +1, 0, -1$ at the $R(t) = 0$ creation, or equivalently when the development angle $\theta = 0$. The result is fundamental because it shows that all Friedmann solutions begin in the state of flat space ($q = 1/2$) and progressively depart from it as the model unfolds for $k =$ either $+1$ or -1. This is the origin of the famous flatness problem. Why is the universe still nearly flat if k does not equal zero (where $q = 1/2$ always)? The density would then be precisely the closure density. The solution is best made by using the development angle formalism and expanding the relevant equations in powers of θ, taking the limit of q as θ approaches zero. The "flatness problem" is developed extensively in Longair's lectures.

---

[6] During the discussion of this point in the lecture as delivered, I suggested that the students try this route to the Mattig equation, but that the route was very difficult. It had been also emphasized that this equation was the turning point in the theoretical development of the subject, separating the subject into two epochs, one before 1958 and the other after. This is because, prior to 1958, each of the equations we have just developed, and will develop in future lectures, were expressed as Taylor expansions in powers of z, no closed form being known, for example, for equation (1.36).

I had also discussed that none of the heroes of the subject had been able to write this most fundamental equation of the subject that was valid for abitrarily large values of z. Reading the pre-1958 literature you cannot help but be struck that the equations are all in series expansions. An example is the classic textbook by McVittie. All of the relevant equations in the 1956 edition of his General Relativity and Cosmology were in series expansions, valid only for small z. But in the 1965 edition all of the relevant equations are in the two closed forms due to Mattig (1958, 1959). Among the last Journal papers showing the pre-1958 style are those of Davidson (1959a, b, 1960) on the redshift-magnitude relation, the count-brightness test, and the angular diameter test.

In the lecture I said that the degree of difficulty might be appreciated by noting that none of the pioneers either could not or, in any case, did not derive the closed form of the $R_0 r$ equation – not Heckmann, nor Tolman, Robertson, McVittie, Bondi, Hoyle, etc. During the campaign to write Part 3 of the HMS (Humason, Mayall, & Sandage 1956) paper, I had also tried with no success.

Longair was astounded at this discussion because he assigns the problem regularly to students and says they have no difficulty in deriving equation (1.36). We all then generally agreed that what is easy in Cambridge all becomes more difficult the further from Cambridge the attempt is made.

1.6.4 Volumes in Curved Space as Functions of z and q_0

The volume contained within coordinate "distance" r for the $k = +1$ case is given by (1.16), which can be written in a more transparent way as

$$V(r) = \frac{4}{3}\pi R^3 r^3 \left[\frac{3}{2}\frac{\arcsin r}{r^3} - \frac{3}{2}\frac{\sqrt{1-r^2}}{r^2} \right], \tag{1.38}$$

where the term in brackets is the correction to $(4/3)\pi(Rr)^3$ due to curvature.

The corresponding equation for the $k = -1$ case is

$$V(r) = \frac{4}{3}\pi R^3 r^3 \left[\frac{3}{2}\frac{\sqrt{1+r^2}}{r^2} - \frac{3}{2}\frac{\operatorname{arsinh} r}{r^3} \right]. \tag{1.39}$$

These two equations, together with the Mattig equations (1.36) and (1.37) for the $r = f(z, q_0)$, solve the problem parametrically, giving the volume $V = g(z, q_0)$ in terms of the observables z and q_0 alone.

To appreciate the content of these equations and their deviation from the Euclidean geometry of the $k = 0$ case, consider their series expansion using the appropriate substitutions of (1.36) and (1.37) into (1.38) and (1.39). This procedure gives

$$V(z, q_0) = \frac{4}{3}\pi \left(\frac{cz}{H_0} \right)^3 \left[1 - \frac{3}{2}(1 + q_0)z + \ldots \right], \tag{1.40}$$

for both $k = +1$ and $k = -1$. Of course, in practice, the exact equations, valid for arbitrarily large z values, are to be used as we do in Lecture 2.

1.6.5 The Special Cases of $q_0 = 0$ and $q_0 = 1/2$

For the special case of $q_0 = 1/2$ (the Euclidean flat geometry) it is especially easy to write the exact $V(z)$ equation directly rather than in parametric form. Solution of the Friedmann equation via equation (1.35) (see Problem 1.3) gives

$$R_0 r = \frac{2c}{H_0\sqrt{1+z}} \left[\sqrt{1+z} - 1 \right], \tag{1.41}$$

which, when put into $V(Rr) = (4/3)\pi(Rr)^3$ gives

$$V(z) = \frac{32\pi c^3}{3H_0^3(1+z)^{\frac{3}{2}}} \left[\sqrt{1+z} - 1 \right]^3. \tag{1.42}$$

The $q_0 = 0$ case is more complicated in view of the complication of equation (1.39) for $k = -1$. The solution is kept in parametric form by using equation (1.39), into which are substituted

$$R_0 r = \frac{c}{H_0} \left(\frac{z}{1+z} \right) \left(1 + \frac{z}{2} \right), \tag{1.43}$$

and

$$r = \left(\frac{z}{1+z}\right)\left(1+\frac{z}{2}\right), \tag{1.44}$$

noting that

$$R_0 = \frac{c}{H_0}, \tag{1.45}$$

all of which follow from the solution to the Friedmann equation for $q_0 = 0$ (see problem 1.3).

Problem 1.3. Prove equations (1.41) for $q_0 = 1/2$, and (1.43)–(1.45) for $q_0 = 0$. Hint: Use the Friedmann equation (1.22) for $k = 0$ (i.e. $q_0 = 1/2$) to find that $R_i(t) = R_0(t_i/t_0)^{2/3}$, and then use the Lemaître equation (1.34) and the metric equation (1.35) to find r.

For the $q_0 = 0$ case, note that $R = ct$ is the solution for $R(t)$, (i.e. there is no mass, and therefore the deceleration is zero). Then use equation (1.34) and (1.35) to find r and Rr. Hint: recall that arsinh $r = \ln[r + (r^2+1)^{0.5}]$ is an identity.

1.6.6 Comparison of Volumes Enclosed Within Coordinate "Distances" for Three Cases

The "distance" as a function of redshift used (incorrectly) by Hubble, although it is what intuition would say, was

$$rR_0 = \frac{cz}{H_0}. \tag{1.46}$$

Table 1.1. Comparison of Rr values

		$(rRH_0)/c$	
z	Hubble	$q_0 = 0$	$q_0 = 1/2$
0.1	0.1	0.095	0.093
0.2	0.2	0.183	0.174
0.3	0.3	0.265	0.247
0.4	0.4	0.343	0.310
0.5	0.5	0.417	0.367
1	1	0.750	0.586
2	2	1.333	0.845
4	4	2.400	1.106

On the other hand the proper and exact coordinate distance corresponding to the $q_0 = 1/2$ and 0 cases are given by equations (1.41) and (1.43), respectively. All three "distances" are similar for small z but differ drastically for large z as shown by Table 1.1. These, appropriately put into the relevant equations for the volume enclosed within "distance" r, show why $V(z)$ differs for different q_0 values. This is the essence of the volume-redshift test.

1.6.7 The Friedmann Equation with a Cosmological Constant

Finally we simply write the Friedmann equation if the cosmological constant is not put to zero. This parameter appears as a constant of integration in the field equations of relativity, and has a formal significance of a repulsive force increasing with distance. It was originally introduced to provide a "static" solution to the field equations. A repulsive force is needed to counter the attractive force of gravity (in the absence of motion), even in the case of Newtonian cosmology. However, although a formal equilibrium can be achieved by balancing gravity with repulsion, the equilibrium is unstable (Eddington), so in the absence of motion, this solution is unphysical. Nevertheless, for certain "problems" with the time scale (i.e. if H_0^{-1} is shorter than the "known" age of the universe; see Lectures 7 ,8, and 9 on the time-scale 4th test) the phenomenon can be saved by introducing Λ.

Here we simply write the equations without proof. The proofs follow the formalisms set out in the previous sections of this lecture.

Clearly, the Friedmann equation becomes

$$\ddot{R} = -\frac{GM}{R^2} + \frac{1}{3}\Lambda c^2 R, \tag{1.47}$$

where the coefficient of 1/3 and the factor of c^2 to the repulsion term containing Λ is by convention. Note that the units of Λ are $(\text{length})^{-2}$.

The equations corresponding to equations (1.24) and (1.22) respectively are

$$\frac{\dot{R}^2}{R^2} = \frac{8\pi G\rho}{3} + \frac{1}{3}\Lambda c^2 - \frac{kc^2}{R^2}, \tag{1.48}$$

and

$$\frac{\dot{R}^2}{R^2} = -\frac{2\ddot{R}}{R} + \Lambda c^2 - \frac{kc^2}{R^2}. \tag{1.49}$$

The equation for curvature corresponding to equation (1.10) and (1.27) is

$$\frac{kc^2}{R_0^2} = H_0^2\left[(2q_0 - 1) + \Lambda\frac{c^2}{H_0^2}\right]. \tag{1.50}$$

The density for any q_0 value, corresponding to equation (1.28), is

$$\rho_0 = \frac{3H_0^2}{4\pi G}\left[q_0 + \frac{1}{3}\frac{\Lambda c^2}{H_0^2}\right]. \tag{1.51}$$

If we define a parameter Ω such that

$$\rho_0 = \frac{3H_0^2}{8\pi G}, \quad \text{or} \quad \Omega_0 = \frac{8\pi G\rho}{3H_0^2}, \tag{1.52}$$

then

$$\Omega_0 = 2q_0 + \frac{2}{3}\frac{\Lambda c^2}{H_0^2}. \tag{1.53}$$

Putting this into (1.50) gives the curvature in terms of Ω as

$$\frac{kc^2}{R_0^2} = H_0^2(\Omega_0 - 1) + \frac{1}{3}\Lambda c^2. \tag{1.54}$$

The solutions of the Friedmann equation for various values of Ω are, of course, classical, leading to the famous "coasting" solutions of Lemaître (1950 for a popular account of his 1927 model for example; or Bondi 1952) with their increased time scales.

References

Alpher, R.A., Bethe, H.A, & Gamow, G. 1948, Phys. Rev. **73**, 803

Alpher, R.A., Follin, J.W., & Herman, R.C. 1953, Phys. Rev. **93**, 1347

Alpher, R.A. & Herman, R.C. 1948, Phys. Rev. **74**, 1737

Alpher, R.A. & Herman, R.C. 1950, Rev. Mod. Phys. **22**, 153

Berry, M. 1976, *Principles of Cosmology and Gravitation* (Cambridge: Cambridge University Press)

Binggeli, B., Sandage, A., & Tammann, G.A. 1988, Ann. Rev. Astron. Astrophys. **26**, 509

Bondi, H. 1952, *Cosmology* (Cambridge: Cambridge University Press)

Burbidge, E.M., Burbidge, G.R., Fowler, W.A., & Hoyle, F. 1957, Rev. Mod. Phys. **29**, 547

Clayton, D.D. 1968, 1983 *Principles of Stellar Evolution and Nucleosynthesis* (Chicago: Univ. Chicago Press)

Davidson, W. 1959a, MNRAS **119**, 54

Davidson, W. 1959b, MNRAS **119**, 665.

Davidson, W. 1960, MNRAS **120**, 271

Eggen, O., Lynden-Bell, D., & Sandage, A. 1962, ApJ **136**, 748

Heller, M., Flin, P., Golda, Z., Maslanka, K., Ostrowski, M., Rudnicki, K., & Sierotowicz, T. 1989, Acta Cosmologica **16**, 87

Hoyle, F. 1953, ApJ Suppl. **1**, 121

Hoyle, F., & Sandage, A. 1956, PASP **68**, 301

Humason, M.L., Mayall, N.U., & Sandage, A. 1956, AJ **61**, 97

Lemaître, G. 1927, Ann. Soc. Sci. Bruxelles, **47A**, 49

Lemaître, G. 1931, MNRAS **91**, 483

Lemaître, G. 1950, *The Primeval Atom* (New York: D. Van Norstrand Co.)

Mattig, W. 1958, Astron. Nach. **284**, 109
Mattig, W. 1959, Astron. Nach. **285**, 1
McVittie, G.C. 1962, ed. of *Problems of Extragalactic Research* (New York: Macmillian)
McCrea, W.H. & Milne, E.A. 1934, Quart. J. Math. (Oxford Ser.), **5**, 73
McCrea, W.H. 1953, Reports on Progress in Physics, **16**, 321
Penzias, A.A. & Wilson, R.W. 1965, ApJ **142**, 419
Robertson, H.P. 1929, Proc. Nat. Acad. Sci. **15**, 822
Robertson, H.P. 1935, ApJ **82**, 284
Robertson, H.P. & Noonan, T.W. 1968, *Relativity and Cosmology* (Philadelphia: Sanders Co.)
Sandage, A. 1961b, ApJ **134**, 916
Sandage, A. 1962, ApJ **136**, 319
Sandage, A. 1975, in *Galaxies and the Universe, (Stars and Stellar Systems, Vol 9)*, eds. A. Sandage, M. Sandage, & J. Kristian (Univ. Chicago Press), Lecture 19
Sandage, A. 1988, Ann. Rev. Astron. Astrophys. **26**, 561
Sandage, A., Tammann, G.A., & Hardy, E. 1972, ApJ **172**, 253
Shapley, H. 1943, *Galaxies* (Philadelphia: Blakiston Co.), Lecture 7
Walker, A.G. 1936, Proc. London Math. Soc. 2nd Ser. **42**, 90
Weinberg, S. 1972, *Gravitation & Cosmology* (New York: Wiley), p. 5
Will, C. M. 1992, *Theory and Experiment in Gravational Physics* (Cambridge: Univ. Cambridge Press)
Wilson, T.L. & Rood, R.T. 1994, Ann. Rev. Astron. Astrophys. **32**, in press
Yahil, A., Sandage, A., & Tammann, G.A. 1979, in *Physical Cosmology*, Les Houches Summer School, ed. R. Balian, J. Audouze, D.N. Schramm (Amsterdam: North-Holland), p. 127
Yahil, A., Sandage, A., & Tammann, G.A. 1980, ApJ **242**, 448

2 The First Test: The Count–Magnitude Relation: Theory and Practice I (The Early Phase 1920–1938)

2.1 Theory of the Count-Magnitude Relation in Parametric Form

Equations (1.38) and (1.39) for $V(Rr)$ in Lecture 1, together with (1.36) and (1.37) for $R_0 r = (z, q_0)$ and $r = g(z, q_0)$ as parametric equations give the volume enclosed in a redshift distance z for a space with deceleration parameter q_0. Recall that q_0 is related to the curvature in units of H_0^2. These equations would be useful if we had galaxy counts that are complete within a given volume. This requires the seemingly impossible operational condition that every galaxy be counted over the complete luminosity function, for all surface brightnesses. A more practical proposition is to obtain galaxy counts that are complete to a given *apparent magnitude*, but again including galaxies of all surface brightnesses.

Of course, the two concepts of a volume-limited and a flux-limited sample are fundamentally different, and that difference must be accounted for by the theory. Lecture 1 was concerned with the volume-limited case. In the present lecture we develop the method of the count-*brightness* test in all its complexity. The theory is then compared with the galaxy count data available to 1936 from Hubble's observational program to determine the "curvature". Lecture 3 continues the count-brightness test into the modern era.

2.1.1 $V(z, q_0)$ Transformed to $V(m, q_0)$

As said, we obtained in Lecture 1 the equations for the volume, $V(r)$, enclosed in coordinate distance r. The equation for $V(r)$ was then thrown into the observables of z and q_0 using the Friedmann equation to give $V(z, q_0)$. Therefore, if we can obtain the relation between observed flux (changed by tradition to astronomical magnitude) and z and q_0 we finally end with the volume expressed as a function of m and q_0 alone. This is the *count-magnitude* relation we seek.

The fundamental equation relating bolometric flux, l, received at the Earth (in units of erg sec^{-1}cm^{-2}) over all wavelengths is

$$l_{\text{bol}} = \frac{L_{\text{bol}}}{4\pi (R_0 r)^2 (1+z)^2}, \tag{2.1}$$

due to Robertson (1938), where $R_0 r$ is the "coordinate" distance of the galaxy at the time light is received at the Earth, given by the Mattig equation (1.36) in Lecture 1.

For many years there was controversy over this equation as to whether the exponent for $(1 + z)$ should be 2 or 3, but the rigorous derivation (late in the game as it was played in 1938) by Robertson settled the issue.

The observed *apparent* luminosity, l, and the total *absolute* luminosity, L, are bolometric luminosities, not the observed heterochromatic luminosities which require an extra $(1+z)$ factor due to band-width shrinkage caused by the redshift. [This may have been the origin of the controversy over equation (2.1).] This band-width term due to the spectrum being stretched through the fixed band-pass of the detector is the non-selective $(1+z)$ factor in the modern definition of the K term discussed in section 3.1.1e of Lecture 3[7, 8].

A heuristic appreciation of equation (2.1) can be had by noting that the area of a sphere centered on the source at a coordinate distance $R_0 r$ at the time of photon reception is $4\pi(R_0 r)^2$. [Recall that $R_0 r = R_0 \sin(d/R_0)$, where d is the "astronomers operationally measured distance"; see equation (1.14)]. In the absence of a redshift, the received flux, l, which is the energy per unit area per unit time (corrected to outside the atmosphere), is L_{bol} divided by the area of the aforementioned sphere. This is equation (2.1) except for the two factors of $(1+z)$. One factor comes because every photon is degraded in energy by $(1+z)$ due to the redshift, no matter what its cause. The second $(1+z)$ factor is due to the dilution in the rate of photon arrival due to the stretching of the path length in the travel time if the expansion is real. This second factor would not be present if the redshift were not caused by the Friedmann-Lemaître expansion but to some unknown physical cause. The test for the reality of the expansion (Lecture 6) makes use of the difference in the predicted surface brightness caused by the appropriateness of either one or two factors of $(1+z)$.

Combining equation (2.1) with equation (1.36), which is

---

[7] Hubble's definition and subsequent use of the K term (both in 1936 and 1953) neglects the band-width shrinkage relative to the rest frame spectrum, and consequently his analysis of both the count-magnitude data and the redshift-magnitude relation is in error by one factor of $2.5\log(1+z)$ in the "correction to the magnitudes due to the effects of redshift". Hubble's incorrect neglect of this band-width shrinkage is not discussed in extant accounts of the history of the count-brightness test as it was set out in his classic 1936 and 1937 reports. The error came to light only in a redevelopment of the theory of the K correction made for the discussion of the redshift-magnitude relation (Lecture 5) by Humason et al. (1956). Hubble's error was discovered independently by Hoyle at the same time. It is a common error even in the modern literature and must be assiduously guarded against at all times.

[8] Throughout these lectures I write as an astronomer, not as Malcolm Longair who, lecturing as a physicist, does not employ the K term in the way we do here. This contrast between the speech of an astronomer and a physicist was often amusing during the lectures as given, but should not confuse the students in seeking an accommodation between the same concepts as presented from both viewpoints in this book. After mastering the difference in the use of units and methods using the language of an astronomer, a physicist should be able to read the classic cosmological literature of the 1920-1950's as written by observational astronomers.

It is to recall these methods that the approach here is directed. Just as physicists might find in these lectures a dictionary for the past, the astronomers that attended the lectures found Longair's presentations a dictionary for the future. Although our language is different, we suspect that our universe is the same, needing only the dictionary to make a translation. Part of the purposes of these lectures is to provide the astronomer-half of the dictionary.

$$R_0 r = \frac{c}{(H_0 q_0^2)(1+z)} \left\{ q_0 z + (q_0 - 1)\left[-1 + \sqrt{1 + 2q_0 z}\right] \right\}, \tag{2.2}$$

gives the exact equation for the redshift-apparent magnitude relation (leading to the theory of the Hubble diagram in Lecture 5) as

$$m_{\mathrm{bol}} = 5 \log q_0^{-2} \left\{ q_0 z + (q_0 - 1)\left[-1 + \sqrt{1 + 2q_0 z}\right] \right\} + \mathrm{const.} \tag{2.3}$$

The constant contains the absolute magnitude of the object in question and the Hubble constant relating distance to redshift. Recall the definition of astronomical magnitude needed to change equation (2.1) and (2.2) into (2.3) is $m = -2.5 \log l + \mathrm{const.}$

However, bolometric magnitudes are never observed in practice because of the finite bandwidth of all detectors, giving the heterochromatic photometric systems such as for the "standard" broadband photometric systems of B, V, R, I, H, K, L, etc. Equation (2.3) is changed to give the predicted heterochromatic magnitude of a galaxy at redshift z in bandpass i as

$$m_i = m_{\mathrm{bol}} + K_i(z,t) + E(z, q_0) + \mathrm{constant}, \tag{2.4}$$

where K is the correction to m_{bol} for the technical effects of redshift (section 3.1.1e of Lecture 3) as a function of redshift and the evolution of the *shape* of the spectral energy distribution (SED) during the light-travel time, and E is the evolutionary change of absolute luminosity in the light travel time at redshift z for the geometry of space determined by q_0 (note that the look-back time depends on *both* z and q_0). The evolutionary correction is discussed in Lecture 5 (section 5.5e). The constant term is the zero-point difference between bolometric magnitudes and the m_i band-pass magnitudes at zero redshift. It is the "bolometric correction" in the standard literature.

2.2 $N(m, q_0)$ Is Degenerate to z in First Order While $N(z, q_0)$ Is Not

An appreciation of the z dependence of the exact equation for $N(z, q_0)$ for different space curvatures can be had by expanding the relevant equations in Lecture 1 in powers of z to give, to first order, as in equation (1.40),

$$N(z, q_0) = \frac{4\pi c^3 z^3}{3H_0^3} \left[1 - \frac{3}{2}(1 + q_0)z + \mathrm{O}(z^2) + \ldots\right], \tag{2.5}$$

showing a *first order* z dependence for the count-volume relation for different q_0. Note particularly the *sign* of the effect. At a given z, the volume (and therefore the counts for a homogeneous and isotropic distribution), is smaller the larger is q_0, varying as $\mathrm{d}N \sim -1.5 \mathrm{d}q_0 z$ for small z. This is the same sense as the area of a spherical cap in the 2-space of the surface of a sphere. The area of the cap is smaller than the Euclidean area with the same radius, the ratio increasing with larger curvature (i.e. smaller radii of curvature).

On the other hand, the sign of the change of apparent *magnitude* with z for different q_0 is in the opposite sense. Expanding the Mattig equation (2.3) in powers of z gives the famous approximation to the bolometric Hubble diagram as

$$m_{\rm bol} = 5 \log z + 1.086(1 - q_0)z + {\rm O}(z^2) + {\rm const}, \tag{2.6}$$

derived first by a Taylor series expansion in the entire pre-1959 literature, i.e. before the exact Mattig solution was known [footnote 6 of Lecture 1 concerning equation (1.40)]. Examples of the Taylor expansion development are in Heckmann (1942), Robertson (1955), McVittie (1956), Davidson (1959a,b; 1960) and undoubtedly others. This expansion of the exact Mattig (1958) equation, showing the correspondence between the theory in the pre-Mattig era compared with the post-Mattig era was made by Mattig himself (1958, his equation 15), by Sandage (1961, 1962, 1988), and by many others in a variety of contexts.

The opposite sign of the q_0 dependence in equation (2.6) (the fact that the apparent luminosities are *brighter* at a given z for larger q_0, whereas the volumes are smaller) is the reason that the effect of curvature cancels to first order in the $N(m, q_0)$ relation, *making* $N(m)$ *independent of* q_0 for "small" z. In contrast, note that it is linearly dependent on q_0 to first order in z in $N(z, q_0)$ [equation (2.5)]. This was shown by the initial numerical calculation (Sandage 1961) where the relative insensitivity of the $N(m)$ test to curvature was shown "empirically" in that way. The analytical proof via series expansion was first demonstrated by Brown and Tinsley (1974) in a derivation repeated here to illustrate the various dependences.

We use the exact Robertson equation (2.1) for $l = f(L, R_0 r, z)$ with the exact equation for $R_0 = g(z, q_0)$ from (1.36) or (2.2). Expanding (2.2) gives

$$R_0 r(1 + z) = \frac{cz}{H_0}\left[1 + \frac{1}{2}z(1 - q_0) + {\rm O}(z^2) + \ldots\right], \tag{2.7}$$

which, with the small quantity x defined as

$$x = \left(\frac{H_0}{c}\right)\left(\frac{L}{4\pi l}\right)^{1/2}, \tag{2.8}$$

when put into (2.7) and noting that to first order $x = z$, and that

$$z \sim x\left[1 - \frac{x}{2}(1 - q_0)\right],$$

when put into equation (2.5) for $N(l)$, the count-brightness relation, $N(l)$, gives to first order in x (or z)

$$N(l) \sim \left(\frac{L}{4\pi l}\right)^{3/2}(1 - 3z + \ldots). \tag{2.9}$$

This is independent of q_0 to *first order* in z, which was to be proved.

This, plus the direct numerical calculations (Sandage 1961), show that the count-*magnitude* relation is a poor vehicle (being nearly independent of q_0 locally) by which to *see* space curvature directly as in the straightforward Gauss experiment, but indeed, because of this, the $N(m)$ is a powerful means by which to determine details of the spatial distribution and especially galaxy evolution in the look-back time (Lecture 3). Said differently, deviations of the data from the homogeneous case of $N(m) \sim 0.6m$ cannot be due to curvature to first order but must be due to inhomogeneous spatial distributions and/or either luminosity or density evolution.

Display of the dependence of $V(z, q_0)$ and also the first order degeneracy of $V(m, q_0)$ on q_0 is made with Figs. 1 and 2 of Sandage (1988). A calculation of $N(m, q_0)$ for the case of a delta luminosity function, which we have assumed to this point, is in the archive literature (Sandage 1961, Fig. 4). The realistic case of a broad luminosity function is the subject of Lecture 3.

2.3 Hubble's 1926–1934 Galaxy Count Program

Galaxy counts by many astronomers had been made since the turn of this century in an effort to establish the relation of the "white nebulae" to the Galaxy. A nearly definitive paper by Seares (1925) established fundamental results such as (a) the Galactic latitude dependence of the distribution, (b) the existence of the dusty plane of the Milky Way and the "zone of avoidance", (a name later given by Hubble), and (c) a rediscovery of the North Galactic anomaly first pointed out by Humboldt (1869) and now called the Local Supercluster after de Vaucouleurs (1956). Seares' "nebular" counts were made using Mount Wilson 60-inch reflector plates which had been taken to produce the great Mount Wilson Catalog of 139 Selected Areas (Seares, Kapteyn, & van Rhijn 1930).

This massive study by Seares of the galaxy distribution was the last of a long series of earlier studies on galaxy distribution begun by Humboldt in the 1850's, Proctor (1869), Hinks (1911), Fath (1914), Hardcastle (1914), Reynolds (1920, 1923a,b), Wirtz (1923, 1924a,b), and undoubtedly others. Many of these works struck at the heart of the local galaxy distribution problems, solving them in principle but not in practice. Hubble's massive program begun in 1926, and published as data (Hubble 1931, 1934, 1936) and then with analysis (Hubble 1936, 1937) was the definitive study until the beginning of the current era with the Lick count program of Shane & Wirtanen (see Shane 1975 for a review). Shapley's Harvard 18th mag survey (see Shapley 1942, 1961 for a review) was too inhomogeneous to give quantitative results on $N(m)$.

The most important result of the early bright galaxy counts to $m \sim 13$ was that the number of galaxies brighter than apparent magnitude m increases as $\log N(m) = 0.6m + \text{constant}$ [e.g. equation (1.1)] when averaged over a "large enough" solid angle. For example, the Shapley-Ames $N(m)$ data to m_{pg} with its all-sky survey showed beyond doubt that the coefficient of the

$N(m)$ distribution is $0.6m$ (Shapley & Ames 1932; Shapley 1942, 1961), a result obtained earlier by Hubble (1926) with less homogeneous material.

Recall that the coefficient of 0.6 is the theoretical expectation for a homogeneous distribution in the mean, no matter what form or how wide is the luminosity function as long as (1) that function is independent of distance, (2) space is Euclidean, and (3) the objects being counted are in fact distributed with constant density outward (see von der Pahlen 1937, Bok 1937 for derivations of this most important result, which is not immediately intuitively obvious).

The fact that the coefficient in $\log N(m)$ is observed to be 0.6 for the bright counts (first to $m \sim 12$ and then extended to $m \sim 17$, Hubble 1926, Table XVIII) was the needed proof that galaxies provide a fair sample of the universe when large enough volumes are surveyed. This means that the averaging over large enough solid angles ($>$ 1 steradian) smooths the clusterings and voids to define a cosmologically important average density.

Said differently, the fact that $\mathrm{d}N(m)/\mathrm{d}m = 0.6$ means that there is no larger hierarchy beyond clusters of galaxies to contend with, and therefore that galaxies and their clusters at the intersections of the sheets separating the voids are the natural objects with which to map the geometry of space. A basic higher hierarchical structure would show itself by a deviation, in the large, of the 0.6 slope, that is not in fact observed (Sandage, Tammann, & Hardy 1972).

This does not mean that galaxies are distributed as uniformly as grains of sand on the beach. Rather, that averages over moderately large solid angles recover the signal for homogeneity in the mean counts.

Indeed, studies of the size of the voids (e.g. Schuecker and Ott 1991) show that by a redshift of $\sim$ 10,000 km s^{-1} the void structure has begun to repeat itself, the distribution averaged over spaces larger than that approaching homogeneity.

One of the stringent arguments of the 1930's between Hubble and Shapley concerned this question of homogeneity. Shapley repeatedly pointed out the tendency of galaxies to cluster and to form filaments in projection on the sky. Hubble argued for large-scale homogeneity based on the slope of $\mathrm{d}N(m)/\mathrm{d}m = 0.6$ when averaged "over the sky". Of course, for Hubble's program to determine space curvature from the counts to mean anything, homogeneity in the large was a requirement. As in many arguments, both protagonists had a grain of the truth. Both were right and both were wrong. The difference between Hubble and Shapley was simply one of scale.

Fig. 2.1. Einstein at a lecture given in 1931 at the offices of the Mount Wilson Observatory (Carnegie Institution of Washington).

2.4 Hubble's Program to Find Space Curvature from Galaxy Counts

During an extended visit to Southern California in 1931 Einstein lectured on general relativity at the offices of the Mount Wilson Observatory of the Carnegie Institution of Washington. The famous photograph (Fig. 2.1) taken during this lecture shows his query as to whether the contracted Riemann curvature tensor for the universe as a whole, R_{ij}, can equal zero.

Hubble was in the audience and perhaps realized then the possibility of using very deep galaxy counts to measure volumes for a Gauss-like program of experimental geometry by determining from galaxy counts how volumes deviate from their Euclidean measure, increasing as r^3 or not. In any case, whether stimulated by Einstein's lecture or not, Hubble began a massive observational program with the Mount Wilson 60 and 100-inch telescopes on galaxy counts to very faint magnitudes to extend the range of the $N(m)$ curve then known.

By 1934 Hubble had completed an intermediate count-"depth" survey to m_{pg} magnitude of 20, made by counting galaxies on 1184 photographic plates taken at random positions but on a grid covering the northern hemisphere visible from Mount Wilson. The total area of the sky covered by these plates was only about 300 square degrees, or only about 0.1 steradian. (Each plate covered an area encompassed by about 1/2 degree on a side, which is 0.25 square degree per plate).

From the count data Hubble (1934, 1936) calculated 5 normal $\log N, m$ points, defining his $N(m)$ count-magnitude relation at the $m_{\rm pg}$ magnitudes of 18.47, 19.0, 19.4, 20.0, and 21.03. There was no discussion in the 1934 paper of the cosmological aspects of space curvature or, indeed, no mention, even as a preamble, as to what was to come two years later in the attempt to find the "Gaussian curvature" via the volume test. (Perhaps, then, the 1931 Einstein lecture was not the driving influence for the beginning of this classic experiment that became so grand in scope by 1936).

What is clear is that Hubble teamed with Tolman in $\sim$ 1934 in a detailed discussion of "theoretical expectations" in the use of the observations to brute-force *measure* the curvature via comparisons of volumes at different distances – the Gauss protocol (Hubble & Tolman, 1935). The 1935 paper formed the basis of Hubble's (1936) analysis of the 1934 data, to which, by 1936, he had added the two faintest $N(m)$ data points at $m = 20.0$ and $m = 21.03$. To these he also added three further points due to Mayall (1934) at $m_{\rm pg} = 18.1$, 18.8, and 19.1. The quoted magnitudes here are on the system of the observations themselves, not yet corrected "for the effects of redshift", incorrect as these K corrections later proved to be, now discussed.

Hubble made corrections to his observed magnitudes, based on his version of the K term (see Lecture 3, section 3.1.1e). His K term was composed both of the technical effect to the measured magnitudes of shifting the energy distribution through the (fixed) filter band pass, (accounting for the shape of the SED), but contrary to modern practice he also included a term (or terms) for the "effects of redshift". By modern practice these are now included into the theory via the Robertson equation (2.1) above, the two factors of $(1 + z)$ for the "energy" and the "number" effect.

Hubble left open the possibility of a non-real expansion by giving *two versions* of the "correction for the effects of redshift". He included either one or two factors of $2.5 \log(1 + z)$ depending on whether the redshift was or was not due to a real expansion (section 2.1.1), but again, incorrectly, neglected the non-selective band-width term of the K correction.

From his analysis, Hubble reached the astounding conclusion that the data, corrected as if the expansion were real, using two factors of $(1 + z)$, could not support a true expansion explanation for the redshift unless an "impossibly" small radius of curvature for non-Euclidean space was adopted.

The argument is interesting, and his conclusion would have been "correct" (neglecting his error in the K correction and in the absence of luminosity evolution) if his data had been adequate. But in fact his magnitude scale gave too bright magnitudes for $m > 17$, see next section.

His method of analysis is intuitively confusing to the modern reader who has been trained to correct the *theoretical* expectations in deriving the $N(m)$ curves of different curvature values by putting the two $(1+z)$ terms into the Robertson equation (2.1) rather than by "correcting" the observations as Hubble did. Because of this difference, a confusion even of the *sign* of the curvature effect is sometimes made when reading Hubble's 1936 analysis. An outline of Hubble's reasoning is as follows. Tedious details follow in section 2.5.3.

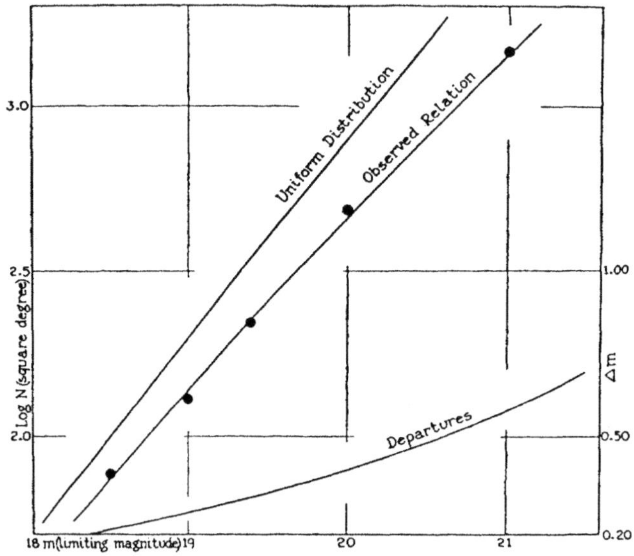

Fig. 2.2. Hubble's final $N(m)$ curve from Fig. 16 of *Realm of the Nebulae* and from his archive paper in ApJ **84**, 517 (1936).

Hubble (1936) assumed that a uniform distribution of galaxies in space (*when corrected for the effect of the curvature*) must obey $\log N(m) = 0.6m_{\rm c} + \text{const}$ where the corrected magnitudes, $m_{\rm c}$, are related to the *observed* magnitudes m_0 by

$$m_{\rm c} = m_0 - A(m_0).$$

Here, A is the correction to the observed magnitudes required to force the observed slope of the $N(m_0)$ curve, which was about 0.5 for the 1936 data as seen in Fig. 2.2, to the slope of 0.6 required by Hubble's assumption of homogeneity. Said differently, given the observed $N(m_0)$ distribution, the A correction converts $N(m_0)$ into $N(m_{\rm c})$ such that the "corrected" counts have a $dN(m_{\rm c})/dm_{\rm c}$ slope of 0.6. Clearly, A can be determined from the observations as the correction that is needed to produce a 0.6 slope on the "corrected" data. Hubble calls the $A(m_0) = f(m_0)$ function the "departure", shown as the bottom curve in Fig. 2.2 taken from Hubble (1936).

Expressed as the magnitude corrections to m_0, A is composed of (1) the effects of redshift on magnitude via the K term (the selective "technical" component plus the band-width term, if it had been used), and either one or two factors of $(1+z)$ depending on the reality of the expansion, and (2) the effect of space curvature which is the "residual" that Hubble sought. The curvature factor is the correction to the leading $(R_0 r)^3$ term in equation (1.38) in Lecture 1, which is

$$f(R_0 r) = \frac{3}{2}\left(\frac{\arcsin r}{r^3} - \frac{\sqrt{1-r^2}}{r^2}\right), \tag{2.10}$$

which Hubble writes explicitly, as we do here.

The principal problem now is that both the K correction and the curvature correction in equation (2.10) are functions of distance $R_0 r$ (and therefore z), whereas Hubble had only counts to various *apparent magnitudes*. Hence Hubble had to make some assumption about the redshift, and therefore the mean distance, of galaxies at a given apparent magnitude. Hubble (1926) had already determined the mean absolute magnitude of an "average" galaxy to be $M_{\mathrm{pg}} = -15^{\mathrm{m}}.2$, and had by that time obtained, with Humason, a reliable mean redshift-apparent magnitude relation for field galaxies (Hubble & Humason 1934, and Lecture 5 later). Hence magnitudes could be converted to redshifts for average galaxies, and all that remained was an assumption of how the distances $R_0 r$ varies with redshift z.

The intuitive assumption was for a simple redshift-distance relation of the form

$$R_0 r = \frac{cz}{H_0}, \tag{2.11}$$

or

$$r = \frac{cz}{R_0 H_0}, \tag{2.12}$$

which Hubble adopted (see section 2.5.3 for the exact relation in curved space).

What remained for Hubble was to apply the corrections "for the effects of redshift" to the data points in Fig. 2.2 (either one or two factors of $1+z$) to account for the "departure",A, in all factors *except* the curvature correction, and then to determine the curvature term $f(R_0 r)$ needed to explain the value of A. Note that the $(1+z)$ and the $f(R_0 r)$ factors *had to ultimately be expressed in apparent magnitudes*, which is all that the count data could provide anyway, redshifts being unknown.

In this way, Hubble made the five data points in Fig. 2.2 brighter by one or two redshift factors (expressed in magnitudes), moving them to the left. He then had to bring the "redshift corrected data points" back to the $0.6m$ line by the curvature term which acts to make the magnitudes fainter [they move again to the right depending on the value of $f(R_0 r)$ from equation (2.10)].

When Hubble applied the magnitude corrections for both the "energy" effect and the "number" effect, the data points moved so far to the left that a very large $f(R_0r)$ curvature correction was needed to move them back to the imposed $0.6m$ slope condition. This gave an "unsatisfactorily" small radius of curvature. But if only the "energy" effect correction was made, the $f(R_0r)$ term was less, and the radius of curvature was large enough to satisfy Hubble's intuition. Hence he concluded that the expansion may not be real (Hubble 1937, 1953).

2.5 Why Hubble's Program Failed

2.5.1 Systematic Errors in Hubble's Magnitude Scale

It became known in the 1940's (Baade 1944) that the faint apparent magnitude scale of the Selected areas in the standard Mount Wilson Catalog of 139 Selected Areas was in need of revision, generally starting at about $m_{\text{pg}} = 16$. The Seares scale in the Selected Areas was the basis of the magnitude scale used by Hubble.

Correction to the scale in the three selected areas of SA 57, SA 61, and SA 68 was the subject of a fundamental paper by Stebbins, Whitford, and Johnson (1950) where the errors to the old Mount Wilson scale became as high as ~ 1.5 magnitudes at $m_{\text{pg}}(\text{Seares}) \sim 20$.

The error is present in all the Selected Areas tested. A sampling of typical corrections to the standard B system for four Selected Areas is shown in Fig. 2.3 obtained in an unpublished investigation of photoelectric photometry using the Mount Wilson Hooker reflector in the 1980's as part of the Mount Wilson Halo Mapping project (Sandage 1983).

Because the effects of space curvature are so subtle, as we have seen earlier, magnitude errors of this size, in fact, destroyed the test even if Hubble had (1) used the exact equations set out here and in Lecture 1, (2) if his K terms had been correct (i.e. if he had used a valid SED and had not neglected the band-width term), and (3) if luminosity evolution in the look-back time could have been negligible.

2.5.2 The Error in Hubble's K Term

Hubble's formulation and error of the K correction to observed magnitudes for the effects of redshift was composed of three terms. (1) By redshifting an adopted spectral energy distribution for a standard galaxy through the filter spectral response, the change of the received flux due to a non-flat SED is calculated. This is the selective term. (2) The stretching of the received SED by the redshift, each wavelength redshifted by $(1+z)$ causes the received band-width to be reduced by $(1+z)$ from what it would have been in the absence of redshift. This is the non-selective *band-width term*, so often emphasized herein as *neglected by Hubble.* (3) As said in the last section, Hubble included one

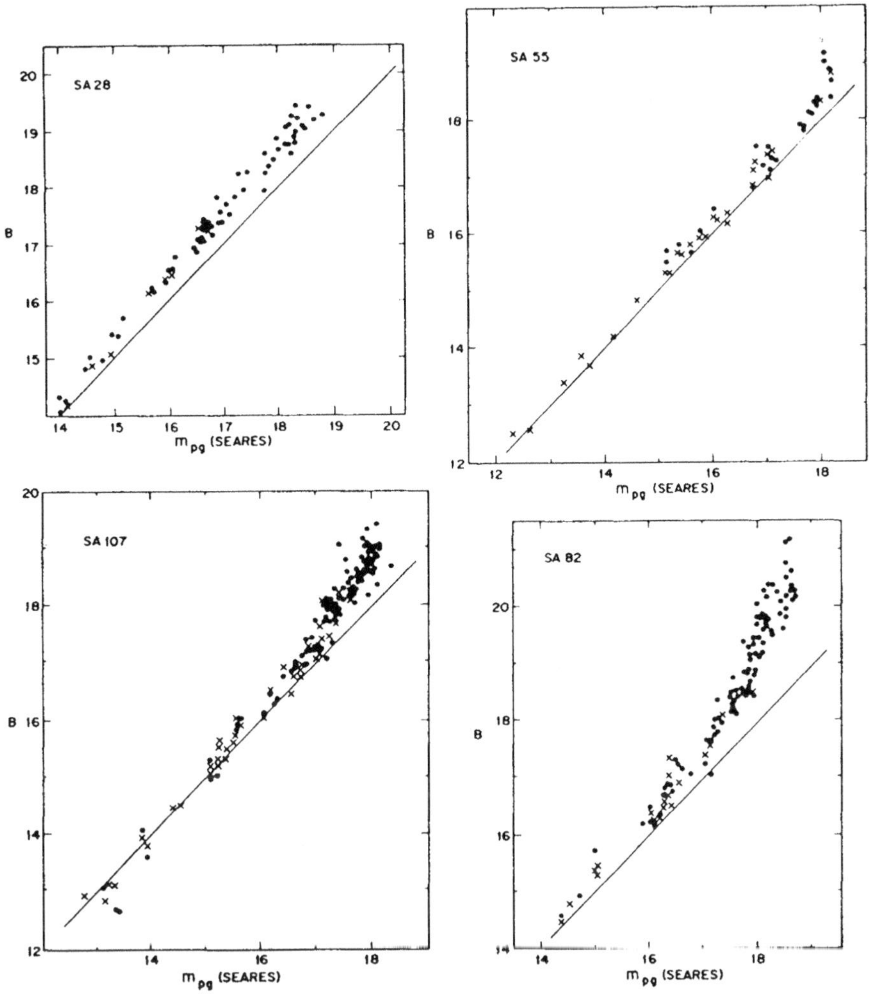

Fig. 2.3. Corrections to the Seares *et al.* magnitudes in four Selected Areas determined by photoelectric photometry in the 1960's. From Sandage (1983).

or two terms of $(1+z)$ in the correction to the received luminosity according to whether the expansion is real or not. Also as said, we now take these two terms into the theory via the fundamental Robertson (1938) equation (2.1) that connects the expected $N(m, q_0)$ and $m(z, q_0)$ relations.

Hubble's different practice of including this third item for the "energy" and the "number" effects in the K correction rather than in the prediction from theory as we do here is not an error and leads to no error in itself. Rather the error in principle was his neglect of the band-width term. The other error in practice was his assumption of a blackbody spectrum for the SED (of temperature 6000° K) to shift through the filter response. First, galaxy spectra are not black body, and second, an effective black body color temperature of M31 is closer to 4000° K than to 6000° K (Greenstein 1938)[9].

The modern definition of the K correction, expressed in magnitudes, following Humason, Mayall, and Sandage (1956), and repeated in Oke and Sandage (1968), is

$$K = 2.5\log(1+z) + 2.5\log\frac{\int_0^\infty S(\lambda)I_0(\lambda)\mathrm{d}\lambda}{\int_0^\infty S(\lambda)I_z(\lambda)\mathrm{d}\lambda}, \tag{2.13}$$

where $S(\lambda)$ is the filter transmission function and I_0 and I_z is the SED at zero and at z redshift. Note that the "energy effect"and the "number effect" are missing from this equation – because they are included in equation (2.1).

2.5.3 Hubble's Definition of "Distance" Before the Mattig Equation

The details of Hubble's analysis are found in his profound (as to the scope of the inquiry and the grandeur of the quest for a world model there is no other description) 1936 paper, which follows from the development by Tolman in his text book (Relativity, Thermodynamics, and Cosmology, 1934) as condensed in Hubble and Tolman (1935). We sketch in somewhat more detail than in section 2.4 the method of analysis.

The volume enclosed within coordinate distance r, in a curved space, both here [equation (1.16)] and in Hubble (1936) and Hubble and Tolman (1935), is

$$V(R_0 r) = \frac{4\pi}{3}(R_0 r)^3\left[\frac{3}{2}\left(\frac{\arcsin r}{r^3} - \frac{\sqrt{1-r^2}}{r^2}\right)\right], \tag{2.14}$$

or,

$$V(R_0 r) = \frac{4\pi}{3}(R_0 r)^3\ f(R_0 r), \tag{2.15}$$

---

[9] Hubble did not take criticism easily. When Greenstein (1938) published his note from Yerkes on the observed color temperature of M31 that was lower than Hubble had assumed in 1936, Hubble was not pleased. Greenstein came to CalTech in 1948 from Yerkes to begin the astronomy department at that Institution, with which Hubble was also connected through the Mount Wilson-Caltech initial agreement to operate Palomar. These men did not become close colleagues. There was a coolness. I saw it first hand during the five years spent as a Caltech graduate student from 1948 to 1953 and as Hubble's observing assistant from 1950 to 1953.

Greenstein had previously recommended me to Hubble in the summer of 1949, and I had begun working for him on galaxy counts using 48-inch Schmidt plates in early 1950. Greenstein had sent me up from the campus to the Mount Wilson offices, following Hubble's request for a student apprentice. Whom would Greenstein have sent had they been friends?

where the second term in brackets in (2.14), named $f(R_0r)$ in (2.15) from (2.10), is due to curvature.

This expands, as in equation (1.18), to

$$V(R_0r) = \frac{4\pi}{3}(R_0r)^3\left[1 + \frac{3}{10}kr^2 + O(r^4)\right]. \tag{2.16}$$

Note the *sign* of the second order correction in kr^2 compared with the sign of the correction using the "distance": determined by astronomical measurement [Lecture 1, section 1.6.4, comparing equations (1.18) and (1.19)].

The problem now becomes one of relating r and R_0r to the redshift which, in principle, is the only observable, or in practice, relating z with m, where m is more easily observed and which was the only parameter generally known in 1936. As we have often seen earlier, Hubble (see the discussion of Table 1.1 in Lecture 1) assumed that

$$R_0r = \text{“astronomical } d\text{”} = \frac{cz}{H_0}. \tag{2.17}$$

However, the correct precept is

$$r = \sin\frac{d}{R}, \tag{2.18}$$

from Lecture 1 [section 1.4 and equations (1.14), and (1.15)].

Hubble's other assumption concerning "luminosity distance" was

$$l_{\text{bol}} = \frac{L_{\text{bol}}}{4\pi(R_0r)^2} = \frac{L_{\text{bol}}H_0^2}{4\pi c^2 z^2}, \tag{2.19}$$

using equation 2.17 for the luminosity distance. [The correct method in the modern theory is to use the Mattig equation for R_0r and then substituting into the Robertson equation (2.1).] Hubble then transformed equation (2.17) into magnitudes as

$$m_{\text{bol}} = (m_{\text{obs}} - \Delta m) = M - 5 + 5\log(R_0r),$$

where Δm contains all the corrections needed to change bolometric magnitudes to the m_{pg} band and further to correct for the "effects of redshift". Hence

$$\Delta m = \text{B. C.} + K_{\text{H}}(z), \tag{2.20}$$

where $K_{\text{H}}(z)$ is Hubble's K correction that includes the technical selective term calculated by shifting the spectrum through the filter band pass (and as said so often before, which should have included the band-width stretching factor as well) plus either one or two factors of $2.5\log(1+z)$ depending on the reality of the expansion, as explained previously.

Hubble then assumed that

$$(R_0 r)^3 = A\ 10^{[0.6(m_{obs} - \Delta m - K)]}\ f\left(\frac{cz}{H_0 R_0}\right), \tag{2.21}$$

where the $f(z, R_0)$ term is the curvature term in equation (2.14) and (2.15).

Hubble then translated the above equations into

$$\log N(m) = 0.6\left[(m_{obs} - \Delta m) + \frac{10}{6}\log f(z, R_0)\right], \tag{2.22}$$

valid for a homogeneous distribution of galaxies in space.

He then operates on equation (2.22), knowing the $N(m)$ values from his counts, m_{obs} from his photometry, and Δm from equation (2.20) for the K term (selective plus energy and/or number terms). The only unknown in equation (2.22) is the curvature term $f(R_0)$ from which he determines R_0 for the two cases of either applying both the energy and number effect term or only one of them. From the analysis, Hubble concluded he must apply only one of these terms, because otherwise the needed curvature $f(R_0 r)$ correction would have been too large to suit his intuition. This then lead to his famous conclusion that the expansion "may not be real".

Where was his error? It was not in the relatively minor problems of not using exact equations for $R_0 r$ [eqs. (1.36) and (1.37)] or the exact Mattig redshift-distance relation [eq. (2.3)] rather than equations (2.11), (2.12), and (2.18). These equations reduce to the same asymptotic values at low redshifts, where we now know Hubble's $N(m)$ data apply. (For Hubble's count data, $\langle z \rangle$ is hardly larger than ~ 0.4.)

Rather, the main differences between Hubble's analysis and any modern discussion rests on (1) the large errors present in the 1936 Mount Wilson Catalog of 139 Selected Areas magnitude scales (section 2.5.1) and (2) the value of the K corrections, properly defined (section 2.5.2). It will be of some considerable historical interest to repeat the analysis using Hubble's 1936 count data with the modern magnitude scales, K corrections, and the exact Mattig equations set out here.

But finally, all the above has been developed as if we can analyze the data using a mean absolute magnitude for all galaxies that enter the $N(m)$ counts, i.e. by neglecting the vast spread in intrinsic luminosity of galaxies given by the luminosity distribution function. The next lecture develops the practical method to obtain a more realistic $N(m, q_0)$ prediction using the full machinery of the luminosity function with the Mattig equations and a Friedmann universe with $\Lambda = 0$.

References

Baade, W. 1944, ApJ **100**, 137
Bok, B.J. 1937, *Distribution of Stars in Space* (Chicago: Univ. Chicago Press), pp. 26-37
Brown, G.S. & Tinsley, B.M. 1974, ApJ **194**, 555
Davidson, W. 1959a, MNRAS **119**, 54
Davidson, W. 1959b, MNRAS **119**, 665
Davidson, W. 1960, MNRAS **120**, 271
de Vaucouleurs, G. 1956, Vistas Astron. **2**, 1584
Fath, E.A. 1914, AJ **28**, 75
Greenstein, J.L. 1938, ApJ **88**, 605
Hardcastle, J.A. 1914, MNRAS **74**, 699
Heckmann, O. 1942, *Theorien der Kosmologie* (Berlin: Springer-Verlag)
Hinks, A.R. 1911, MNRAS **71**, 588
Hubble, E. 1926, ApJ **64**, 321
Hubble, E. 1931, Pub. Astron. Soc. Pac. **43**, 284
Hubble, E. 1934, ApJ **79**, 8
Hubble, E. 1936, ApJ **84**, 517
Hubble, E. 1937, *The Observational Approach to Cosmology* (Oxford: Clarendon Press)
Hubble, E. 1953, MNRAS **113**, 658
Hubble, E. & Humason, M.L. 1934, Proc. Nat. Acad. Sci. **20**, 264
Hubble, E. & Tolman, R.C. 1935, ApJ **82**, 302
Humason, M.L., Mayall, N.U., & Sandage, A. 1956, AJ **61**, 97
Humboldt, W. v. 1869, *Cosmos* **4**, 25 (New York: Harper Brothers)
Mattig, W. 1958, Astron. Nach. **284**, 109
Mayall, N.U. 1934, Lick Obs. Bull. **16**, 177 (No. 458)
McVittie, G.C. 1956, *General Relativity and Cosmology*, 1st ed. (London: Chapman & Hall.)
Oke, J.B. & Sandage, A. 1968, ApJ **154**, 21
Proctor, R.A. 1869, MNRAS **29**, 337
Reynolds, J.A. 1920, MNRAS **81**, 129, 598
Reynolds, J.A. 1923a, MNRAS **83**, 147
Reynolds, J.A. 1923b, MNRAS **84**, 76
Robertson, H.P. 1938, Zs. f. Ap. **15**, 69
Robertson, H.P. 1955, Pub. Astron. Soc. Pac. **67**, 82
Sandage, A. 1961, ApJ **133**, 355
Sandage, A. 1962, in *A Memorial Tribute to H.P. Robertson*, J. Soc. Appl. Math. **10**, 737-801
Sandage, A. 1983, in *Kinematics, Dynamics, and Structure of the Milky Way*, ed. W.L.H. Shuter (Dordrecht: Reidel), p. 315
Sandage, A. 1988, Ann. Rev. Astron. Astrophys. **26**, 561
Sandage, A., Tammann, G.A., & Hardy, E. 1972, ApJ **172**, 253
Schuecker, P., & Ott, H.-A. 1991, ApJ **378**, L1
Seares, F. 1925, ApJ **62**, 168
Seares, F.H., Kapteyn, J.C., & van Rhijn, P.J. 1930, *The Mount Wilson Catalog of Photographic Magnitudes in Selected Areas 1-139* (Carnegie Inst. Pub. 402)
Shane, C.D. 1975 in *Galaxies and the Universe*, eds. A. Sandage, M. Sandage, & J. Kristian (Chicago: Univ. of Chicago Press), Chapter 16
Shapley, H. 1942 and 1961, in *Galaxies*, 1st and 2nd editions (Blakiston and/or Harvard Univ. Press), Chapter 6
Shapley, H., & Ames, A. 1932, Har. Ann. **88**, No. 2

Stebbins, J., Whitford, A.E., & Johnson, H.L. 1950, ApJ **112**, 469
Tolman, R.C. 1934, *Relativity, Thermodynamics, and Cosmology* (Clarendon: Oxford Univ. Press)
von der Pahlen, E. 1937, *Lehrbuch der Stellarstatistik* (Leipzig: J. A. Barth), p. 427f.
Wirtz, C. 1923, Medd. Lunds Obs. Ser II. No. 29
Wirtz, C. 1924a, Astron. Nach. **222**, 33
Wirtz, C. 1924b, Astron. Nach. **223**, 123

3 The Count–Magnitude Relation: Theory and Practice II (The Modern Attempts)

3.1 The Realistic $N(m)$ Test

The machinery developed in Lectures 1 and 2 is for the unrealistic case where the spread in the absolute luminosities of the galaxies has been neglected. The distribution in absolute magnitude was replaced there by some appropriate mean luminosity. This was the assumption made by Hubble, and was also the basis on which the theory of $N(m, q_0)$ using the Mattig equations for $R_0 r$, r, and $m(z, q_0)$ was developed (Sandage 1961) soon after the Mattig (1958, 1959) solutions had been found.

However, the only realistic procedure is to calculate the expected $N(m, q_0)$ by solving the fundamental equation of stellar statistics for the predicted $A(m)$ values by integrating over an appropriate set of luminosity functions, one for each galaxy type. In the usual notation, $A(m)$ is the number of objects (galaxies in this case) per unit area on the sky between apparent magnitude $m - \mathrm{d}m/2$ and $m + \mathrm{d}m/2$. The integral count, $N(m)$, is obviously

$$N(m) = \int_0^m A(m)\mathrm{d}m,$$

where $N(m)$ is the number of galaxies per square degree *brighter* than apparent magnitude m.

The purpose of this lecture is to develop a method of solving the fundamental equation using the non-Euclidean volume elements set out in Lecture 1 for any given q_0, for any adopted set of luminosity functions, for any adopted mixture of Hubble morphological types, for any adopted K correction to the magnitudes for the technical effects of redshift, and for any given assumed evolution (either density or luminosity) in the look-back time corresponding to the relevant redshift.

The problem is exceedingly complex because of the plethora of parameters. With so many degrees of freedom (the luminosity function, the morphological mix, the density function with its evolution with look-back time, and the luminosity variation again with look-back time, etc.) the solution is not robustly unique. Any observed $N(m)$ count distribution can, in principle, be fitted by varying any or all of these parameters.

However, a very strong constraint is the concomitant predicted and observed *redshift distribution*, $N(z)$, for a sample that is complete to a given apparent magnitude, or, the more easily obtained differential redshift distribution for a survey that is complete within the magnitude interval from m_1 to m_2. Because of the importance of this constraint, we necessarily are also concerned here with the calculation of the $N(z, q_0)$ redshift distribution (section 3.3). A comparison is then made between the predicted and the observed $N(m, q_0)$ and $N(z, q_0)$ distributions. Illustrations are given showing the effect of changing the various parameters on the predictions for both $N(m, q_0)$ and $N(z, q_0)$. In this way, agreement can be achieved between the predictions and the observations (section 3.4) imitating (perhaps) uniqueness.

For pedagogical reasons the method of calculation to be set out below is painfully elementary, made by numerical methods using a spread sheet. The advantage is that the roles of the separate functions required in the integral are made transparent. The method was invented by Kapteyn to deal with the problem of star counts in the Galaxy (Schouten 1918, Bok 1937, Mihalas and Binney 1981). Seares (1931) also solved the problem in a similar way. The Kapteyn solution is also set out by von der Pahlen (1937).

The scheme is discussed here with sufficient tedium to prepare the intuition for the formal, modern, powerful analytical methods described by Longair for the same problem. The subject is fundamental enough that the problem and its solution must be a part of the student's culture. Although the two approaches, one here by the m, log(distance) table and the other as the Longair, Cambridge analytical tripos route, seem so different, they are in fact the same. One appeals to the intuition while the other appeals to the elegance of a formalism in the limit of infinitesimal distance intervals $\mathrm{d}r$, treating the distribution functions as continua.

3.1.1 Theory of the $N(m, q_0)$ Calculation

3.1.1a The Fundamental Equation of Stellar Statistics

The familiar integral equation, valid in Euclidean geometry, for the number of objects, $A(m)\mathrm{d}m$, per square degree at apparent magnitude m in interval $\mathrm{d}m$, if the absolute magnitudes are distributed by a luminosity function $\phi(M)\mathrm{d}M$, which is the number of objects per unit volume at absolute magnitude M in interval $\mathrm{d}M$, is

$$A(m)\mathrm{d}m = \frac{4\pi}{41,253}\int_0^\infty r^2 D(r)\phi(M)\mathrm{d}M\mathrm{d}r. \tag{3.1}$$

The integration is made over all distances r. The apparent magnitude, m, the absolute magnitude, M, and the distance, r, are related by

$$m = 5\log r - 5 + M + A(r). \tag{3.2}$$

The distance r is in parsecs[10], and $A(r)$ is the absorption at distance r. (In the case of cosmology, A can be any correction to the apparent magnitudes due to say the K term for the effects of redshift, and/or a correction for luminosity evolution for example).

[10]Equation (3.2) simply defines the zero point of the absolute magnitude scale such that an object at 10 parsecs has the value of M equal to m. In reading the early literature (pre-1925) cognizance should be taken that the "absolute magnitudes" defined then were based on $M = m$ at a distance of 1 parsec.

Problem 3.1. Show that the number 41,253 in equation (3.1) is the number of square degrees over the sky given by $4\pi(180/\pi)^2$. This, and other numbers are part of a game played with any group of astronomers to test their past education. Can you identify and show the origin of such numbers as 206265, 4.74, 3.26, 0.4343, $3.16 \cdot 10^7$, $3.09 \cdot 10^{18}$, $2 \cdot 10^{33}$, $1.5 \cdot 10^{13}$, $4 \cdot 10^{33}$, $6.7 \cdot 10^{-8}$, etc. All are important for back-of-the envelope calculations in astronomy.

Other important bits of astronomical culture include knowing, for example, the angle subtended by your thumb when held at arms length, or by your fist, or by your hand with all fingers and thumb extended. Using your hand in this way to measure angles you can tell the altitude of the sun and therefore, for example, in an afternoon, how long to sunset knowing that the sky turns at 15° per hour.

Find, by means you should devise, the angle subtended by your thumb at arms length, which, of course, is an invariant from birth to death. With this knowledge, determine the approximate angular diameter of the sun by attempting to block it out with your thumb and seeing the ratio of size of thumb to the sun. Before you try the experiment, what is your intuition concerning the relative angular size of your thumb at arms length and the sun? Most non-astronomers, and indeed many astronomers have a bad intuition before trying this observation. Careful with the sun – it is bright. Try instead the full moon. Are the sun and moon the same angular size?

The meaning of equations (3.1) and (3.2) is that the volume contained in a shell at distance r of thickness $\mathrm{d}r$, whose inner and outer boundaries are at $r - \mathrm{d}r/2$ and $r + \mathrm{d}r/2$ has a volume of $4\pi r^2 \mathrm{d}r$ reduced to a solid angle of one square degree. Only objects with absolute magnitude M in absolute magnitude interval $M + \mathrm{d}M/2$ and $M - \mathrm{d}M/2$ at distance r, will be observed at apparent magnitude m in apparent magnitude interval $m + \mathrm{d}m/2$ and $m - \mathrm{d}m/2$.

Treating each shell at different distances in the same way (at a larger distance r_i, a brighter absolute magnitude, M_i, is required to give the same apparent magnitude at the observer), and summing the contribution of each shell at all distances gives the total number of objects per square degree that will be observed with apparent magnitude m in interval $\mathrm{d}m$. The number of objects per unit volume at absolute magnitude M in $\mathrm{d}M$ is given by the luminosity function $\phi(M)\mathrm{d}M$. The density of objects at distance r is the density function $D(r)$. Hence all terms in equation (3.1) are understood.

3.1.1b Solution via the m, $\log r$ Table

As said, a method of solution with great flexibility is via the $m, \log r$ table that sets out a spread sheet (Bok 1937, von der Pahlen 1937, Mihalas and Binney 1981). Such a table is shown in Fig. 3.1 for the case of Euclidean volume elements.

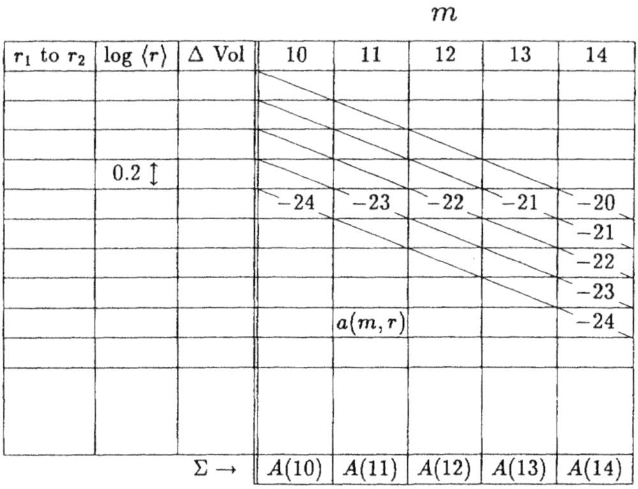

Fig. 3.1. An m, log(distance) table for Euclidean volume elements.

The discrete annular shells are separated in distance by equal intervals of $\Delta \log r = 0.2$. Because of the factor of 5 in equation (3.2) for $\log r$, the absolute magnitudes at the center of the shell intervals are separated by one magnitude. This defines the granularity of the numerical calculation with this interval in $\log r$. Finer integrations can, of course, be made by using smaller $\Delta \log r$ intervals.

The spread sheet in Fig. 3.1 has vertical columns with $\log r$ increasing downward. Each column is for a different apparent magnitude, marked along the top. Lines of absolute magnitude thread as diagonals through the diagram, following equation (3.2), taking account of any absorption, A, (or the K term etc. for the equivalent problem in cosmology at large z) if needed.

The volume enclosed between the shells of inner and outer radius r_i and r_{i+1} is the same for all boxes in a given row. Its value is

$$\Delta \mathrm{Vol} = \frac{4}{3}\pi(r_2^3 - r_1^3), \tag{3.3}$$

or, in the limit,

$$\mathrm{d}V = 4\pi r^2 \mathrm{d}r,$$

over the whole sky. The volume elements calculated for each individual shell via equation (3.3) are to be listed in column 3 of Fig. 3.1.

Each box can now be filled with a number that is the product of (1) the number of objects of the absolute magnitude corresponding to the particular box, found from the adopted luminosity function per unit volume at that absolute magnitude, (2) the volume from column 3, and (3) the density of objects at the particular distance. In the case of galaxies, this may include the effects of density and/or luminosity evolution with time corresponding to the light-travel-time for the distance (redshift) appropriate for that box. As developed in the next sections, all distances can be expressed as redshifts for given q_0 values, and therefore the light-travel time (section 5.5 of Lecture 5) can be calculated. Clearly, the entire calculation is to be made in redshift space.

Summing the columns in Fig. 3.1 gives the $A(m)$ values at the bottom of each. Summing the $A(m)$ values horizontally (across the bottom) to a any given limiting apparent magnitude m, gives $N(m)$. This is the number of objects per square degree brighter than apparent magnitude $m + \Delta m/2$. This is the count-magnitude relation which we seek.

3.1.1c A General $m, \log r$ Table for Cosmology

An $m, \log z$ spread sheet valid for cosmology with non-Euclidean geometry is more complicated than Fig. 3.1 because the volume element between r_i and r_{i+1} does not increase with radial coordinate r as in equation (3.3). Nor does the apparent magnitude vary with M and r as in equation (3.2), but rather in the more complicated way derived in Lectures 1 and 2. We must use the $V(z, q_0)$ equation of Lecture 1 together with the $R_0 r(z, q_0)$, the $r(z, q_0)$, and the $m(M, z, q_0)$ Mattig equations.

However, the problem is even more complicated because of the effects of redshift on the apparent magnitudes, which is the $K(z)$ correction term. Moreover, $K(z)$ is different for each galaxy type because the spectral energy distributions (SED) differ amongst the types. Therefore, any realistic calculation of $N(m)$ for field galaxies must use a set of $K(z, \text{galaxy type})$ corrections and an assumed fraction of types in the mix at all distances. Because each Hubble galaxy type will also have different evolutionary changes (luminosity and/or density evolution), the adopted luminosity function must be galaxy type and evolution dependent.

To make the calculation we therefore need:

a. The volume elements $V(z, q_0)$ from Lecture 1,
b. The relation $m(z, q_0)$ between redshift and apparent magnitude (i.e. the equation for the Hubble diagram) from the Mattig equation,
c. The K correction for the effects of redshift for every galaxy type and evolutionary assumption,
d. The mix of galaxy morphological types at every redshift,
e. The luminosity function for every galaxy type and redshift.

This list embraces an enormous parameter space giving a wide range for variations in the parameters in attempts to match actual $N(m)$ count data. The ingredients that enter the prediction of $N(m)$ are not unique without constraints based on other considerations (e.g. as said, the redshift distribution for the same galaxies used for the count-brightness data). This lack of uniqueness explains the plethora of papers in the literature, varying this and that to make a fit to observed data but with the conclusions, in general, lacking acceptance.

However, to demonstrate the method we go part way along the path paved with these necessities, showing how to construct an $m, \log z$ spread sheet for a given q_0 from items (a) and (b) to which the complication of items (c) - (e) can be added at will to each individual box of the spread sheet at given redshift and galaxy type.

3.1.1d The General Parametric Equations of the Problem

For completeness and for ease of reference we repeat the relevant equations for the volume elements $V(r)$ and $V(z, q_0)$ from the previous lectures. We then apply them for the two special cases of $q_0 = 0$ and $q_0 = 1/2$ using the Mattig equations for $m(z, q_0)$ to obtain $V(m, q_0)$, first without K correction. (The examples can be generalized to any value of q_0.)

$V(r)$ is obtained by integrating the Robertson-Walker metric in equation (1.1) of Lecture 1 for the space part of that metric, giving

$$\left.\begin{aligned} V(r) &= \frac{4}{3}\pi R_0^3 r^3 && \text{for } k = 0, \\ V(r) &= \frac{4}{3}\pi R_0^3 r^3 \left[\frac{3}{2}\frac{\arcsin r}{r^3} - \frac{3}{2}\frac{\sqrt{1-r^2}}{r^2}\right] && \text{for } k = +1, \\ V(r) &= \frac{4}{3}\pi R_0^3 r^3 \left[\frac{3}{2}\frac{\sqrt{1+r^2}}{r^2} - \frac{3}{2}\frac{\operatorname{arsinh} r}{r^3}\right] && \text{for } k = -1. \end{aligned}\right\} \quad (3.4)$$

These are changed to functions of z and q_0 from the Mattig equations of

$$\left.\begin{aligned} R_0 r &= \frac{c}{H_0 q_0^2 (1+z)}\left\{q_0 z + (q_0 - 1)\left[-1 + \sqrt{1 + 2q_0 z}\right]\right\}, \\ \frac{kc^2}{R_0^2} &= H_0^2(2q_0 - 1) \\ r &= \frac{\sqrt{2q_0 - 1}}{q_0^2(1+z)}\left\{q_0 z + (q_0 - 1)\left[-1 + \sqrt{1 + 2q_0 z}\right]\right\}. \end{aligned}\right\} \quad (3.5)$$

The special cases of interest, because they are the least messy to calculate, are

$$\left.\begin{aligned} R_0 r &= \frac{2c}{H_0}\left(1 - \frac{1}{\sqrt{1+z}}\right), \\ V(R_0 r) &= \frac{4}{3}\pi R_0^3 r^3, \\ m(\text{obs}) &= M(z=0) + K(z) + E(z) + 5\log\left[2(1 + z - \sqrt{1+z}\right] \\ &\quad + 43.89, \end{aligned}\right\} \quad (3.6)$$

for $q_0 = 1/2$. The constant of 43.89 in the magnitude equation is valid if the Hubble constant is 50 km s^{-1} Mpc^{-1}, found by evaluating the equation in the limit of z approaching zero and requiring that $m - M + 5 = 5\log r$ with the ordinary astronomical distance in parsecs.

The similar equations for $q_0 = 0$ are

$$\left.\begin{aligned} R_0 &= \frac{c}{H_0}, \\ r &= \frac{z}{1+z}\left(1+\frac{z}{2}\right), \\ V(R_0 r) &= \frac{4}{3}\pi R_0^3 r^3 \left[\frac{3}{2}\frac{\sqrt{1+r^2}}{r^2} - \frac{3}{2}\frac{\operatorname{arsinh} r}{r^3}\right], \\ m(\mathrm{obs}) &= M(z=0) + K(z) + E(z) + 5\log z\left(1+\frac{z}{2}\right) + 43.89. \end{aligned}\right\} \quad (3.7)$$

Note again that Hubble (1926) used the intuitive equation

$$R_0 r = \frac{cz}{H_0}, \quad (3.8)$$

at all redshifts, which is an approximation valid only in the limit of low z.

Problem 3.2. Prove that equation (3.8) is the $z \to 0$ limit of equations (3.6) and (3.7).

In constructing an $m, \log z$ table for any value of q_0, the general method is to find the z_i and z_{i+1} values for the inner and outer boundary shells that correspond to the apparent magnitude values such as $11^{\mathrm{m}}\!.0$, $12^{\mathrm{m}}\!.0$, $13^{\mathrm{m}}\!.0$ etc. for values of the corrected absolute magnitude, $M + K(z) + E(z)$, that differ by 1 mag. This requires solving the Mattig $m(M, z, q_0)$ Hubble diagram equation (i.e. in reverse) for z, given m and $M + K(z) + E(z)$. This equation, repeated from Lecture 1, is

$$m_{\mathrm{bol}}(\mathrm{obs}) = M_{\mathrm{bol}}(z=0) + 5\log\frac{1}{q_0^2}\left\{q_0 z + (q_0 - 1)\left[-1 + \sqrt{1 + 2q_0 z}\right]\right\} + 43.89. \quad (3.9)$$

[The bolometric magnitudes can be replaced by heterochromatic magnitudes if we use $M_i(z=0)$ in photometric band i at zero redshift and the observed apparent magnitude in band i at redshift z, corrected for the effects of redshift, i.e. use $m_i(\mathrm{observed}) - K(z) - E(z)$ instead of simply the observed m.]

In constructing an $m, \log z$ table for $q_0 = 0$ (or $q_0 = 1/2$), we would follow the following steps.

A. Solve the magnitude equation, i.e. the fourth equation of (3.7) [or the third equation of (3.6)] for the run of z values that correspond to 1 mag intervals in $m_{\mathrm{bol}} - M_{\mathrm{bol}}$, or $m(\mathrm{obs}, z) - [M(z=0) + K(z) + E(z)]$ in heterochromatic magnitudes. This solution will give the z_i and z_{i+1} values for the inner and outer boundaries of the shells that give apparent magnitudes say $11\overset{m}{.}0$, $12\overset{m}{.}0$, $13\overset{m}{.}0$ etc. for particular assigned values of $M + K(z) + E(z)$.

B. Using these z_i and z_{i+1} values, calculate the corresponding r_i and r_{i+1} and $R_0 r_i$ and $R_0 r_{i+1}$ values from the explicit equation (3.5) for any arbitrary q_0. Note that the special cases for $q_0 = 1/2$ and 0 are equations (3.6) and (3.7).

C. Use these specific r and $R_0 r$ values for the specific apparent magnitudes used in item A. Each $R_0 r$ value corresponds to a specific z value. The family of these z values are used to calculate the volumes enclosed in each z value from equations (3.4) for any general q_0 value, or from equations (3.6) and (3.7) for the special cases of $q_0 = 1/2$ or 0, respectively.

D. Subtract the inner volume from the outer volume to obtain the volume to be used for the particular relevant box in the $m, \log z$ table corresponding to apparent magnitude m between $m + 0.5$ and $m - 0.5$. These are the volume elements of the particular shells to be entered in column 3 of the table equivalent to Fig. 3.1.

E. Note that each box now has an assigned volume element calculated as just described, an apparent magnitude corrected for the effects of redshift, and a corresponding absolute magnitude $M(z = 0) + K(z) + E(z)$. Note also that each box corresponds to a known mean redshift at the center of the box, bracketed by the redshifts at the inner and outer shell boundaries calculated in step A.
To fill the box we must multiply the volume element by the number of galaxies per unit volume at absolute magnitude $M+K(z)+E(z)$ obtained from the luminosity function, where $E(z)$ is defined by an assumed luminosity evolution. If there is density evolution, a density factor as a function of redshift, $D(z)$, must also be included.

F. As before, summing the columns vertically gives the predicted $A(m)$ values. Summing this sum horizontally at the bottom of the table gives the predicted $N(m)$, *but only for the galaxy type for which the adopted* $K(z)$ *applies.*

G. Repeat steps (A) - (F) for each galaxy type; the $K(z)$ correction for the effects of redshift depend on the SED and hence on Hubble type (see the next section). Hence, the m, $\log z$ table calculated in the way just described (for any fixed q_0) must be used in series, once for each galaxy type using a mean $K(z)$ correction for that type.

The predicted $N(m)$ distribution for a particular mix of Hubble types in the general field is the sum of the individual type-specific $N(m)$ predictions, weighted by an adopted ratio of Hubble types in the total mix. Because of the known change of the percentages of Hubble types with environment (Hubble and Humason 1931; Dressler 1980; Postman and Geller 1984) the mix ratio also depends on environment. Data on the mix in typical environments have been published by Pence (1976), Tinsley (1980), and Ellis (1986) among others. These references give an entrance to the extensive literature.

Each or all of the parameters can be tweaked to obtain agreement of the "theory" with the data. Clearly, the $N(m)$ test tells us more about luminosity and density evolution than about space curvature because the dependence on q_0 turns out to be nearly miniscule (Lecture 2, section 2.2; see also section 3.4), whereas the effect of evolution in the $E(z)$ term is at least linear in z.

3.1.1e The $K(z)$ Correction

The definition of $K(z)$, following Humason, Mayall, and Sandage (1956) and Oke and Sandage (1968), which is consistent with its application in concert with the fundamental Robertson $l(L, z, R_0 r)$ in equation (2.1) of Lecture 2 [rather than the way Hubble discussed the redshift-magnitude data where he included the $(1 + z)$ term either once or twice] is, as in equation (2.13) of Lecture 2,

$$K = 2.5 \log(1 + z) + 2.5 \log \frac{\int_0^\infty S(\lambda) I_0(\lambda) \mathrm{d}\lambda}{\int_0^\infty S(\lambda) I_z(\lambda) \mathrm{d}\lambda}. \tag{3.10}$$

The first term is the bandwidth term (neglected by Hubble). The second is the "selective" term due to the shifting of the SED, called $I_z(\lambda)$, through the filter response, denoted by $S(\lambda)$ in the equation.

The calculation was made by Oke and Sandage (1968) using SED measured by them for eight giant E galaxies and for the center of M31. These data gave $K(B)$, $K(V)$, and $K(R)$ corrections for giant E galaxies to redshifts of $z = 0.28$, which, at the time, was the extent of most of the cluster data (Humason, Mayall, & Sandage 1956) except for that of Minkowski (1960) at $z = 0.46$.

However, because the spectral energy distributions upon which these K corrections were based were for the central regions of E galaxies, i.e. $(A/D) < 0.05$, where A is the measuring aperture and D is the isophotal diameter of the galaxy to an isophote of say 25 mag arcsec^{-2}, the Oke/Sandage $K(z)$ corrections refer to too red an SED (the centers of E galaxies are redder than the outside) needed for a "total" galaxy magnitude.

Whitford (1971) produced more relevant K corrections based on scans of five giant E galaxies over the wavelength interval from 3400Å to 11,000Å and for a large fraction of the total light ($A/D = 0.6$). His resulting SEDs are more representative of distant cluster E galaxies for which the measuring apertures (Sandage 1972a, 1972b, 1973; Sandage & Hardy 1973) were large enough to cover most of the image. Said differently, the A/D ratio used by Whitford for his standards were close to the multiple measuring apertures used by Sandage in the photometry of cluster galaxies. A similar study by Schild & Oke (1971) gave similar $K(z)$ results for the B, V, and R photometric pass bands.

Hubble's K correction, besides adding the $(1+z)$ terms for the "effects of redshift" as part of "K" (which we do not do), and neglecting the bandwidth stretching term, were for a *blackbody* SED of 6000 K temperature which he redshifted through his filter bands to calculate the selective term. His final adopted values were K_{pg}(Hubble) $\sim 4z$ for "motion" and $K(z) \sim 3z$ for "no motion" , but again note that these contain the "number" and/or "energy" effect terms, whereas the modern definition of $K(z)$ does not. If Hubble had not added these effects of redshift to his K term, but if he *had* added the bandwidth term, he would have obtained $K_{pg} = 2z + 2.5 \log(1+z)$ (i.e. adding here the neglected bandwidth term and subtracting the energy and number effects to conform to modern practice, but retaining his assumption of a 6000 K blackbody spectrum). Hence, Hubble's final K_{pg} term would have been, on the precepts of modern usage but with Hubble's assumption of the SED, $K_{pg} = 3z$ at small z. We now compare this with the modern value calculated using realistic (observed) SEDs with the result that $K_B = 5z$ at low z. Hence, Hubble would have underestimated the $K(z)$ term due to his assumption of too hot an effective SED.

A definitive calculation using modern SEDs such as summarized by Yoshii & Takahara (1988) for galaxies of each morphological type has been made by Coleman, Wu, & Weedman (1988). Their adopted SEDs are based on IUE moderate-to-far UV spectra combined with modern spectral scans in the optical. For z smaller than 0.3, their calculated K corrections, consistent with the definition by Humason, Mayall, & Sandage (1956) and Oke & Sandage (1968) are $K_B = 5.0z$, $K_V = 2.3z$, and $K(R) = 1.0z$ for giant E galaxies. They give extensive tables for other galaxy types as well and for redshifts reaching $z = 2$ where the low redshift linear approximations, given here as examples, are to be replaced by the tables.

Using the modern K corrections we can test if the SED of E galaxies have changed due to evolution of the stellar content in the look-back time. Such tests using broadband $B - V$ and $V - R$ colors were made by Kristian, Sandage, & Westphal (1978) showing no color evolution at the level of less than $0\overset{m}{.}04$ in $B - V$ in the look-back time to at least redshifts of $z = 0.4$. The data are shown in Fig. 3.2. A similar test with the same null result was made using higher resolution SEDs by Wilkinson and Oke (1978). These studies followed earlier pioneer work on SEDs as a function of redshift by Oke (1971) and by Crane (1975, 1976).

It is important to emphasize that different K corrections must be applied for different galaxy types. It then follows that in any sample of mixed types where such tests as the $N(m)$ counts or the $m(z)$ Hubble diagram are being used, the number ratios of the mixture of types must be known (item d of section 3.1.1c).

3.1.1f Galaxy Luminosity Functions

The luminosity function for each galaxy type in the mix must be known before the individual $N(m)$ count-magnitude predictions even for a given type can be calculated by the methods of this lecture.

A review of the evolving ideas concerning this function is in Binggeli, Sandage, & Tammann (1988) where it is emphasized that the apparent dis-

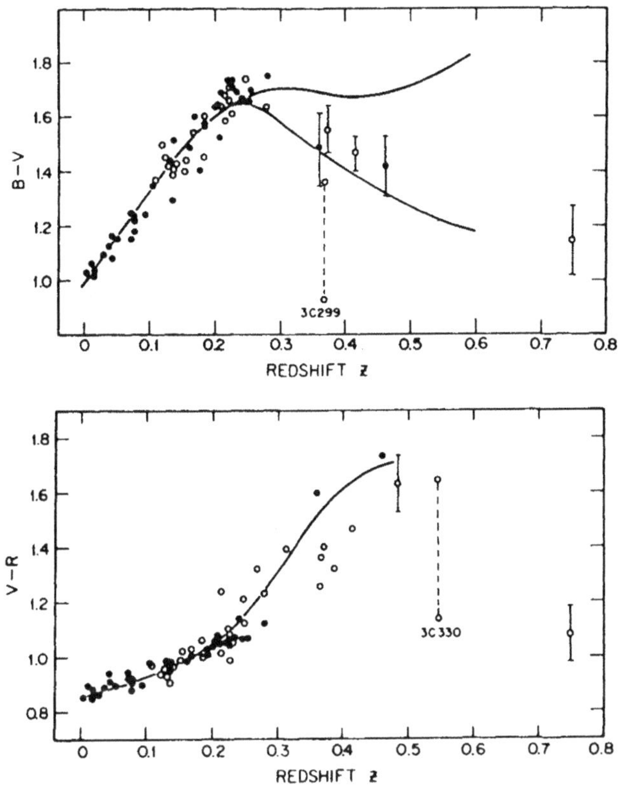

Fig. 3.2. Color as a function of redshift showing no color evolution in the look-back time. The diagram is from Kristian, Sandage, & Westphal (1978).

parate views of Hubble and Zwicky on the form of this function are actually not so disparate. Hubble's near log-normal (Gaussian in magnitudes) distribution applies to spirals (Sandage, Binggeli, & Tammann 1985), whereas Zwicky's exponential increase in numbers with decreasing luminosity applies to the dwarf irregular and dwarf E galaxies (the dE types).

A concept for the form for a "general" luminosity function has been set out by Schechter (1976) as a modification of an original proposal by Abell (1975). The precept that a "general" function exists has been widely used, often incorrectly. The concept is too simple in practice where predictions of $N(m)$ and $N(z)$ that are correct in principle are required. Each galaxy type has its own form of the luminosity function.

The type-specific functions must be added in different proportions in different environments according to the morphological type-density relation (Hubble & Humason 1931; Dressler 1980), giving a different "total" luminosity function for each specific environment (Jerjen, Tammann, & Binggeli

Fig. 3.3. Binggeli's (1987) cartoon of Schechter suppressing the details of the "general" luminosity function.

1992). This, of course, is fatal to the concept of a "general" luminosity function. *Type-specific* luminosity functions are required, and these cannot generally be summed with a fixed type ratio to produce a "general" luminosity function that has universal application.

The problem is illustrated in Fig. 3.3 due to Binggeli (1987), showing that the details of the problem are buried under foot with the Schechter function. In particular, the analysis by Jerjen *et al.* shows that the value of L^*, central to the Schechter formulation, varies with the morphological mix in specific environments.

The "total" luminosity function in the intermediate density environment of the Virgo cluster has been determined for each morphological type in a study of a complete catalog of Virgo cluster members (Binggeli, Sandage, & Tammann 1985, BST). The resulting composite luminosity function obtained by summing each type-specific function is in Fig. 3.4 (BST 1988 op. cit.). Further discussions of the type-specific luminosity functions in different environments, as they sum to a total in each case, are given in Ferguson & Sandage 1991), with a sample of the results shown in Fig. 3.5.

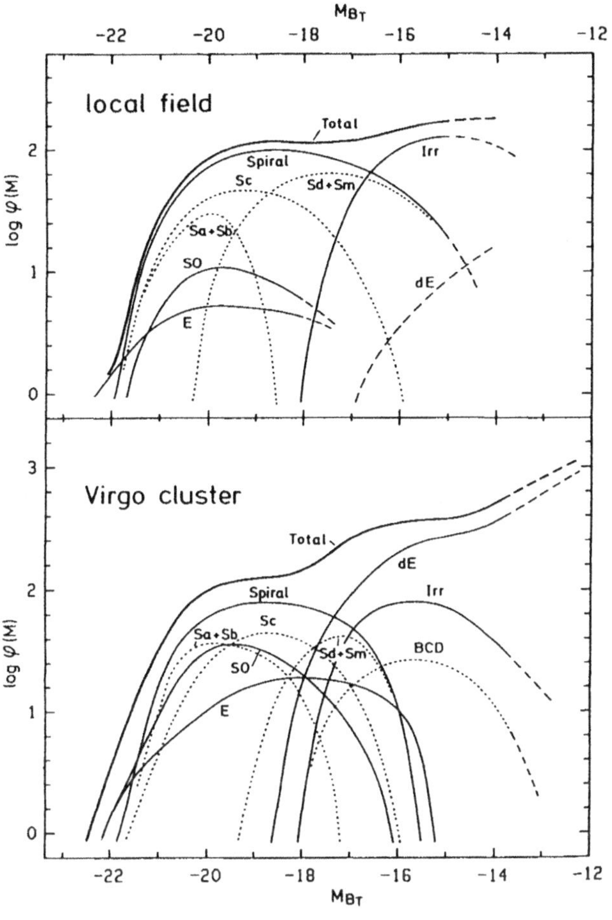

Fig. 3.4. The type-specific luminosity functions combined into a "total" luminosity function for the Virgo cluster and for an "average" place in the general field. Diagram from Binggeli, Sandage & Tammann (1988).

3.2 The $N(m)$ Data

3.2.1 The Early Count Data

An important early $N(m)$ count-magnitude analysis was the 1926 study by Hubble that followed many earlier, less complete studies of galaxy distribution by Hardcastle, Fath, Wirtz, Seares, and others discussed in Lecture 2. Hubble showed beyond doubt that the number of galaxies counted per unit area on the sky increased as 0.6 dex per unit change in the apparent magnitude, valid at least to $m_{pg} = 17$. Later, in an extension of this early study to fainter magnitudes, Hubble (1934) had carried the count-magnitude data to the limit of the Mount Wilson 100-inch telescope at $m = 21$, and again found no edge to the distribution.

Modern data on galaxy counts to $m_{pg} = 19$ are shown in Fig. 3.6 taken from a review (Sandage, Tammann, & Hardy 1972, STH). The purpose of the study was to counter a claim by de Vaucouleurs (1970, 1971) and by Haggerty & Wertz (1971, see also Haggerty 1971, and Wertz 1971) that the

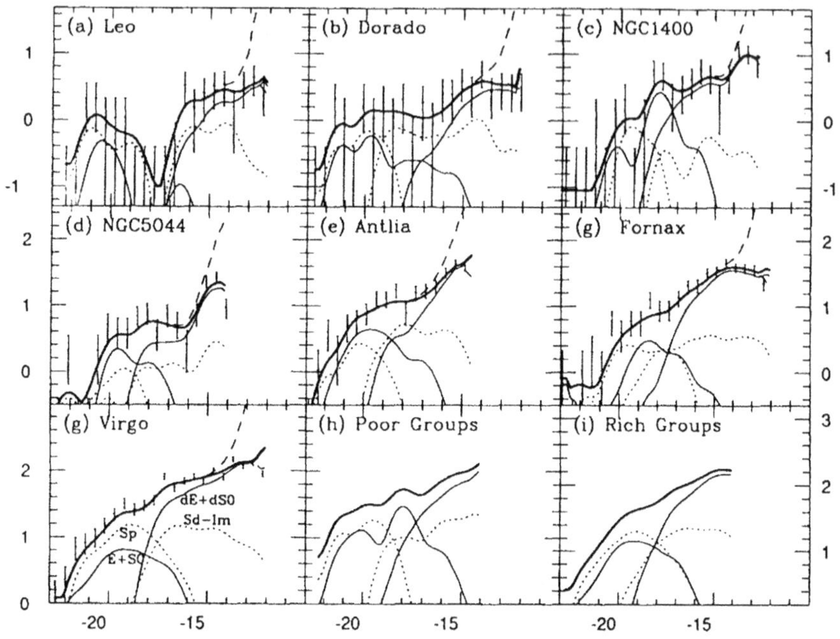

Fig. 3.5. The individual luminosity function of various galaxy groups compared with the combined function for the Virgo Cluster. Relative $\log \Phi(M)$ is the ordinate. Diagram from Ferguson & Sandage (1991).

distribution of galaxies defined a hierarchy, as set out by Charlier (1908, 1922) with a hierarchical scaling factor (see STH for definitions) of 1.7, shown as a dot-dashed line in the diagram.

The hierarchical model is clearly disproved by Fig. 3.6. The observed data are inconsistent with any thinning rate; the slope of the count-magnitude correlation remains at $d\log N(m)/dm \sim 0.6$. Nevertheless, the data show the well known very local (i.e. $v < 2000$ km s^{-1}) North Galactic Hemisphere density anomaly given by the offset of the North Galactic Cap counts from the South Galactic Cap data alone. Both *slopes* are $d\log N(m)/dm \sim 0.6$, but there is an offset in the zero point due to the north galactic cap density enhancement by about a factor of 2 (see Yahil, Sandage, & Tammann 1980).

The second point of the diagram shows the depth to which the counts reached in the era just before the very extensive count data began in modern times since 1970, discussed in the next section.

Data by Mayall (1934), Holmberg (1958), and Shane & Wirtanen (1954, 1967) are shown. Note that at $m = 19$ the slope of the count-magnitude relation has fallen short of 0.6 due to the effects of space curvature[11]. (The

[11] As discussed in Lecture 2, the principle reason for Hubble's conclusion that the expansion is "probably not real", is that the pre-1950 magnitude scales used by

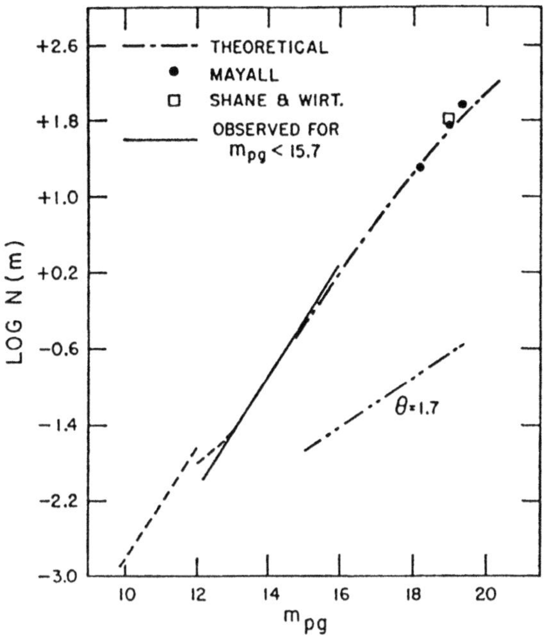

Fig. 3.6. The bright $N(m)$ count-magnitude relation from extant data available to 1972. The diagram is from Sandage, Tammann, & Hardy (1972).

systematic errors in the magnitude scales used in the studies shown in Fig. 3.6 are relatively minor to at least $m = 17$).

3.2.2 Modern Counts and the Blue Excess

Many modern studies exist on the $N(m)$ relation to very faint magnitudes that themselves are on a correct zero point and the Pogson magnitude scale. Among the first is by Tyson & Jarvis (1979) shown in Fig. 3.7 where several important features of the problem are displayed. (1) The effects of space curvature are shown beginning at $J = 17^{\mathrm{m}}$ in the decrease of slope from 0.6. (2) The number of stars per unit area becomes less than the number of galaxies at about $J = 22^{\mathrm{m}}$. (There are problems with the summary plot

all astronomers fainter than about $m_{\mathrm{pg}} = 17$ were incorrect [except the International Polar Sequence which was fine to its limit at $m_{\mathrm{pg}} \sim 20$. The Polar Sequence, adopted internationally is set out in Transactions of the IAU, Vol. 1, Table I of the report of Commission 25, Seares (1922) as author].

This can be seen by comparing the ordinate values in Fig. 3.6, which has approximately the correct magnitude zero point (the Shane/Wirtanen point was determined on the modern system) at a given m_{pg} with Hubble's $N(m)$ data shown as Fig. 2.3, as discussed in Lecture 2. This comparison shows that fainter than $m_{\mathrm{pg}} \sim 17$, Hubble's zero point is too bright, and the magnitude error increases progressively with faintness, with the consequences discussed in Lecture 2.

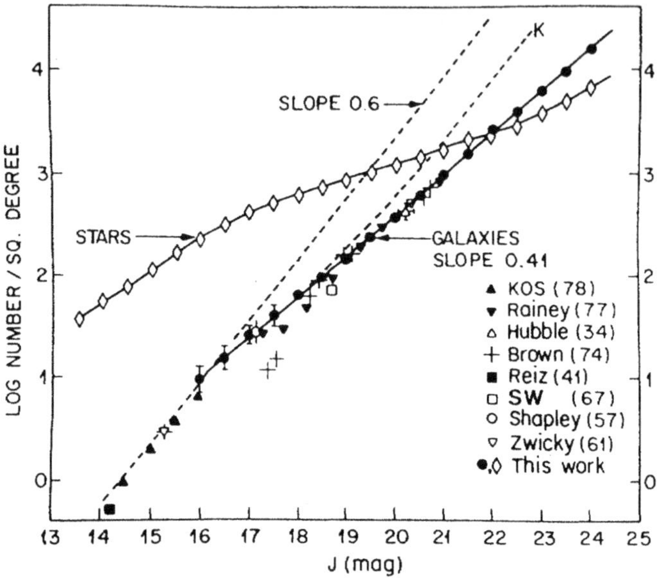

Fig. 3.7. The count-magnitude diagram from Tyson & Jarvis (1979).

here, among which is the apparent agreement of Hubble's 1936 points with the modern data, whereas we know that Hubble's magnitude scale was at least $1^{\rm m}_{\cdot}5$ too bright at his $m_{\rm pg} = 21$, but never mind, the diagram was the first of what is now a very long list of modern $N(m)$ studies, some of which we now cite to gain access to the extensive literature. The Hubble data should be reevaluated. This is left as a problem for the students.)

Counts with the Russian 6-m telescope, with comparison with the predictions from theory were given by Karachentsev & Kopylov (1978). Counts to $J = 25^{\rm m}$ with the AAT telescope in Australia were set out (also with a review) by Peterson *et al.* (1979) where again a comparison is made with the predicted counts with and without luminosity evolution in the look-back time.

A summary to 1986 is given by Ellis (1986) whose diagram of the $N(m)$ counts in the blue is reproduced here as Fig. 3.8. The striking feature is the enormous excess in the observed numbers compared with the predictions of no evolution models. This excess has continued to be found in all more recent counts since the summary by Ellis, and is the current problem not yet understood.

A comprehensive study and review by Yoshii & Takahara (1988) summarizes the faint count data to $B_J = 25^{\rm m}$, again showing the excess of the counts over the no evolution predictions. The excess was discovered by Kron (1980) where he also showed that the faintest galaxies in the count survey are blue, providing, it is widely believed, a clue for the explanation.

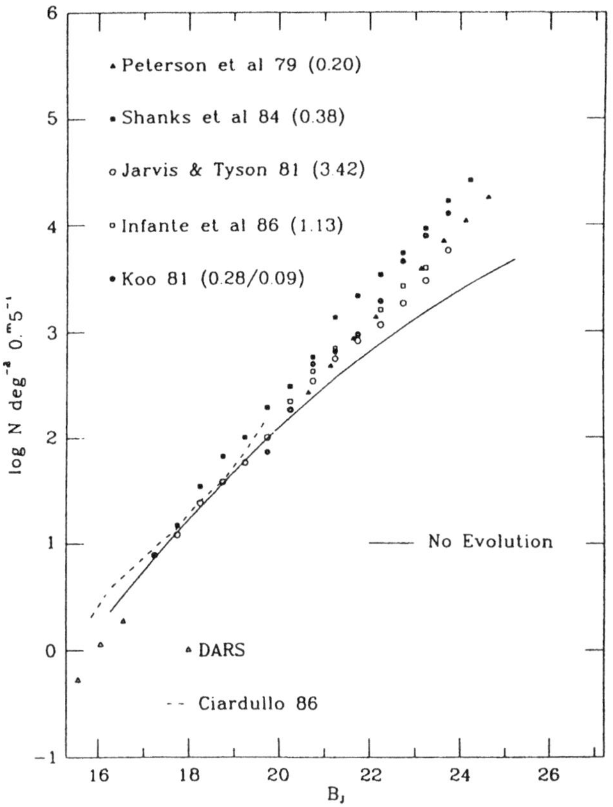

Fig. 3.8. Number counts from Fig. 2 of Ellis (1986) from IAU Symposium 124.

The data to 1990 in the three wavelength bands of B, r, and K have been summarized by Koo & Kron (1992), which we set out here as Fig. 3.9. Note that the excess in the actual counts over the no evolution expectation occurs in the B and the r band but not the K band, which again must be an important clue to an explanation.

A discussion of faint counts in B and I magnitudes is by Lilly, Cowie, & Gardner (1991) showing the effects of luminosity evolution vs. density evolution in the look-back time. Their Fig. 8 is reproduced here as Fig. 3.10, again showing the excess as in the data by Ellis (Fig. 3.8), Koo & Kron (Fig 3.9), Yoshii & Takahara (1988, their Figs. 3, 8, and 9, not shown here).

The most complete discussion of the possibilities of the predictive model such as has been set out earlier in this lecture is the discussion by Cole, Treyer & Silk (1992) where they compare the $N(m)$ distribution with the observed $N(z)$ and thereby put limits on possible explanations of the Kron blue excess in the faint counts. But before we can discuss this method to constrain the explanations, we must first show how to calculate the number-

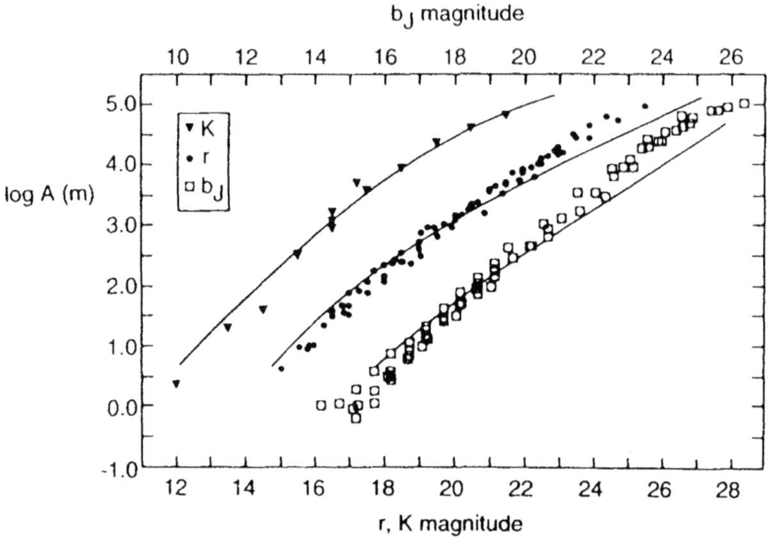

Fig. 3.9. The $N(m)$ count data in three bands discussed by Koo & Kron (1992) showing the excess in the blue data but no excess over the standard model in the K band to their magnitude limit.

redshift, $N(z, m)$, distribution from the standard model for any arbitrary apparent magnitude interval.

3.3 Calculation of the N(z,q_0) Distribution

The problem is to calculate the expected *redshift distribution* of the galaxies in a list that is complete within the apparent magnitude interval of m and $m + \Delta m$ (between say $m = 12$ and $m = 13$). To do so requires summing the number of galaxies *over all distances* that contribute to that apparent magnitude interval, i.e. over all absolute magnitudes at *each* distance that give apparent magnitudes in m to $m + \Delta m$. The calculation requires finding the relevant volume elements, and with them, using the luminosity function to compute the number of galaxies at each distance that can contribute to the magnitude interval m to $m + \Delta m$ in a way similar to the calculation of $N(m)$ via an m, $\log r$ table.

Using Euclidean volume elements as illustration, it is evident from the discussion in section 3.1.1c that the calculation of the expected $N(v)\Delta v$ "velocity" distribution at low redshift (i.e. the number of galaxies at velocity v in velocity interval Δv) is given by

$$N(v)\Delta v = \omega v^2 \Delta v H_0^{-3} \int_{-\infty}^{\infty} \phi(M) D(v) \mathrm{d}M, \tag{3.11}$$

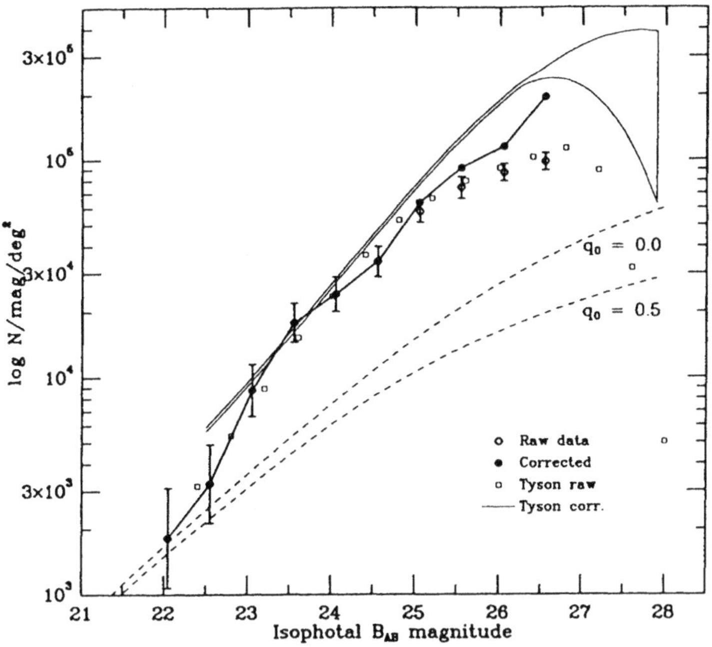

Fig. 3.10. Number counts from Lilly, Cowie, Gardner (1991, their Fig. 8) showing the great excess of the faint counts compared with the standard model.

(cf. Yahil, Sandage, & Tammann 1980) where ω is the solid angle sampled by the catalog, H_0 is the Hubble constant (in this approximation the "distance" is $r = cz/H_0$), $\phi(M)$ is the luminosity function, and $D(v)$ is the density function. The relation between the variables is $m = M + 5\log v + 16.51$, where v is in km s^{-1} and which assumes a Hubble constant of $H_0 = 50$ km s^{-1} Mpc ^{1}. In this local approximation, the $N(v)$ distribution is independent of q_0 because the volume elements are Euclidean. In the general case of large redshift and arbitrary q_0 we must use the complications of the non-Euclidean volume elements and the Mattig equation for the distance $R_0 r$ as developed in section 3.1. This final complication is made for the diagrams shown in the last section 3.4 of this lecture.

3.3.1 Calculation Via an m, $\log z$ Table

The solution of (3.11) can, as in section 3.1.1b, be done using the spread sheet formalism of the $m, \log r$ table (Fig. 3.1), but now where the volume elements are calculated using the redshifts z and $z + \Delta z$ rather than r and $r + \Delta r$ in column 1. Again each box of the spread sheet at apparent magnitude m_i corresponds to a particular absolute magnitude $M = m - 5\log cz - 16.51$, and, in this Euclidean approximation has its characteristic volume element

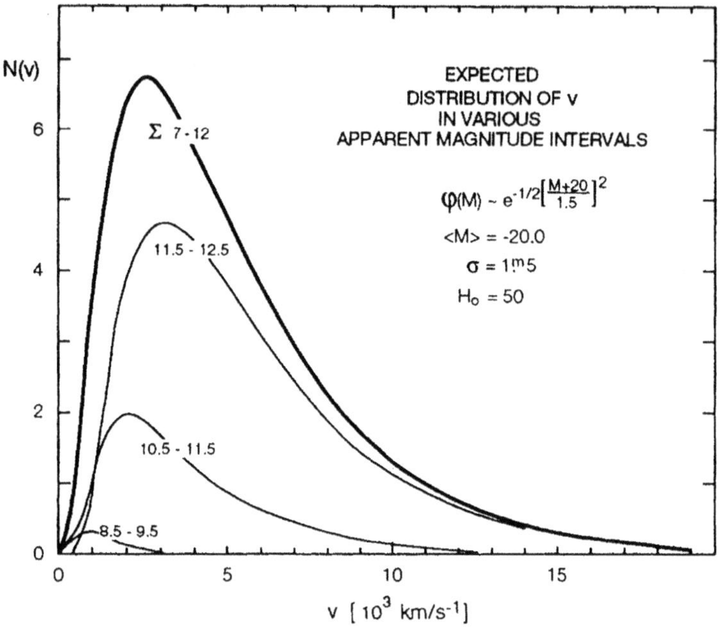

Fig. 3.11. Illustrative calculation of the expected count-redshift distribution for four different flux-limited samples assuming $\sigma = 1^{\mathrm{m}}\!.5$ for the luminosity function. The assumed parameters are $H_0 = 50$ and $\langle M \rangle = -20^{\mathrm{m}}$. Euclidean geometry is assumed.

$$\Delta V = \frac{1}{3}\pi\Big(\frac{c}{H_0}\Big)^3(z_i^3 - z_j^3), \tag{3.12}$$

between redshifts of z_i and z_j. These redshifts define the inner and outer boundaries of the Euclidean shell when the relevant absolute magnitude at the distance of the center of the box is M_i when the apparent magnitude is m_i.

The entries in each box are the volume elements, so calculated, multiplied by the luminosity function per unit volume at the appropriate absolute magnitude. Again, these box numbers are the number of galaxies in the redshift interval z and $z + \Delta z$ listed in column 1 that contribute to the number of galaxies between apparent magnitude m and $m + \Delta m$. The $N(z)\Delta z$ value at the mean z of the particular box is found by dividing the number in that particular box by the redshift interval (the range) listed in column 1.

An illustration of this calculation using a Gaussian luminosity function with $\langle M \rangle = -20^{\mathrm{m}}$, a dispersion of $1^{\mathrm{m}}\!.5$, $H_0 = 50$, and a constant space density is in Fig. 3.11 for the four apparent magnitude intervals of $8^{\mathrm{m}}\!.5$ to $9^{\mathrm{m}}\!.5$, $10^{\mathrm{m}}\!.5$ to $11^{\mathrm{m}}\!.5$, $11^{\mathrm{m}}\!.5$ to $12^{\mathrm{m}}\!.5$, and from 7^{m} through 12^{m}. This last distribution would, for example, be the predicted velocity distribution of galaxies in the Revised Shapley-Ames Catalog (Sandage & Tammann 1987) where redshifts are known for the entire sample that is nearly complete to $m = 12.5$.

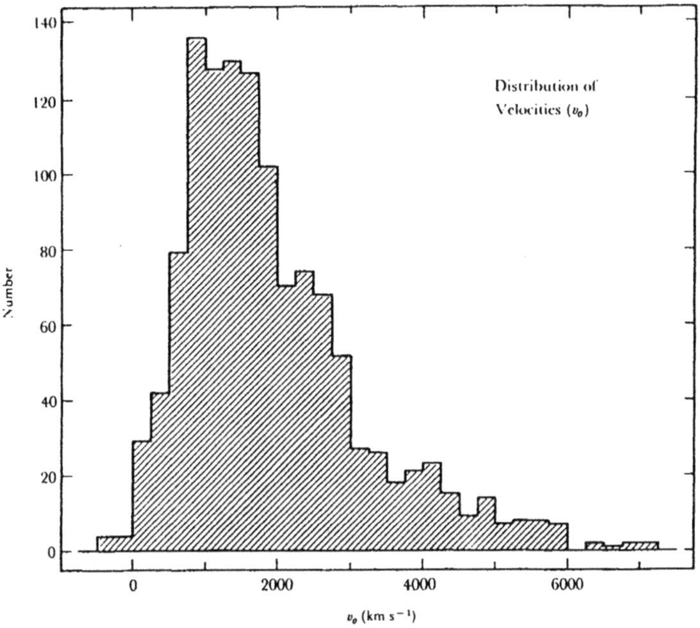

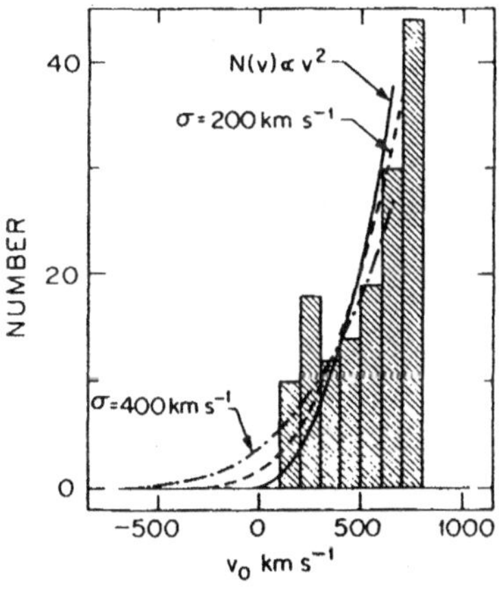

Fig. 3.12. (Top) Observed distribution of velocities in the flux-limited sample of the Revised Shapley-Ames Catalog. Diagram is from Fig. 2 of the RSA. (Bottom) From Fig. 5 of Sandage (1978).

The shape of the individual distributions in Fig. 3.11 shows the effect of incompleteness in a magnitude-limited catalog at large redshifts. At the small redshifts the $N(v)$ distribution rises at first strictly as v^2. This is the part of the nearby space that has been completely sampled in the magnitude-limited catalog. Said differently, the catalog defines a strict distance-limited sample at the smallest distances, and the $N(z)$ distribution rises directly as the volume. That this is in fact the case for the Shapley-Ames catalog, is shown by the

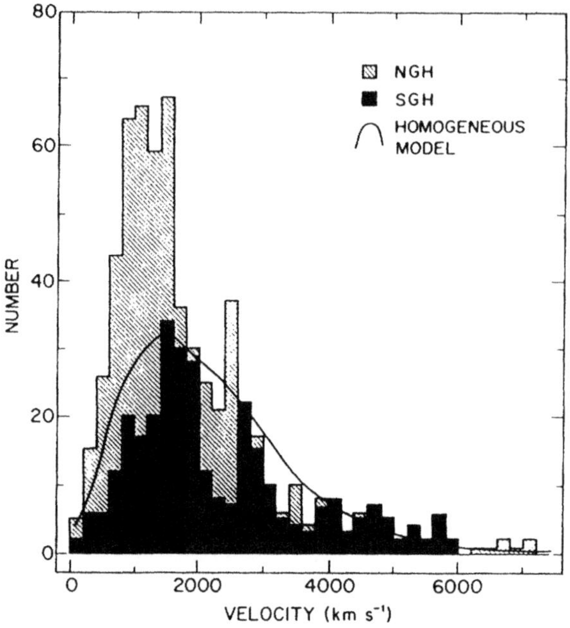

Fig. 3.13. The $N(v)$ distribution from Yahil *et al.* (1980), from their Fig. 5.

actual velocity distribution at low velocities, discussed elsewhere (Sandage 1978, his Fig. 5), and shown here as the second panel of Fig. 3.12.

The expected $N(v)$ distribution for larger velocities has a maximum and then declines with a long tail that approaches zero in a way determined by the *bright* end of the luminosity function. This is the region where the observational selection due to the imposed apparent magnitude limit progressively cuts away the faint end of the luminosity function.

Fig. 3.12 shows the observed $N(v)$ distribution of the RSA Catalog (Sandage & Tammann 1987, Fig. 2) which has the characteristic shape of the distributions in Fig. 3.11. Note, however, that the calculated distribution in Fig. 3.11 for the magnitude interval between $m = 7$ and $m = 12$ continues to higher velocities (up to 16,000 km s^{-1}) whereas the observed distribution in Fig. 3.12 continues only to 6,000 km s^{-1}. This is a consequence of the parameters we chose here for the pedagogical calculation in Fig. 3.11. If we had used a fainter $\langle M \rangle$ value, a better match could have been obtained such as we obtained from a more exact analysis (Yahil, Sandage, & Tammann 1980), shown in Fig. 3.13.

Note again the extreme flexibility we can achieve in the calculation via the $m, \log z$ table approach. Changes in the luminosity function as a function of morphological mix, in arbitrary density functions (for example, accounting for sheets and voids), in the $K(z)$ correction as a function of galaxy type, in the evolutionary correction $E(z, t)$, etc. can be incorporated easily into the calculation, all of which will be evident in the next section.

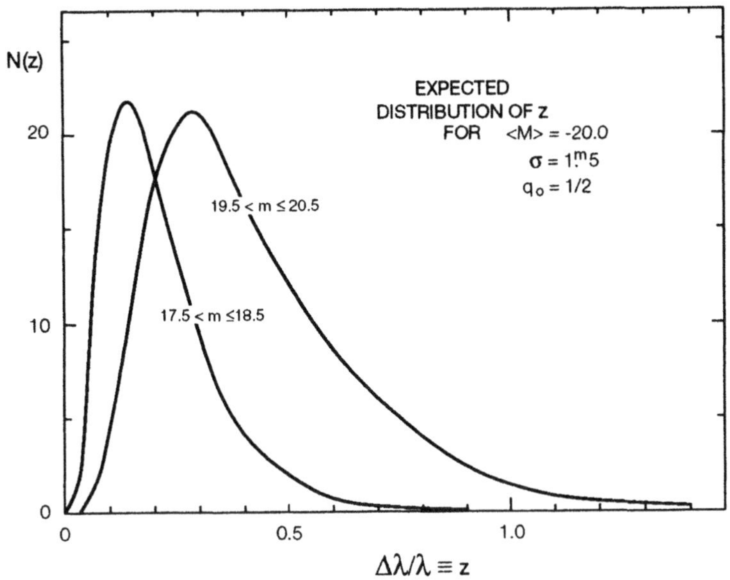

Fig. 3.14. Illustrative calculation of the shape of the $N(z)$ distributions for the magnitude intervals of m = 17.5 to 18.5 and 19.5 to 20.5 assuming the same parameters as in Fig. 3.11.

Fig. 3.11 was calculated from a Euclidean $m, \log z$ table using the volume elements from equation (3.3). As discussed earlier, at large redshifts we must use the appropriate $m, \log z$ table calculated as in section 3.1.1c taking into account the curvature if q_0 is not 1/2, i.e. the appropriate cosmological $m, \log z$ table must be used.

Figure 3.14 shows a calculation of $N(z)$ (with arbitrary absolute numbers) for an apparent magnitude interval from $m = 17.5$ to 18.5 and $m = 19.5$ to 20.5 for $q_0 = 1/2$, $\langle M \rangle = -20^{\mathrm{m}}$, and a dispersion to a Gaussian luminosity function of $1\overset{\mathrm{m}}{.}5$. Note again the characteristic shape and the shift toward larger redshifts for fainter magnitudes. If the more realistic value of $\langle M \rangle = -19^{\mathrm{m}}$ had been used (compare Fig. 3.11 with Fig. 3.13), the scale of the redshifts (the abscissa) would have been reduced by about a factor of 2, consistent with the real observations.

Figure 3.15 shows a sample of the available data, here from Colless *et al.* (1990).

We note several things. (1) The number of galaxies with large redshifts, say with $z > 0.4$, is very small. Hence, sampling *field* galaxies to faint magnitudes is not the way to study the universe at large redshifts. (2) Not shown explicitly here, but shown in the next section, is the fact that the theoretical $N(z)$ distributions are insensitive to differences in q_0 compared with either luminosity evolution or density evolution. Hence, the $N(m)$ count data and the $N(z)$ redshift distribution data tell us almost nothing about space curvature,

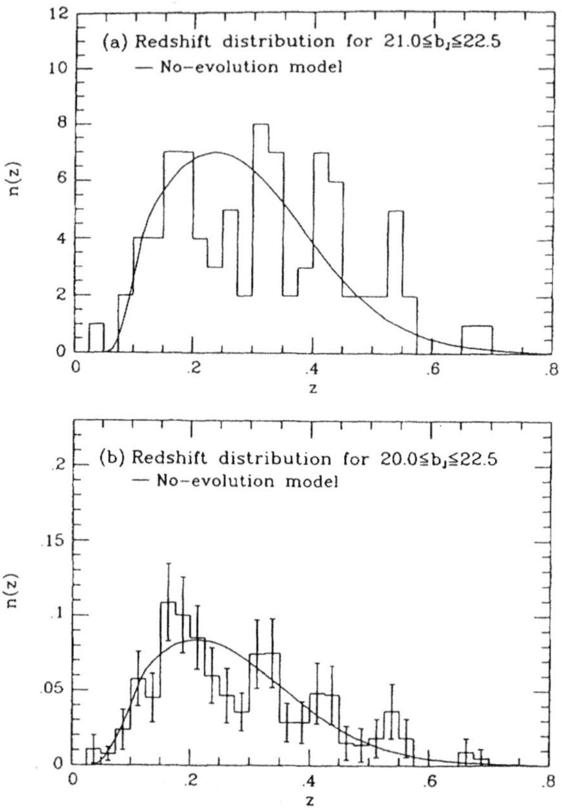

Fig. 3.15. The observed $N(z)$ distribution function of Colless *et al.* (1990).

but rather are sensitive indicators of density and/or luminosity *evolution*, which of course is useful for evolutionary studies.

The method of studying the predicted $N(m)$ and the $N(z)$ distributions compared with the observations, and seeing the effect of varying q_0, the luminosity and density evolution functions has begun to be used in the literature as a most powerful technique to understand the Kron blue excess in the $N(m)$ counts.

3.4 Comparing the $N(m, q_0)$ and the $N(z, q_0)$ Distributions for Consistency

The machinery developed so far provides the method to calculate the expected $N(m, q_0)$ and $N(z, q_0)$ distributions for any assumed values of the relevant parameters. In a remarkable summary paper Cole, Treyer, & Silk (1992) aim to discover the reason for the blue excess in the B_J counts, its lack in the K counts, and the concomitant observed $N(z)$ distributions. Their method is to range over much of the relevant parameter space, changing the parameters until a match is had with the observations.

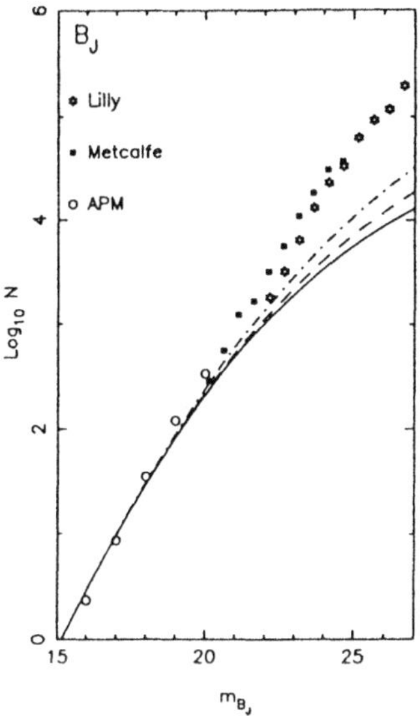

FIG. 1a

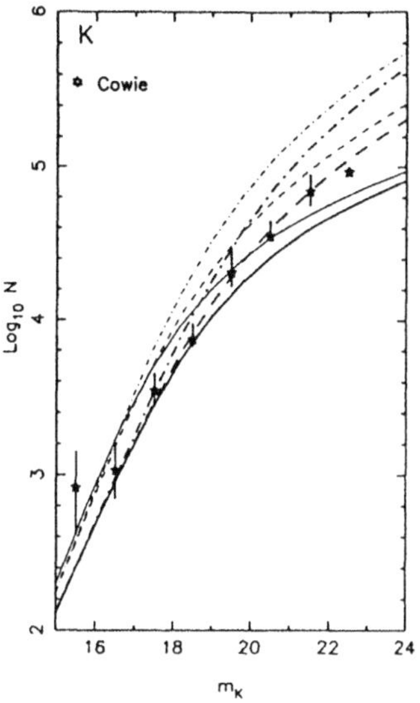

FIG. 1b

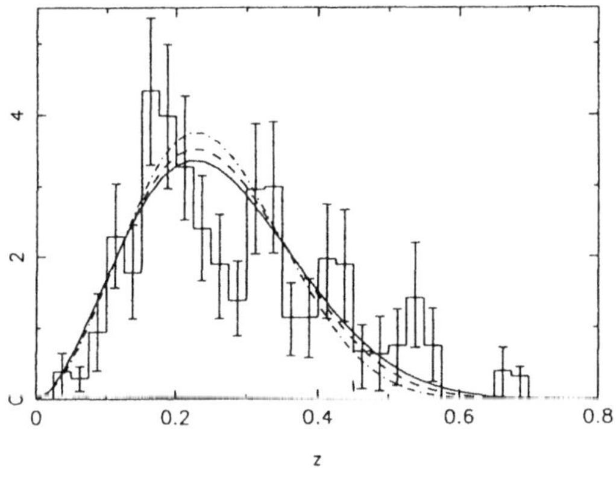

FIG. 1c

FIG. 1.—(*a*) Blue (B_J-band) differential number counts. The stars show the faint counts of Lilly et al. (1991), the squares the faint counts of Metcalfe et al. (1990), and the open circles the brighter galaxy counts of the APM galaxy survey (Maddox et al. 1990b). (*b*) Infrared (K-band) differential number counts. The stars show the faint counts of Cowie et al. (1991). The three curves in (*a*) and the heavy curves in (*b*) show the predictions of the no-evolution model detailed in § 4.1 for three cosmological models: flat, $\Omega_0 = 1$ (*solid line*); open, $\Omega_0 = 0.1$ (*dashed line*); and flat with $\Omega = 0.1$ and $\Lambda = 3(1 - \Omega_0)H_0^2$ (*dot-dash line*). The three light curves in (*b*) show the effect on the K-band counts of passive evolution modeled using the stellar population synthesis program of Bruzual (1983). (*c*) Redshift distribution of galaxies with apparent B_J magnitudes in the range $20.0 < m_{B_J} < 22.5$. The histogram shows the observed distribution of Colless et al. (1990) and Broadhurst et al. (1988). The three curves show predictions of the no-evolution model detailed in § 4.1 for three cosmological models: flat, $\Omega_0 = 1$ (*solid line*); open, $\Omega_0 = 0.1$ (*dashed line*); and flat with $\Omega = 0.1$ and $\Lambda = 3(1 - \Omega_0)H_0^2$ (*dot-dash line*).

Fig. 3.16. Figure 1a, b, c of Cole, Treyer, & Silk (1992) with their Fig. caption attached.

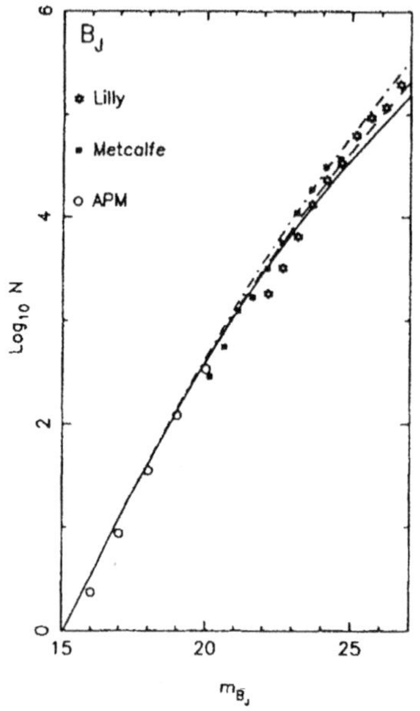

FIG. 2a

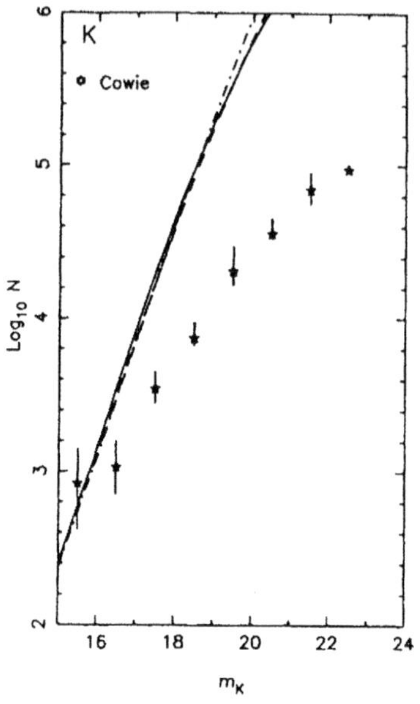

FIG. 2b

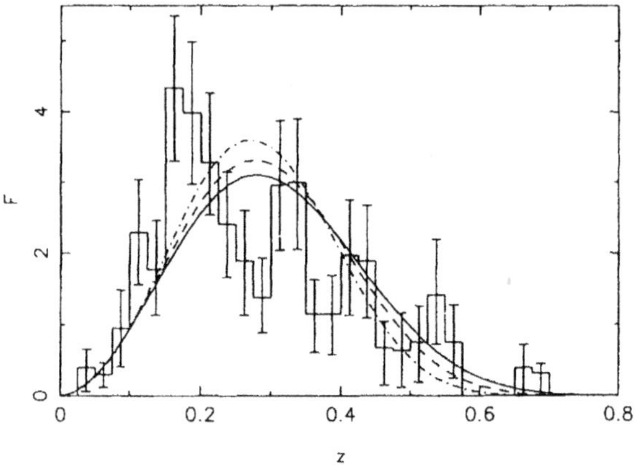

FIG. 2c

FIG. 2.—(*a, b*) As in Figs. 1*a* and 1*b*, but for models with number density evolution of the form $\phi^* \propto (1+z)^4$. (*c*) Redshift distribution of $20 < m_{B_J} < 22.5$ galaxies. As in Fig. 1*c*, but for models with number density evolution of the form $\phi^* \propto (1+z)^4$.

Fig. 3.17. Figure 2a, b, c of Cole, Treyer, & Silk (1992).

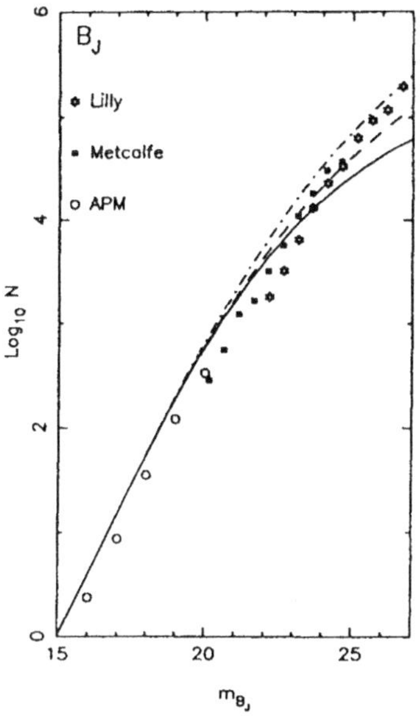

FIG. 3a

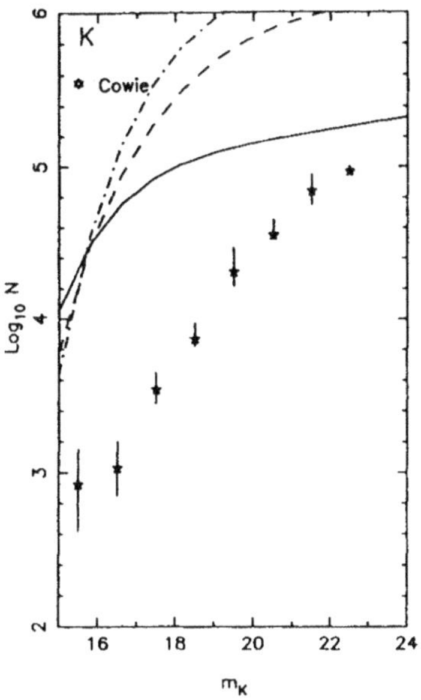

FIG. 3b

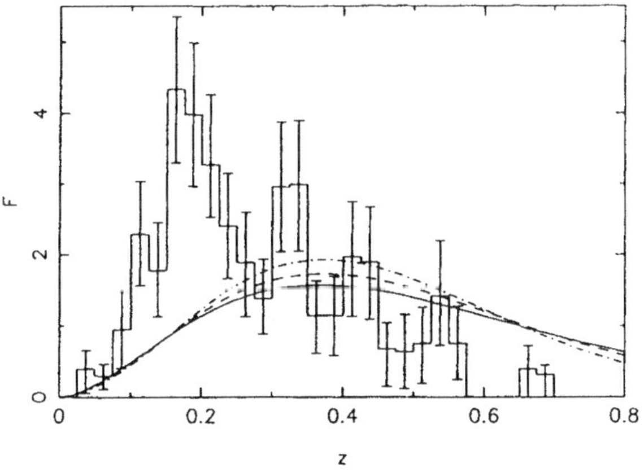

FIG. 3c

FIG. 3.—(*a, b*) As in Figs. 1*a* and 1*b*, but for models with luminosity evolution of the form $L^* \propto (1+z)^4$. (*c*) Redshift distribution of $20 < m_{B_J} < 22.5$ galaxies. As in Fig. 1*c*, but for models with luminosity evolution of the form $L^* \propto (1+z)^4$ assumed for each morphological type.

Fig. 3.18. Figure 3a, b, c of Cole, Treyer, & Silk (1992).

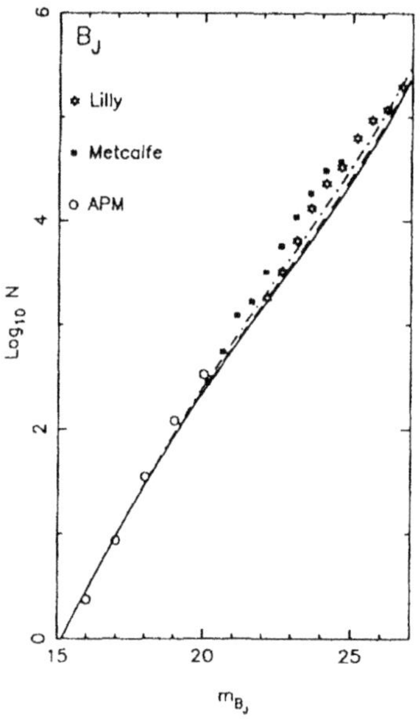

FIG. 4a

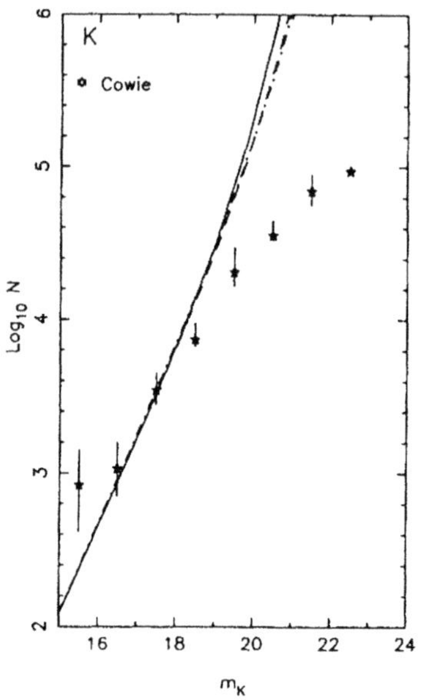

FIG. 4b

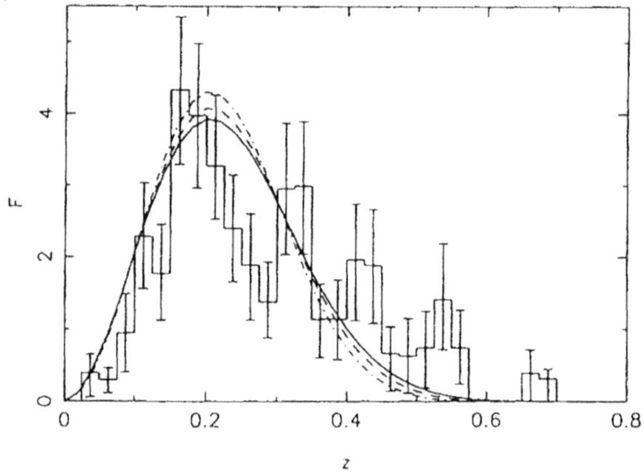

FIG. 4c

FIG. 4.—(*a, b*) As in Figs. 1*a* and 1*b*, but for models with a faint-end slope of the luminosity function which has been taken to evolve as $\alpha \propto (1 + z)^{1.5}$. (*c*) Redshift distribution of $20 < m_{B_J} < 22.5$ galaxies. As in Fig. 1*c*, but for models with a faint-end slope of the luminosity function, which has been taken to evolve as $\alpha \propto (1 + z)^{1.5}$.

Fig. 3.19. Figure 4a, b, c of Cole, Treyer, & Silk (1992).

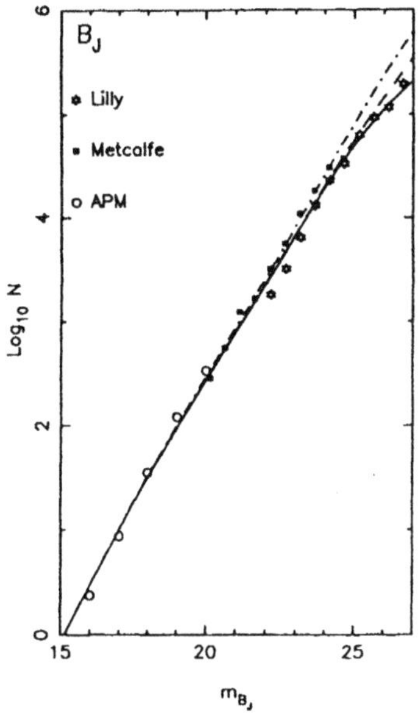

FIG. 5a

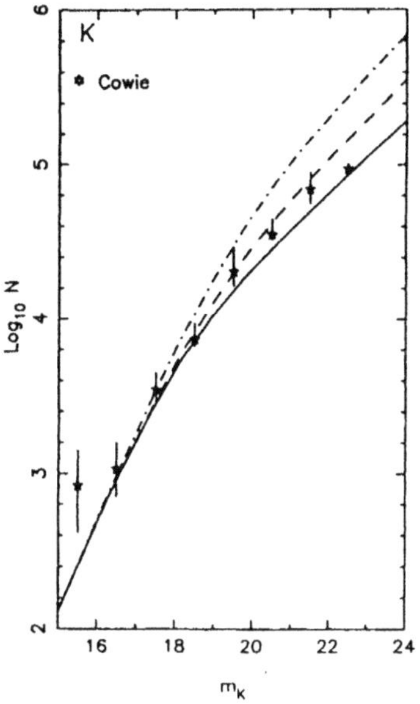

FIG. 5b

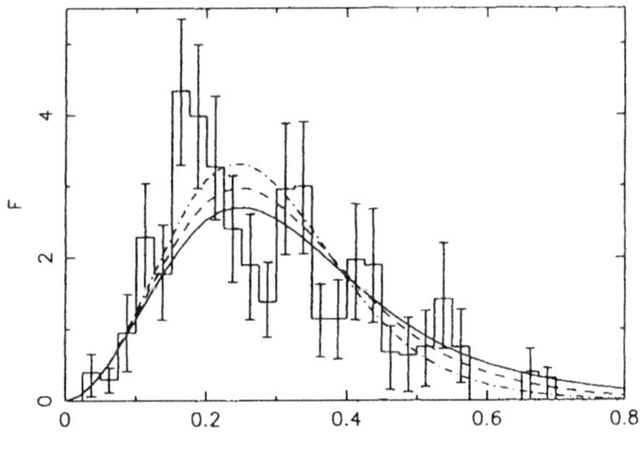

FIG. 5c

FIG. 5.—(*a*, *b*) As in Figs. 1*a* and 1*b*, but the curves are for a model containing a population of flat-spectrum galaxies at redshifts $z \gtrsim 0.7$ as described by eq. (4.4). (*c*) Redshift distribution of $20 < m_{B_J} < 22.5$ galaxies. The curves are for a model containing a population of flat-spectrum galaxies at redshifts $z \gtrsim 0.7$ as described by eq. (4.4).

Fig. 3.20. Figure 5a, b, c of Cole, Treyer, Silk (1992).

Figure 3.16 from Cole *et al.* (1992, their Fig. 1) shows the observed $N(m)$ values using the data of Lilly *et al.* (1991) as star symbols, Metcalfe *et al.* (1991) as squares, and the bright APM galaxy survey (Maddox *et al.* 1990) as open circles. The three continuous curves in panels a, b, and c are predictions of models calculated by methods similar to those set out in these Lectures for (1) the no evolution model; i.e. no density or luminosity evolution using a $q_0 = 1/2$ geometry (the solid line), $q_0 = 0.05$ (the dashed line), and $q_0 = 0.1$ with a cosmological constant of $\Lambda = 3(1 - \Omega_0)H_0^2$ necessary to make space flat (the dot-dashed line). The three light lines in Fig. 3.16 for the K band counts of Cowie *et al.* (1991) show the effects of passive evolution on $N(K)$, modeled using the stellar population synthesis program of Bruzual (1983). (Passive evolution means the burning down of the main sequence and the moving up of the giant branch as a function of time.) See the caption of Cole *et al.* (1992) (Fig. 3.16 here) for details.

The $N(z)$ distribution in the third panel shows the combined redshift data of Colless *et al.* (1990), and Broadhurst *et al.* (1988), together with the predicted expected $N(z)$ distribution for the same three parameters shown in the curves in panel a. The difference between various values of q_0 is nearly negligible, justifying the statement in the preceding section. Clearly, the $N(B)$ count data cannot be accommodated by this simple set of input parameters, although the $N(z)$ data show a fairly good fit; the $N(z)$ distribution is relatively insensitive to the *no evolution* models that only adjust the geometry and/or the cosmological constant.

The conclusion is amplified in Fig. 3.17 where the same data are shown but now with theoretical curves made by adopting a density evolution in the look-back time in the form of $(1+z)^4$. The B_J count curve can be well fitted, but the $N(K)$ count curve cannot. Again, the redshift distribution, $N(z)$, is insensitive to q_0.

Figure 3.18 shows the comparison of the same data with predictions for *luminosity* evolution of the form $L^* \sim (1+z)^4$ assumed for each morphological type. The $N(B)$ counts, with the blue excess, are well fit, but the $N(K)$ counts and the $N(z)$ distributions are not.

Figure 3.19 shows the effect of forcing the slope of the faint end of the luminosity function to evolve as $\alpha = (1+z)^{1.5}$. None of the three observed distribution functions are fit particularly well.

Figure 3.20 introduces a *new popoulation* of galaxies appearing at redshifts $z > 0.7$ with the properties described in Cole *et al.* (1992, their section 4.4) in the numbers and with the luminosity function described there. The curves are for the the same geometrical parameters described for Fig. 3.16. A good fit can be made with this mix to all three distribution functions.

The main point is that one must seek consistency between the $N(m)$ and $N(z)$ distribution functions with any assumed evolution in the look-back time if the observations are useful in constraining the models with and without evolution. What has emerged to date in the many papers that survey the problem (see Cole *et al.* 1992 for detailed references) is that the $N(m)$ count test, so prominent in the early history of the subject (Hubble 1936), *is nearly insensitive to the intrinsic space curvature*, but is, nevertheless, very powerful for discovering the evolution, both in density (or new types of galaxies in earlier times) and/or the luminosity evolution in the look-back time. Hence, the $N(m)$ test for the "world model" , dreamt about for so long by pre-1960 cosmologists, is impotent in the original purpose for which it was devised.

References

Abell, G.O. 1975, in *Galaxies and the Universe*, eds. A. Sandage, M. Sandage, & J. Kristian (Chicago: Univ. Chicago Press), Chapter 15

Binggeli, B. 1987, in *Nearly Normal Galaxies*, ed. S. Faber (New York: Springer-Verlag), p. 195

Binggeli, B., Sandage, A., & Tammann, G.A. 1985, AJ **90**, 1681

Binggeli, B., Sandage, A., & Tammann, G.A. 1988, Ann. Rev. Astron. Astrophys. **26**, 509

Bok, B.J. 1937, *Distribution of Stars in Space*, (Chicago: Univ. Chicago Press) p. 17-18

Broadhurst, T.J., Ellis, R.S., & Shanks, T. 1988, MNRAS **235**, 827

Bruzual, A.G. 1983, ApJ **273**, 105

Charlier, C.V.L. 1908, Ark. Mat. Astr. Pys. **4**, No. 24

Charlier, C.V.L. 1922, Medd. Lunds Obs. No. **98**

Cole, S., Treyer, M.A., & Silk, J. 1992, ApJ **385**, 9

Coleman, G.D., Wu, C.-C., & Weedman, D.W. 1988, ApJ Suppl. **43**, 393

Colless, M.M., Ellis, R.S., Taylor, K., & Hook, R.N. 1990, MNRAS **244**, 408

Cowie, L.L., Gardner, J.P., Wainscoat, R.J., & Hodapp, K.W. 1991, ApJ, submitted

Crane, P. 1975, ApJ **198**, L9

Crane, P. 1976, ApJ **206**, L133

de Vaucouleurs, G. 1970, Science **167**, 1203

de Vaucouleurs, G. 1971, Pub. ASP. **83**, 113

Dressler, A. 1980, ApJ **236**, 351

Ellis, R. 1986 in *Observational Cosmology*, IAU Symp. 124, eds. A. Hewitt, G. Burbidge, & L.Z. Fang (Dordrecht: Reidel), p. 367

Ferguson, H., & Sandage, A. 1991, AJ **101**, 765

Haggerty, M.J. 1971, Physica **50**, 391

Haggerty, M.J. & Wertz, J.R. 1971, Preprint

Holmberg, E. 1958, Medd. Lunds Obs. Ser. II, No. 136

Hubble, E. 1926, ApJ **64**, 321

Hubble, E. 1934, ApJ **79**, 8

Hubble, E. 1936 ApJ **84**, 517

Hubble, E., & Humason, M.L. 1931, ApJ **74**, 43

Humason, M., Mayall, N.U., & Sandage, A. 1956, AJ **61**, 97

Jerjen, H., Tammann, G.A., & Binggeli, B. 1992, in *Morphological and Physical Classification of Galaxies*, eds. G. Longo *et al.* (Dordrecht: Kluwer), p. 17

Karachentsev, I.D., & Kopylov, A.I. 1978, IAU Symp. No. 79, p. 339

Koo, D.C., & Kron, R.G. 1992, Anu. Rev. Astron. Astrophys. **30**, 613

Kristian, J., Sandage, A., & Westphal, J.A. 1978, ApJ **221**, 383

Kron, R.G. 1980, ApJ Suppl. **43**, 305

Lilly, S.J., Cowie, L.L., & Gardner, J.P. 1991, ApJ **369**, 79

Maddox, S.J., Sutherland, W.J., Efstathiou, G., Loveday, J., & Peterson, B.A. 1990, MNRAS **247**, 1P
Mattig, W. 1958, Astron. Nach. **284**, 109
Mattig, W. 1959, Astron. Nach. **285**, 1
Mayall, N.U., 1934, Lick Obs. Bull. No. 458
Metcalfe, N., Shanks, T., Fong, R., & Jones, L.R. 1991, MNRAS **249**, 498
Mihalas, D., & Binney, J. 1981, *Galactic Astronomy* (San Francisco: Freeman & Co.), Chapter 4
Minkowski, R. 1960, ApJ **132**, 908
Oke, J.B. 1971, ApJ **170**, 193
Oke, J.B., & Sandage, A. 1968, ApJ **154**, 21
Pence, W. 1976, ApJ **203**, 39
Peterson, B.A., Ellis, R.S., Kibblewhite, E.J., Bridgeland, M.T., Hooley, T., & Horne, D. 1979, ApJ **233**, L109
Postman, M., & Geller, M. J. 1984, ApJ **281**, 95
Sandage, A. 1961, ApJ **133**, 355
Sandage, A. 1972a, ApJ **178**, 1
Sandage, A. 1972b, ApJ **178**, 25
Sandage, A. 1973, ApJ **183**, 711
Sandage, A. 1978, AJ **83**, 904
Sandage, A., Binggeli, B., & Tammann, G.A. 1985, AJ **90**, 1681
Sandage, A., & Hardy, E. 1973, ApJ **183**, 743
Sandage, A., & Tammann, G.A. 1987, *A Revised Shapley-Ames Catalog of Bright Galaxies* (Carnegie Institution of Washington: Second ed.)
Sandage, A., Tammann, G.A., & Hardy, E. 1972, ApJ **172**, 253
Schechter, P. 1976, ApJ **203**, 297
Schild, R., & Oke, J.B., 1971, ApJ **169**, 209
Schouten, W.J.A. 1918, Dissertation: *On Determination of the Principal Laws of Statictical Astronomy* (Groningen)
Seares, F.H. 1922, Trans. IAU, Vol. 1 (London: Imp. College Bookstall), p. 71
Seares, F.H. 1931, ApJ **74**, 268
Shane, C.D., & Wirtanen, C.A. 1954, AJ **59**, 285
Shane, C.D., & Wirtanen, C.A. 1967, Lick Obs. Pub. **22**, 1
Tinsley, B.M. 1980, ApJ **241**, 41
Tyson, J.A., & Jarvis, J.F. 1979, ApJ **230**, L153
von der Pahlen, E. 1937, *Lehrbuch der Stellarstatistik* (Leipzig: Barth), p. 400f
Wertz, J.R. 1971, ApJ **164**, 227
Whitford, A. E. 1971, ApJ **169**, 215
Wilkinson, A., & Oke, J.B. 1978, ApJ **220**, 376
Yahil, A., Sandage, A., & Tammann, G.A. 1980, ApJ 242, 448
Yoshii, Y., & Takahara, F. 1988, ApJ **326**, 1

4 The Second Test: Angular Size = $f(z, q_0)$

4.0 Introduction

If a standard rod were to exist such that its length was the same everywhere, its observed angular diameter would decrease as the inverse of the distance if the geometry is Euclidean. If the intrinsic diameter were known, the observed angular diameter could be used to determine the distance (projection effects complicating the experiment).

In curved space, where the curvature acts as a lens, the angular diameter of a standard rod will decrease closely as the inverse distance at small distances, but will deviate from d^{-1} at large distances due to the effect of the curvature. The deviation can, in principle, be used to measure the sign of the curvature, and the concomitant radius of curvature (Hoyle 1959).

The principle of this test for space curvature has been known at least since Tolman's (1930, 1934) early discussions, shortly after it became apparent that the metric line element must be non-static (Tolman 1929, 1930), necessitated by the redshift-distance relation if the redshift is due to a true expansion (Lemaître 1927, 1931; Robertson 1928; Hubble 1929). However, the stumbling block was always then, and is yet today, the necessity to find a suitable standard rod whose properties (static and/or evolutionary over time) are known. This assurance must be had before any putative "standard rod" can be used to mark, and thereby to measure, the geometry.

Various standard rods have been suggested, such as angular diameters of first ranked cluster galaxies (Sandage 1972a), the average angular separations of the first n galaxies in rich glusters (Hickson 1977a,b; Hickson & Adams 1979, Bruzual & Spinrad 1979), the separation of the components of double radio sources (Miley 1968, 1971; Kapahi 1975, 1985, 1987, 1989; Daly 1994), the sizes of compact radio sources (Kellermann 1993) amongst others, many of which we discuss in this lecture.

But we first discuss the much easier proposition that locally – i.e., at distances (a) where the look-back time is small and therefore any evolutionary change with time (redshift) can be neglected, and (b) the effects of space curvature are negligible – the redshift-distance relation is, in fact, linear. The evidence is clear that the angular size decreases as d^{-1}, using several types of objects including individual galaxies, clusters of galaxies, double lobed radio galaxies, and compact radio sources.

4.1 The Very Local Redshift-Angular Diameter Test

A prime question set in the 1930's at the outset of observational cosmology was the form of the redshift-distance relation (Lecture 5). Is it either linear if the expansion is real or is it parabolic in the static (but highly unphysical) metric of de Sitter (1917, 1933), (also Eddington 1923 for a derivation, and Sandage 1975 for a discussion)?

In more recent times a different static theory also requires a quadratic redshift-distance relation (Segal 1975, 1976, 1980, 1993). Multiple claims of verification have been made by Segal using highly selected observational data. He finds the data we discuss in this and the following lecture inconclusive.

Hubble & Humason (1931) first showed that the apparent magnitude-redshift correlation for galaxies with redshifts as high as 20,000 km s^{-1} has very little dispersion when the data are restricted to cluster galaxies. The observed slope of the apparent magnitude-redshift relation at $\mathrm{d}\log m/\mathrm{d}\log cz = 5$ using cluster galaxies is itself a powerful proof that the redshift-distance relation is linear.

The early data of Hubble & Humason gave the first indication that first-ranked cluster galaxies were restricted in the spread of their intrinsic luminositites. All subsequent work, set out by a series of observational programs in the 1970's (Lecture 5) showed this indeed to be true.

Because first-ranked cluster galaxies define a good standard *candle* it seemed likely that they also might define a standard metric *rod*. The notion follows from Hubble's (1926) demonstration that the apparent magnitude and the angular diameter are highly correlated for galaxies of a given type. Although this shows only that their surface brightness has a statistically stable mean with small dispersion, the inference could reasonably be drawn that their linear diameter is also a quantity of small enough dispersion to be useful for relative distance estimates. The inference was that such galaxies are built on a fixed model by the unknown physics of their formation.

But even if the diameters of first-ranked cluster galaxies would provide a *standard rod*, considerable operational difficulties exist in defining an angular diameter of objects with strong gradients of surface brightness across their images. Ways must be devised to define a proper angular diameter.

The expectation that first-ranked cluster galaxies provide at least a first order test was demonstrated in an experiment (Sandage 1972a) using plate material for galaxy clusters obtained with the 200-inch telescope. The work was a by-product of the search for new distant clusters for the Palomar redshift program in the 1950's.

The observational material was of very uniform quality, obtained to a standard isophotal level, care being taken with exposure time and photographic development procedures to give a sky background of uniform density. Similar plate material was also obtained, again under uniform conditions, with the Palomar 48-inch Schmidt telescope for the nearer clusters out to redshifts of 0.08. The combined material covered the redshift range from the Leo group with $z = 0.0023$ to the cluster containing the radio source 3C295 at a redshift of $z = 0.46$ (Minkowski 1960).

Figure 4.1 shows the result of the angular diameter of the image of the first ranked cluster galaxy in 66 clusters and groups, measured to an isophote of about 23^{m} arcsec^{-2} in the B band. No K corrections have been applied; they are small over the redshift range of most of the data that range from

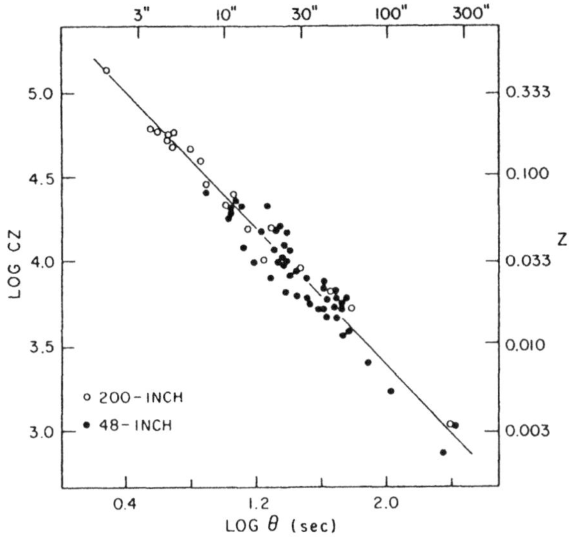

Fig. 4.1. The observed angular size-redshift relation for 66 clusters & groups from Fig. 2 of Sandage (1972a).

very local to redshifts of $z \sim 0.1$. The purpose of this demonstration was not to determine q_0, where refinements of all kinds are needed (set out in the next section) but rather to test the first-order dependence of angular diameter on redshift.

The line in Fig. 4.1 is the expectation of a z^{-1} redshift dependence in Euclidean space if the redshift-distance relation is linear. A squared relation is denied by the data such as is predicted in the static theories of de Sitter (1917, 1933) and Segal (1976).

The same conclusion has been reached in all other tests using the other candidate standard rods listed in section 4.0. Data and analysis in the papers of Hickson (1977a,b), Hickson & Adams (1979), Miley (1968, 1971), Kapahi (1975, 1985, 1987, 1989), will be discussed in the next section after setting out the theory of the second-order term depending on q_0 in the diameter-redshift relation. The cited papers are all concerned with determining q_0 from the data at high redshift, the analysis of which requires the theory. But we shall see in section 4.5 that the first order z^{-1} term is recovered in each of these experiments, similar to the result of the experiment set out in Fig. 4.1.

4.2 General Theory of Angular Diameter = $f(z, q_0)$ Valid for All Redshifts

The treatment here follows the initial suggestion by Hoyle (1959), extending Hubble and Tolman (1935), but using the exact equations of Mattig for $R_0 r$, where the distinction between observations of metric angular diameters compared with isophotal diameters was shown to be decisive (Sandage 1972a).

Consider a "standard" rod of linear (metric, not isophotal) dimension y at a redshift z in a curved space with deceleration parameter q_0, Hubble constant H_0, and zero cosmological constant. We require the general equation for the angular diameter that will be observed at different redshifts in the absence of evolutionary effects, i.e., y is assumed not to be a function of look-back time.

Recall from Lecture 1 that the space-time Minkowski metric that is invariant to coordinate frames consistent with special relativity, is

$$ds^2 = c^2 dt^2 - du^2, \tag{4.1}$$

where du^2 is the curved isotropic, homogeneous space-like part whose most general form, as proved by Robertson (1929, 1935), Tolman (1929), and Walker (1936), is

$$du^2 = R(t)^2 \left[\frac{dr^2}{1 - kr^2} + r^2 d\theta^2 + r^2 \sin^2\theta d\phi^2\right], \tag{4.2}$$

where $R(t)$ is the scale factor that, in an expanding manifold, is a function of time (Lecture 1, section 1.6).

The time dependent $R(t)$ function is given by the solution of the Friedmann equation, which, with the Lemaître equation for R_0/R_1 and the Mattig equation for $R_0 r$, solves the problem.

Two light rays from the ends of a rod perpendicular to the sight line, travelling toward the observer from a source that is a distance rR_1 away from the observer at the time of light emission, converge toward the observer with a separation angle of

$$\theta = \frac{y}{R_1 r}, \tag{4.3}$$

where r is the constant co-moving radial coordinate. (Without loss of generality we have orientated the rod such that $dr = d\phi = 0$ in the metric. The $dr = 0$ condition means that the rod is put perpendicular to the direction of the observer).

The angle in equation (4.3) does not change during the travel time, even as the distance increases from $R_1 r$ to $R_0 r$ due to the expansion. Using both the Lemaître equation

$$R_0 = R_1(1 + z),$$

and $R_0 r = f(z, q_0)$ from the Mattig equation with equation (4.3) gives the angular size observed by us at the time of light reception as

$$\theta = \frac{y(1 + z)}{R_0 r}, \tag{4.4}$$

giving, in all generality,

$$\theta(\text{metric}) = \left(\frac{yH_0}{c}\right) \frac{q_0^2(1+z)}{q_0 z + (q_0 - 1)\left[-1 + \sqrt{1 + 2q_0 z}\right]}. \tag{4.5}$$

Hoyle (1959) was the first to realize that equation (4.5) (or in what must have been its equivalent; he gives no details) has a minimum for all $q_0 > 0$. For $q_0 = 1/2$ this minimum is at $z = 5/4$. Hoyle emphasized that a deviation from a strict z^{-1} dependence at large z is a test of the lens effect of space curvature.

Problem 4.1 Show that equation (4.5) for all q_0, and also in the equations for "tired light" (section 4.3), each reduce to $\theta = yH_0/cz$ for small z, which is the intuitive Euclidean solution because "distance" $= cz/H_0$.

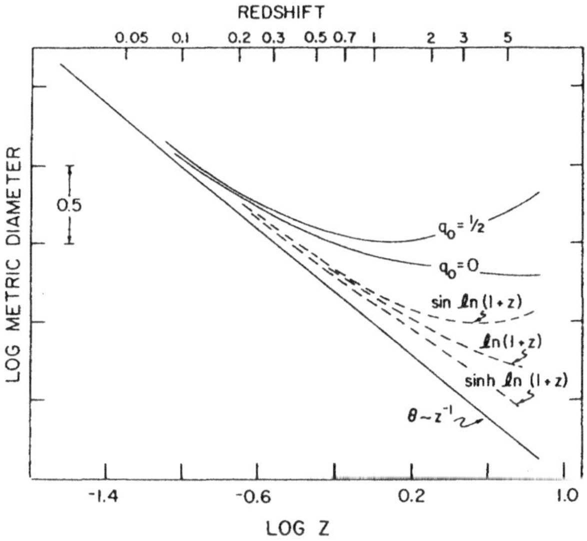

Fig. 4.2. Theoretical angular size-redshift relation from Fig. 11 of Sandage (1988c).

The two special simple cases for $q_0 = 0$ and $q_0 = 1/2$ give

$$\theta = \left(\frac{yH_0}{c}\right) \frac{(1+z)^2}{\left(z + \frac{z^2}{2}\right)}, \tag{4.6}$$

for $q_0 = 0$, and

$$\theta = \left(\frac{yH_0}{2c}\right) \frac{(1+z)^{3/2}}{\left(\sqrt{1+z} - 1\right)}, \tag{4.7}$$

for $q_0 = 1/2$ using the relevant equations for $R_0 r$ (section 1.6.5 of Lecture 1).

Figure 4.2 shows the predicted run of angular size with redshift for the $q_0 = 0$ and $q_0 = 1/2$ cases compared with the Euclidean z^{-1} case. Again, all standard models with mass (i.e. with $q_0 > 0$, and for a cosmological constant of zero) have a minimum in the observed metric diameter at large redshifts. No such minimum exists for the steady state model ($q_0 = -1$) or for the predictions of an *ad hoc* "tired light" model described in the next section.

Problem 4.2. Prove that in the steady state model ($q_0 = -1$ at all cosmic times), the minimum angular size is reached only asymptotically as the redshift approaches infinity. Hint: recall that the metric scale factor increases such that $R = e^{H_0 t}$, and then apply the machinery that relates Rr, R, z, and t from the previous lectures to the steady state function for $R(t)$.

4.3 Predictions for a "Tired Light" Model

The implications of a truly expanding model seem so bizzare, both in the 1930's and yet today, that alternative explanations of the redshift are still sought, sometimes in the professional, but most often in the underground literature. A perennial favorite is some form of the *ad hoc* hypothesis that photons interact with matter along the transit path, losing energy in some fashion in direct proportion to the distance travelled (Zwicky 1929, 1957; LaViolette 1986). There are observational reasons why this is untenable; energy loss must be accompanied by scattering, whereas the images of galaxies at high z show no sign of blurring. Nevertheless, it is interesting to derive the angular size-redshift relation for one version of the tired light suggestion and to show that the predicted relations lie between the standard cosmological models with q_0 between 0 and 1/2 and the Euclidean z^{-1} relation, seen already in Fig. 4.2.

For the "tired light" case, evidently

$$1 + z = e^{\frac{H_0 R_0 r}{c}}, \tag{4.8}$$

or,

$$R_0 r = \frac{c}{H_0} \ln(1 + z), \tag{4.9}$$

as a guess for the "coordinate" distance for flat space. For curved space we must use the relations between "interval" distance u and "coordinate" distance $R_0 r$ derived in Lecture 1 [equations (1.14) and (1.15), requiring the interval distance to be $R \sin(RrH/c)$ or $R \sinh(RrH/c)$ for $k = +1$ and -1, respectively]. Hence, in the case where the curvature factor is $HR/c = 1$ we have the three possibilities

$$\theta = \frac{yH_0}{c} \frac{1}{\ln(1 + z)} \qquad \text{for } k = 0 \text{ (flat space)} \tag{4.10}$$

$$\theta = \frac{yH_0}{c} \frac{1}{\sin[\ln(1+z)]} \quad \text{for } k = +1 \tag{4.11}$$

$$\theta = \frac{yH_0}{c} \frac{1}{\sinh[\ln(1+z)]} \quad \text{for } k = -1, \tag{4.12}$$

shown in Fig. 4.2.

4.4 The Difference Between Metric and Isophotal Diameters

The theory in section 4.2 refers to angular diameters of a rigid rod that does not expand with the scale factor, and where the sharp edges of the rod can be observed. We refer to such sizes as "metric" diameters. A second type of diameter refers to the size of a luminous object as measured to a given isophotal level such as say 25^{m} arcsec^{-2}.

The two types of diameters differ as a function of redshift. Tolman (1930, 1934) was among the first, if not the first, to derive the relation that the bolometric *surface brightness* of objects in an expanding manifold become fainter with redshift by the factor of $(1+z)^4$. A heuristic derivation is as follows.

In an expanding manifold the metric surface brightness is found by dividing the observed (received) flux by the (metric) angular area. The equations for both quantities as a function of redshift in an expanding manifold have already been derived in these lectures, the results of which are repeated here.

The Robertson (1938) equation (2.1) in Lecture 2 for the received *bolometric* flux, l_{bol}, is

$$l_{\text{bol}} = \frac{L_{\text{bol}}}{4\pi(R_0 r)^2(1+z)^2}, \tag{4.13}$$

where $R_0 r$ is the coordinate distance of the source at the time of light reception, and the two factors of $(1+z)$ are due to the "energy" and the "number" effects, assuming the expansion is real. If it is not, but rather due to some unknown physical process in a stationary universe, only one factor of $(1+z)$ applies.

The $R_0 r$ factor in equation (4.13) is identical to the same factor in equation (4.4). Hence, dividing (4.13) by equation (4.4) squared, which is proportional to the angular metric area subtended by the source at light reception, gives the mean surface brightness as

$$\langle \text{SB} \rangle \equiv \frac{l_{\text{bol}}}{\theta^2} \sim \frac{1}{(1+z)^4}. \tag{4.14}$$

This is the famous Tolman result (Lecture 6).

One of the many consequences of equation (4.14) is that as the redshift increases, the *isophotal* radius of a galaxy becomes smaller relative to the *metric* radius. This is because the position of a numerically fixed isophotal level (say 25^{m} arcsec^{-2}) moves progressively toward the center (in parsecs) due to the decrease in the surface brightness by the $(1+z)^4$ factor. The ratio of metric to isophotal diameter as a function of redshift is derived as follows.

We consider only E galaxies, which is the practical case in the application of the test. Hubble's (1930) intensity profile

$$I(\theta) = \frac{I(0)}{\left(1+\frac{\theta}{a}\right)^2}, \tag{4.15}$$

is the useful approximation, where a is the "core" radius, i.e. the radius where the surface brightness falls to 1/4 of the central value. Clearly a is a metric rather than isophotal radius because it is a radius determined by a ratio of fluxes, both of which contain the $(1+z)^4$ factor, which therefore cancels.

At angular radii θ that are large compared with a, the ratio of radii for which the observed fluxes have a ratio of $I(\theta_1)/I(\theta_2)$ is $(\theta_2/\theta_1)^2$ by equation (4.15). Hence, the radius θ_1 where the SB is down by a factor of $(1+z)^4$ from the SB at radius θ_2 is such that the ratio of θ_1 to θ_2 is $(1+z)^2$. From this it follows that

$$\theta(\text{metric}) = \theta(\text{isophotal})\ (1+z)^2. \tag{4.16}$$

Note, however, that the isophotal level to be used for both the galaxy at rest and the identical (standard) one under redshift *must first be reduced to the same proper (rest) wavelength by the $K(z)$ term* as in Lecture 3 (section 3.3.1e). Without the $K(z)$ correction, equation (4.16) is valid only when using *bolometric* intensities. As this is never done, *attention must be paid to the $K(z)$ correction when analyzing actual data* (Sandage 1972a).

The distinction between isophotal and metric diameters and the effect on the analysis of data was first made by Stock & Schücking (1957).

Problem 4.3 Show that one of the consequences of equation (4.16) is that isophotal diameters will not show the minimum in the θ-redshift relation at large redshifts as is present for metric diameters. Hint: divide equations (4.5), (4.6), and (4.7) by $(1+z)^2$ and follow the consequences.

4.5 Samples of Angular Size–Redshift Data

Figure 4.1 shows only the first order angular size-redshift relation for small redshifts such that the distinction between isophotal and metric diameters and the difference caused by space curvature can be ignored. The purpose there was only to show the inverse first power dependence of angular size on redshift, not the second order (q_0) term.

The attempt to determine q_0 from the test requires observing similar objects to high redshifts. Because both quasars and radio galaxies were early observed at high enough redshifts to make the test (i.e., $z > 0.5$), much initial work on the problem was done using double-lobed radio sources. The obvious

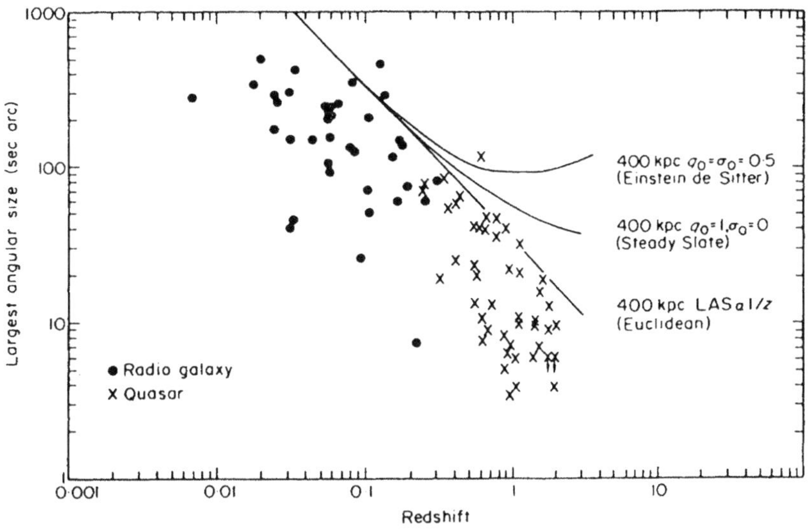

Fig. 4.3. Miley's (1971) relation between angular sizes of double lobed radio sources and redshift.

problem with orientation of the radio jets to the sight line was solved either by considering upper envelope fits to the angular size-redshift correlations (for example Legg 1970; Miley 1971), or taking median values of the angular diameters at each redshift (Kapahi 1985, 1987, 1989). A review of the early literature is given elsewhere (Sandage 1988c, section 8.3).

The data and analysis of both Miley (1968, 1971) and Legg (1970) showed decisively that the angular separation of the radio lobes do not have the expected minimum predicted by the standard models for q_0 between 0 and 1/2. Rather, they continue to decrease approximately as z^{-1} over the entire redshift range from $z = 0$ to $z = 2$.

Figure 4.3 from Miley (1971) shows a typical correlation using the upper envelope fits that compensate for the orientation. Figure 4.4 from Kapahi (1987) using median values shows the same linear decrease as z^{-1} with increasing redshift. These data showed either that the standard model leading to Fig. 4.1 is wrong in second order[12] of z, or that (a) evolution of radio lobe separations with epoch, or (b) a dependence of linear separation with radio power (see next section) is required. The second-order test for geometry then becomes *ad hoc* if *arbitrary* evolution corrections must be applied, based only on the necessity to make the data conform with the theory.

The same null effect of no observed minimum was found in several well constituted optical experiments. Figure 4.5 shows the data of Hickson &

[12]It is to be noted again how badly the Segal (1976) squared, size-redshift prediction, based on his static, *ad hoc*, model fails already in first order.

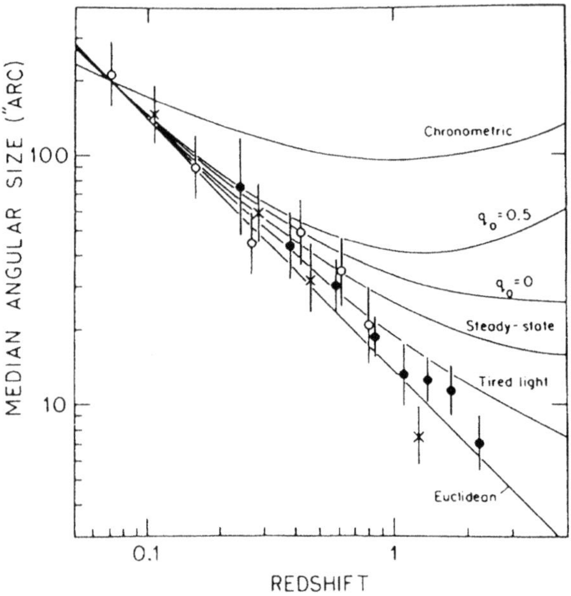

Fig. 4.4. Kapahi's (1987, his Fig. 7) data for the median angular sizes (arcsec) and redshift for his radio-source sample compared with the predictions of the standard model for various q_0 values (without evolution). The Segal "chronometric" model violates the data at all redshifts.

Adams (1979) and of Bruzual & Spinrad (1978) following the brilliant initiative of Hickson (1977a, 1977b) of using the "mean angular separations", suitably defined, of the first N galaxies in clusters at different redshifts. Note that Hickson's definition of mean angular separation clearly gives a *metric* size.

The relevant theoretical curves of Fig. 4.1 are superposed on the data in Fig. 4.5. Although the redshift range is not as high for these optical cluster data as for the radio lobe data in Figs. 4.3 and 4.4, yet again no convincing evidence for a start of a turn up is seen. Hickson & Adams (1979) invoke evolution of cluster sizes with look-back epoch to reconcile the data with the second-order terms of the standard theory, but then the test for q_0 again becomes *ad hoc*.

A similar optical study was made by Djorgovski & Spinrad (1981) but now using the angular size of the first ranked cluster galaxy itself, defined as a Petrosian *metric* size (see Kron's Lecture 6, this volume; also Lecture 6 here and Sandage & Perelmuter 1990 for properties of Petrosian diameters). The result (their Fig. 3), shown here as Fig. 4.6, again shows an excellent first order term of z^{-1}, but again fails to show the second order q_0 turn-up. As before, evolution must be introduced to "save" the standard model.

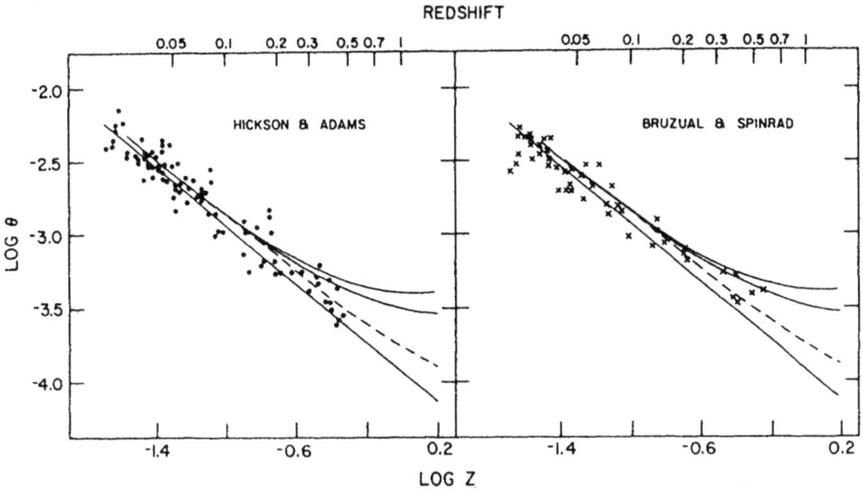

Fig. 4.5. Data for the "mean" angular separation of galaxies in great clusters as a function of redshift from two data sets and analyses. Diagram from Sandage (1988c, Fig. 13). The z^{-1} limiting relation is the lower straight line. For low redshifts the relation is strictly linear, as in Fig. 4.1.

4.5.1 Selection and Evolutionary Effects

Evolution is a sharp-edged sword whose introduction may or may not be appropriate to save the phenomonon in the same way a scimitar has been used in history to force a point of view. Before claiming evolution with redshift in any astronomical data set it is necessary to identify observational selection effects that can imitate changes with redshift but which are due to incomplete sampling. The problem is central to many statistical astronomical settings. A most important problem for this lecture series is the effect of selection bias on the determination of the Hubble constant (Lectures 8 and 10).

As an introduction to the general case set out in Lecture 10, we illustrate the dilemma of a choice between evolution, observational selection, or an incorrect second-order theory. The illustration uses the radio data for the angular size distributions in Figs. 4.3 and 4.4, but the principles apply to a large number of other astronomical situations as well.

The early data such as used by Miley (Fig. 4.3) were for radio sources chosen from radio source catalogs that were made complete to a given radio flux level within a particular frequency interval. In current language, the catalogs were flux-limited in contrast to the much more difficult task (but much more important task) of making a catalog complete over all absolute radio power levels *in a given volume* of space. The difference between a flux-limited catalog and a volume-limited catalog is profound. This difference is the basis of all corrections for observational selection bias to observed data. It is the subject of Lecture 10.

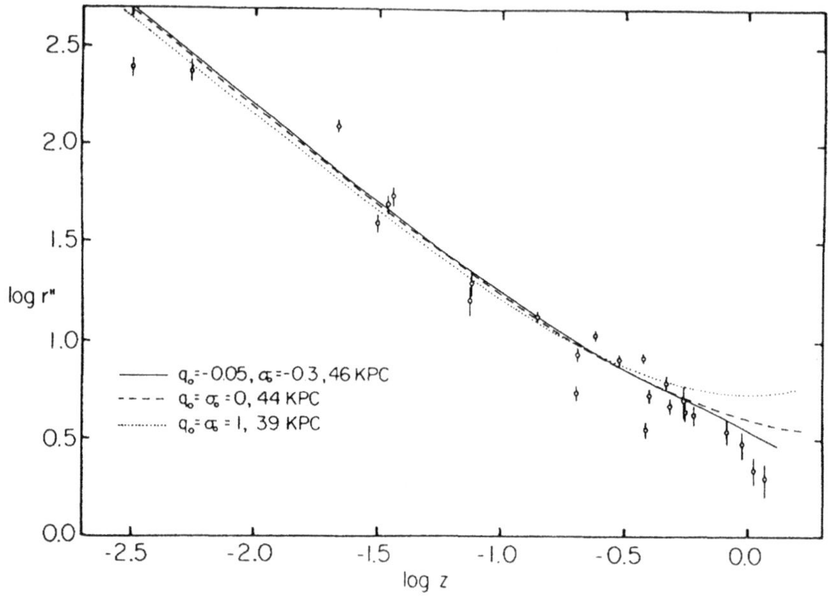

Fig. 4.6. The angular size-redshift relation from Djorgovski & Spinrad (1981), their Fig. 3. A linear redshift-distance relation in Euclidean space (without evolution) is the lower curve.

The effect of flux limitation is illustrated in Fig. 4.7 from an early study of the range of absolute radio power for galaxies and quasars optically identified, primarily from the Cambridge 3C and 4C and the Parkes catalogs (Sandage 1972b). From the known redshifts and the apparent radio flux, the redshift-distance (using $H_0 = 50$ km s^{-1} Mpc^{-1}), and therefore the absolute radio power (in ergs per sec), $L_{\rm R}$, could be calculated. The increase of $L_{\rm R}$ with increasing redshift, varying as $L_{\rm R} \sim z^2$, could have been interpreted as a real increase with look-back time until it is realized from the position of the 1, 3, and 10 Jansky flux lines that it is the flux limitation of the catalogs that produces the observed effects in Fig. 4.7, not any real effect in nature. The correlation is the result of a very broad luminosity function, sampled with normalization (total number) factors that increase with the increasing volumes at larger redshifts, growing as $z^2 dz$.

This observational selection effect due to the flux limitation of the catalogs dominates the distribution of points in Fig. 4.7. In any flux-limited sample of objects those at the largest distances have the highest absolute luminosities because the intrinsically fainter sources are denied entry into the catalog due to the fact that the grasp of the catalog does not include them. The remaining objects have ever increasing intrinsic luminosity (power) the higher is the redshift. Why is this important?

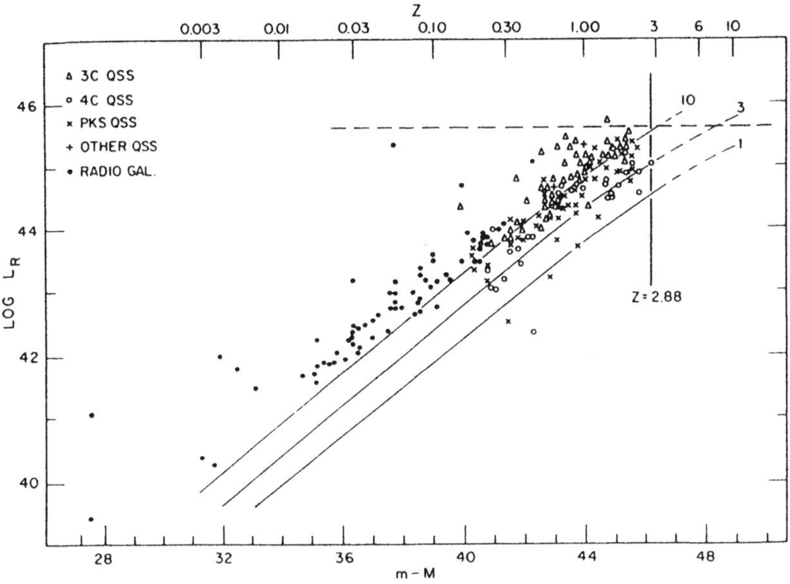

Fig. 4.7. The relation from *flux-limited* data between absolute radio power, L_R, and redshift for identified radio sources from various catalogs. Diagram is from Fig. 7 of Sandage (1972b).

Return to Figs. 4.3 and 4.4. Because galaxies at high redshifts have higher intrinsic luminosities than those at small redshifts *in these catalogs* (not in nature), the deviation of the data from the expected lines of q_0 between 0 and 1/2 relative to the theoretical model could be explained if the intrinsic linear lobe separations of radio sources were smaller at high absolute radio power due to the unknown physics of the double radio galaxies. We would then not need to invoke secular evolution in the look-back time; the deviation of the data from the standard model would, in that case, be due to selection bias, and not real. We would then not be working with a "standard rod" but with one whose apparent properties change with redshift. In that case the test for q_0 fails.

Although in various disguises of the same fundamental problem, this is the basic premise in all problems of observational selection bias.

Three explanations are possible from the biased data in Figs. 4.3 and 4.4 from flux-limited catalogs. Either (1) there is true size evolution in the look-back time, (2) there is a physical (true) correlation between linear lobe separation and absolute radio power, and/or (3) the second-order cosmological theory of section 4.2 leading to Fig. 4.2 is wrong.

An obvious way of separating possibilities (1) and (2) is to add a fainter sample and see the bias properties change in ways that can distinguish between the cases. The method is the subject of Lecture 10. Its application to the problem of the Hubble constant is set out elsewhere (Sandage 1988a, b, 1994a, 1994b). An impressive analysis by Kapahi (1987, 1989) was made of the radio lobe problem in the following way.

Suppose we add a much fainter sample to Fig. 4.7, reaching say to a flux level of 10 mJy. This is 1000 times fainter than used in that diagram. The radio sources would then fill the region between the upper 10 Jy line and that fainter flux limit line, but will still hug the fainter flux limit line, again sloping upward at large redshifts but now at a level 1000 times fainter in absolute power at a given redshift.

The new data would then be extensive enough to permit a subdivision of the sample into galaxies at various redshifts but in a narrow range of radio power. (The data are not extensive enough for such a test in Fig. 4.7, i.e. a horizontal cut does not encompass a large enough range in redshift.) Study of the lobe separation in such a subsample would reveal a real evolutionary change of linear lobe separation with look-back time, because any dependence on absolute luminosity is supressed in that subsample (being all of the same intrinsic power). Study of a second subsample, restricted to narrow *redshift* bins, would reveal any effect of increased absolute radio power on lobe separation at a given redshift (i.e. constant look-back time) thereby eliminating the time-dependent evolutionary effect in evaluating the power dependence.

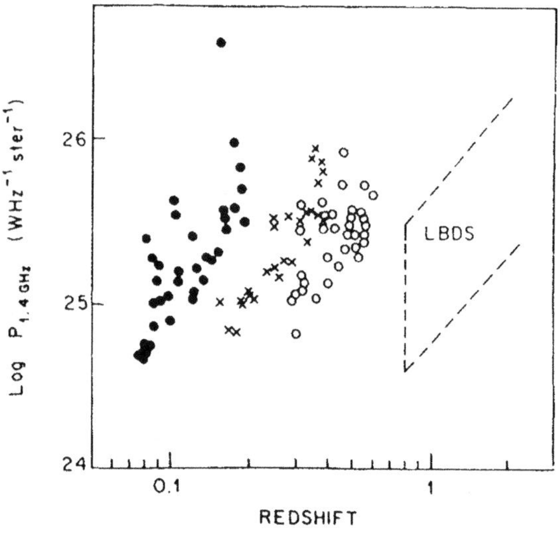

Fig. 4.8. Absolute radio power from four source catalogs used by Kapahi (1987) to control observational bias problems in the angular size-redshift program.

Kapahi (1987, 1989) made these two tests by adding fainter and fainter samples such that his complete sample has a range of apparent flux of a factor 10^4 and a range in redshift from $z = 0.05$ to $z > 1$. Characteristics of the radio objects in the four source catalogs used by Kapahi are shown in Fig. 4.8. From this analysis he concludes that there is both size evolution in the look-back time[13] (which is the dominant effect) and a much smaller effect of size on intrinsic power in the sense that the larger is the power, the *larger* is the intrinsic (linear) size. Note that this latter effect is in the opposite sense needed, alone, to explain the deviation of the data from the models in Figs. 4.3 and 4.4. Kapahi's (1989) conclusion is that the small dependence of linear size, l, on absolute power, P, is approximated by $l \sim P^{0.3\pm0.1}$ (larger size for higher power), and the large dependence on epoch by $l \sim (1+z)^{-2.5\pm0.5}$ for $q_0 = 0$, and $l \sim (1+z)^{-3\pm0.5}$ for $q_0 = 1/2$ (smaller size at earlier epochs), presumably consistent with an idea of pressure confinment by an intergalactic medium that becomes less dense with cosmic time due to the expansion. Of course, the derived epoch dependence is based on the assumption that the standard theory giving the curves in Fig. 4.1 is correct.

The conclusion by Kapahi (1987) summarizes the present state of the test for q_0 using double lobe radio sources. He writes: "The angular size-redshift relation is important for observational cosmology as it provides valuable information on the evolution with epoch of the linear sizes of extragalactic radio sources and can be used for discriminating against some models of the Universe. The physics of the evolution of individual sources is, however, far from being understood. There is therefore little hope at this stage of using the relation as a test of the geometry of the Universe." Perhaps Daly's (1994) work is progress in this direction, but it is too early to tell.

4.6 Kellermann's Data on Compact Radio Sources

Kapahi's conclusion, with its caution, is universally accepted. However, suppose there were a class of objects that, without correction of any kind, either for evolution or for physical effects, would follow one of the curves in Fig. 4.1. If we had found such objects early in the decades of the 1960 and 70's rather than the situation in Figs. 4.3–4.6, we would have taken the test as a determination of q_0.

Such a class of objects may have been discovered by Kellermann (1993) using compact, single component, radio sources at all redshifts whose angular sizes have been measured by the intercontinental very large base line radio telescope (VLBA). Kellermann defined the angular size to be that diameter

[13] A size dependence, if true, is extremely important for physical cosmology. Presumably, it would give information on a combination of two effects, (1) the physical evolution of the radio source itself with time (but only if the life times are long compared with the look-back time, otherwise, if the life time of individual radio sources is short, say 10^8 years, we do not see this time-dependence by sampling various redshift where the travel time is long compared with 10^8 years). (2) Hence, the more likely product of the size dependence with look-back time is information on the density of the intergalactic medium with the epoch of cosmic time. If such evolution exists, and if the intergalactic density is important in determining the lobe separation, the method suggested by Daly (1994) to determine q_0 using lobe separation fails in its first formulation.

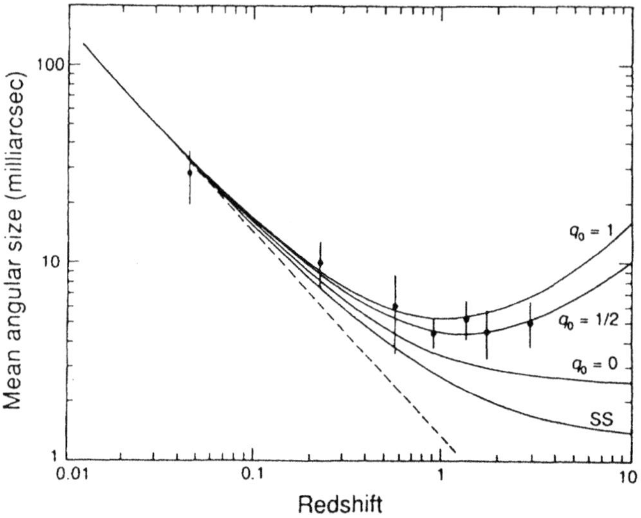

Fig. 4.9 Diagram by Kellermann (1993) of the angular size-redshift relation for compact radio sources.

where the surface brightness is a fixed fraction (1/100) of the central surface brightness. Clearly, this is a metric diameter rather than an isophotal one because of the *ratio* definition; *each* surface brightness is decreased by the Tolman $(1+z)^4$, and the ratio is invariant with z.

Figure 4.9 is Kellermann's famous diagram. If evolution of the linear size with epoch, and if the dependence of size on radio power, could both be shown to vanish using some presently unknown argument, either from a well founded theory or by experiment, the conclusion concerning the intrinsic geometry of space would be powerful from the diagram[14]. But again evolution and selection effects cloud the conclusion despite the extreme, almost mystical beauty of Fig. 4.9 for cosmologists, hoping for a sign that the standard model of the lens effect of space-time curvature can actually be manifest.

[14]Kellermann argues that the diameter, so defined, is very close to the central engine of the embedded radio sourse. Its linear diameter, referring to such a compact region of the source, is likely to be impervious to any effect of the intergalactic medium which is expected to be nearly vacuous compared to the matter densities in the radio source at the positions measured by Kellermann. (The objection to Daly's method based on possible effects caused by the intergalactic medium would not apply here.) From these precepts Kellermann argues that evolutionary effects on the compact diameters , so close to the center, might be negligible.

References

Bruzual, A.G., & Spinrad, H. 1978, ApJ **220**, 1
Daly, R.A. 1994, ApJ **426**, 38
de Sitter, W. 1917, MNRAS **78**, 3
de Sitter, W. 1933, *The Astronomical Aspects of the Theory of Relativity*, Univ. Calif. Pub. in Math. (Berkeley: Univ. Calif. Press), Vol. 2, No. 8. Section 35, eqs. 79 & 80
Djorgovski, S. & Spinrad, H. 1981, ApJ **251**, 417
Eddington, A.S. 1923, *Mathematical Theory of Relativity* (Cambridge: Cambridge Univ. Press), Section 70
Hickson, P. 1977a, ApJ **217**, 16
Hickson, P. 1977b, ApJ **217**, 964
Hickson, P., & Adams, P.J. 1979, ApJ **234**, L91
Hoyle, F. 1959, in *Radio Astronomy*, IAU Symp. 9, ed. R.N. Bracewell, (Stanford [Calif.]: Stanford Univ. Press), p. 529
Hubble, E. 1926, ApJ **64**, 321
Hubble, E. 1929, Proc. Nat. Acad. Sci. **15**, 168
Hubble, E. 1930, ApJ **71**, 231
Hubble, E., & Humason, M. 1931, ApJ **74**, 43
Hubble, E., & Tolman, R.C. 1935, ApJ **82**, 302
Kapahi, V.K. 1975, MNRAS **172**, 513
Kapahi, V.K. 1985, MNRAS **214**, 19P
Kapahi, V.K. 1987, in *Observational Cosmology*, IAU Symp. 124, eds. A. Hewitt, G. Burbidge, L.-Z. Fang (Dordrecht: Reidel), p. 251
Kapahi, V.K. 1989, AJ **97**, 1
Kellermann, K.I. 1993, Nature **361**, 134
La Violette, P.A. 1986, ApJ **301**, 544
Legg, T.H. 1970, Nature **226**, 65
Lemaître, G. 1927, Ann. Soc. Sci. Bruxelles **47**, 49
Lemaître, G. 1931, MNRAS **91**, 483
Miley, G. 1968, Nature **218**, 933
Miley, G. 1971, MNRAS **152**, 477
Minkowski, R. 1960, ApJ **132**, 908
Robertson, H.P. 1928, Phil. Mag. **5**, 835
Robertson, H.P. 1929, Proc. Nat. Acad. Sci. **15**, 822
Robertson, H.P. 1935, ApJ **82**, 284
Robertson, H.P. 1938, Zs. f. Astrophys. **15**, 69
Sandage, A. 1972a, ApJ **173**, 485
Sandage, A. 1972b, ApJ **178**, 25
Sandage, A. 1975, in *Galaxies and the Universe*, eds. A. Sandage, M. Sandage, & J. Kristian (Chicago: Univ. Chicago Press), Chapter 19
Sandage, A. 1988a, ApJ **331**, 583
Sandage, A. 1988b, ApJ **331**, 605
Sandage, A. 1988c, Ann. Rev. A&A **26**, 561
Sandage, A. 1994a, ApJ **430**, 1
Sandage, A. 1994b, ApJ **430**, 13
Sandage, A. & Perelmuter, J.-M. 1990, ApJ **350**, 481
Segal, I. 1975, Proc. Nat. Acad. Sci. **72**, 2473
Segal, I. 1976, *Mathematical Cosmology and Extragalactic Astronomy* (New York: Academic Press), p. 104
Segal, I. 1980, Proc. Nat. Acad. Sci. **77**, 10
Segal, I. 1993, Proc. Nat. Acad. Sci. **90**, 4798 (Section 4c)

Stock, J., & Schücking, E. 1957, AJ **62**, 98
Tolman, R.C. 1929, Proc. Nat. Acad. Sci. **15**, 297
Tolman, R.C. 1930, Proc. Nat. Acad. Sci. **16**, 511
Tolman, R.C. 1934, *Relativity, Thermodynamics, & Cosmology* (Oxford: Clarendon), p. 467
Walker, A.G. 1936, Proc. London Math. Soc. 2nd Ser. **42**, 90
Zwicky, F. 1929, Proc. Nat. Acad. Sci. **15**, 773
Zwicky, F. 1957, in *Morphological Astronomy* (Berlin: Springer-Verlag), Sections 30/31

5 The Third Test: The Redshift–Distance Relation Based on the $m(z, q_0)$ Hubble Diagram

5.0 Introduction

This lecture is concerned with the meaning of the empirical evidence for a redshift-distance relation. We treat its many aspects, beginning with the theoretical expectations of the nature of a linear velocity field. The history is then followed to the convoluted discovery phase from the mid-1920's to the early 1930's, the confirmation phase from 1931 to 1936, the completion of the Humason-Mayall northern redshift survey of Shapley-Ames bright galaxies in 1956 including a small sample of great clusters, the extension to many directions and using many clusters in the first Palomar phase to the mid-1970's, the completion of an all-sky redshift program for the complete set of Shapley-Ames galaxies by an extension of the observing program to the south in the late 1960s, the extension of the Hubble diagram to large redshifts by the search for distant clusters and measurements of their redshifts at Palomar in the mid-to late-1970's, the use of these clusters to determine q_0 (both theory and practice) and why that route to the space curvature (q_0) is temporarily stalled, and finally a discussion of how close to the Local Group the expansion begins (i.e., the location of the nearby zero-velocity surface).

5.1 Expected Form

5.1.1 Four Claims

Four claims are in the current literature concerning the form of the redshift-distance relation.

1. A linear relation is required by the standard model, as shown in section 5.1.3. It is the form suggested by Hubble (1929a), was proved by Hubble & Humason (1931) using galaxies in great clusters, and is required by all the modern data when properly corrected for observational selection bias. A redshift-distance relation, expressed as $\Delta\lambda/\lambda_0 = r^n$ is a linear relation if $n = 1$ precisely.
2. A radical theory by Segal (1982) where the universe does not expand but where the redshift is due to a new concept based on a non-conventional precept of space-time requires $n = 2$ at all distances.
3. A claim for an expanding universe but where the expansion varies approximately as $n = 2$ locally but with $n = 1$ beyond $cz = 5000$ km s^{-1} is made by de Vaucouleurs & Peters (1986), Giraud (1985, 1986a, b, c), Tully (1988), and others in their analyses of flux-limited samples. That the claim is an artifact of an analysis where the effect of observational selection bias at every redshift is assumed to have an effect smaller than it in fact does, is shown in Lecture 10. The case is made in tedious detail in the general literature (Sandage 1988a, b 1994a, b; Federspiel, Sandage, & Tammann 1994); Lecture 10 gives an overview.

4. Because the consequences of an expanding universe seem so bizarre, a proposition was floated from the beginning (Zwicky 1929) that the expansion is not real. The photons lose energy by some process in their flight toward the observer. The details by known physics are not specified, but it must be by interactions with intervening matter in such a way that a redshift is caused by energy loss that is exponential with distance. The law would obviously be $(1+z) = e^{(HRr/c)}$.

   Zwicky did not follow up his initial suggestion with rational physics although he continued to write (Zwicky 1957) as if the suggestion was viable. Critics still point out that any scattering process in which energy is exchanged (causing a redshift) will cause scattering of the photon beam, leading to increased fuzziness of galaxy images with increasing distance, which is not observed. Nevertheless the "tired light" postulate remains in the modern literature (see La Violette 1986).

5.1.2 Properties of a Linear Velocity Field

It can be shown that a linear velocity field (velocity increasing directly with distance from any arbitrary particle in the field), has the following three properties, each crucial for the cosmological problem.

1. It is the only velocity field that preserves the shape of configurations as time proceeds, as proved in section 5.1.3. A non-linear velocity field will, over time, cause an initially uniform density field to become non-uniform. If the velocity increases with distance more slowly than linearly (the consequence of which would be that the point of reference in the manifold would then not be unique rather than just being any old point, all seeing the same field from any point if and only if the field is linear), then faster particles closer to the reference point will overtake the slower movers at some future time, causing an increase in the density outward. The opposite is true if the velocity increases faster than linear with distance.
2. There is no preferred position in the velocity field. Every position views the same linear increase in velocity toward or away from it and with the same rate seen from any other arbitrary place. To show this, transpose yourself to any other place in the field from whatever place you are now, subtract the velocity vectors in the appropriate way and see that the field is again linear as seen from the new position (Heckmann 1942). Furthermore it has the same velocity-to-distance ratio, i.e., Hubble constant, everywhere. This is a most beautiful property, possessed only by a linear velocity-distance relation. It shows the ultimate Copernican democracy.
3. A linear field permits an initial singularity. It is a necessary but not sufficient condition for a singularity. Consider an expanding field where an object twice as far from an observer as another is moving away at twice the speed. Reverse the velocity vectors. The object twice the distance now approaches the observer and has twice the distance to travel as the first, but it is moving twice as fast. Therefore, if in free fall, both objects reach

the observer at the same time. The same is true for all particles in the manifold. However, the actual case as to whether there is a singularity depends upon the dynamics. $R(t)$ need not be zero at $t = 0$ in some of the Friedmann models of the standard theory (Robertson 1933), i.e., the particles are, in general, not in "free-fall" there will be acceleration or deceleration with time). The dynamics is as given by the Friedmann equation.

5.1.3 Theoretical Expectations of Linearity from the Standard Model

(1) Proof Through Similar Triangles

If a geometrical configuration is to remain similar to itself upon expansion, the expansion velocity field must be linear. Proof can be made using similar triangles. Consider a triangle with two sides of length r_1 and r_2 at some cosmic time t. Fix an observer at the apex of the triangle formed by r_1 and r_2. Let the triangle expand, keeping the observer fixed in the manifold. By item (2) in the last section, fixing this point and observing the expansion elsewhere in the manifold loses no generality. Every point in the manifold appears to an observer placed there to be the center of the expansion.

If the triangle is to remain similar to itself upon expansion, the ratio of r_1 to r_2 must be the same at all times. Call that ratio H. Therefore, regardless of the past history of the expansion, i.e., independent of the nature of the $R(t)$ variation set out in Lecture 1, we must have

$$r_1 = Hr_2 \tag{5.1}$$

at all times.

However, because r_1 and r_2 change with time due to the expansion, the end points of the triangle at these distances from the observer at the apex will show velocities relative to the apex. These are $\mathrm{d}r_1/\mathrm{d}t$ and $\mathrm{d}r_2/\mathrm{d}t$, called $v_1(t)$ and $v_2(t)$ at any particular arbitrary time. The time derivative of equation (5.1), evaluated at a particular time (H is the same for all points on the triangle at this time and for all the space of which the triangle is a part, i.e., over the entire manifold), gives therefore

$$v_1 = Hv_2 \tag{5.2}$$

at this time. Combining the two equations (neglecting any light-travel time over the manifold) gives

$$\frac{r_1}{r_2} = \frac{v_1}{v_2} = H.$$

Hence,

$$\frac{v_1}{r_1} = \frac{v_2}{r_2} = H,$$

or, for any arbitrary distance i,

$$v_i = H r_i, \tag{5.3}$$

i.e., the velocity-distance relation is linear.

(2) Proof via the Expanding Balloon Analogy

An analogy that is part of the culture, due perhaps first to Eddington (1933, Chapter III), is to consider the kinematic properties of dots placed on the surface of an expanding balloon. Air is pumped into the balloon in *any arbitrary fashion.* As shown below, the history of how the radius, $R(t)$, changes with time is irrelevant to the instantaneous existence of a velocity-distance relation that is the same as seen from any arbitrary dot on the surface at any arbitrary time.

Assume that the balloon is spherical. (An elliptical balloon will show a different expansion rate, H, in different directions.) Pick an arbitrary dot on the surface as origin, O. Measure distances, r_i, along the surface to any number of dots elsewhere on the surface. Call two of these, dots 1 and 2, with distances from O of D_1 and D_2.

Now imagine looking at the surface from the center of the balloon (a dimension denied a 2-space creature confined to the surface, but the mathematicians of the 2-space culture would understand the concept of the "radius of curvature" of their non-Euclidean space).

Name the angle between two vectors drawn from the center to the surface between O and point 1 θ_1. Name the similar angle between O and point 2 θ_2.

With this construction, clearly

$$D_1 = \theta_1 R(t),$$

and

$$D_2 = \theta_2 R(t),$$

where $R(t)$ is the radius of the balloon at time t.

Differentiation of both equations relative to time gives the instantaneous velocities of dots 1 and 2 as seen from O. Noting that in a *radial* expansion (of the spherical surface) the angles θ_1 and θ_2 do not change with time, clearly then,

$$\frac{\dot{D}_i}{D_i} = \frac{\dot{R}}{R} = H, \tag{5.4}$$

all arbitrary spots, i, satisfying the relation. Hence, H is a constant over the entire surface at any given cosmic time, providing, again, that we neglect the travel time of information over the surface. Therefore, as before,

$$\dot{D}_i \equiv v_i = H D_i,$$

which is a linear velocity-distance relation.

(3) Proof from the Metric

From Lecture 1 it was shown that the operational distance, l, as measured by an astronomer using practical means at the telescope, is related to the co-moving coordinate "distance", r, of the metric by

$$l = R(t) \int_0^r \frac{dr}{\sqrt{1 - kr^2}}. \tag{5.5}$$

Differentiate with respect to time, recalling that r is a constant for all time for a particular galaxy, giving

$$\dot{l} = \dot{R}(t) \int_0^r \frac{dr}{\sqrt{1 - kr^2}}, \tag{5.6}$$

where the definite integral is the same in both equations. Dividing the two equations gives

$$\frac{\dot{l}}{l} = \frac{\dot{R}}{R} = \text{constant} = H, \tag{5.7}$$

which again is a linear velocity-distance relation.

(4) Lemaître's 1927 Prediction and Discovery

Many of the central ideas of practical cosmology were set out by Lemaître (1927) in his fundamental paper on non-static solutions to the Einstein field equations. However, the paper was published in a rather inaccessible journal. It was brought to Eddington's (1930) attention by Eddington's research student, G.C. McVittie, later a leader in the field in his own right (McVittie 1956, 1965). At Eddington's suggestion a translation from the French was published in the Monthly Notices (Lemaître 1931).

Lemaître's equation 23 shows that a linear velocity-distance relation is the prediction of his non-static theory. He was aware of Hubble's (1926a) estimate that the mean absolute magnitude of field galaxies was $M_{\mathrm{pg}} = -15\overset{\mathrm{m}}{.}2$ on Hubble's distance scale at the time[15]. By combining Hubble's mean absolute magnitude for galaxies with Slipher's measured redshifts, in Eddington (1923, *Mathematical Theory of Relativity*, section 70), Lemaître calculated and set out as his equation 24 the rate of expansion (as the ratio between velocity and distance) of $H = 627$ km s^{-1} Mpc^{-1}. The remarkable irony is that this calculation occurred two years before Hubble's (1929a) discovery. History is more complicated than the simplest explanations of it.

[15]The modern value of M_{B}^* in the Schechter function is $-20\overset{\mathrm{m}}{.}7$ (Tammann, Yahil & Sandage 1979) based on $H = 50$ km s^{-1} Mpc^{-1}, giving a ratio of $5\overset{\mathrm{m}}{.}5$ or a factor of 12.6 between the modern distance scale with $H_0 = 50$ and Hubble's original distance scale.

(5) Robertson's 1928 Prediction and Discovery

H.P. Robertson (1928), in a similarly remarkable paper communicated to the Royal Society by Eddington and published in Philosophical Magazine, also set out some of the observational consequences of relativistic cosmology. As there is no mention of Lemaître (1927), we can be certain that Robertson (and by inference, Eddington) did not know of Lemaître's central paper at the time, but he does use the coordinates employed earlier by Lemaître (1925) in a discussion of the de Sitter *static* solution. Later Robertson (1933, p. 67) in his famous 1933 Reviews of Modern Physics paper does credit Lemaître with fine priority for the non-static 1927 Lemaître solution and the discussion of its consequences.

Near the end of the 1928 Phil. Mag. paper, Robertson states "— we should expect a correlation $v \sim cl/R$" [marked as his equation 17] "between the velocity v, the distance l and the radius [R] of the observable world. Comparing the data given by Hubble (1926a) concerning the value of l for the spiral nebulae with that of Slipher (see Eddington 1923, p. 162 for Slipher's Table of Radial Velocities) concerning the corresponding radial velocities, we arrive at a rough verification of (17) and a value of $R = 2 \times 10^{27}$ cm".

Robertson's value of R, when placed in his equation 17, gives $H = c/R =$ 461 km s^{-1} Mpc^{-1}, again before Hubble's 1929 announcement[16].

[16]These facts are particularly vivid for the present writer. H.P. Robertson became Professor of Physics at the California Institute of Technology in 1949, replacing R.C. Tolman as the resident cosmologist. Tolman had passed to his reward in the summer of 1948. The initial group of graduate students in the new department of astronomy at Caltech had the privilege (and the necessity) of attending Robertson's lectures on mathematical physics, and in that way we each gained an acquaintance with him in 1950-1952, especially because he took a great interest in assigning most difficult problems and grading them himself in formidable detail. From 1952 on, as a new employee of the Mount Wilson and Palomar Observatories, I had a more detailed opportunity to talk with him on a number of occasions on the Caltech campus.

In 1959-1960 I was attempting to throw the four observational tests of "world models" into operational forms in the manner of Bridgeman (1927, 1936). The requirement was that the specific questions to be asked have operational answers found by practical means at the telescope.

With a late draft of what later became a Journal paper (Sandage 1961a), I asked Robertson to grade the manuscript as he had done with the students nearly a decade before. He did and in the process told me of his 1928 paper with its prior "suggestion", as he put it, of the linear redshift-distance relation. He also said that he had discussed the result with Hubble in 1927/8, but did not know if Hubble ever remembered the conversation. I had not heard of this before, although I had worked as an assistant to Hubble from 1950 through 1953 on the the redshift-distance relation.

Robertson was without rancour concerning any claim of priority – he was the most fair-minded man any of us knew. To the end he held Hubble in high regard, dedicating his mid-1950's review (Robertson 1955) to Hubble's memory.

Both the Lemaître and the Robertson pre-Hubble value of H are within the range of Hubble's value of $H = 500$ km s^{-1} Mpc^{-1} set out in his 1929 announcement paper.

5.2 The Early Observations 1927–1936

5.2.1 Early Search for the "de Sitter Effect"

Before the discovery by Friedmann that a time-dependent scale factor to the metric could satisfy the relativity field equations, and the further development of that idea by Lemaître, static solutions of the equations had been found. Three such solutions for the gravitational potentials, g_{ij}, in the metric were found, and it was proved later by Tolman (1929) that these three and only these such solutions exist. The solution found by Einstein (1917) has the familiar g_{ij} coefficients of equation (1.12) (Lecture 1) but with a fixed (i.e., static) scale factor. A second solution is that of Minkowski space-time, where the four components of the metric tensor are 1,1,1,−1, and the third solution found by de Sitter (1916a,b, 1917) has the metric

$$ds^2 = \frac{dr^2}{1 - \frac{r^2}{R^2}} + r^2(d\theta^2 + \sin^2\theta d\phi^2) - (1 - \frac{r^2}{R^2})dt^2. \tag{5.8}$$

Although all three solutions are static, the third de Sitter solution contains the remarkable term $g_{44} = (1 - r^2/R^2)$ as the coefficient of the time interval dt^2. This means that a clock placed at the observer with $dr = d\theta = d\phi = 0$ will keep a different time than an identical clock placed at some distance r in the mainfold because the timelike interval depends on the distance r. The consequence would be that the timelike interval becomes smaller for larger r, meaning that clocks would appear to slow down depending on the distance. The effect would manifest itself as a redshift, increasing with distance. This is the de Sitter effect in a static universe.

The predicted redshift-distance relation follows by noting that the ratio of times for a clock at the observer and at distance r is $\sqrt{1 - r^2/R^2}$, obviously giving

$$\frac{\lambda_1}{\lambda_0} \equiv 1 + \frac{\Delta\lambda}{\lambda_0} = \sqrt{1 - \frac{r^2}{R^2}}, \tag{5.9}$$

or, expanding for small r, giving the squared dependence of redshift on distance as

$$\frac{\Delta\lambda}{\lambda_0} = \frac{1}{2}\frac{r^2}{R^2}, \tag{5.10}$$

to first order.

The problem was, however, more complicated. A second property of the de Sitter solution is that particles placed in the manifold have a tendency to accelerate, leading to more terms in equation (5.10) (Eddington 1923, section 70; de Sitter 1933, equations 79 and 80).

This second effect was not appreciated by the observers and by some theoreticians in the 1920's, and the quest for the "de Sitter" effect was a search for a quadratic relation, or in less specific terms, a search for any relation between redshift and distance, but the motive was always to test for the predicted effect in equation (5.10).

A feeling for the prevalent climate can be had by reading the literature of the 1920's. Particularly revealing are the attempts to "determine the curvature of de Sitter's space-time" using measured radial velocities of astronomical objects.

Before galaxies were seriously considered as markers of the space in the early 1920's, data on globular clusters were used. The ideas and the data started fermenting together, seen, for example, in the series of papers by Silberstein (1924a,b, 1925) and the reply to them by Lundmark (1924). Silberstein used globular clusters; Lundmark used many different types of objects including galaxies. Armed with Slipher's redshifts, both derived values of de Sitter's radius of curvature based on the "de Sitter redshift effect".

In his 1924 paper, Lundmark plots (his Fig. 5) relative distances of galaxies against Slipher's measured radial velocities. His galaxian distances were based on his own measurements of apparent diameters, determined from Lick Observatory Crossley plates "from the supposition that the apparent angular dimensions and total magnitudes of the spiral nebulae are only dependent on distance" (a totally modern view now 70 years old!: van der Kruit 1986; Sandage 1993a,b). His conclusion was "we find that there may be a relation between the two quantities (v and r), although not a very definite one. If this phenomenon were due to the [de Sitter] curvature of spacetime..." etc. He then proceeds to discuss Silberstein's idea of finding the de Sitter effect in a *static* universe from the radial velocity data.

Also of considerable interest is the paper by Hubble (1926a). Near the end of that paper Hubble derives the radius of curvature of an Einstein *static* universe based on the mass density of galaxies. He used the theoretical treatment in the well known text book of the time by Haas (1924). This curvature, although having nothing to do with the de Sitter effect, nevertheless is still grounded in a *static* model as late as 1926, four years after the Friedmann dynamic solutions.

Based on the current histories of these events which state that "Hubble discovered the expansion in 1929", it will strike the reader as remarkable that Hubble was still interpreting his 1929 velocity-distance plot as the likely discovery of the de Sitter *static* case.

Of particular interest in the period just before 1929 is Lundmark's (1925) long paper on "The Motions and Distances of Spiral Nebulae". There, among the landmark results such as the separation of novae into "upper class" (supernovae in modern terms) and "lower class" (normal novae), he attempted to correlate galaxy distances with Slipher's redshifts. As said, Lundmark had derived relative galaxy distances based on angular diameters in his 1924 paper. In this more complete 1925 paper, Lundmark writes: "A rather definite correlation is shown between apparent dimensions and radial velocity in the sense that the smaller and presumably more distant spirals have a higher space motion". Lundmark should have stopped there. If he had, his case for shared credit with Hubble for the ultimate step from speculation to the real beginnings of the modern velocity-distance relation would have been more secure. But analyzing the data beyond its capabilities, Lundmark, a pure Baconian empiricist, fitted an arbitrary second degree polynomial of the form $v = a + br + cr^2$ to his distances and Slipher's redshifts, unfortunately deriving a negative value for the c coefficient. Lundmark concluded therefrom that "one would scarcely expect to find any radial velocity larger than 3000 km s^{-1} among the spirals" (Lundmark 1925, p. 867).

Hubble (1936b, Chapter 5), in his short history of the period emphasizes Lundmark's shot in the foot as the last attempt to find a $v - r$ relation before 1929. Hubble also relates the earlier attempts to understand the "solar motion" relative to the nebulae by the introduction of a standard K term (a general expansion or contraction of a system of objects, independent of their distance) by Wirtz (1918, 1921) who followed Truman's (1916) unsuccessful attempt to find a simple dipole solar motion solution. Hubble states that Wirtz's (1925) final discussion implied to Wirtz that de Sitter's prediction of a redshift-distance effect (in a static universe) had been verified. This was the tenor of the times. These papers just preceded the seminal papers of Lemaître (1927) and Robertson (1928) with their recrudescence of the Friedmann (1922, 1924) time-dependent dynamic model of a truly expanding universe.

5.2.2 The Announcement

(a) Hubble 1929

Following Hubble's (1925, 1926b, 1929b) three papers on the Cepheid distances to NGC 6822, M33, and M31 he began accumulating data on the apparent magnitudes of what he considered to be the brightest resolved stars in nearby galaxies. He had already used the first of these data in the 1926 paper as the basis to estimate that $\langle M_{\rm pg} \rangle = -15\overset{\rm m}{.}2$ for the average galaxy, based on $M_{\rm pg} = -6\overset{\rm m}{.}3$ for the absolute magnitude of the brightest stars[17].

[17]During the initial phases of correcting Hubble's distance scale to the modern scale when Baade (1952) doubled the distance to M31, Hubble told me that his 1936 scale, based on brightest star absolute magnitudes of $M_{\rm pg} = -6\overset{\rm m}{.}3$, was severely criticized in the opposite direction by the Mount Wilson spectroscopists (as being much too bright). They had told him that no star was known with absolute magnitude brighter than about $M_{\rm pg} = -3^{\rm m}$. The same story was told me by Paul Merrill about 1955, a Mount Wilson astronomer critical of Hubble but

We now know that the objects used by Hubble were not the brightest stars but are HII regions (Sandage 1956 in HMS, Appendix C; 1958).

With the apparent magnitudes of these brightest "stars" in hand, and with the assumption that $M_{pg} = -6^{m}_{.}3$ for them, Hubble had distances that were independent of angular diameters and/or total apparent galaxian magnitudes. These had been the criteria previously used by Wirtz, Lundmark, and Dose (1927). It was this set of new data on distances that Hubble (1929a) used to correlate with the known velocities in his discovery paper.

But Hubble had one additional most crucial bit of information not available to Lundmark (1925). In the paper just preceding Hubble's famous announcement there is a short note by Humason (1929) reporting an enormous (for the time) redshift $v = 3779$ km s^{-1} for NGC 7619 in a small group of E galaxies. Humason's third sentence reads "About a year ago Mr. Hubble suggested that a selected list of fainter nebulae... be observed to determine, if possible, whether the absorption lines in these objects show large displacements toward longer wave-lengths, as might be expected *on de Sitter's theory of curved space-time.*" (Emphasis mine.) Note again the absence of mention of either Friedmann's or Lemaître's expanding model. (There is no doubt that Hubble worked closely with Humason in the writing of Humason's note. Humason always credited Hubble in this regard. Hence, in 1929 Hubble apparently did not know of the truly expanding models from the Friedmann-Lemaître theory).

We now come to two most interesting things about Hubble's announcement paper. The linear velocity-redshift relation is clearly visible in Hubble's Fig. 1, certainly to a much greater degree than the previous discussions by Lundmark or of Silberstein. The first notable omission is that Hubble does not plot Humason's large redshift of NGC 7619 which would have strengthened his case enormously. But, to be sure, he had no independent distance for NGC 7619 because it is an E galaxy with no "brightest stars" of the kind he had used. Hubble does state that extrapolating the relation he had obtained from the spirals at velocities less than 1200 km s^{-1} out to Humason's NGC 7619 redshift gives "the absolute magnitude at such a distance as $-17^{m}_{.}65$, which is of the right order for the brightest nebulae in a cluster" (known from some early knowledge of the luminosity function in clusters; Hubble & Humason 1931). But, indeed, it was Humason's crucial discovery of the large redshift that showed immediately a new phenomenon that was in clear violation of Lundmark's unfortunate quadratic prediction. Humason's results surely gave Hubble knowledge for the first time that he was on to something.

The second interesting thing concerns what Hubble thought he had discovered. The last several sentences of Hubble's paper are:

---

who, nevertheless, at that late date admitted the incorrect spectroscopic absolute magnitudes of the brightest stars assumed in the 1930's. The modern calibration of the upper limit to stellar luminosity in large galaxies is $M_{pg} = -10^{m}$ (Hubble & Sandage 1953 corrected to Baade's distance scale; Sandage & Tammann 1974, 1982b; Humphreys 1983; Sandage & Carlson 1985; Sandage 1986a).

"The outstanding feature, however, is the possibility that the velocity-distance relation may represent the de Sitter effect, and hence that numerical data may be introduced into discussions of the general curvature of space[18]. In the de Sitter cosmology, displacements of the spectra arise from two sources, an apparent slowing down of atomic vibrations and a general tendency of material particles to scatter. The latter involves an acceleration and hence introduces the element of time." [Hubble apparently partially misunderstood the theory here because his "element of time" *is* the slowing down of clocks as a function of distance, caused by the scandalous mixed coordinate in g_{44} of the metric.] "The relative importance of these two effects should determine the form of the relation between distances and observed velocities; and in this connection it may be emphasized that the linear relation found in the present discussion is a first approximation representing a restricted range in distance".

What Hubble meant by the last phrase is not quite clear, but we can guess that he was not convinced that the relation *should* be linear. Rather, he knew by this time of the quadratic prediction of the empty de Sitter model [equation (5.10)], and may have thought that the data he presented defined the relation only near the origin of a real parabola (Lundmark's quadratic attempt but with a positive coefficient for the squared term) that could be approximated by a linear relation over short distance near the observer.

(b) Shapley 1929

Stunned by Hubble's paper, Shapley (1929) made an attempt to dilute the priority of the discovery by a fresh discussion using Lundmark's (1924, 1925) methods based on a correlation of angular diameters (and indirectly on apparent magnitudes) with redshift, a method also used earlier (but not shown explicitly) by Lemaître (1927) and Robertson (1928). But these latter two investigations were evidently as unknown to Shapley as they were to Hubble because there is no mention of them in Shapley's paper.

Shapley began his note by: "In one of the first general considerations of the velocities of extragalactic nebulae, I pointed out some ten years ago the systematic recession of these objects on both sides of our Galaxy and the fact that the speed of spiral nebulae is dependent to some extent upon apparent brightness indicating the relation of speed to distance ...". Shapley then correlated known radial velocities with dex $(0.2\mathrm{m} - 2.0)$ which is a measure of relative distance if there is a well defined mean absolute magnitude for the galaxies.

The most astounding feature of Shapley's paper is the lack of mention of Hubble's prior discovery. Shapley does reference this paper but only to attribute to Hubble the determination of apex and the size of the sun's *dipole*

[18]The last phrase of this sentence is somewhat ingenuous in view of the fact that Silberstein and Lundmark had indeed introduced numerical data into their discussions of the "general curvature of space" by the de Sitter effect.

motion toward $A = 277°$, $D = 36°$ with a dipole amplitude of 280 km s^{-1}, i.e., what we would now call the "solar motion relative to the centroid of the Local Group due predominantly to Galactic rotation". Shapley claims "except for the important result for the faint object NGC 7619, for which Humason has recently found a velocity of 3800 km s^{-1}, the relation of speed to apparent magnitude [shown in Shapley's new figure] is much the same as in 1918 (sic)." An inspection of this paper by Shapley and Betz (1919, section VII) to which Shapley refers shows a vast difference between this claim and Hubble's discovery[19].

(c) de Sitter 1930

De Sitter (1930a, b) wrote two papers on various observational matters concerning redshifts, using diameters instead of magnitudes. As with Shapley, he also failed to afford a proper priority to Hubble's announcement. Hubble, sensing that his methods and result using "brightest stars" would be diluted by the flavor of de Sitter's language, wrote a strongly worded objection privately (correspondence in the Huntington Library) to de Sitter, to which a friendly reply mentioning the acceptance of priority was received in return.

However, the matter was never fully resolved publicly. In his later Hitchcock lectures, de Sitter (1933) fails to mention Hubble's discovery or to reference Hubble's 1929 announcement, although he does say that "During the last few years the limits of our neighborhood have been enormously extended by the observations of extragalactic nebulae, made chiefly with the 100-inch telescope at Mount Wilson", with no attribution to either Hubble or Humason. And, for what it is worth as an indication of his possible recognition of the importance of Hubble's 1929 initial correlation of redshift and distance,

[19]Shapley made the same type of claim of priority in another grand moment in science with Baade's (1952) discovery of the factor of two error in Hubble's distance scale. After hearing Baade lecture on the result at the 1952 Rome meeting of the IAU (Report of Commission 28) "Prof. Shapley then asked Dr. Baade if he could describe in a little more detail how he (Dr. Baade) had arrived at the error in the zero-point of the classical Cepheids. Dr. Baade offered two arguments in support of this conclusion. – ".

Upon returning from the Rome meeting Shapley (1953) published a long paper, again in the Proceedings of the National Academy of Sciences, where his principal mention of Baade was simply "Dr. Baade presented several of the arguments for revision [of the Cepheid zero-point] at the Rome meeting of the International Astronomical Union, September 1952", when, in fact, Baade had presented much more, leading to the total concept of a revised distance scale, doubling Hubble's scale.

Baade was furious in a way which, as his graduate student from 1950 to 1953, I had never seen in him either before or after. It was a case of Shapley writing, not quite as if he had invented it all, but also with a flavor of not denying that he had not.

Because these Lectures are aimed at students just starting their careers, the moral is that (1) eventually everything works for the good (i.e., "time wounds all heels"), and (2) scientific morality is worth an infinite sum compared with false priority.

de Sitter does write the relation (unnumbered equations on his page 155) as $dr/dt = hr$, using, for the first time, the letter h for the coupling constant of what is now known as the Hubble constant. In a copy of the published Hitchcock lectures that belonged to Hubble, the few sentences on this page containing this equation have been marked in pencil, almost certainly by Hubble.

5.2.3 The 1930's Confirmations and Extensions

(a) Hubble & Humason 1931, 1934

Hubble immediately set in motion a comprehensive observational campaign with Humason to confirm if his preliminary 1929 result was real and to extend the first suggestions of a linear velocity-distance relation to larger redshifts. In perhaps the most important observational paper of this period, important because it removed all doubts as to the raw facts, Hubble & Humason (1931) wrote remarkably about a number of topics in "The Velocity-Distance Relation Among Extra-Galactic Nebulae". Humason had determined 40 new redshifts, 26 of which for galaxies in 8 groups and clusters. Hubble had determined apparent magnitudes and types for many galaxies in each group and cluster and so could construct rough luminosity functions from which the "mean apparent magnitude" of galaxies in the aggregate was determined.

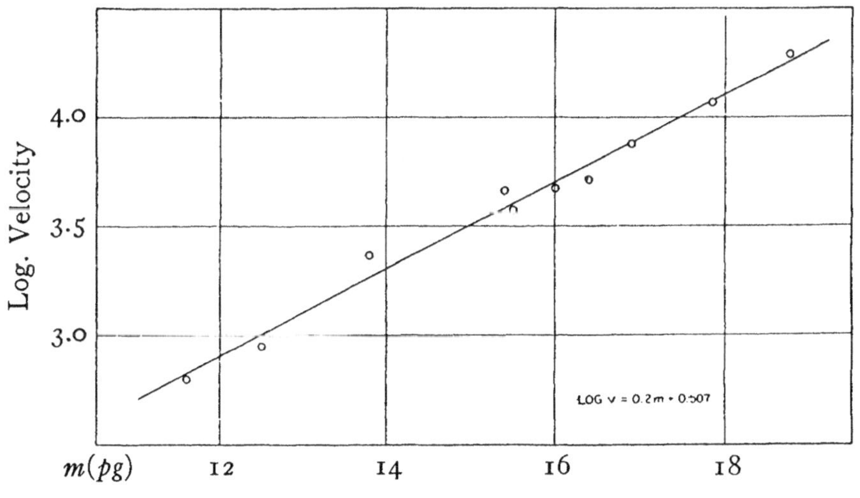

Fig. 5.1. Hubble ($m, \log v$) diagram from Fig. 4 of Hubble & Humason (1931). Note that the magnitudes (abscissa) are the mean of the entire luminosity function in each cluster, not the mag of the brightest or the "n^{th}" brightest member.

The most remarkable result was the large redshift of 19,700 km s^{-1} for a cluster galaxy in the Leo group. The next largest redshift in the 1931 paper was for the brightest galaxy in the Ursa Major I cluster with a redshift of 11,700 km s^{-1}. Redshifts of this size had never been seen before, and the result was clearly spectacular, especially because the linear velocity-distance relation, expressed as $\log v = 0.2m + \text{const.}$, was satisfied beyond all doubts as early as 1931. Those who claim differently even today (Segal 1993) ignore the record and the most extensive evidence consistently obtained over the past 65 years.

The redshift-"distance" relation, plotted as an apparent magnitude-log velocity diagram (i.e. the Hubble diagram) obtained just two years after the "discovery" paper is reproduced here as Fig. 5.1. The slope of the line is 0.2 which is the value required for a linear velocity-distance relation. (Recall the log-log nature of the ordinate and abscissa; the slope here defines the exponent n in $v \sim r^n$. The slope of 0.2 requires $n = 1$ precisely.)

Figure 5.1 was the most potent and the earliest proof of linearity despite the fact that only 10 points define the relation. The work over the next 40 years was directed to extending the redshift range and enlarging the sample to test the universality of the relation in various directions and to search for deviations from a smooth expansion flow.

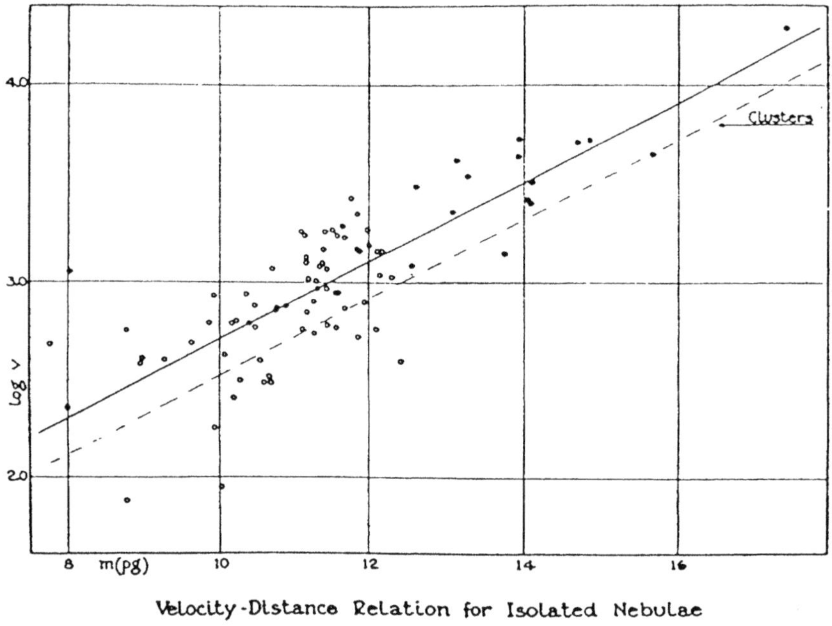

Fig. 5.2. Hubble diagram for field galaxies from Hubble & Humason (1934).

Progress was reported in a short paper by Hubble & Humason (1934) which dealt with the redshifts and magnitudes of field galaxies. The result, using new redshifts of 35 field galaxies observed by Humason, when combined with all known previous redshifts by Slipher, Pease, Sanford, and others, gave 94 galaxies in the data set. The Hubble diagram for this field galaxy sample is shown as Fig. 5.2 taken from Fig. 1 of their paper. The solid line was put through the data with a slope of 0.2. The dashed line is the cluster data from Fig. 5.1. It is fainter, due to a selection effect of using a mean luminosity-function-magnitude for the clusters and the individual magnitudes for the field galaxies. The redshift range of these data is to about 10,000 km s^{-1}. (Compare with Fig. 5.15 in section 5.6 that uses modern data.)

(b) Humason and Hubble 1936

In separate papers, Humason (1936) and Hubble (1936a) carried the determination of redshifts and magnitudes to more field galaxies and to several new remote clusters, the largest redshift of which was 42,000 km s^{-1} for the Bootis cluster. The Hubble diagram for 10 great clusters available from the 100-inch program in 1936 is shown in Fig. 5.3 taken both from the Humason and the Hubble papers (it is the same in both). This is also Fig. 14 of Hubble's (1936b) book "The Realm of the Nebulae".

Figure 5.3 is the final extent of the Mount Wilson 100-inch program for clusters. Humason repeatedly tried for redshifts of still fainter known clusters. At the time, all had been found by accident in fields photographed for other reasons with the small-field, large-aperture reflectors whose areas were generally less than 45 arc minutes on a side. Many of the clusters had been found at Mount Wilson with the 60 and 100-inch telescopes, but not all. For example, Baade (1928) had discovered the important Ursa Major No. 1 cluster with the Hamburg 40-inch reflector.

Humason could not obtain redshifts for clusters fainter than Ursa Major No. 2. He often tried the Hydra cluster (redshift 60,000 km s^{-1}, $z = 0.2$) but he was not certain enough of his 100-inch results to be convinced, waiting for the 200-inch telescope (expected in 1938) with which he eventually measured Hydra in 1950.

The slope of the line in Fig. 5.3 is 0.2, showing again a linear redshift-distance relation. The plotted magnitudes are for the fifth ranked galaxy in the cluster, a departure from the convention of "mean" magnitudes which Hubble & Humason had employed in 1931 (Fig. 5.1). Note also that the magnitudes have been corrected for K dimming as the K corrections were known in 1933 (see Lecture 3).

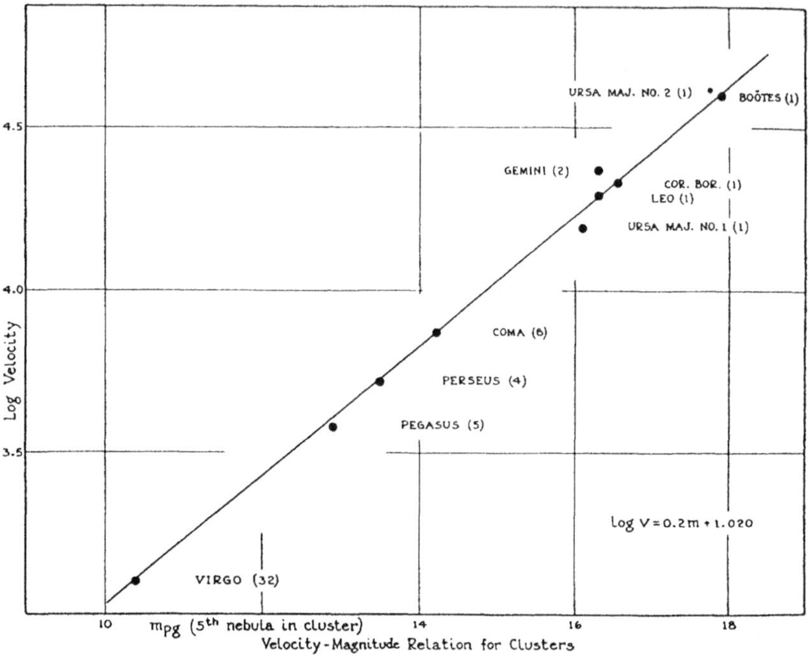

Fig. 5.3. The 1936 Hubble diagram for 10 great clusters from Hubble (1936a) and also from Humason (1936). This turned out to be the extent of the Mount Wilson redshift 100-inch Hooker telescope program. The diagram was not extended to larger redshifts until the program was resumed after the Second World War with the Palomar 200-inch telescope in 1949.

5.3 The Palomar Program 1953–1980

5.3.1a Using Known Clusters 1950–1953

A new start in defining the redshift-distance relation in multiple directions and larger distances was begun as soon as the 200-inch telescope was commissioned in November 1949. A report of progress to 1953 and a prospectus for the future was made by Hubble (1953) in his May 1953 Darwin Lecture before the Royal Astronomical Society, four months before he passed to his reward.

Only 11 of the standard clusters had been studied with the new data as reported by Hubble. The photometric work had been started by the present writer again from scratch, based on modern photometric standards. New data for the previous redshifts were also obtained by Humason using the superior spectrograph and camera made for the prime focus of the 200-inch. The photometry was tied to photoelectric standards in Selected Areas previously measured by Stebbins, Whitford, and Johnson (1950, SWJ) and by W.A. Baum. These sequences in SA 57, 61, and 68 replaced the standards for these areas in the Mount Wilson Catalog of 139 Selected Areas (Seares,

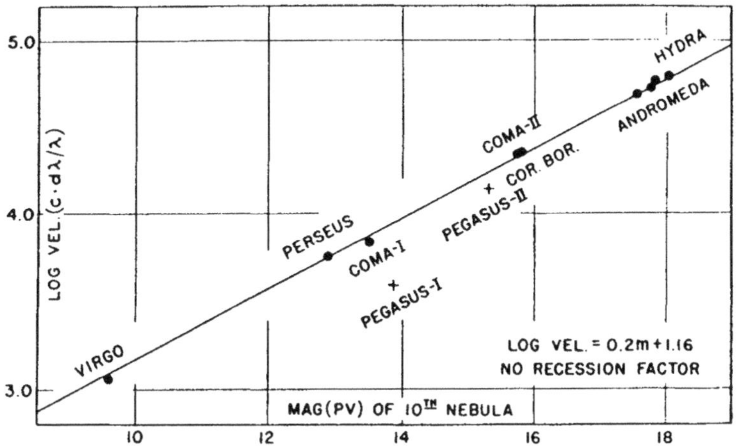

Fig. 5.4. Hubble diagram from Hubble's (1953) Darwin Lecture showing the first post-war results from the 200-inch telescope. Note the "no recession" factor for the "correction of the magnitudes for the effects of redshift".

Kapteyn, & van Rhijn 1930) used previously by Hubble in the 1930-1936 campaign. Recall that Hubble's 1936 $N(m)$ data had been calibrated by his early photographic transfers to the Mount Wilson sequences in SA 57 and the North Polar Sequence, now revised by SWJ and Baum.

The resulting redshift-magnitude diagram is in Fig. 5.4 taken from the Darwin Lecture. It is based on the apparent magnitude of the 10th brightest member galaxy. The magnitudes in that diagram are corrected for the selective part of the K term (he still forgot the bandwidth term, an omission discovered only in 1956) and, as was Hubble's procedure, also have been corrected only for one of the two factors of $2.5\log(1+z)$ for the "energy" effect but not the "number" effect (Lecture 3). This is evident by the notation "no recession factor" in Fig. 5.4. *Hence, as late as 1953 Hubble was still questioning if the expansion is real.* We have seen earlier that he began doubting the expansion in the mid-1930's, based on his analysis of the galaxy counts (Lectures 2 and 3).

5.3.1b HMS 1956

The results of the long, mid-1930's, redshift campaign on Shapley-Ames field galaxies by Humason at Mount Wilson and Mayall at Lick, as continued at Palomar and at Lick after the Second World War, were combined with Humason's Palomar campaign on clusters and Pettit's photoelectric photometry, and published in 1956 (Humason, Mayall, & Sandage 1956, hereafter HMS). Many of the clusters in Humason's new list had been discovered in the first years of the 48-inch Schmidt Palomar Sky Survey. Included in HMS was also

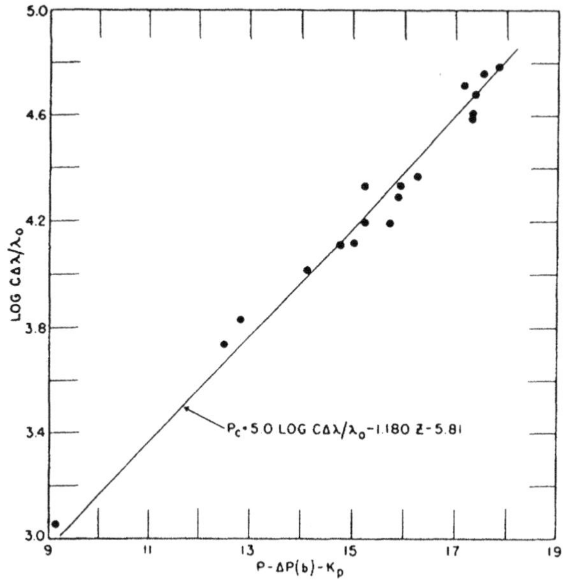

Fig. 5.5. The Hubble diagram in P for 18 clusters from HMS (1956). The data were used to attempt a determination of the second order (q_0) term in search for deceleration of the expansion.

a discussion of the apparent magnitudes of the program galaxies, corrected to a well defined fraction of the "total" light by considering a standard luminosity profile and integrating it to different fractional angular diameters (HMS, Appendix A).

Photometric and redshift data were given for 18 great clusters, from which the Hubble diagram in the P photometric band is reproduced in Fig. 5.5. The magnitudes in this diagram have been corrected only for the selective and the band-width K term (Lecture 3). The two effects of redshift (energy and number) were put into the theory (not the observations) via the Robertson (1938) equation as discussed earlier. The "second order" deceleration term (q_0) could finally be looked for with these data[20].

Although a start on the general redefinition of the Hubble diagram, the 1956 cluster data were not so much more extensive than the result in Fig. 5.3 where the 1936 campaign had ended. The maximum redshift was only 0.2 (for the Hydra cluster), compared with 0.16 from the earlier campaign. The main advances were (1) the photometric data had been tied to the Stebbins,

[20]On two occasions Hubble had written that deviations of the data from a slope of $d \log v/dm = 0.2$ at large redshifts would be a measure of the past history of the expansion by providing a mapping of the $R(t)$ function over the look-back time (he did not use these terms). However, on both occasions (Hubble 1938, 1953) he stated that if the data points deviate upward from a line of slope 0.2 it would indicate *acceleration* which in fact is incorrect. Galaxies at large redshift are seen as they were when light left, hence a higher redshift for them than a line of 0.2, would show higher redshift *in the past* compared to the present. This sense is deceleration.

Whitford, & Johnson (1950) and the Baum photoelectric standards on a true Pogson scale and a well defined zero point, and (2) adequate procedures were in place to define either "total" magnitudes or fractional magnitudes related to aperture ratios relative to the galaxy diameter. These definitions of galaxy magnitudes were independent of the apparent galaxy size, removing a previous systematic error that depended on distance. (Previous measurements often included a larger fraction of a galaxy's light for galaxies of small angular diameter compared with large).

5.3.1c 1970–1975

Many new clusters of galaxies had been systematically found and cataloged (Abell 1958, 1975) from the Palomar Sky Survey plates, providing the principal observing list in the north for the extended program in the 1970's. A series of eight papers was published in this period (Sandage 1972a, b, c; 1973a, b, c; 1975; Sandage & Hardy 1973). These results, together with Minkowski's (1960) large-redshift cluster containing the radio source 3C 295, provided the Hubble diagrams shown in Figs. 5.6 (from Sandage 1972b) and 5.7 (from Sandage & Hardy 1973) which also contain nearby groups. All the bells and whistles of the necessary corrections (K term, Galactic absorption, Bautz-Morgan contrast effect, and population luminosity function effect) are put in the photometry shown in these diagrams.

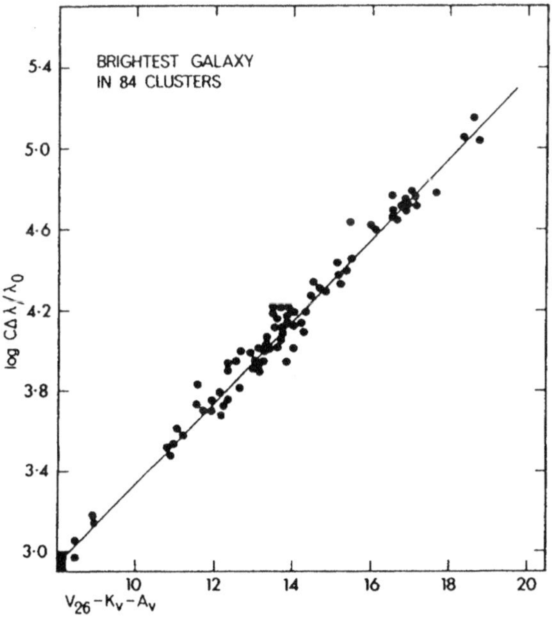

Fig. 5.6. Hubble diagram in the V photometric band showing the progress to 1972 in the extension of the data to higher redshifts (from Sandage 1972b). The black box in the lower left is the extent of the data available to Hubble (1929a) in his discovery announcement. The line has the imposed slope of d log $z/\mathrm{d}V = 0.2$ required if the redshift-distance relation is linear.

The black box in the lower left corner of Fig. 5.6 is the extent of Hubble's data in his 1929 announcement paper. The data in these two diagrams refer to the first-ranked cluster or group member.

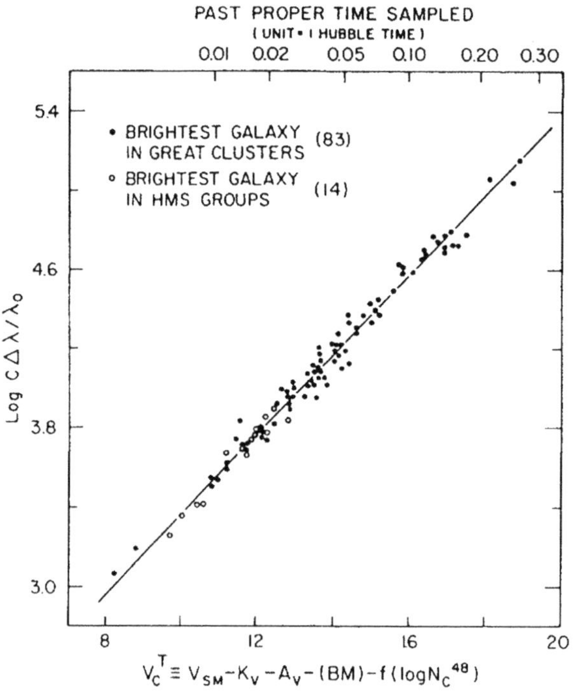

Fig. 5.7. Hubble diagram from Sandage & Hardy (1973) for first-ranked cluster E galaxies with all the corrections to the magnitude data included (K term, Galactic absorption, BM contrast correction, and luminosity function normalization). The line has an imposed slope of d $\log z/dV = 0.2$.

5.3.2 Extension to Large Redshifts with New Clusters

Minkowski's (1960) redshift of $z = 0.46$ was the largest known redshift up to the mid-1970's. Figure 5.8, taken from Sandage, Kristian, & Westphal (1976, SKW), shows the progressive redshift limits of the various previous programs. The black circles are additional clusters, all fainter than Abell's limit that we had found in a special search program with the 48-inch Palomar Schmidt on III-aJ plates, for which SKW determined redshifts using new instrumentation at Palomar. Despite the considerable extension beyond Humason's previous $z = 0.2$ wall at the Hydra cluster, no redshifts larger than $z = 0.46$ for 3C 295 had been obtained even in this new search and seizure program. To find q_0 it was imperative to find more distant clusters and to determine their redshifts.

Two new search programs were then begun, one by Gunn & Oke (1975) whose used random fields photographed with the 200-inch, and the other by Kristian, Sandage, & Katem (1974, 1978) based on identifying clusters containing 3C radio sources that heretofore were "empty fields" when searched

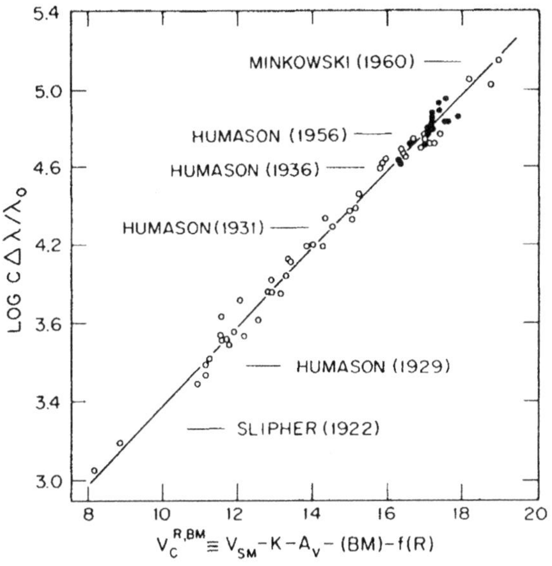

Fig. 5.8. Hubble diagram from Sandage, Kristian, and Westphal (1976) showing new clusters (black dots) as the early result of a program to find and measure more distant clusters. The various limits to earlier programs are identified by their authors.

at bright limits. In this second program we knew where to look because of the known radio position.

The result of each independent program was that clusters were found to redshifts as large as $z = 0.75$, and Hubble diagrams were constructed from the redshifts and the concomitant photometry. The results of the second search (Kristian, Sandage, & Westphal 1978) are discussed in section 5.5 after we set out the theory of the Hubble diagram at arbitrarily large redshifts.

5.4 Expected Hubble $m(q_0, z)$ Diagram at Arbitrarily Large Redshift

(a) The Standard Friedmann-Robertson-Walker-Mattig Model

Mattig's (1958) solution for rR_0 for arbitrarily large z in Lecture 2 (section 2.1.1), when combined with the Robertson (1938) equation for the observed *bolometric* flux l of an object with intrinsic *bolometric* flux L at redshift z and coordinate distance rR_0 at the time of reception as

$$l = \frac{L}{4\pi(R_0 r)^2(1+z)^2},$$

when changed to Pogson magnitudes by $m = -2.5 \log l + \text{const.}$, gives the (q_0, z) Hubble diagram in bolometric magnitudes as

$$m_{\text{bol}} = 5 \log q_0^{-2} \Big\{ q_0 z + (q_0 - 1)\big[-1 + \sqrt{1 + 2q_0 z}\big] \Big\} + \text{const.} \tag{5.11}$$

Note that for $q_0 = 1$,

$$m = \text{const.} + 5\log z$$

for all z. This also is the asymptotic value at small z for all q_0, seen from the series expansion of (5.11) which is

$$m_{\text{bol}} = 5\log z + 1.086(1 - q_0)z + O(z^2) + \text{const.}, \tag{5.12}$$

near the limit of $z = 0$. Note again, as in Lecture 2, that $m \sim 5\log z$ for all z was assumed by Hubble and all others in the early work on the redshift-distance relation.

Bolometric magnitudes are never observed. Changing to observed magnitudes in a particular wavelength band i requires only a change in the constant in equation (5.11). This is the "bolometric correction". It can be determined from spectral energy distributions for given objects observed at small z. For any other redshift, the observed heterochromatic magnitude, $m_i(z)$, [aside from the distance factor which is the log term in equation (5.11)] is converted to the $m(z = 0)$ proper wavelength system by the $K(z)$ term (Lecture 3) which (it cannot be repeated often enough) consists of (1) the selective term to take account of the different "effective wavelength upon redshifting" (actually it is an integral over the SED), and (2) the bandwidth term $2.5\log(1 + z)$ to account for the stretching of the spectrum.

The final effect on the observed heterochromatic magnitude of a redshifted galaxy is the change of absolute luminosity due to evolution of the stellar content of the galaxy in the look-back time. Call this factor $E(z)$.

In summary, the observed heterochromatic magnitude of a galaxy at redshift z, corrected for the technical effects of redshift if the detector has a fixed bandpass, is

$$\begin{aligned} m_i = {} & M_i - K_i(z) - E_i(z) \\ & + 5\log q_0^{-2}\left\{q_0 z + (q_0 - 1)\left[-1 + \sqrt{1 + 2q_0 z}\right]\right\} + \text{const.} \end{aligned} \tag{5.13}$$

Again, the two interesting special cases of $q_0 = 1/2$, where

$$(R_0 r)(1 + z) = \frac{2c}{H_0}\left(1 + z - \sqrt{1 + z}\right),$$

and $q_0 = 0$, where

$$(R_0 r)(1 + z) = \frac{cz}{H_0}(1 + 0.5z),$$

give

$$m = M - K(z) - E(z) + 5\log\left[2\left(1 + z - \sqrt{1 + z}\right)\right] + \text{const.}, \tag{5.14}$$

for $q_0 = 1/2$, and

$$m = M - K(z) - E(z) + 5\log z(1 + \frac{z}{2}) + \text{const.}, \tag{5.15}$$

for $q_0 = 0$.

Inspection of any and all of these equations shows that the sign of the deviations of the family of $m(z, q_0)$ curves is in the sense that as z becomes larger for a given m for increasingly positive q_0 the curves deviate toward brighter magnitudes at a given z (Mattig 1958; Sandage 1961a, Fig. 13). Hence, because larger q_0 means larger *deceleration*, an upward curving locus of points in a Hubble diagram when plotted with $\log z$ as ordinate and m as abscissa with bright luminosity to the left, shows deceleration.

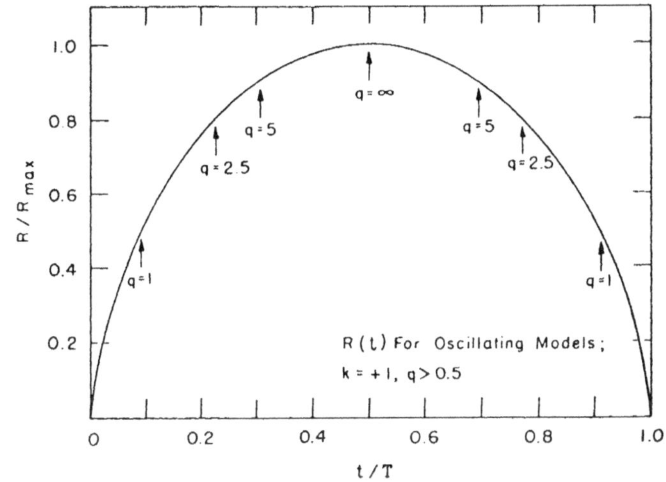

Fig. 5.9. Variation of q_0 along the $R(t)$ curve for the $k = +1$ geometry. Diagram from Sandage (1961b).

To fix ideas, it is of interest to realize that q_0 changes continuously along the $R(t)$ curves of the Friedmann solutions. Figure 5.9 shows q_0 at various times for the $k = +1$ geometry where the $R(t)/R(\max)$ curve is a cycloid (Lecture 1).

Problem 5.1. (a) Show that for the $k = +1$ geometry, $q_0 = 1/2$ at $R(t) = 0$ and increases to infinity at $R(\max)$. Derive the relation between q_0 and time for this geometry. (b) Show that $q_0 = 1/2$ for all times for the $k = 0$ geometry. (c) Show that q_0 starts at $1/2$ for $R(t) = 0$ and monotonically decreases toward zero at large times for the $k = -1$ geometry. (d) Draw a family of $R(t)$ curves for $q_0 < 1/2$, each passing through the (R, t_0) point for the "present", and show thereby the variation of the $R(t) = 0$ point on the time axis for different q_0 values, preparing the way for Lecture 7. Draw the similar $R(t)$ family, again all passing through the $t_0 =$"present" point, for $q_0 > 1/2$.

Problem 5.2. Show that for the steady state model where H is not a function of time, (a) $R \sim e^{Ht}$, (b) $q_0 = -1$, and (c) the (m, z) redshift-magnitude relation is

$$m = \text{const.} + 5 \log z(1+z).$$

Inherent in Fig. 5.9 and in the consequences of problem 5.1 is the "flatness" mystery. As we see in the next section, q_0 is observed to be very close to 1 (i.e. $-1 < q_0 < 2$ are the current observational limits). With q_0 changing so radically in the $k = +1$ geometry from 1/2 to infinity, if space is closed why then is q_0 so close to 1 now, implying by Fig. 5.9 that we are only about 1/10 the way to R_{max}. However, the problem of "fine tuning" disappears if $k = 0$ and therefore $q_0 = 1/2$ exactly, always. It is at this point that the new idea of inflation (see Narlikar & Padmanabhan 1991) in the early universe enters practical cosmology.

(b) Predicted m(z) For the Tired Light Model

As developed in Lecture 2, the equivalent of the Robertson equation for $l = f(L, z, rR)$ is

$$l = \frac{L}{4\pi(Rr)^2(1+z)}, \tag{5.16}$$

with only one power of $(1+z)$ in the denominator.

Problem 5.3. Show that the equation of the predicted Hubble diagram in the tired light model using $1 + z = e^{(HrR/c)}$, and therefore

$$Rr = \frac{c}{H} \ln(1+z) \tag{5.17}$$

from Lecture 4, is

$$m = \text{const.} + 5\log(1+z) + 5\log[\ln(1+z)]. \tag{5.18}$$

Show that this "tired light" equation imitates the $q_0 = +1$ standard model to first order in z, and therefore that the Hubble diagram test is not a test between tired light and the standard expanding model.

5.5 The Observational Search for q_0 Via the Hubble Diagram

The purpose of the programs to extend the Hubble diagram to large redshifts (section 5.3.2) was to determine q_0 by the spread of the family of Hubble diagram curves [equation (5.13)] for different q_0 values at high z. Although the analysis of both the Gunn-Oke (1975) and the Kristian, Sandage, & Westphal (1978) independent data sets differ in detail (undoubtedly due to the different ways of treating the aperture correction – a problem not yet looked into adequately, although it is the opinion of this writer that the Gunn-Oke adoption of a fixed metric aperture of only 16 kpc at which to define the measured apparent magnitude and then applying "growth curve slope corrections" is unnecessarily complicated), both are consistent with showing that q_0 lies between -1 and $+2$, and more likely between 0 and 1.

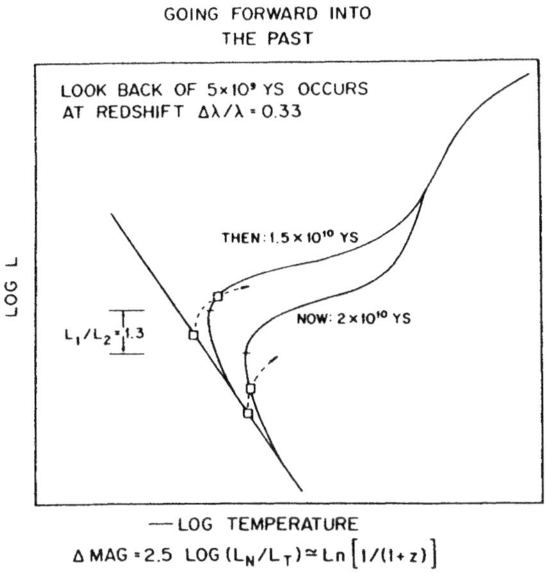

Fig. 5.10. HR diagram illustrating evolution of E galaxy luminosity in the look-back time. The approximate change of absolute magnitude in the look-back time is written at the bottom of the abscissa. Diagram from Sandage & Tammann (1983).

It was early realized (HMS 1956, Appendix C; Sandage 1961b, 1963) that evolution of the stellar content of galaxies must have occurred in the look-back time, and must be accounted for by the $E(z)$ term in the Hubble diagram equations. Detailed and elaborate models were eventually made to approximate the effect (examples are by Tinsley 1968; 1972a, b; 1976; 1977a, b; 1980). However the simplest approach to the evolutionary correction is to calculate the change of luminosity of the main sequence termination point in the HR diagram as a function of time and to combine this with the look-back time as a function of redshift. The method is shown in Fig. 5.10.

Problem 5.4. Derive the general equation for the look-back time (LBT) for any q_0 as a function of z. Show that the special cases reduce to

$$\mathrm{LBT} = \frac{T_0 z}{1+z}, \tag{5.19}$$

for $q_0 = 0$, and

$$\mathrm{LBT} = T_0 \left[1 - \frac{1}{(1+z)^{3/2}}\right], \tag{5.20}$$

for $q_0 = 1/2$, where T_0 = the age of the universe. Express these equations in terms of the present value of the Hubble constant for each q_0 value.

Problem 5.5. Show that the look-back time for tired light is

$$\mathrm{LBT} = \frac{\ln(1+z)}{H_0}. \tag{5.21}$$

In Fig. 5.10, L_{N} means "luminosity now"; L_{T} means "luminosity then". It is almost trivial, and is left as an exercise, to show that the difference in luminosity of the turn-off points between N and T, when combined with the look-back time for the $q_0 = 0$ model gives the approximate correction without considering light from the giant branch. With this approximation, $E(z) = 2.5 \log(L_{\mathrm{N}}/L_{\mathrm{T}}) \sim \ln[1/(1+z)]$. From this, the correction is $E(z, t) \sim 0^{\mathrm{m}}.07$ per 10^9 years (when $T_0 = 15 \times 10^9$ years). The detailed calculation is, of course, more complicated than simply the ratio of main sequence termination points because both the main sequence stellar luminosity function near the turn-off and the luminosity function along the red giant branches must be taken into account for both the N and the T states. But remarkably, the simple technique of considering only the change of luminosity of the main sequence termination point gives nearly the same answer as that of the detailed models.

The effect on the Hubble diagram of this correction for evolution is shown in Fig. 5.11 applied to the data determined by Kristian, Sandage, & Westphal (1978). This comparison shows that the deviation of the points from the theoretical Mattig expectation line for $q_0 = 0$ in the top panel of Fig. 5.11 with no evolutionary correction disappears in the bottom panel where evolution is included.

The smallness of the separation of the family of curves for different q_0 values is shown in Fig. 5.12, taken from Narlikar (1983, 1993). Because of this smallness and the uncertainty of the precise evolutionary correction, depending on unknown details of the stellar population of E galaxies in the look-back time, a different route to q_0 is necessary. The time scale test set out in Lectures 7, 8 and 9, based on the value of the Hubble constant and the age of the Galactic globular clusters, is free from many of these uncertainties.

Finally, the recent K-band Hubble diagrams discussed by Longair in his lectures are of great interest in this classical approach to the q_0 problem because of the very large redshifts reached at $z > 2$ using radio galaxies. Special efforts were made to prove that the absolute K magnitudes are not a function of absolute radio power, and therefore that a selection effect similar to that discussed in Lecture 4 (radio power *of the sample* increases with redshift due to the flux-limited selection effect) does not occur. Again, as in the optical diagrams in the previous figures, the apparent q_0 value is between 0

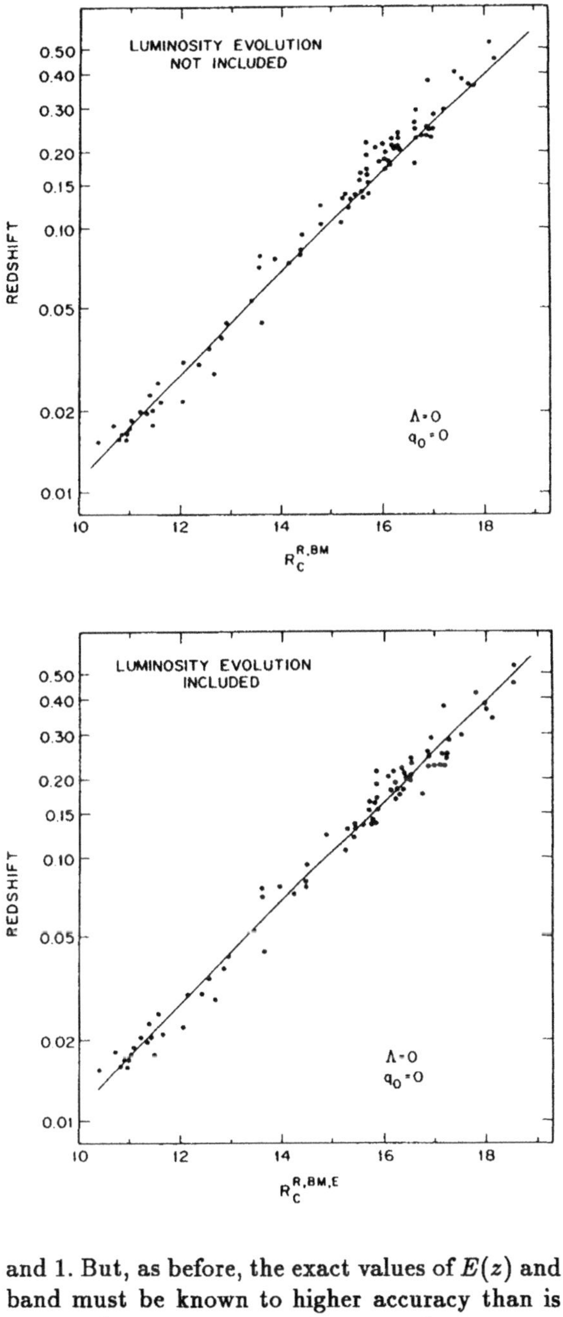

Fig. 5.11. Hubble diagram using KSW (1978) data with and without the evolutionary correction of Fig. 5.10. Diagram from Sandage & Tammann (1983). The abscissa shows fully corrected magnitudes in the R (near red) photometric band.

and 1. But, as before, the exact values of $E(z)$ and $K(z)$ in the K photometric band must be known to higher accuracy than is probably now available to correct the apparent q_0 to its true value.

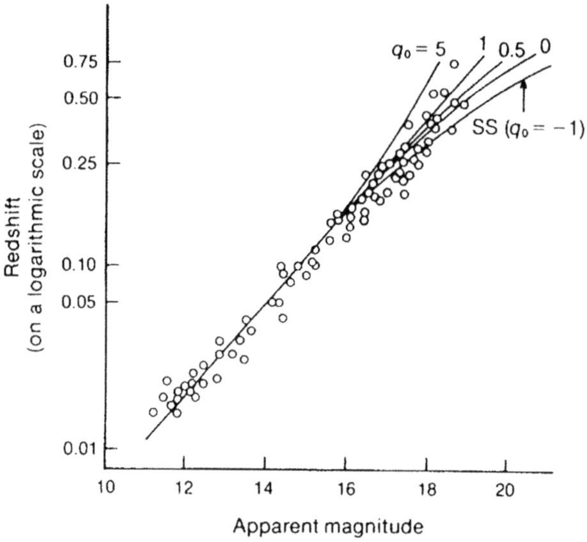

Fig. 5.12. Narlikar's (1983, 1993) plot of the Hubble diagram with q_0 curves superposed, using the 1978 data of KSW.

5.6 The Local Velocity Field

We close with a return to the local region from which we can set limits on the random and possible streaming motions about a noiseless, linear, redshift-distance relation. This most controversial subject is treated here only in outline. At the outset, to state a conviction that is, at the moment, a minority position, this writer is convinced that the only large-scale, non-Hubble, systematic motions relative to the centroid of the Local Group that have been identified with certainty are (1) an infall (actually a retarded expansion) of about 220 km s^{-1} toward the center of the Virgo complex, and (2) the dipole signal of about 600 km s^{-1} of the Alpher-Herman microwave background radiation. This latter is composed of the Virgo infall vector and a motion of the local Virgo complex (the region to $v < 2000$ km s^{-1}) almost in bulk toward the direction defined by their vector sum as in Fig. 5.13, taken from Tammann & Sandage (1985).

A discussion of the unreality of any non-cosmological motions except these two (besides, of course, the non-systematic, i.e. random component) is given elsewhere (Federspiel, Sandage, & Tammann 1994, FST). In particular, we see no evidence for the Great Attractor, in agreement with the initial analysis of Mathewson, Ford, and Buchhorn (1992).

These two certain non-cosmological motions (the retarded expansion of us from Virgo and the dipole of the Alpher-Herman radiation) are small compared to the cosmological expansion itself, appearing only as a small perturbation relative to the Machian frame. Otherwise all the Hubble diagrams shown here would be less well defined. In particular, the CMB dipole is caused

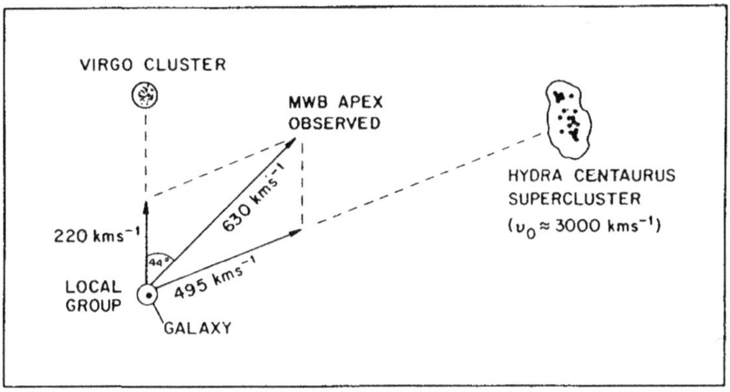

Fig. 5.13. Vector diagram showing the "infall" (actually retarded expansion) of the Local Group toward the Virgo cluster center plus the "infall" toward the Hydra-Centaurus region resulting in the observed dipole motion toward the warm pole of the microwave background. (Tammann & Sandage 1985)

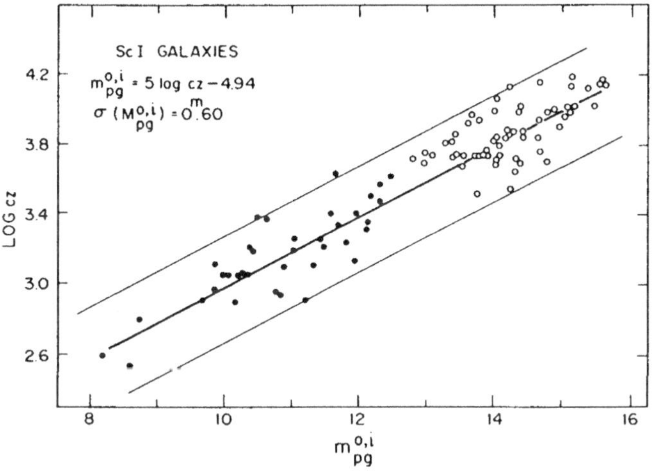

Fig. 5.14. Hubble diagram for Sc I galaxies (Sandage & Tammann 1975).

by a nearly bulk motion of the Local Supercluster (out to perhaps 5000 km s^{-1}) carrying the Local Group with it (FST). Within this relatively local region, the Hubble expansion occurs, giving the well-defined Hubble diagrams for galaxies within it (except for the immediate direction of the Virgo cluster where the scatter and the velocity dispersion is high).

That the effect of the CMB dipole motion within the Local Supercluster is so small is shown by the lack of a *step* in the Hubble diagrams at the "edge" of the Local Supercluster (again FST 1994, their Figs. 14-19). In second approximation, where the shear over the Local Supercluster is taken

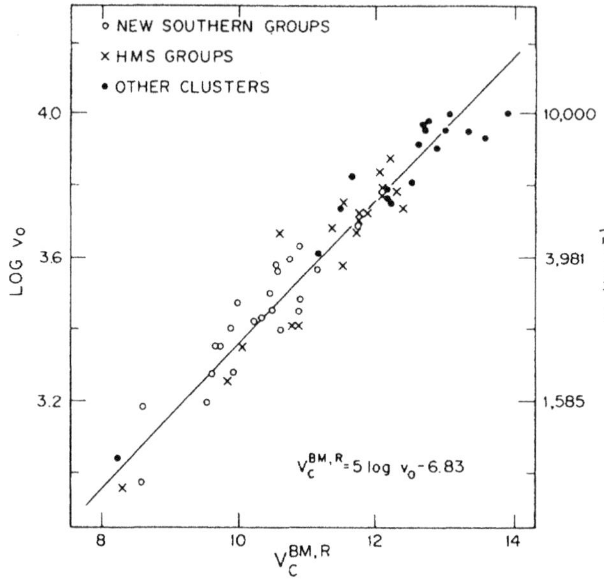

Fig. 5.15. Hubble diagram for nearby groups and clusters (Sandage 1975).

into account, the effect is very small. The lack of an effect can be made quantitative.

The Hubble diagram for Sc I galaxies (Sandage & Tammann 1975) to 10,000 km s^{-1} is shown in Fig. 5.14 where the scatter about the line of slope d $\log cz/dm = 0.2$ is independent of direction[21] and where there is no systematic deviation of the data from the $\log cz \sim 0.2m$ line, except for the effects of observational selection bias (Sandage 1988a).

A different example using nearby groups and clusters to 10,000 km s^{-1} is shown in Fig. 5.15 (Sandage 1975) which combines data from the northern and southern hemisphere. The southern RSA redshifts were completed in a separate campaign (Sandage 1978). Analysis of these data show no Rubin-Ford effect nor any other large-scale streaming motions at the level of about 500 km s^{-1}.

We also know that the very local region just beyond the Local Group (i.e., to 500 km s^{-1}) is an order of magnitude quieter about the cosmological flow than a 500 km s^{-1} random motion. Study of Hubble diagrams shows

[21]We believe the interpretation of data of the same kind as Tammann and I used in our Sc I experiment, as carried out by Rubin *et al.* (1976), is flawed. A systematic motion was claimed which is the so-called Rubin-Ford effect. Selection effects are suspected, leading to the difference in the conclusions of Rubin *et al.* from those reached in the initial experiment (Sandage & Tammann 1975, Appendix). This selection effect, caused by density inhomogeneities, has been called the Fall-Jones (1976) effect in the later literature.

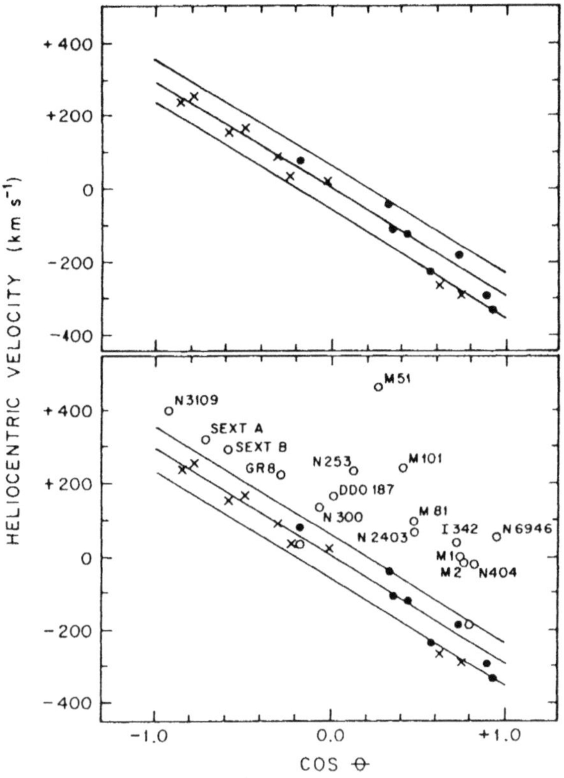

Fig. 5.16. Dipole solution for the solar motion relative to known Local Group members (top). Nearby galaxies are added (bottom) that are not Local Group members, showing the beginning of the expansion (the added galaxies stand above the upper Local Group limit line). Diagram from S86b.

a mean random motion of less than $\sim$ 60 km s^{-1} (Sandage 1986b, 1987). This is small enough to be able to detect the deceleration of the expansion caused by the mass of the Local Group. The argument is as follows. We begin by asking for the distance from the centroid of the Local Group where the expansion starts.

The top part of Fig. 5.16 shows the heliocentric velocity of galaxies that have traditionally been assigned membership in the Local Group on the basis of distance alone ($d < 1$ Mpc). The abscissa is the cosine of the angle between the direction of the object and an apex position (solved for from the data) at Galactic coordinates near $l = 90°$, $b = 0°$ (the most tedious details are in S86b). The amplitude of the dipole (a drift of the sun relative to the Local Group members) is 300 km s^{-1} composed of the rotation of the sun about the Galactic center plus the motion of the Galaxy relative to the centroid of the Local Group. The dots in Fig. 5.16 are Local Group galaxies that are expected to move independently of the Galaxy. Crosses are satellites of the Galaxy which, nevertheless, follow the adopted solar motion toward $l = 97.2°$, $b = -5.6°$. This is the apex of the dipole solution shown as the ridge line.

The dispersion about the mean line in the top panel of Fig. 5.16 is $\sigma = 60$ km s^{-1}.

Plotted in the bottom panel are nearby galaxies whose distances are known well enough (from Cepheids, brightest stars, etc.) to place them beyond the traditional boundary of the Local Group. The new feature in the bottom panel compared to the top is that every one of the galaxies beyond the traditional boundary of the Local Group has a positive residual velocity relative to the ridge line of the Local Group dipole motion. This, of course, is the signature of the beginning of the expansion.

Entrance to the literature of the dynamics of the Local Group can be gained through Lynden-Bell (1981, 1982). A simpler, but more idealistic model that displays the basis of a calculation of the expectation of the size of the deceleration of the local expansion field due to the mass of the Local Group is in the reference from which Fig. 5.16 has been taken (Sandage 1986b).

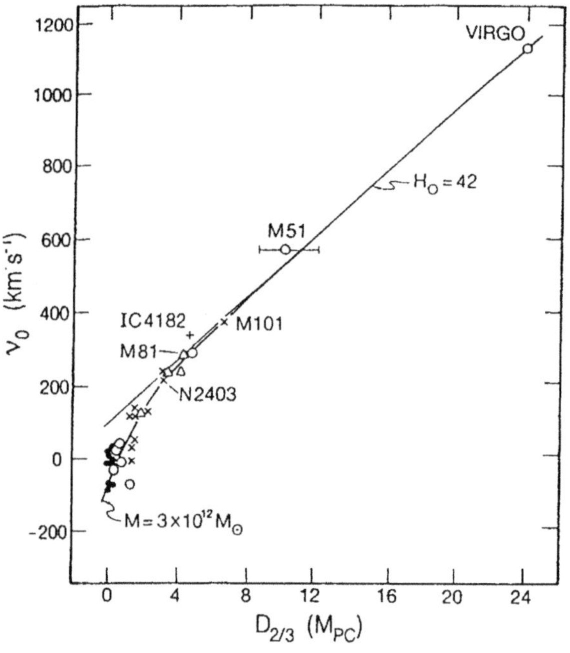

Fig. 5.17. Deceleration seen in the very local Hubble diagram caused by the pull due to the mass of Local Group. These data update those in Sandage (1986b, 1987).

Cepheid distances are known for most of the galaxies in Fig. 5.16, permitting the Hubble diagram to be plotted as velocity against Cepheid (mostly) distance, with the result shown in Fig. 5.17. The Virgo cluster distance has been put at 24 Mpc, based on the supernova distance, calibrated via SN 1937C in IC 4182 (Sandage *et al.* 1992). The distance to M51 is based on the apparent magnitude of the brightest resolved stars, and is the most uncertain of the set.

A deceleration has clearly been detected from these data. The simplest model based on a uniform distribution of matter outside the Local Group (i.e., neglecting the granular effect of such nearby groups as that containing NGC 342, and Maffei 1 and 2, and also the NGC 300 South Polar Group nearly in the opposite direction) gives the calculated curve as the expected effect of the deceleration caused on the local expansion field by the Local Group acting as a point for a total Local Group mass of 3×10^{12} solar masses.

Problem 5.6. Show by a back-of-an-envelope calculation that a velocity deviation from a no-deceleration line at a distance of 1 Mpc, seen in the data in Fig. 5.17, can be the result of a force on a test particle at that representative distance caused by the mass of the order of magnitude shown in Fig. 5.17 operating over a Hubble time.

The second feature to note from Fig. 5.17 is the very tight adherence of the points to the curve, showing a velocity *dispersion* of the order of only $\sim$ 50 km s^{-1}. This and other similar evidence demonstrates that the local region ($v <$ 500 km s^{-1}) is extraordinarily quiet, as was to have been shown.

References

Abell, G.O. 1958, ApJ Suppl. **3**, 211
Abell, G.O. 1975 in *Galaxies and the Universe*, ed. A. Sandage, M. Sandage, J. Kristian (Chicago: Univ. Chicago Press), p. 601
Baade, W. 1928, Astr. Nach. **233**, 66
Baade, W. 1952, Trans. IAU VIII (Cambridge: Cambridge Univ. Press), p. 397
Bridgeman, P.W. 1927, *The Logic of Modern Physics*, (New York: MacMillan)
Bridgeman, P.W. 1936, *The Nature of Physical Theory* (Princeton: Princeton Univ. Press)
de Sitter, W. 1916a, Mon. Not. RAS **76**, 699
de Sitter, W. 1916b, Mon. Not, RAS **77**, 155
de Sitter, W. 1917, Mon. Not. RAS **78**, 3
de Sitter, W. 1930a, Proc. Nat. Acad. Sci. **16**, 474
de Sitter, W. 1930b, Bull. Astron. Inst. Netherlands, Vol. 5, No. **185**, 157
de Sitter, W. 1933, *The Astronomical Aspect of the Theory of Relativity*, Univ. Calif. Pub. in Math. (Berkeley: Univ. Calif. Press), Vol. 2, No. 8
de Vaucouleurs, G., & Peters, W.L. 1986, ApJ **303**, 19
Dose, A. 1927, Astron. Nach. **229**, 157
Eddington, A.S. 1923, *The Mathematical Theory of Relativity* (Cambridge: Univ. Cambridge Press)
Eddington, A.S. 1930, MNRAS **90**, 668
Eddington, A.S. 1933, *The Expanding Universe* (Cambridge: Univ. Cambridge Press)
Einstein, A. 1917, *Sitzungsberichte d. Preussischen Akad. d. Wissenschaften zu Berlin*, p. 142

Fall, S.M., & Jones, B.J.T. 1976, Nature **262**, 457
Federspiel, M., Sandage, A., & Tammann, G.A. 1994, ApJ **430**, 29
Friedmann, A. 1922, Zs. f. Physik **10**, 377
Friedmann, A. 1924, Zs. f. Physik **21**, 326
Giraud, E. 1985, A&A **153**, 125
Giraud, E. 1986a, ApJ **301**, 7
Giraud, E. 1986b, ApJ **309**, 312
Giraud, E. 1986c, A&A **174**, 23
Gunn, J.A., & Oke, J.B. 1975, ApJ **195**, 255
Haas, A. 1924, *Introduction to Theoretical Physics* (London: Constable & Co.)
Heckmann, O. 1942, *Theorien der Kosmologie* (Berlin: Springer-Verlag)
Hubble, E. 1925, ApJ **62**, 409
Hubble, E. 1926a, ApJ **64**, 321
Hubble, E. 1926b, ApJ **63**, 236
Hubble, E. 1929a, Proc. Nat. Acad. Sci. **15**, 168
Hubble, E. 1929b, ApJ **69**, 103
Hubble, E. 1936a, ApJ **84**, 270
Hubble, E. 1936b, *Realm of the Nebulae* (New Haven: Yale Univ. Press)
Hubble, E. 1938, *Observational Approach to Cosmology* (Oxford: Oxford Univ. Press)
Hubble, E. 1953, Mon. Not. RAS **113**, 658
Hubble, E. & Humason, M.L. 1931, ApJ **74**, 43
Hubble, E. & Humason, M.L. 1934, Proc. Nat. Acad. Sci. **20**, 264
Hubble, E. & Sandage, A. 1953, ApJ **118**, 353
Humason, M.L. 1929, Proc. Nat. Acad. Sci. **15**, 167
Humason, M.L. 1936, ApJ **83**, 10
Humason, M.L., Mayall, N.U., & Sandage, A. 1956, AJ **61**, 97
Humphreys, R.M. 1983, ApJ **269**, 335
Kristian, J., Sandage, A., & Katem, B.N. 1974, ApJ **191**, 43
Kristian, J., Sandage, A., & Katem, B.N. 1978, ApJ **219**, 803
Kristian, J., Sandage, A., & Westphal, J.A. 1978, ApJ **221**, 383
La Violette, P.A. 1986, ApJ **301**, 544
Lemaître, G. 1925, *Note on de Sitter's Universe*, J. Math & Physics (MIT) **4**, 188
Lemaître, G. 1927, Ann. Soc. Sci. Bruxelles **47A**, 49
Lemaître, G. 1931, Mon. Not. RAS **91**, 483
Lundmark, K. 1924, Mon. Not. RAS, **84**, 747
Lundmark, K. 1925, Mon. Not. RAS, **85**, 865
Lynden-Bell, D. 1981, Observatory **101**, 111
Lynden-Bell, D. 1982, in *Astrophysical Cosmology*, eds. H.A. Bruck, G.V. Coyne, M.S. Longair (Rome: Specola Vaticana), p. 85
Mathewson, D.S., Ford, V.L., & Buchhorn, M. 1992, ApJ **389**, L5
Mattig, W. 1958 Astr. Nach. **284**, 109
McVittie, G.C. 1956, *General Relativity and Cosmology* (London: Chapman & Hall), 1st edition
McVittie, G.C. 1965, ibid, 2nd edition
Minkowski, R. 1960, ApJ **132**, 908
Narlikar, J. 1983, *Introduction to Cosmology*, 1st ed. (Boston: Jones & Bartlett), p. 363
Narlikar, J. 1993 *Introduction to Cosmology*, 2nd ed. (Cambridge: Univ. Cambridge Press), p. 346
Narlikar, J.V., & Padmanabhan T. 1991, Ann. Rev. A&A **29**, 325
Robertson, H.P. 1928, Phil Mag. **5**, 835
Robertson, H.P. 1933, Rev. Mod. Phys. **5**, 62

Robertson, H.P. 1938, Zs. f. Ap. **15**, 69
Robertson, H.P. 1955, Pub. ASP **67**, 82
Rubin, V.C., Ford, W.K., Thonnard, N., Roberts, M.S., & Graham, J.A. 1976, AJ **81**, 687
Sandage, A. 1958, ApJ **127**, 513
Sandage, A. 1961a, ApJ **133**, 355
Sandage, A. 1961b, ApJ **134**, 916
Sandage, A. 1963, ApJ **138**, 863
Sandage, A. 1972a, ApJ **173**, 485
Sandage, A. 1972b, ApJ **178**, 1
Sandage, A. 1972c, ApJ **178**, 25
Sandage, A. 1973a, ApJ **180**, 687
Sandage, A. 1973b, ApJ **183**, 711
Sandage, A. 1973c, ApJ **183**, 731
Sandage, A. 1975, ApJ **202**, 563
Sandage, A. 1978, AJ **83**, 904
Sandage, A. 1986a, in *Luminous Stars and Associations in Galaxies*, IAU Symp. 116, ed. C.W.H. de Loore, H.J. Willis, P. Laskarides (Dordrecht: Reidel), p. 31
Sandage, A. 1986b, ApJ **307**, 1
Sandage, A. 1987, ApJ **317**, 557
Sandage, A. 1988a, ApJ **331**, 583
Sandage, A. 1988b, ApJ **331**, 605
Sandage, A. 1993a, ApJ **402**, 1
Sandage, A. 1993b, ApJ **404**, 419
Sandage, A. 1994a, ApJ **430**, 1
Sandage, A. 1994b, ApJ **430**, 13
Sandage, A. & Carlson, G. 1985, AJ **90**, 1464
Sandage, A. & Hardy, E. 1973, ApJ **183**, 743
Sandage, A., Kristian, J. & Westphal, J.E. 1976, ApJ **205**, 688
Sandage, A., Saha, A. Tammann, G.A., Panagia, N., & Macchetto, F.D. 1992, ApJ **401**, L7
Sandage, A., & Tammann, G.A. 1974, ApJ **191**, 603
Sandage, A., & Tammann, G.A. 1975 ApJ **197**, 265 (Appendix on the "Rubin-Ford effect")
Sandage, A., & Tammann, G.A. 1982a in *Astrophysical Cosmology*, eds. H.A. Bruck, G.V. Coyne, & M.S. Longair (Vatican City: Pont. Acad. Sci.), p. 23
Sandage, A., & Tammann, G.A. 1982b, ApJ **256**, 339
Sandage, A., & Tammann, G.A. 1983 in *Large Scale Structure of the Universe, Cosmology, and Fundamental Physics*, eds. S. Setti & L. Van Hove (Geneva: ESO/CERN), p. 127
Seares, F. H., Kapteyn, J.C., & van Rhijn, P.J. 1930, *The Mount Wilson Catalog of Photographic Magnitudes in Selected Areas 1-139* (Washington: Carnegie Institution of Washington), Pub. 402
Segal, I. 1982, ApJ **252**, 37
Segal, I. 1993, Proc. Nat. Acad. Sci. **90**, 4798
Shapley, H. 1929, Proc. Nat. Acad. Sci. **15**, 565
Shapley, H. 1953, Proc. Nat. Acad. Sci. **39**, 349
Shapley, H. & Betz, M.B. 1919, ApJ **50**, 107
Silberstein, L. 1924a, Nature **113**, 350, 602, 818
Silberstein, L. 1924b, Mon. Not. RAS **84**, 363
Silberstein, L.1925, Mon. Not. RAS **85**, 285
Stebbins, J., Whitford, A.E., & Johnson, H.L. 1950, ApJ **112**, 469
Tammann, G.A. & Sandage, A. 1985, ApJ **294**, 81

Tammann, G.A., Yahil A. & Sandage, A. 1979, ApJ **234**, 775
Tinsley, B.M. 1968, ApJ **151**, 547
Tinsley, B.M. 1972a, ApJ **173**, L93
Tinsley, B.M. 1972b, A&A **20**, 383
Tinsley, B.M. 1976, ApJ **203**, 63
Tinsley, B.M. 1977a, in *Expansion of the Universe*, IAU Colloq. 37, (Paris: CNRS), p. 223
Tinsley, B.M. 1977b, ApJ **211**, 621
Tinsley, B.M. 1980, ApJ **241**, 41
Tolman, R.C. 1929, Proc. Nat. Acad. Sci. **15**, 297
Truman, O.H. 1916, Popular Astronomy **24**, 111
Tully, B. 1988, Nature **334**, 209
van der Kruit, P. 1986, A&A **157**, 230
Wirtz, C. 1918, Astron. Nach. **206**, 109
Wirtz, C. 1921, Astron. Nach. **215**, 349
Wirtz, C. 1925, Scientia **38**, 303
Zwicky, F. 1929, Proc. Nat. Acad. Sci. **15**, 773
Zwicky, F. 1957, *Morphological Astronomy* (Berlin: Springer-Verlag)

6 Is the Expansion Real?

6.1 History

(a) A Litany of Doubts

As discussed in previous lectures, there has never been total acceptance that the expansion is real. We have seen that three versions of this apostasy are (1) Hubble's reticence, based on a less than cautious belief in his $N(m)$ count data and their interpretation, (2) Zwicky's obduracy in his *ad hoc* speculation on "tired light", and (3) Segal's mysticism in his radical speculation that his new hypothesis on the nature of space-time leads to a static universe which nevertheless shows a redshift. Segal's model is a faint echo of the "de Sitter effect", also in a static universe, but is chimeric.

Refutation of Hubble's reticence has come from proof that the apparent magnitude scales and zero points of the pre-1950 photometry fainter than $m_{\rm pg} \sim 17$ are progressively too bright relative to the Pogson definition of $m = -2.5 \log l + \text{const}$. The corrections of the pre-1950 scales to the modern reliable scales sometimes reach $\sim 1^{\rm m}_{.}5$ at $m_{\rm pg} = 21$ (Fig. 2.3 of Lecture 2). If Hubble could have applied these (then unknown) corrections to his data, he could have eliminated his concern that the "correction to the observed magnitudes for the number effect" spoils the interpretation of the size of the space curvature.

Refutation of Zwicky's position relies on the clear requirement that if "tired light" exists, the process involving energy loss in the path must be accompanied by scattering of the beam, blurring the images of distant galaxies. Observation shows that this does not occur. Furthermore, the energy loss must be independent of wavelength if agreement is to be had with the observational data where the redshift, $\Delta\lambda/\lambda(\text{lab})$, has been shown to be independent of wavelength (see below) over 19 octaves of the spectrum. No known scattering process has energy losses with so constant a fractional wavelength shift.

Refutation of Segal's radical model has been made in Lectures 4 and 5 using data on the angular diameters and apparent magnitudes of galaxies at different redshifts. These data clearly show that the redshift-distance relation is linear. Segal's model requires that the redshift should increase as the square of the distance. Segal's claims are based on analyses of unsuitable samples that suffer selection biases, improperly corrected for. The general problem of observational selection bias is the subject of Lecture 10.

(b) Constraints on the Data if the Expansion is Real

If the redshifts are due to a true change of the metric scale factor with time, then space is expanding carrying the galaxies with it (Lemaître 1927, 1931; Eddington 1933). The three robust observational consequences of such an expansion are (1) the redshift-distance relation must be linear to maintain homogeneity (section 5.1 of Lecture 5), (2) the measured redshift must be independent of wavelength, and (3) the surface brightness of "standard" galaxies must vary with redshift as $(1+z)^4$, which is the subject of this Lecture.

That the redshift-distance relation is linear has been discussed often in these pages (Lectures 4 and 5). An early proof that the second requirement is satisfied was given by Wilson (1949) using emission lines in the Seyfert galaxy NGC 4151 over a wavelength range of about one octave (3425 – 6583 Å). A more stringent test was made by Minkowski and Wilson (1956) using the optical emission lines from the Cygnus A radio source.

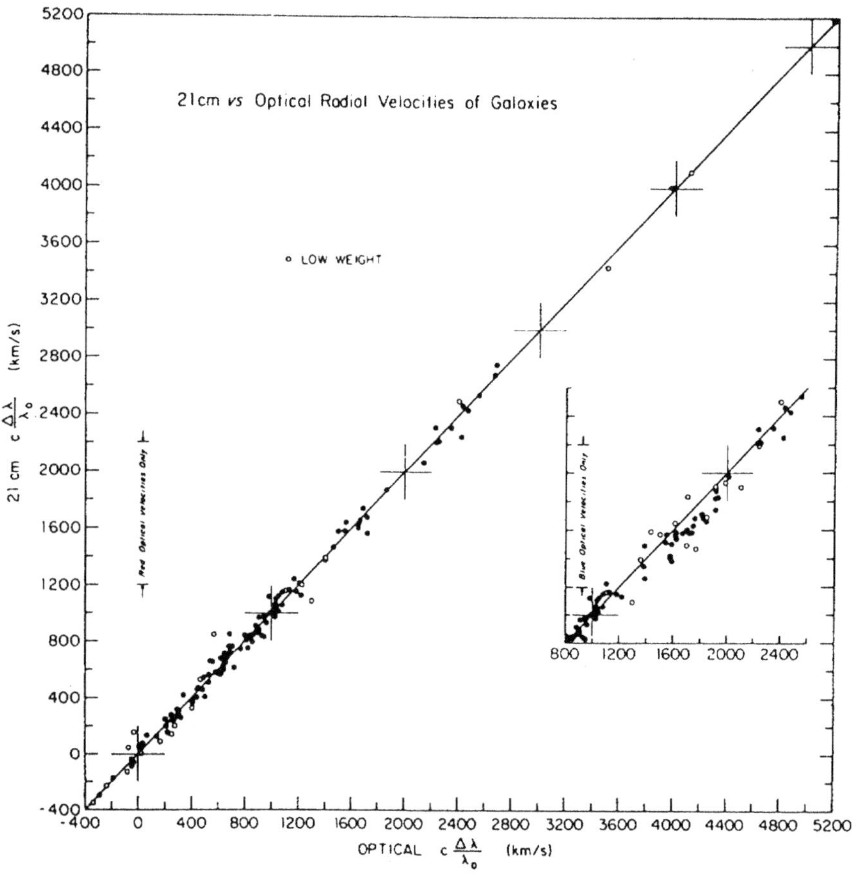

Fig. 6.1. Comparison of 21 cm and optical redshifts for the same galaxies. Diagram from Roberts (1972). Red and blue optical data are shown separately.

The test for constancy of $\Delta\lambda/\lambda_0$ over 19 octaves was made by McCutcheon and Davies (1970) and by Roberts (1972) by comparing the optical redshifts of galaxies, mostly from the Humason/Mayall redshifts known at that time from the HMS (1956) catalog, with H I 21 cm redshifts. Agreement between the two kinds of redshifts was excellent. The analysis by Roberts is shown in Fig. 6.1.

The third constraint, and by far the most powerful as a test between the expanding and the static models is via the Tolman surface brightness test, mentioned in passing in Lecture 2, and derived in some detail in Lecture 4 [section 4.4, equation (4.14)]. To show how the test discriminates between a model with true expansion and a static manifold, recall the basis of the test.

The Robertson (1938) equation for the expanding case, set out as equations (2.1) and (4.13), has two factors of $(1+z)$ in the denominator, giving the observed flux at distance R_0r from a source of intrinsic luminosity L, as $l = L/[4\pi(R_0r)^2(1+z)^2]$, which, when divided by the apparent (angular) surface area, θ^2, seen by the observer as $\theta^2 = y^2(1+z)^2/(R_0r)^2$ [from the square of equation (4.14)] gives the dependence of surface brightness on redshift as $(1+z)^{-4}$, stated before. This is the proof given by Tolman (1930, 1934 equation 189.5) but with a different notation.

If the manifold is not expanding the Robertson equation would have only one factor of $(1+z)$ from the "energy" effect. The same procedure would then predict that *SB* would decrease only as $(1+z)$. The difference in the predictions by three factors of $(1+z)$ is large even at only moderate redshifts and is the basis of the test.

Although the test has been known since 1930, operational difficulties have prevented its practical application until recently. These difficulties concern an appropriate definition of mean surface brightness averaged over a *metric* angular diameter. This lecture concerns the means to overcome these difficulties at the telescope.

6.2 Preliminaries to the Tolman Surface Brightness Test

(a) Three Operational Problems to be Solved in Realizing the Test

(1) The main difficulty with using galaxies for the test is that the *SB* varies by a factor of $\sim$ 1000 over the face of suitable galaxies (E types). This makes ambiguous most definitions of the galaxy size over which to calculate a stable mean surface brightness. (2) Furthermore, any angular diameter that is to be used must be a metric rather than an isophotal diameter (Lecture 4 for definitions and comparisons). (3) The third problem concerns the variation of a mean *SB*, however defined, with the absolute magnitude of the galaxy in question. We will show later in a series of diagrams that the $\langle SB \rangle$ of E galaxies is a strong function of absolute magnitude (and therefore of linear radius as well). Because the test is based on a comparison of *SB* of "identical" objects at different redshifts, we must devise ways to overcome the absolute magnitude correlation. The $\langle SB \rangle = f(M) = g(R)$ functions to be used to reduce the data

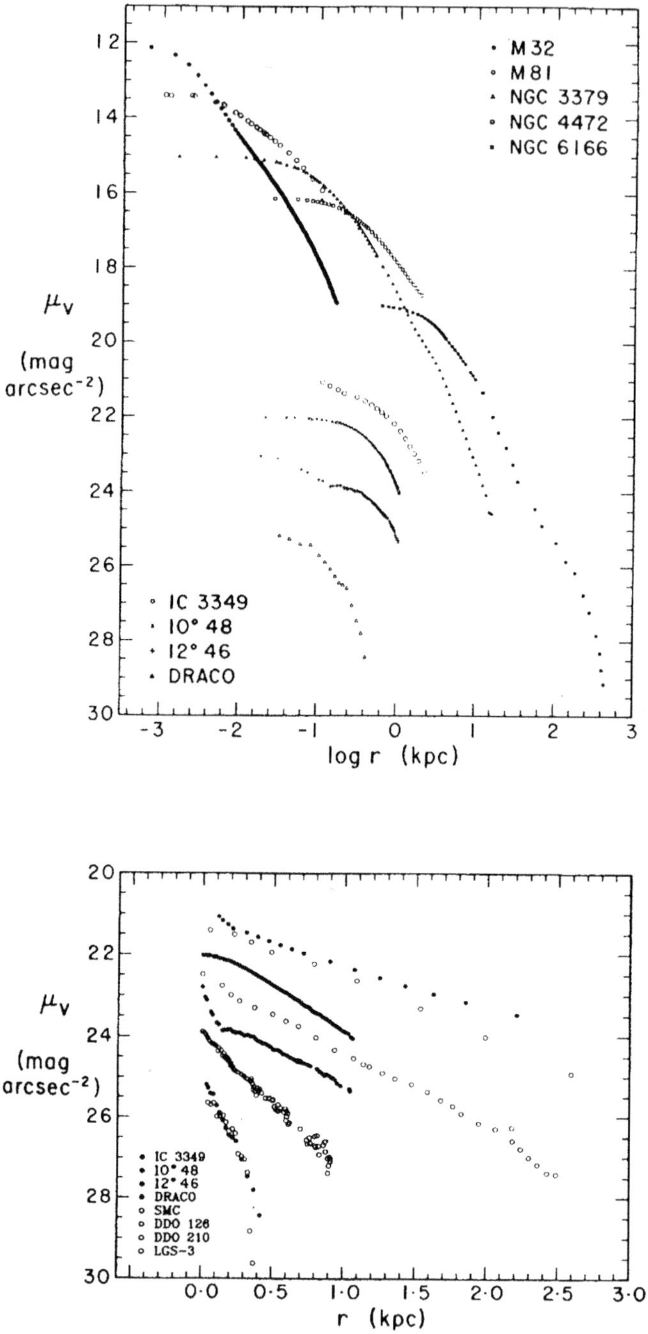

Fig. 6.2. Summary of $SB(r)$ variations of E galaxies as a function of absolute radius, and consequently with absolute luminosity. Diagram from Kormendy (1987).

to standard parameters are *a priori* unknown and must be determined. The problem is illustrated in the summary diagram of Fig. 6.2 from Kormendy (1987).

The first part of this lecture addresses the problem of an operationally sound definition of a metric rather than an isophotal diameter. The middle third is concerned with a method to reduce the mean *SB* values to an adopted absolute magnitude standard (or its equivalent "standard" radius). The final third shows a trial application to extant data in the literature.

(b) Metric vs. Isophotal Angular Radii

Isophotal diameters are easiest to measure. They are defined as the angular diameter of a galaxy enclosed within a fixed value of a *SB* isophote such as 25^{m} per square arc second. This type of size is what the eye, for example, sees as an apparent edge to the image on a photographic plate or print. For example, the apparent diameter of galaxy images on the original Palomar Sky Survey blue plates corresponds to an isophote of about $22^{\mathrm{m}}_{\cdot}5$ per square arcsec in the *B* photometric band (see HMS 1956, Appendix A). This is about $0^{\mathrm{m}}_{\cdot}6$ fainter than the *SB* of the air glow of the night sky at a dark site. Easy as isophotal diameters are to measure, it is metric diameters that we need for the test.

The most intuitively satisfying concept of a metric size is the "effective" diameter defined as that diameter which encloses half the "total" light. The total light is difficult to measure. The procedure is to construct a growth curve giving the increase in integrated intensity with increasing aperture. The curve is then extrapolated to "infinite" radius. The asymptotic magnitude level is defined as the total light. The "effective" diameter is that diameter where the integrated light via the growth curve is $0^{\mathrm{m}}_{\cdot}75$ fainter than the "total" magnitude.

For galaxies where only a few aperture measurements exist, the extrapolation requires use of a standard template curve such as in HMS (1956, Appendix A), as later constructed from much larger samples (Sandage 1972; de Vaucouleurs & de Vaucouleurs 1964) and most recently by many others using very large samples (e.g. Burstein *et al.* 1987).

The values of "effective" diameters depend sensitively on the extrapolation to infinity. A slight error in the intensity level of the asymptote makes an appreciable error in r_e. A far cleaner way is to devise some use of the *SB* profile that is independent of the concept of a total magnitude.

Such an operationally clean definition of a metric diameter was proposed by Petrosian (1976). Its strong merits were first realized by Kron (c. 1980; see also his Lecture 6 in this volume) who brought the concept to the attention of Spinrad. Its first use was by Djorgovski and Spinrad (1981) in their attempt to apply the angular size test (Lecture 4). The Tolman test is only a single step beyond their analysis.

(c) Properties of Petrosian Sizes

Although Petrosian's paper contains a definition of his metric radius, $r(\eta)$, it is operationally obscure, having been devised for a different purpose. Kron's insight showed that an equivalent definition is that $r(\eta)$ is the radius at which the ratio of the surface brightness averaged over the area inside that radius to the surface brightness at that radius is a particular assigned number, which can be expressed in magnitudes. For instance, the Petrosian $\eta = 2$ (mag) radius is the radius at which the average *SB* inside it is 2^{m} brighter than the *SB* at that radius.

It is shown in problem 6.1 that this definition is equivalent to

$$\eta(\mathrm{mag}) = 2.5 \log[2 \mathrm{d} \log r / \mathrm{d} \log L(r)], \tag{6.1}$$

which, expressed as the slope, $\mathrm{d} \log r / \mathrm{d} \Delta \mathrm{mag}$, of the growth curve is,

$$\eta(\mathrm{mag}) = 2.5 \log(5 \mathrm{d} \log r / \mathrm{d} \Delta \mathrm{mag}). \tag{6.2}$$

[The growth curve is the locus of the *increase* in brightness (as Δ mag) with increasing $\log r$ determined from aperture photometry.]

Problem 6.1. From equation (6.2) with its definition of the slope of the growth curve, prove that the radius, x, at which the growth curve has a particular value is that radius where the ratio of the *SB* averaged over the image interior to x to the *SB* at that radius is equal to that value. Show that this provides the operational method to determine values of Petrosian radii using data from an observed luminosity profile or from aperture photometry plus knowledge of the *SB* at various points in the image. [If stuck, non-Cambridge students (see footnote 6 of Lecture 1) can consult the solution as set out in several places in the literature: Petrosian 1976; Djorgovski & Spinrad 1981; Sandage & Perelmuter 1990a, section II.]

One of the advantages of working with the η function is the felicity with which the radius at which $\eta = 2$ mag (say) can be obtained from the observed data by a clean operational procedure. We need not know the central intensity of the profile which, in any case, is generally unmeasurable because of the finite spatial resolution of the telescope, the detector, and the circumstances (i.e., seeing for example). With the Petrosian formalism we need only determine the total flux coming from any finite area of the image that is larger than the resolution limit. Dividing the observed flux from this area by the area gives the *SB* averaged over this area, no knowledge needed of $I(0)$ at the center. (Of course, from the definition of the Petrosian η parameter we do need to know the *SB* at the edge of the area so as to form η).

Other metric radii such as the "core" radius, a, in the Hubble (1930) profile equation, or the ratio of core radius to the "tidal" radius in Oemler's (1976) modified Hubble profile, depend on knowledge of the unknowable "central surface brightness" and are therefore impractical.

We now divert for a few paragraphs to discuss again the photometric program of Gunn and Oke (1975) mentioned in Lecture 5, section 5.5. These authors used a fixed metric radius of 16 kpc within which to determine the received flux. To find the angular radius corresponding to 16 kpc they used a redshift distance calculated from the linear redshift-distance relation and particular values of the Hubble constant and q_0.

By adopting a fixed metric size rather than a size that is ratioed to a percentage of the total image, Gunn and Oke measure a smaller percentage of the total light for intrinsically large galaxies and a larger percentage for intrinsically small galaxies. The effect is that the magnitudes within their standard size refer to *different parts of the growth curve*, and therefore define a different percentage of the total light.

In the language of Petrosian radii, the η values at the "standard" radius are different for each galaxy. They correct for this inconsistency by correlating the slope of the growth curve at their 16 kpc place with absolute magnitude, M, calculated from the redshift and their measured apparent magnitude. They then apply the correlation of the growth-curve slope value with M as a correction to their partial magnitudes in their test for q_0 from the resulting Hubble diagram.

However, from their observations, they could have reduced the data more cleanly by determining the flux within a radius at which a particular value of the growth-curve slope occurs, i.e., at a fixed Petrosian radius. From that, they could have used the Petrosian formalism in their reduction. (Of course, in that case, corrections for cluster richness and contrast effects would be needed, which they argue are not required in their procedure, but our contention here is that these problems are then simply hidden, not dealt with.) We mentioned in Lecture 5 that the difference in the q_0 values obtained in the Gunn/Oke experiment and in that of Kristian, Sandage, and Westphal (1978) undoubtedly lies in this different treatment of galaxy apparent magnitudes.

Returning now to the main theme of understanding the properties of Petrosian diameters, an appreciation of how the value of η grows with radius for regular and not so regular *SB* galaxy profiles can be had by numerical integration of any available $I(r)$ profile. Such integration can give the run of *SB* averaged over a circle of radius r, divided by the *SB* at that r, i.e., the run of η with r.

A calculation of the η function using the profile of the E galaxy NGC 3379 (de Vaucouleurs & Capaccioli 1979) is shown in Fig. 6.3, taken from Sandage & Perelmuter (1990a, hereafter SPI). The data themselves rather than the smoothed profile were used for the numerical integration. The zero point of the $SB(r)$ profile, in B mag per square arcsec, are shown along the left hand ordinate. The zero point of the resulting calculated growth curve, $m_B(r)$, is along the outside scale on the right hand ordinate. The calculated η curve, marked $\eta(r)$, has the value in magnitudes along the inner scale on the right

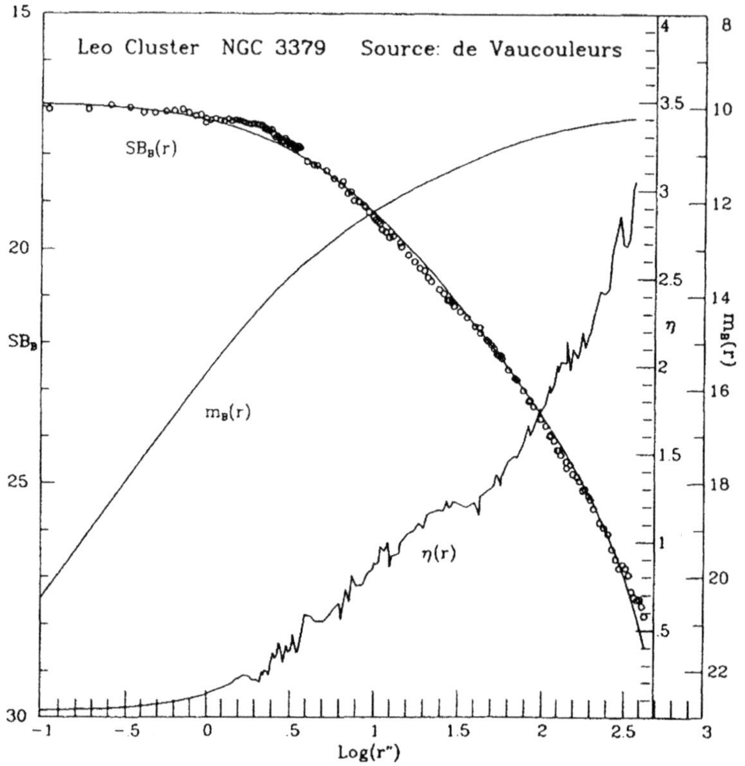

Fig. 6.3. Numerical calculation of the Petrosian η function for NGC 3379. The adopted data for the $B(r)$ profile are from de Vaucouleurs and Capaccioli (1979). The magnitude values for the growth curve and for the $\eta(r)$ function are along the right hand ordinate. Diagram from Sandage & Perelmuter (1990a).

hand ordinate. The abscissa is the log radius in arc seconds along the major axis.

It is obvious from the $\eta(r)$ curve how to read the angular (metric) radius corresponding to any desired η(mag) value. The noise on the $\eta(r)$ curve is due to the small deviations from smoothness of the data points from the continuous profile.

The systematic deviation of the $\eta(r)$ trajectory from a monotonic rise magnifies the deviation of the $SB(r)$ data points from the smooth standard $SB(r)$ curve. For example, the near standstill in the $\eta(r)$ near $\log r'' = 1.5$ reflects the dip in the $SB(r)$ data below the smooth curve.

This sensitivity to the deviations of the profile is seen in the more extreme case of NGC 1272 in Fig. 6.4. The standstill starting near $\log r'' = 1$ is pronounced. Clearly, extended envelopes can be more easily seen in the $\eta(r)$ curve than in the $B(r)$ profile itself.

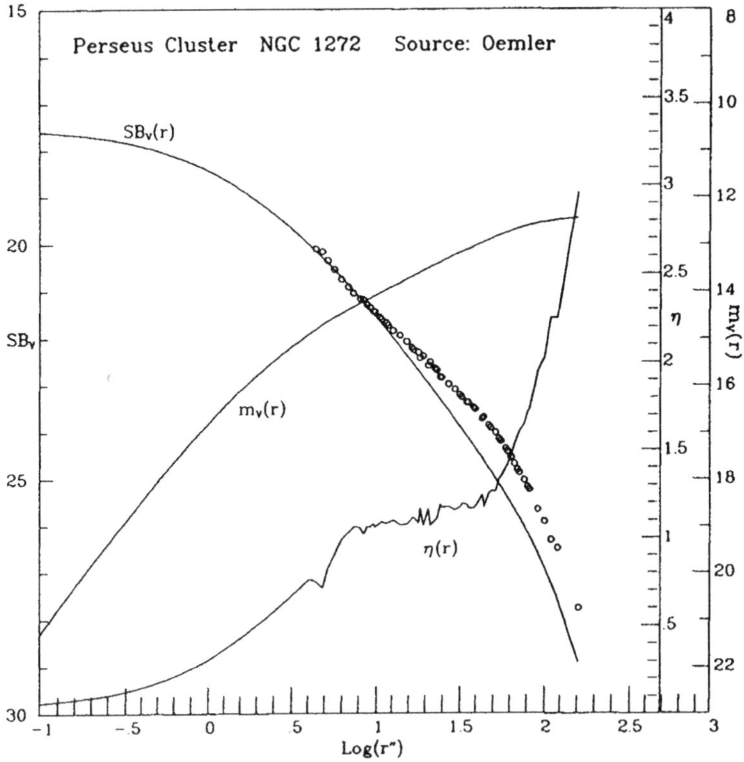

Fig. 6.4. The Petrosian $\eta(r)$ function for NGC 1272. Diagram from Sandage and Perelmuter (1990a).

Because η is defined as a ratio of surface brightnesses, the function is impervious to many systematic effects that affect other measurements. For example, $\eta(r)$ is wavelength independent if there is no variation of color across the galaxy image. Hence, it is also independent of redshift, and obviously also is in no need of the $K(z)$ correction in its selective and bandwidth stretching terms. Said differently, a "standard" galaxy taken to different redshifts, observed in any wavelength frame, will have the same $\eta(r)$ linear values as at zero redshift.

Furthermore, the $\eta(r)$ values are independent of absorption in the sight line and, importantly, are also independent of any change of SB due to luminosity evolution in the look-back time as long as that evolution is independent of radius in the image. For all these reasons, the Petrosian metric radius is a very powerful parameter.

Proof of the wavelength independence of $\eta(r)$ is shown in Fig. 6.5 using data in the B, V, and R pass bands for NGC 4697, taken from Fig. 4 of SPI. The cited literature sources for the profile data are identified in the diagram

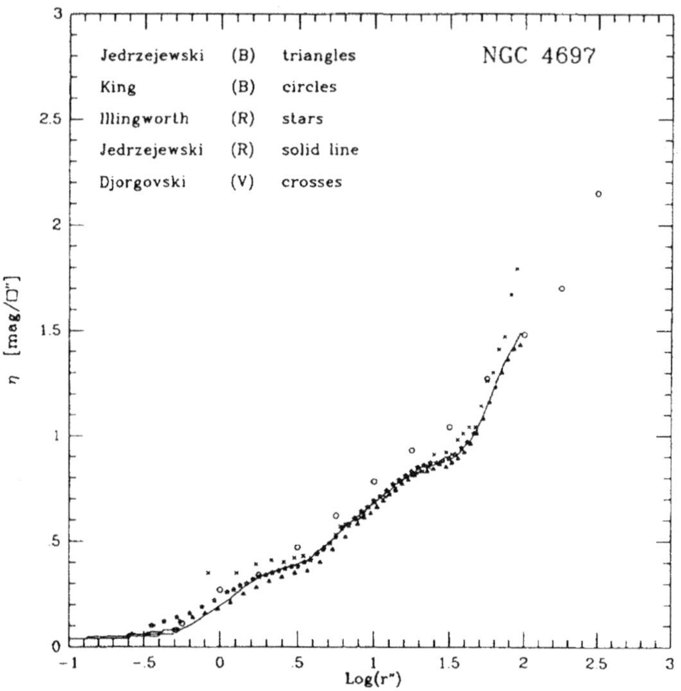

Fig. 6.5. Wavelength independence of the $\eta(r)$ function. Diagram from Sandage and Perelmuter (1990a).

and are listed in the references. The label "Illingworth" refers to Jedrzejewski, Davies, & Illingworth (1988).

The ratio of $\eta(r)$ radii values to the "effective" (half light) radii for various common profile laws in the literature is shown in Fig. 6.6 obtained by numerical calculation. The de Vaucouleurs $r^{0.25}$ profile and three members of the Oemler (1976) modified Hubble profile, identified by the ratios of tidal radius to core radius, are shown. Note that the Hubble (1930) law (not shown) with its far field decay as r^{-2}, has a tidal-to-core radius ratio of $\alpha/\beta = \infty$.

The purpose of Fig. 6.6 is to show that a Petrosian radius of $\eta(r)$ say 2, easily obtainable even from the ground, has a radius $\sim$ 2.5 times larger than the half-light radius. This is important because such large Petrosian η values can be more easily attained at very high redshifts than the effective radii because they are larger and therefore are more completely above the seeing disk.

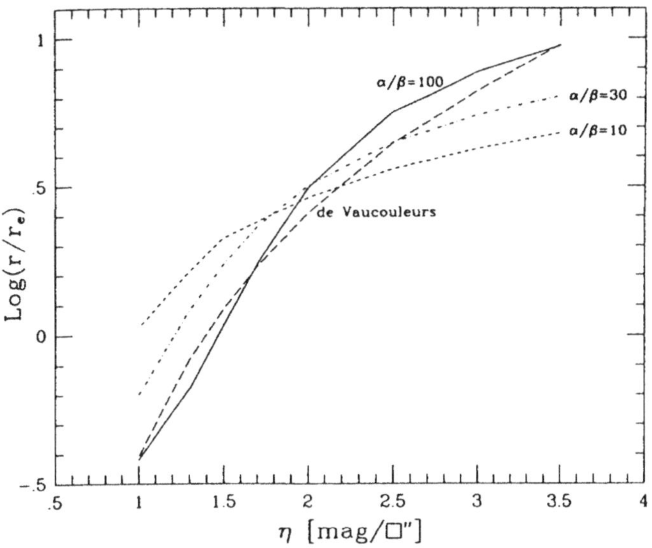

Fig. 6.6. Comparison of the increase of $\eta(r)$ with effective size for three Oemler models with different ratios of "tidal" radius to core radius, and for the $r^{0.25}$ distribution. Diagram from Sandage and Perelmuter (1990a).

6.3 Selective Effects for a Practical Tolman Test

(a) Both SB and $R(\eta)$ are Functions of Absolute Magnitude

An early indication that the surface brightness of giant E galaxies is not the same from galaxy to galaxy was available from the puzzling (for then) direct photographs of NGC 6166 and the two components of the close pair NGC 4782 and 4783 by Burbidge (1962) and Burbidge, Burbidge, and Crampin (1964). These photographs clearly showed that at least some brighter E galaxies have lower mean surface brightness in their central regions (i.e., are less compact) than intrinsically fainter E galaxies.

It was not clear at that time whether there is a general correlation between "average" surface brightness and absolute magnitude. Later work by many people showed there is in fact. These initial discoveries were the first hints of what is now variously called the "manifold of E galaxies" or the "fundamental plane" of E galaxy parameters, exploiting the sometimes sterile technique (no physics involved) of "principal component analysis".

The essence of the $\langle SB \rangle = f(M)$ mean surface brightness-absolute magnitude function was uncovered in an early demonstration by Oemler (1974, 1976), followed by Kormendy (1977, 1980), Strom & Strom (1978a,b,c), Thomsen & Frandsen (1983), and undoubtedly by others[22].

[22]The relation between mean SB, suitably defined, and absolute magnitude for giant E galaxies shown here (Figs. 6.7-6.9 later), discovered in these early papers, con-

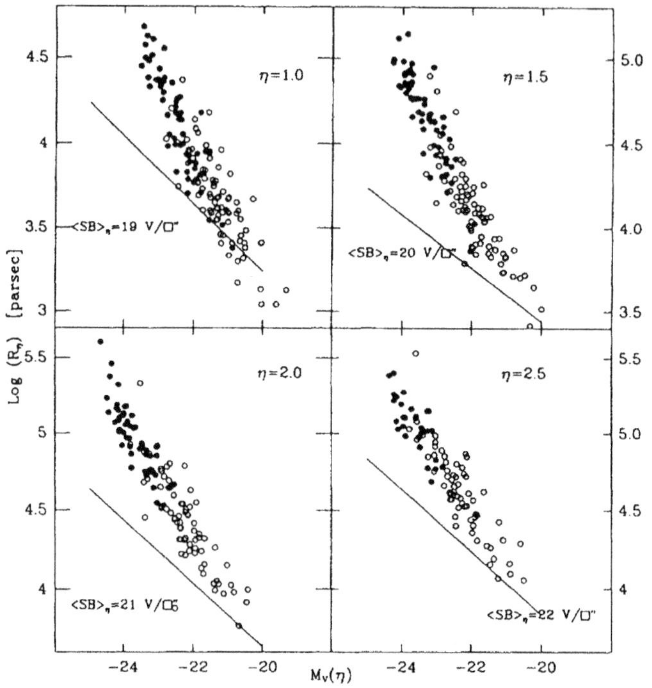

Fig. 6.7. *R* vs. *M* for the first several ranked galaxies in 56 clusters for four η values. Diagram from Sandage and Perelmuter (1991).

An analysis that combined homogeneous data sets for E galaxies in the Virgo, Fornax, and Coma clusters, plus four samples of bright field E galaxies, plus a sample of the brightest 146 galaxies in 56 great clusters was made by Sandage and Perelmuter (1990b, 1991, hereafter SPII and SPIII). The purpose was to set out the operational foundations for the Tolman test using either Petrosian angular radii or "effective" angular radii. The results confirmed the surface brightness correlations obtained in the earlier studies and provided the calibrated corrections needed to reduce $\langle SB \rangle$ data to standard conditions of constant linear radius and/or fixed absolute magnitude. The results are shown in the three summary Figs. 6.7-6.9.

Figure 6.7 shows the variation of Petrosian η linear radii with the absolute magnitudes defined by the light inside these radii. The linear sizes

tains one of the principal components of the "fundamental plane" first discussed by Brosche and Lentes (1982, 1983) for E galaxies, following Brosche (1973) who named the construct "the manifold of galaxies". Djorgovski (1985), Djorgovski and Davis (1987), and independently Dressler *et al.* (1987) renamed the concept the "fundamental E galaxy plane". A review and extension of the results has been made by Peacock (1990) and by Guzmán, Lucey, & Bower (1993), each providing a comprehensive entrance to the literature.

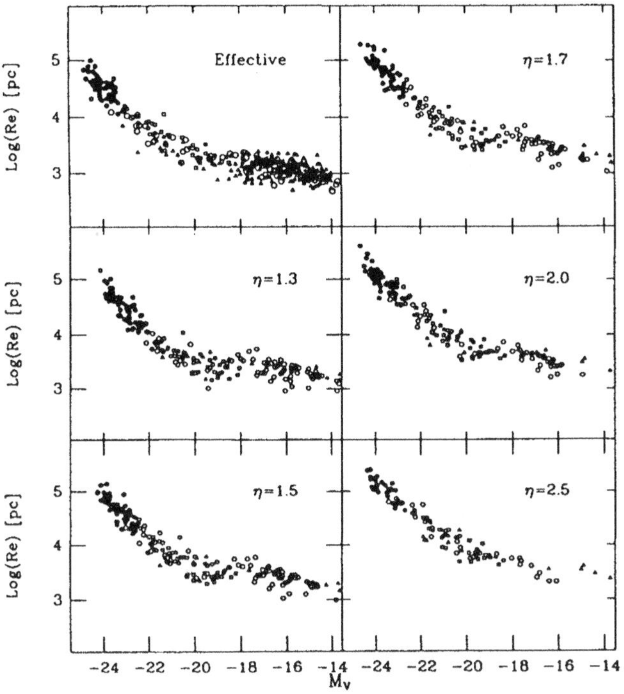

Fig. 6.8. Same as Fig. 6.7 but extending the correlation to $M_V = -14^m$ by adding data from the nearby Virgo, Fornax, and Coma clusters to Fig. 6.7. Diagram is Fig. 2 of SPIII (1991).

and absolute magnitudes are calculated using adopted distances to the three nearby clusters obtained by fundamental means (Lectures 7 and 8). Redshift distances ($H_0 = 50$ km s^{-1} Mpc^{-1}) are used for the field E galaxies and for the more distant great clusters. This Hubble constant is consistent with the adopted distances from redshifts for the field galaxies and the clusters more distant than Virgo, Fornax, and Coma.

The closed circles are for first ranked galaxies in the clusters. The open circles are also for cluster galaxies but of lower rank than the first. The data are from Sandage and Perelmuter (1991, their Table 1). The four different η radii in Fig. 6.7 are for η values of 1, 1.5, 2, and 2.5 mag. Figure 6.6 shows that this range of $\eta(r)$ covers the galaxian image from $\sim 0.5R_e$ to $4R_e$.

The solid line of slope 5 in Fig. 6.7 is the locus of constant surface brightness at the SB_V values marked, ranging from $\langle SB \rangle = 19$ V mag per square arcsec for the $\eta = 1.0$ data at $M_V \sim -20^m$, to $\langle SB \rangle = 22$ (in same units) far out in the image at $\eta(r) = 2.5$ mag, also at $M_V \sim -20^m$. The systematic deviation of the points from the constant SB line shows that the radius grows with absolute magnitude faster than the slope of d log $R/\mathrm{d}M = -0.2$. This is the signature that the SB becomes systematically fainter as the absolute mag-

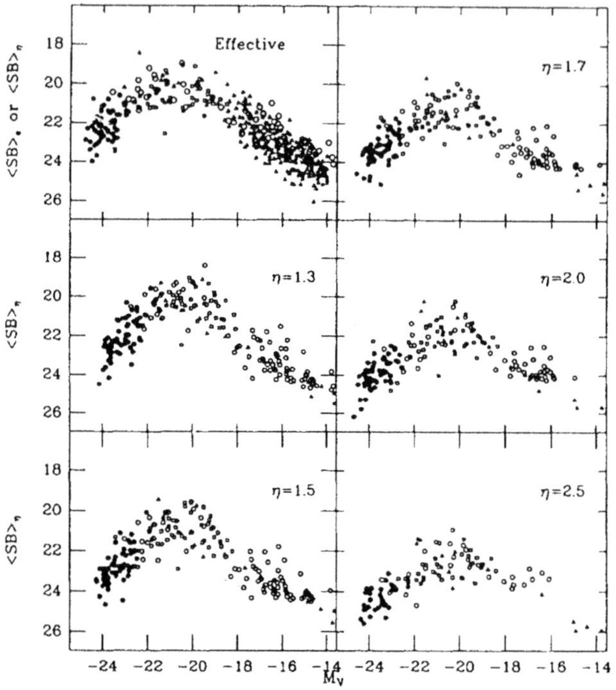

Fig. 6.9. The $\langle SB \rangle = f(M)$ correlation using data in Figs. 6.7 and 6.8. Diagram is from Fig. 4 of SPIII (1991).

nitude becomes brighter. It is the effect seen in the Burbidge's photographs of both the double galaxies NGC 4782/4783 and of the central region of NGC 6166 compared with its satellites that were on the same photographic plate.

Figure 6.8 extends the data in Fig. 6.7 to $M_V = -14^m$ by adding the data from the three nearby clusters of Virgo, Fornax, and Coma from Tables 1, 2, and 3 of SPII. From this it is evident that the nearly constant *SB* relation condition obtains only over the relative narrow absolute magnitude range from -22^m to -19^m where the slope of the R, M relation in Fig. 6.8 is in fact close to $\mathrm{d}\log R/\mathrm{d}M = -0.2$.

The data in Fig. 6.8, changed to surface brightness (units of mag per square arcsec) are shown in Fig. 6.9, taken from SPIII (their Fig. 4) defined at the effective radius and for five values of the $\eta(r)$ radius. The variation of compactness with absolute magnitude is seen in all six panels. The brightest E galaxies near absolute magnitude $M_V = -24^m$ ($H_0 = 50$) have an average surface brightness that is $3\overset{m}{.}5$ fainter than E galaxies at $M_V = -20\overset{m}{.}5$ near the peak of the correlation.

Because of this variation of *SB* with *M*, the Tolman test is considerably more complicated than if all E galaxies would have had identical mean surface brightnesses at all absolute magnitudes. In such an ideal case it would only have been necessary to measure $\langle SB \rangle$ values using any particular value of the Petrosian $\eta(r)$ radius for galaxies at different redshifts and test if the shift faintward of that SB is or is not fainter by the Tolman factor. However, given the systematic variation with M in Fig. 6.9 we must first correct the $\langle SB \rangle$ data to a fixed M_V or, in clusters, measure $\langle SB \rangle$ over a sufficiently large range of M_V for cluster members to obtain a significant portion of the curves in that diagram. The characteristic shape could then be used to test if this shape is mapped isomorphically, but fainter, at the Tolman rate.

No data of this kind are yet available. But using the available data in the literature (see next section), an equivalent procedure can be used to correct the data to "standard" conditions of either fixed radius, or fixed M values at $z = 0$ (i.e., corrected for the K dimming).

(b) Effects of Evolution

Evolution in the look-back time that affects the surface brightness as a function of epoch would compromise the Tolman test. Three types of evolution can be expected.

(1) The simplest is the change of total absolute magnitude of E galaxies due to evolution of the stellar content with time. Such passive evolution, caused by the burning down of the main sequence with time will, in itself, not affect the value of either the effective radius or the Petrosian radii unless the change of luminosity depends on radius. However, the SB itself, averaged over any Petrosian radius, will change directly with evolutionary change and must be accounted for.

(2) Dynamical evolution due to cannibalism may exist in the first few dominant galaxies in rich clusters, changing both the luminosities and the radii. Mergers of small galaxies with the largest will, in general, change the size of the original galaxy and hence its Petrosian radii values. Djorgovski and Spinrad interpreted the failure of their angular size test (Lecture 4, section 4.5.1) to this cause.

A way to account for this type of evolution is to use galaxies over a large range of absolute magnitude in a given cluster. The SB-magnitude locus for each cluster, similar to Fig. 6.9, can then be determined for galaxies fainter than the first few ranked members. We would suppose that cannibalism could be ignored in all but the first few ranked members.

(3) Tidal stripping may occur in the dense clusters, which, like the second effect, may be a problem in the rich clusters. Here the process, if it occurs at all, affects only the outermost parts of the galaxy profile (at least to first order). The inner parts are less vulnerable to stripping, thereby providing a test and a method of correction for such an evolution.

6.4 Result of a 1991 Attempt at the Tolman Test

(a) The Data

No serious attempt has yet been made to perform the Tolman test. Part of the reason is that, until recently, no practical method was available to define a metric angular size that was operationally useful. But with Djorgovski and Spinrad's success (1981, hereafter DS) in demonstrating that Petrosian $\eta = 2$ angular diameters can be measured for first ranked cluster galaxies to redshifts as large as $z = 1$ even from the ground, part of the path to practicality has been cleared.

DS did not discuss the Tolman test using their data. Rather, their aim was to attempt only the angular diameter-redshift test of Lecture 4. As with others before them, that test also failed. Their measured angular diameters show no upturn at redshifts $z > 0.7$ but rather decrease more rapidly than the standard theory predicts, presumably due to evolution in the look-back time (section 4.5.1 of Lecture 4). As said, DS attributed the failure to the evolutionary effect of cannibalism in their first ranked cluster data.

Because of this, taking the next step to make the Tolman test is suspect using their data. We do so here only as an illustration. The analysis is set out simply as a preparation for a full mounting of a definitive effort by some astronomer in the future. This section is, therefore, tutorial despite its apparent proof that the Tolman prediction is satisfied.

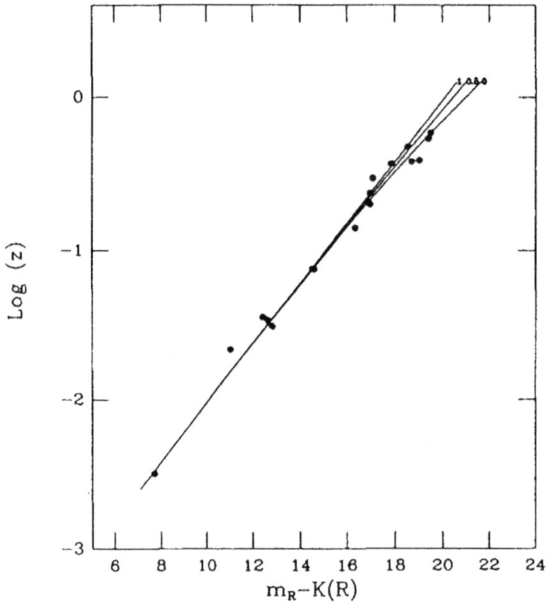

Fig. 6.10. Hubble diagram for the Djorgovski and Spinrad sample with the theoretical curves for three q_0 values superposed. Diagram is Fig. 11 of SPIII (1991).

DS list data for the angular radii for $\eta = 2$ and also list their observed R-band magnitudes for first ranked galaxies in 19 clusters whose redshifts range from $z = 0.003$ to $z = 0.593$. The reduction of these data by the machinery labored here was done in SPIII (their Table 2). The R-band magnitudes were corrected to the rest frame values by the adopted K corrections, $K(R, z)$, from Coleman, Wu, and Weedman (1980). Linear radii were computed from the angular radii by the exact equations in Lectures 1 and 2 for the three cases of q_0 of 0, 1/2, and 1.

That the data for the first ranked cluster galaxies of DS have only a small range of absolute magnitude is shown by the tightness of their Hubble diagram in Fig. 6.10. The theoretical curves for the three q_0 values (for constant absolute magnitude) are also shown. The abscissa is the observed R-band magnitude corrected only for the K term, $K(R)$. The ordinate is log redshift. The conclusion from Fig. 6.10 is that no large luminosity evolution has occurred, consistent with the expectations mentioned before (Lecture 5, section 5.5).

(b) Reduction of the $\langle SB \rangle$ Data to a Standard Linear Radius

Although the range of M_R for galaxies in Fig. 6.10 is evidently small, it is not zero. In view of the strong variation of $\langle SB \rangle$ with both M and R for giant E galaxies shown in Figs. 6.7-6.9, and the variation of M and R among even the first ranked cluster galaxies (i.e., the range shown by the black dots in Figs. 6.7-6.9), we need a procedure to correct the measured $\langle SB \rangle$ of individual galaxies to either a "standard" M or R value.

If we choose to correct the measured $\langle SB \rangle$ values to what would have been measured if the galaxy had a standard value of R(linear) at $\eta = 2$, the correction curve in Fig. 6.11 is to be used. This was calculated in SPIII from an analysis of the 146 first several ranked galaxies in the 56 clusters used in Figs. 6.8-6.9. The symbols in Fig. 6.11 are the same as in these previous figures. Note particularly the range of linear R values of the black dots which are the first ranked galaxies in individual clusters. Open circles are lower ranked cluster members.

To emphasize the problem again, note as in the previous diagrams, that the largest E galaxies have the faintest surface brightness. This is the same trend shown by Kormendy (Fig. 6.2 here) and is also inferred from the correlations made with different parameter projections (that is SB on M, and M on R) in the bright parts of Figs. 6.7-6.9.

(c) The Result

With all parts of the reduction machinery now in place, the results of applying them to the DS sample are shown in Fig. 6.12, taken from SPIII (their Fig. 12). The top left panel shows the surface brightness averaged over the angular diameters corresponding to $\eta = 2$ for 18 of the DS galaxies (not reduced

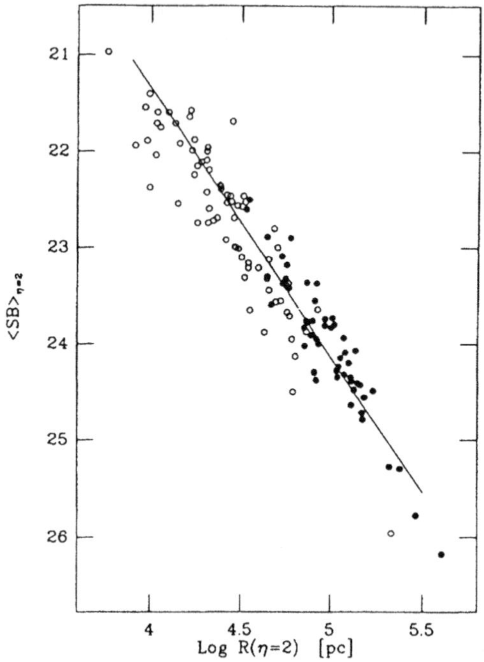

Fig. 6.11. Correlation of $\langle SB \rangle$ with linear radius for the cluster data in Figs. 6.7-6.9. This correlation is used to reduce $\langle SB \rangle$ data to the $\langle SB \rangle$ that would apply at the "standard" R(linear) size. Diagram is Fig. 6 of SPIII (1991).

to fixed linear diameters) plotted vs. $\log(1+z)$. The least squares fit is shown by the solid line. The dashed line has the slope of $\mathrm{d}\langle SB \rangle/\mathrm{d}\log(1+z) = 10$ expected for the Tolman cosmological effect if the expansion is real. Recall that the "tired light" prediction is for a slope of 2.5 which the data clearly violate.

The other three panels show the same data but reduced via Fig. 6.11 to a standard linear radius of $\log R = 5.0$. Figure 6.11 was used only differentially, reading the $\langle SB \rangle$ difference (i.e., which is, then, the correction) to be applied to the calculated $\langle SB \rangle$ data if the galaxy in question has a Petrosian $\eta = 2$ linear radius that is different from $\log R = 5.0$.

To apply the correction via Fig. 6.11 we must know the linear radius of each of the program galaxies. These were calculated by the exact equations of Lecture 4 that relate linear size to angular size, to q_0, and to z. It is assumed that $H_0 = 50$, consistent with the earlier diagrams in this lecture.

The scatter in the three "reduced" panels is smaller than in the unreduced upper left panel. Again, the solid lines are the least squares fits. The dashed lines are the $(1+z)^4$ prediction. The formal result is that the slopes, $\mathrm{d}\langle SB \rangle/\mathrm{d}\log(1+z)$, for the four panels are 8.9 ± 1.5 for the upper "unreduced" panel, 9.9 ± 0.9, 10.8 ± 0.9, and 11.5 ± 0.9 for the q_0 cases of 0, 1/2, and 1 respectively for the other three panels, all data reduced to $\log R = 5.0$ (pc).

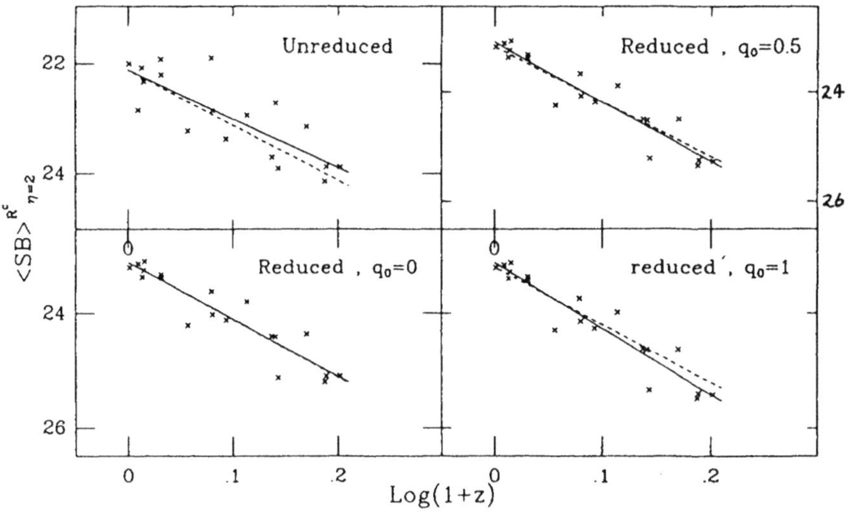

Fig. 6.12. Result of the Tolman test described in the text. Diagram is from Fig. 12 of SPIII (1991).

If this had been a definitive set of data, the conclusion would have been that the expansion is real.

References

Brosche, P. 1973, A&A **23**, 259

Brosche, P., & Lentes, F.-Th. 1982, Mitt. Astron. Gesell. **55**, 116

Brosche, P., & Lentes, F.-Th. 1983, in *Internal Kinematics and Dynamics of Galaxies*, IAU Symp. 100, ed. E. Athanassoula (Dordrecht: Reidel), p. 377

Burbidge, E.M. 1962, ApJ **136**, 1134

Burbidge, E.M., Burbidge, G.R., and Crampin, D.J. 1964, ApJ **140**, 1462

Burstein, D., Davies, R.L., Dressler, A., Faber, S.M., Remington, P.S.S., Lynden-Bell, D., Terlevich, R.J., & Wegner, G. 1987, ApJS **64**, 601

Coleman, G.D., Wu, C.-C., & Weedman, D.W. 1980, ApJS **43**, 393

de Vaucouleurs, G., & de Vaucouleurs, A. 1964, *Reference Catalog of Bright Galaxies* (Austin: Univ. Texas), 1st ed.

de Vaucouleurs, G., & Capaccioli, M. 1979, ApJS **40**, 699

Dressler, A., Lynden-Bell, D., Burstein, D., Davies, R.L., Faber, S.M., Terlevich, R.J., & Wegner, G. 1987, ApJ **313**, 42

Djorgovski, S.B. 1985, PhD Thesis, Univ. Calif. (Berkeley)

Djorgovski, S.B., & Davis, M. 1987, ApJ **313**, 59

Djorgovski, S.B., & Spinrad, H. 1981, ApJ **251**, 417 (DS)

Eddington, A.S. 1933, *The Expanding Universe* (Cambridge: Univ. Cambridge Press)

Gunn, J.E., & Oke, J.B. 1975, ApJ **195**, 255

Guzmán, R., Lucey, J.R., & Bower, R.G. 1993, MNRAS **243**, 517

Hubble, E. 1930, ApJ **71**, 231

Humason, M.L., Mayall, N.U., & Sandage, A. 1956, AJ **61**, 97 (HMS)
Jedrzejewski, R.I., Davies, R.L., & Illingworth, G.D. 1988, AJ **94**, 1508
Kormendy, J. 1977, ApJ **218**, 333
Kormendy, J. 1980, in *Proc. ESO Workshop on Two-Dimensional Photometry*, eds. P. Crane and K. Kjar (Garching: ESO), p. 191
Kormendy, J. 1987, in *Nearly Normal Galaxies*, ed. S.M. Faber (New York: Springer), p. 163
Kristian, J., Sandage, A., & Westphal, J.A. 1978, ApJ **221**, 183
Lemaître, G. 1927, Ann. Soc. Sci. Bruxelles **47A**, 49
Lemaître, G. 1931, MNRAS **91**, 483
McCutcheon, W.H., & Davies, R.D. 1970, MNRAS **150**, 337
Minkowski, R., & Wilson, O.C. 1956, ApJ **123**, 373
Oemler, A. 1974, ApJ **194**, 1
Oemler, A. 1976, ApJ **209**, 693
Peacock, J.A. 1990, MNRAS **243**, 517
Petrosian, V. 1976, ApJ **209**, L1
Roberts, M.S. 1972, in *External Galaxies and Quasi-Stellar Objects*, IAU Symposium No. 44, ed. D.S. Evans (Dordrecht: Reidel), p. 12
Robertson, H.P. 1938, Zs. Ap. **15**, 69
Sandage, A. 1972, ApJ **173**, 485
Sandage, A., & Perelmuter, J.-M. 1990a, ApJ **350**, 481 (SPI)
Sandage, A., & Perelmuter, J.-M. 1990b, ApJ **361**, 1 (SPII)
Sandage, A., & Perelmuter, J.-M. 1991, ApJ **370**, 455 (SPIII)
Strom, S.E., & Strom, K.M. 1978a, AJ **83**, 73
Strom, K.M., & Strom, S.E. 1978b, AJ **83**, 732
Strom, K.M., & Strom, S.E. 1978c, AJ **83**, 1293
Thomsen, B., & Fransen, S. 1983, AJ **88**, 789
Tolman, R.C. 1930, Proc. Nat. Acad. Sci. **16**, 511
Tolman, R.C. 1934, *Relativity, Thermodynamics, & Cosmology* (Oxford: Clarendon), p. 467
Wilson, O.C. 1949, PASP **61**, 132

7 The Fourth Test: Timing (Age of the Galaxy)

7.1 Theory of the Test

The three tests to discriminate between world models discussed in previous lectures are (1) galaxy counts as a function of brightness to measure the deviation of volumes from the Euclidean expectation, (2) magnitudes as a function of redshift at large redshifts to measure the deceleration by changes in the redshift in the look-back time, and (3) angular size as a function of redshift to measure the lensing effect of space curvature. To date, each of these tests have failed because the physics of the multiple evolutionary effects in the look-back time and in different density environments is largely unknown, denying, therefore, their correction.

The fourth test, and the only one immune to the effects of evolution, is the timing test. Here, the "real" age of the universe is compared with a time obtained by extrapolating the current expansion rate back to the time when $R(t) = 0$, assuming no deceleration. A significant actual difference between the extrapolated time and the actual age would signal a deceleration of the expansion if the extrapolated time was longer than the actual age. An *acceleration*, caused by a positive cosmological constant, or occurring naturally in the old steady state model, would be required if the extrapolated time was shorter than the actual age.

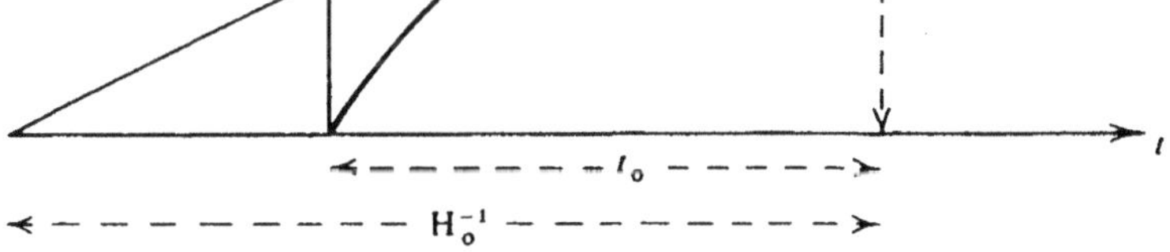

Fig. 7.1. Schematic of the two time scales involved in the timing test. The test consists in comparing the ratio of the two scales of H_0^{-1} and t_0, determining thereby the degree of curvature of the $R(t)$ function. Diagram from Berry (1976).

The particulars of the test are shown in Fig. 7.1. The curved line is the $R(t)$ function from one of the families of solutions of the Friedmann equation. The amount of its bending determines q_0. Hence, if we can determine R and $\dot{R}$ in some way from observation (recall the definition of q_0 as $-\ddot{R}_0 R_0/\dot{R}_0^2$) at the present epoch, shown as t_0 in the diagram, then the world model would follow directly.

From the definition of the Hubble constant as $H = \dot{R}/R$, it is clear that putting a tangent to the curve at the present epoch and extending it backwards to the $R = 0$ abscissa gives a time interval that is the reciprocal of the Hubble constant, marked as such in the diagram. The amount of bending of the $R(t)$ curve at the present epoch, t_0, can be found by the ratio t_0/H_0^{-1}. The essence of the timing test is to measure this ratio.

From the derivations in the previous lectures it is clear that equality of the two scales occurs when $R(t)$ is a straight line, and hence when $q_0 = 0$. For the closure density giving $q_0 = 1/2$, the time ratio is $t_0/H_0^{-1} = 2/3$.

Problem 7.1. (a) Derive the general equation for the time, t_0, since $R(t) = 0$ for any arbitrary value of q_0 in the range of $q_0 = 0$ to q_0 that is arbitrarily large. Use the Mattig machinery of Lectures 1 and 2. Note that the Friedmann solutions change from hypercycloids to cycloids at the critical deceleration of $q_0 = 1/2$. (Said differently, the Euclidean case separates the two kinds of solutions.) (b) Construct a table of t_0/H_0^{-1} values for different q_0 values and determine how accurately one must know both t_0 and H_0 to determine q_0 to within any given accuracy.

(Hints are available in the accessible literature, the easiest of which uses the development angle formalism; Sandage 1961b, but see also 1961a if you are really stuck. Cambridge students are not permitted to use these hints in view of footnote 6 of Lecture 1.)

This lecture is concerned with the determination of the "true" age of the universe which is one of the two required time scales. The next two lectures are concerned with the determination of H_0, giving the second time scale for comparison.

The oldest known objects in the Galaxy are the globular clusters. There are reasons to believe that these are the first aggregates to form in the nascent galaxy (Peebles & Dicke 1968) and hence can be used to age-date the Galaxy itself. It then only remains to know how long after $t = 0$ it took to form galaxies (section 7.6).

The age-dating of the globular clusters depends on an understanding of the H–R diagram.

7.2 The Road to Understanding the H–R Diagram

7.2.1 The Essence of the Dating Method

The most direct method of dating is to determine the amount of time necessary to nuclear-burn a particular amount of hydrogen to helium, knowing the rate of conversion. In a steady state between the rate of generation of energy and its radiation away, this is the luminosity, L, in erg s^{-1}. The problem reduces to determining the amount of H converted at an identifiable point of known L in the H–R diagram.

The problem was not solved until the effects of stellar evolution on the H–R sequences became understood. The centrally important paper in the history is that of Schönberg and Chandrasekhar (1942), where they traced the evolution of a main sequence star as it changed H to He. The calculations showed an impasse due to the impossibility of fitting an isothermal core to a radiative envelope of a standard stellar model when $\sim 10\%$ of the mass of the star had burnt to He in a central, fully mixed core. The impasse is in fact the main sequence termination point, but not identified as such until a termination point was observed in the globular clusters a decade later. This story occupies the first third of the lecture. It contains the dating method, but only in its most distilled essence.

The total energy released in converting a mass MXf of H to He is $0.007Mc^2Xf$, where X is the hydrogen content by mass fraction, M is the total stellar mass, and f is the fraction of that mass converted to He during time t. The generated energy is radiated away at a rate L erg s^{-1}.

The equality

$$E = \int_0^{t_0} L(t)\mathrm{d}t = 0.007Mc^2Xf = \langle L \rangle T \tag{7.1}$$

is obvious, where E is the total energy generated by the H to He conversion. The integral accounts for the variation of the radiation rate, L, due to evolution from an initial start on the zero age main sequence to the identified place in the H–R diagram occupied after MXf grams of H have been converted. Knowing M, X, f, and $\langle L \rangle$ gives the age T. Calculations of the very complicated details by which to refine this early approximation are "simply" aimed at determining $L(t)$ and f.

From equation (7.1) it follows that the order-of-magnitude age of a main sequence termination point if at solar luminosity of $L = 4 \times 10^{33}$ ergs s^{-1}, at a solar mass of 2×10^{33} g, with $X = 0.75$, and with $f = 0.1$ (close to the value of f at the Schönberg-Chandrasekhar limit), is $T \sim 10^{10}$ years.

Fundamentally different views of the nature of the evolution in the H–R diagram and the methods of age-dating prevailed as late as 1945, having their roots in the speculations about the direction of evolution by Russell (1910a, b, 1914). Russell's apparently powerful synthesis was the foundation of thought about secular evolution of stars for three decades. Stars moved down the main sequence after first gravitationally contracting leftward *to* the sequence from the giant branch in the earliest phases. The idea of a main sequence termination point due to stars moving *off* the main sequence was not to come until the 1950's. An example is Bok's (1946) report to the RAS council where he summarizes ideas of evolution as they were understood at that time. As a star burns H to He, the evolution occurs *along* the main sequence, moving up it with time if it is fully convectively mixed (Gamow 1939, Fig. 1; 1940, pages 136 and 150). Bok writes, "as its hydrogen is transformed gradually into helium, the sun will probably expand, grow hotter, and hence more luminous. The time intervals in successive stages decrease rapidly. Following a brief B-star stage, the sun will presumably fall back upon its gravitational energy to end as an inconspicuous white dwarf".

From Bok's review it is clear that even as late as 1946 the debate had not yet been conclusive regarding long vs. short time scales of $10^{12.5}$ (Jeans 1929) or 10^{10} years. Even with the shorter time scale, Bok's way of deriving an approximate evolution time for the sun shows how an argument that holds together logically can be so incorrect because its precepts are wrong. That age-dating method was as follows.

Strömgren's (1933) explanation of the H–R diagram, based on his calculation of stellar opacity as a function of the H and He content, began with a mapping of loci of constant hydrogen content near the main sequence. From this he concluded that the hydrogen mass fraction, X, varies systematically along the main sequence, changing from 91% at dK8 spectral type to 11% at B9 stars as the stars, fully mixed by convection, move up the main sequence as their hydrogen is converted to helium.

From this Bok calculates the time required for the sun to change from a dK5 star to a dG0 star, decreasing its hydrogen content from 56% to 46%. Hence, $f = 0.1$, and $\langle X \rangle = 0.51$, which, with $\langle L \rangle = 4 \times 10^{33}$ ergs s^{-1} in equation (7.1) gives T of the order of 10^{10} years, justifying for Bok the "short" time scale.

The search for the correct value of the hydrogen content, X, of the sun was complicated, taking the better part of three decades until the mid 1950's. Milestones were the papers by Schwarzschild (1946), Keller (1948), Epstein (1950a, b), Oke (1950), Demarque (1967) and his students (Demarque & Percy 1964; Demarque & Heasley 1971). These central papers also provide an entrance to the extensive literature.

The development of these ideas centered on Trumpler's (1925) seminal open cluster paper. An appreciation is set out elsewhere (Sandage 1988).

7.2.2 The H–R Diagram to Be Explained

From an observational point of view, two studies of the highest importance in establishing the properties of the diagram are (1) the initial cluster studies by Trumpler (1925), and (2) the field star studies using spectroscopic parallaxes. We discuss the second development first.

The method of determining absolute magnitudes of stars from their spectra had been discovered and developed by Adams and Kohlschütter (1914) following a series of discoveries of the effect of temperature and pressure on the intensities and shapes of spectral lines in sunspots. As this story has been lost in most histories of the origin of the spectroscopic parallax method, and as the development is a case study of how great discoveries are made, we divert for a moment in its retelling.

In the decade before Saha (1920) developed the theory of ionization equilibrium, Hale, Adams, and the laboratory physicists on the observatory staff began spectroscopic experiments in the physics laboratories of the Mount Wilson Observatory, both on the mountain and in the main office complex in Pasadena. Arc, spark, and later temperature-controlled electric furnace methods were used to excite the spectral lines of many elements. From these experiments it became clear that most of the striking differences seen between the spectrum of the solar disk and of sunspots, discovered earlier by Young, Lockyer, Fowler, and others could be explained as a temperature effect alone, the disk being hotter than the spots.

Hale and Adams had also noticed that the sunspot spectra closely resemble the stellar spectra of K and early M stars along the Secchi spectral sequence, rather than the G-type spectrum of the solar disk.

The work then took a crucial turn when this discovery was coupled with the results of new experimental data on the effect of *pressure* and density on the wings and depths of spectral lines *at fixed temperature*. The experimental results in Pasadena from the electric furnace studies made by varying the pressures and densities were being obtained concurrently with the solar data. With these insights from the laboratory, Adams and Kohlschütter could begin in 1914 to identify *pressure effects* as the explanation of the spectral differences between the sharp and the broad lines in stars of the same temperature as a *luminosity* effect. They could also begin to calibrate the effect, thereby obtaining the intrinsic luminosity of a star *from its spectrum alone*. Hints that some such correlation must exist had, to be sure, come from the prescient observation by Antonia Maury a decade earlier at Harvard that sharp-lined stars exist in the A and F types. Both Maury and Hertzsprung suspected that the so-called cA and cF stars were of high luminosity, but it was for Adams and Kohlschütter to begin the development of a calibration via continuous correlations of line ratios with luminosities that led to the spectroscopic parallax method.

The wondrous connection between sunspot spectra, the distance of stars, the fleshing out of the H–R diagram with its connection to stellar evolution, and the structure of the galaxy was traced many years later by Walter Adams, second director of the Mount Wilson Observatory, in a stunning article on the history of those nascent days of stellar astrophysics when the method was developed. Adams (1939) wrote:

> "In this brief outline of a relatively simple research we have passed from a sunspot to the interior of the atom, thence to the atmosphere of the sun and stars, and so to the problems of stellar constitution and development. That the behavior of certain lines in the spectrum of a sunspot should have any bearing upon the determination of the distance of a star seems at first almost inconceivable, but the successive steps in the relationship are logical and by no means complicated. The investigation forms an excellent illustration of the innumerable inter-

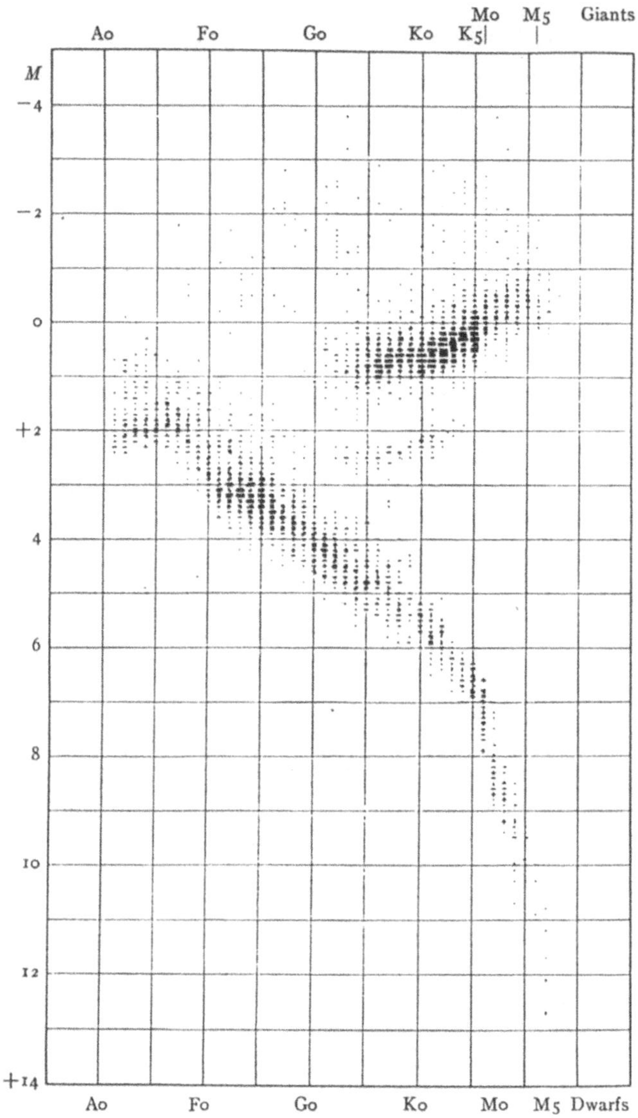

Fig. 7.2. The H–R diagram determined from the Mount Wilson program of spectroscopic absolute magnitudes for 4179 stars. From Adams *et al.* (1935).

connections of the problems of physical science, of the simplicity and directness with which they may be approached, and the effectiveness of bringing to their solution varying interests and a diversity of points of view."

After its discovery in 1914, the method was refined into a practical procedure beginning with a general review by Adams (1916), where he attributes

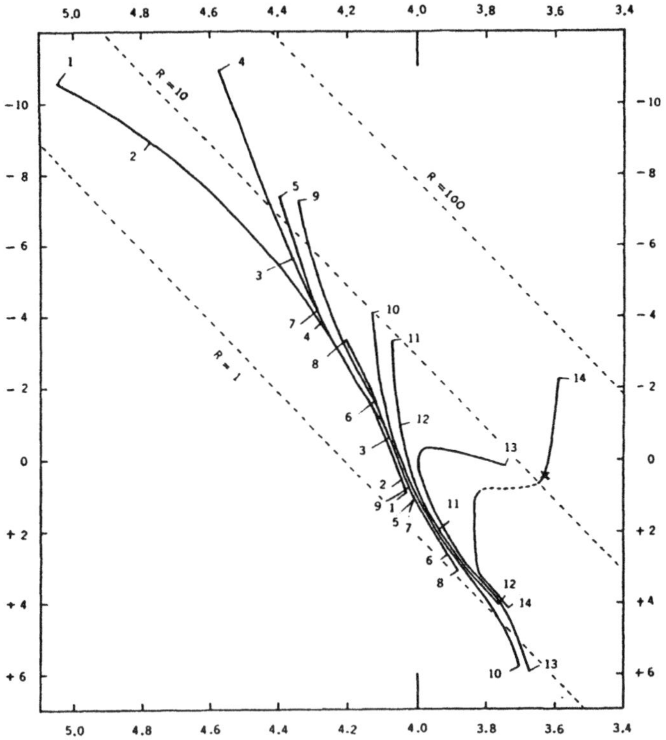

Fig. 7.3. Kuiper's (1936, 1937) composite H–R diagram of open clusters made by combining the observational data of Trumpler. Log T_e and M_{bol} are the coordinates.

much of the initial quantitative work to Kohlschütter. In the major summary by Adams, Joy, Humason, & Brayton (1935) of the 20 years work on the problem, the principal references to the foundations leading to the summary are listed as Adams & Joy (1917), Adams, Joy, Strömberg, & Burwell (1921), and Adams, Joy, & Humason (1926). The first sentence of that paper is:

> "The determination of absolute magnitudes of stars from their spectra has been continued as a major program of research since the method was first developed here [at Mt. Wilson] and reduced to a practical basis in 1916. For stars later than A5 at least, the method has proved eminently satisfactory."

An analysis of the absolute magnitude calibration of the Mount Wilson scale has been given by Blaauw (1963).

The H–R diagram from the 1935 summary, shown in Fig. 7.2, had a profound effect on the subsequent work, leading to the modern views of stellar evolution. Although wrong in the details of the connection (rather the lack of it) of the subgiant sequence near $M_V = 2\overset{m}{.}5$ with the giant branch in the diagram as presented, such items as the Hertzsprung gap between the F and

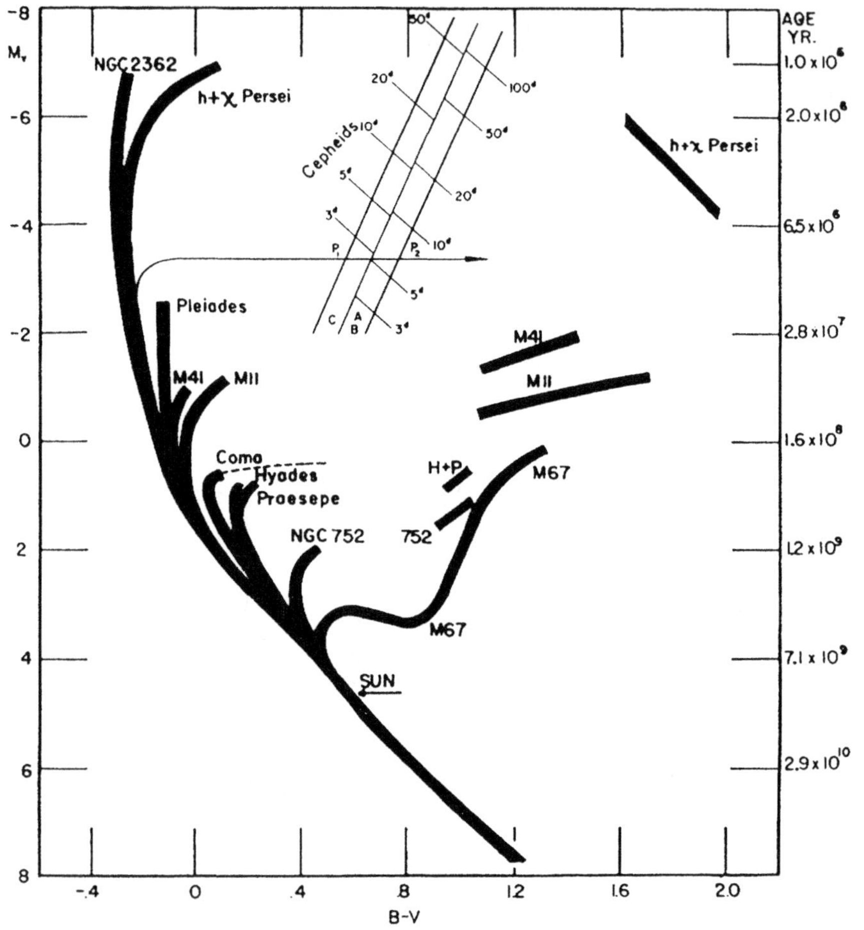

Fig. 7.4. Modern composite color-magnitude diagram of clusters of ages from 10^6 to 10^{10} years (Sandage 1958a). The age scale is that extant in 1957, which remains correct in principle but has been changed in detail.

G stars near $M_V = +1^m$, the ratio of giants to dwarfs (account taken of selection effects), the rarity of stars of higher luminosity than $M_V = 0^m$, and the gap between dK and gK stars showed for the first time those features of the H–R diagram that later proved to be of such importance for stellar evolution.

Consider next the approach through clusters. Hertzsprung had begun this approach to the H–R diagram in his early color studies of the Pleiades. The cluster method was developed decisively by Trumpler (1925) using spectral types rather than color equivalents.

Kuiper (1936, 1937) combined Trumpler's individual cluster H–R diagrams into the composite shown in Fig. 7.3 where he used his empirical tem-

Fig. 7.5. Martin Schwarzschild (c. 1980) who, with Gamow (1945), broke through the Schönberg-Chandrasekhar 10% limit on the way to the giant branch.

perature scale to convert Trumpler's spectral types to the temperature scale of the abscissa. He then interpreted this diagram via Strömgren's (1933) theory of the opacity as a method to estimate the hydrogen content of the various sequences by their positions in Fig. 7.3. The work was further developed into an idea of evolution up the main sequence as the hydrogen was converted to helium in a fully mixed star as we saw from Bok's (1946) summary written a decade later.

This interpretation via Strömgren's theory (Chandrasekhar 1939, Chapter 7), although spectacular and logically coherent, was not correct. While accounting for some of the features of the diagram such as the Trumpler turnups near the bright ends of each of the individual cluster sequences, it did not account for (1) the strong giant branch of Fig. 7.2, (2) the "peculiar" vertical subgiant branch of M67 (cluster number 14 in Fig. 7.3), or (3) the most unusual brighter parts of the H–R diagrams of globular clusters, none of which had main sequences to the limit of the available studies that reached about 2 magnitudes below the horizontal branch (Shapley 1930). These features were made much more conspicuous in later versions of the Kuiper-like diagram by Oke (1955), and by Sandage (1958a) where the giant sequences in the open clusters of h and χ Persei, M41, M11, Hyades, NGC 752, M67, NGC 188, and the globular clusters were seen to form a continuous pattern. An early composite diagram discussed at the Vatican conference (O'Connell 1958) is shown in Fig. 7.4.

Fig. 7.6. Walter Baade (c. 1960), who, on the basis of the H–R diagram of field high-velocity stars, expected the globular cluster main sequence to commence near the equivalent of F5 stars, and who set in motion the program to find it immediately after the commissioning of the Palomar 200-inch telescope in 1949.

The beginning toward the modern explanation of these features, where the subgiant and the giant branches of the diagram are a result of evolutionary tracks of unmixed stars away from the main sequence are contained in the papers of Henrich & Chandrasekhar (1941) and Schönberg & Chandrasekhar (1942, SC) where it was shown that no stable solution of the equations of stellar structure are possible with an isothermal core when that core contains more than $\sim$ 10% of the total mass and where there is no discontinuity of chemical composition between the core and the envelope. The paper by SC was particularly important because it traced the evolution of a star from the zero-age main sequence (not called that at the time) to the SC limit. It was later understood that this explained the Trumpler turn up.

Extending these ideas, Gamow (1944, 1945) proposed the way to break through the SC limit by a contraction of the core of *unmixed* stars. This nascent idea that the evolution of such stars, with their increasing discontinuity of chemical composition between the core and the envelope, was the way to breach the SC limit was developed in full detail by Schwarzschild (Fig. 7.5) in a series of papers beginning in 1952.

In 1950 a parallel observational program had begun based on Baade's (1948) prediction that the main sequence of the globular clusters, never seen in the H–R diagrams of Shapley, would be found near the middle F spectral types. Baade (Fig. 7.6) began the observations to find this main sequence in the first stellar population studies made following the commissioning of the 200-inch Palomar telescope. He assigned the technical problem to his two graduate students (H.C. Arp and the present writer), first as a training exercise at the bright end of the globular cluster H–R diagrams (to test our methods against Shapley's earlier diagrams) using the 60 and the 100-inch telescopes at Mount Wilson. We then pushed to the faint end using photographic plates taken by Baade with the Palomar 200-inch.

7.2.3 Evolution off the Main Sequence: The Globular Cluster Development

The main sequences of M92 (Arp, Baum, & Sandage 1952, 1953), and M3 (Sandage 1953) were found close to where Baade (1948) had predicted them to be from his analysis of the H–R diagram of field high-velocity stars (Oort 1926). The verification of this prediction was one of the first pegs that identified the high-velocity field stars with globular cluster-like stars. As such, it was a crucial step in Baade's continuing development of the population concept in the early 1950's.

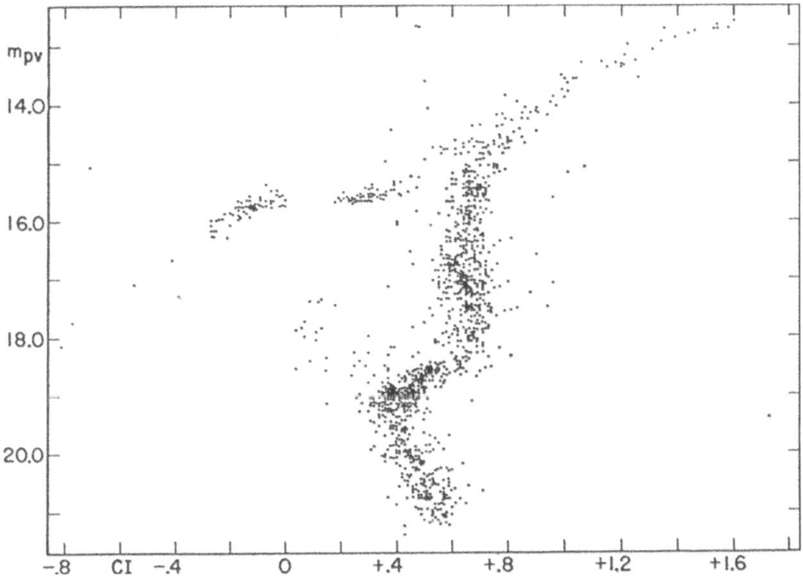

Fig. 7.7. Early color-magnitude diagram of M3 showing the main sequence termination point at apparent magnitude $V \sim 19$. Diagram from Sandage (1953).

The initial color-magnitude diagram for M3, based on the old "International" $CI = m_{\text{pg}} - m_{\text{pv}}$ color system in universal use at that time (Seares 1922 both for definitions and for the data in the North Polar Sequence), is in Fig. 7.7.

The identification of the main sequence termination point with either the SC limit, or something very much like it, was made, following Gamow (1945), by Schwarzschild with his introduction of a gravitationally contracting core after hydrogen exhaustion. The resulting temperature and density increase in the core ignited a hydrogen-burning shell surrounding the (by now) inert helium core. The non-mixed configuration, with a discontinuity in mean molecular weight between the core and the envelope, was the key to breaking the SC limit. With ignition of the shell to burn hydrogen, the star left the main sequence becoming a subgiant (Sandage & Schwarzschild 1952) on its way to the giant branch (Hoyle & Schwarzschild 1955).

An early review by Sears & Brownlee (1963) shows part of the ineffable excitement and ferment we all felt in the 1950's with this new "beginning of the final" understanding of the H–R diagram, bringing as it did its fully evident implications for stellar evolution.

The age-dating then followed immediately from equation (7.1). The mass fraction, Xf, was now known, being approximately the SC value of $Xf = 0.07$. The observed color-magnitude diagram of both M92 and M3 gave the absolute luminosity of the turnoff where this value of Xf applies (but of course only after the cluster distances had been estimated, which then, and yet today, was the remaining thorny problem as we shall see in section 7.3).

Sandage & Schwarzschild (1952) made a first crude estimate of the age of M3 as $T = 3 \times 10^9$ years. We used an absolute magnitude of the horizontal branch based on a fit of the M3 main sequence to the color-magnitude diagram of low velocity population I main sequence stars in the solar neighborhood as determined from trigonometric parallaxes. Baade (1948, 1952) had argued for the appropriateness of that identification, nothing at the time being known either about the low metallicity of the population II stars or the effect of different metallicities on the position of the main sequence. All that was to come a year or two later with (1) the direct observations of the very strange line-weakening in the spectra of individual globular cluster stars (Baum 1952), Roman's (1954) discovery of the UV excess in the high-velocity field F stars, (3) the finding of the same effect in the globular cluster giants (Sandage & Walker 1955), (4) the discussion of the metal-weakening in the composite spectra of globular clusters (Mayall 1946; Morgan 1956, 1959), and (5) the subsequent grouping of the globular clusters into metallicity classes (Deutsch 1955, and Kinman 1959).

Concomitantly, it became understood from theory (Strömgren 1952, Schwarzschild 1958) and from observation (Sandage & Eggen 1959) that the position of the main sequence in the H–R diagram depends sensitively on the metal abundance through a combination of its effect on the rate of energy generation and on the stellar interior bound-free opacity.

But none of that was known in the dating which Schwarzschild and I had done in 1952. Charged with that aspect of the work, I also failed to account for the change of L during the first phases of the evolution from the zero-age main sequence to the termination point, i.e., the $L(t)$ variation in the integral in equation (7.1). Both Schönberg and Chandrasekhar and Cowling (1935) (who had made the first numerical integration of a stellar model), had calculated the $\sim 1^{\mathrm{m}}$ rise in this phase of the evolution. This latter effect caused the age we suggested in 1952 to be too short by a factor of about 1.5 from what it would have been with a "correct" calculation from knowledge of what was indeed known at the time. My incorrect thesis value of the absolute magnitude of the RR Lyrae stars on the horizontal branch at $M_{\mathrm{V}} = 0^{\mathrm{m}}$ (Baade 1952) also shortened the age from what it later became when $\langle M_{\mathrm{V}}(\mathrm{RR\ Lyr})\rangle$ was closer to $+0\overset{\mathrm{m}}{.}7$ (Sandage & Cacciari 1990).

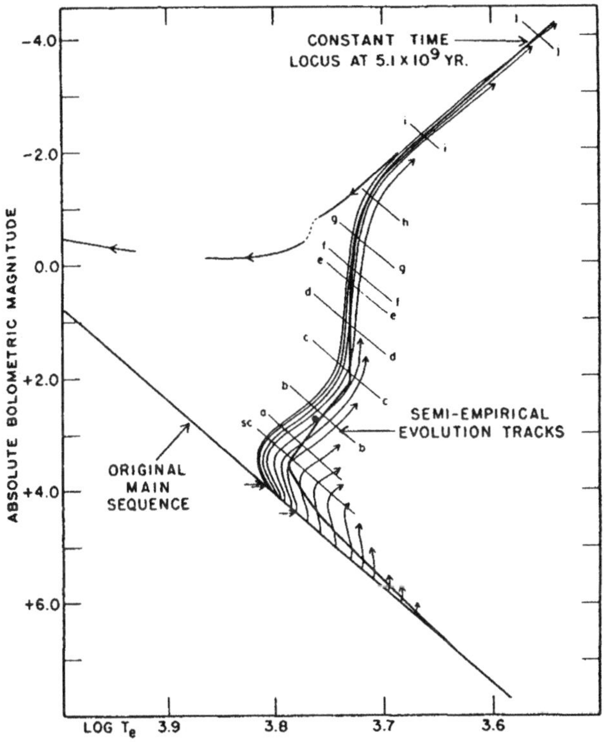

Fig. 7.8. Early schematic of the way the individual evolution tracks for stars of slightly different mass on the main sequence converge in the subgiant and giant region to form the observed locus of constant age, which is the position in the H–R diagram for coeval stars in the cluster. Diagram from Sandage (1957).

The situation as it appeared in 1957 is shown in Fig. 7.8, taken from a review of the observational developments at that time (Sandage 1957). Note again that the horizontal branch is put at $M_{\mathrm{bol}} = 0^{\mathrm{m}}$ there. The calculated age, marked on the giant branch, remained short at 5×10^9 years. This is ~ 3 times smaller than modern ages, due in part to the difference between

M_V(RR Lyr) $= 0^m$ and M_V(RR Lyr) $= +0\overset{m}{.}7$, and also more importantly to the much improved calculation of stellar evolutionary models near the turnoff-point.

A major development in these calculations occurred with the discovery by Simoda & Iben (1968, 1970) that the turnoff luminosity depends not only on the age and helium abundance, Y, but also sensitively on the metal abundance, Z. Their dependence showed that at equal age and Y, the metallicity dependence is $dL_{TO}/d([Fe/H]) = -0\overset{m}{.}13$, a result that has remained in all subsequent dating tables.

An appreciation of the enormous theoretical effort in the making of the relevant stellar models is had by a partial listing of the major grids of age $= f(L_{TO}, Y, Z)$ such as those of Iben & Rood (1970a,b), Demarque, Mengel, & Aizenmann (1971), Hartwick & VandenBerg (1973), Ciardullo & Demarque (1977), Saio (1977), VandenBerg (1983), and Bergbush & VandenBerg (1992), each of which also lists many more references to the vast dating literature.

Convenient linear interpolations of the grids exist with which one can see the sensitivity of the age-dating process to the various input parameters. As examples, an interpolation in the Yale tables of Ciardullo & Demarque (1977) give (Sandage, Katem, & Sandage 1981)

$$\log T(\mathrm{yr}) = 8.319 + 0.41 M_{\mathrm{bol}}(\mathrm{TO}) - 0.15([\mathrm{Fe/H}]) - 0.43(Y - 0.24). \quad (7.2)$$

An interpolation by Buonanno (1986) of the VandenBerg (1983) tables give

$$\log T(\mathrm{yr}) = 8.49 + 0.37 M_V(\mathrm{TO}) - 0.13([\mathrm{Fe/H}]), \quad (7.3)$$

for $Y = 0.23$.

The oxygen enhanced models of Bergbush & VandenBerg (1992) can be interpolated (Sandage 1993c) by

$$\log T(\mathrm{yr}) = 8.441 + 0.39 M_{\mathrm{bol}}(\mathrm{TO}) - 0.10([\mathrm{Fe/H}]), \quad (7.4)$$

for $Y = 0.24$. The smaller dependence on [Fe/H] in eq. (7.4) compared with eqs. (7.2) and (7.3) is due to the oxygen enhancement, effectively increasing the opacity for a given [Fe/H].

The equations show that the two crucial input parameters are M_{bol} (or M_V) of the turnoff point and the metallicity (keeping the initial value of Y fixed). They show that for every $0\overset{m}{.}1$ change in the turnoff luminosity, the age changes by 10%. This, of course, also follows from equation (7.1) for fixed XfM because a change of $0\overset{m}{.}1$ is a factor of 1.1 in L which, therefore, is 10% in the age.

The current uncertainty in the age-dating problem is therefore, effectively, due to uncertainties in the distances to globular clusters. Fig. 7.9 shows a modern normalization of the absolute magnitude scale in Fig. 7.8. The cluster main sequences were fitted to a family of fiducial main sequences, taking into account their different positions in the H–R diagram as a function of metal abundance as these positions were understood both from theory and from

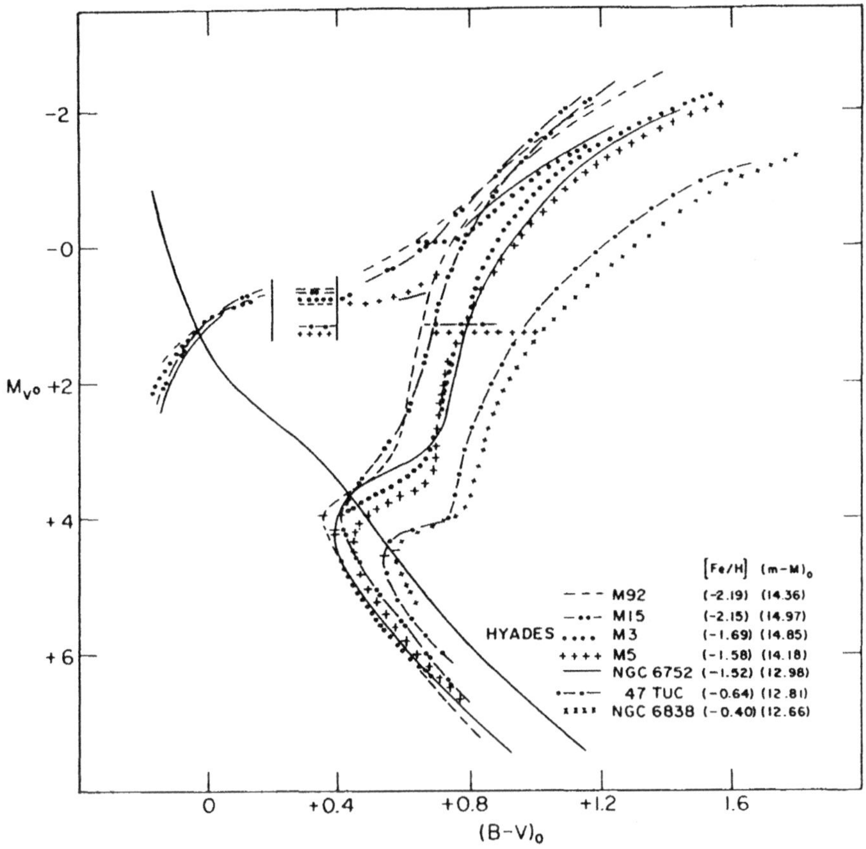

Fig. 7.9. Composite CMD circa 1980 for seven clusters of different metallicities. Diagram from Sandage (1986).

the parallax data available in the 1980's. Note that this fit gave M_V(RR Lyr) values between $+0\overset{m}{.}7$ and $+1\overset{m}{.}0$ along the horizontal branches. With this normalization the various age-dating grids all gave ages between 16 and 20×10^9 years.

Problem 7.2. Show by a transformation of units that a Hubble constant of 75 km s^{-1} Mpc^{-1} gives a reciprocal of $H_0^{-1} = 13.0 \times 10^9$ years. This is one of those convenient numbers to memorize for back-of-the envelope calculations in comparing time scales to scotch the tendentious tenebrous speculations that the cosmological constant is different from zero.

Because the cosmological times derived from the expansion rate via the inverse Hubble constant using H_0 between 50 and 100 (problem 7.1) are all smaller than these long globular cluster ages, the standard model with any deceleration (i.e., with mass) cannot conform to this timing test with a zero cosmological constant if the long globular cluster ages are correct. However, a new calibration of the RR Lyrae absolute magnitudes as a function of [Fe/H] has been made recently, based on the pulsation properties of these stars, combined with an understanding of the Oosterhoff-Arp mean period-metallicity correlation. This calibration, discussed in the next section, is $0\overset{m}{.}3$ brighter than in Fig. 7.9, reducing all cluster ages by $\sim$ 30% when combined with oxygen-enhanced stellar models. With this revised calibration of M_V(RR Lyr), the globular cluster ages are reduced to 14×10^9 years, which with $H_0 = 45 - 50$ km s^{-1} Mpc^{-1}, makes the inverse Hubble constant longer than the age of the Galaxy, withdrawing the need to introduce a cosmological constant to "save the standard model".

With this revised M_V(RR Lyr) calibration, the details of Fig. 7.9 are changed by renormalizing the ordinate using HB luminosities rather than by using main sequence fits which, themselves, have multiple uncertainties. Details of the new RR Lyrae absolute magnitudes as a function of metallicity are as follows.

7.3 Absolute Magnitude of RR Lyrae Stars from Pulsation Properties

An early baffling problem concerning the properties of RR Lyrae stars in globular clusters began when Oosterhoff (1939, 1944; see also van Agt & Oosterhoff 1959) showed that the mean periods of RR Lyrae stars in clusters divided into two nearly discrete groups, one with $\langle P \rangle = 0.55$ days and the other with $\langle P \rangle = 0.65$ days. The mystery deepened when Arp (1955) showed that the groups also divided the clusters by metallicity. The long mean period group contained only clusters of the lower metallicity with [Fe/H] ~ -2.0, while the short period group contained clusters of higher metallicity with [Fe/H] ~ -1.5.

The Oosterhoff groups are not as discrete as once thought, the apparent effect being caused by a combination of the variation of globular cluster morphology with [Fe/H] and a difference in the luminosity level of the HB's of Oosterhoff I and II clusters. When the extreme blue morphology of M13-like clusters is accounted for as compared with clusters such as M3 and NGC 7006, the fundamental Oosterhoff-Arp period-metallicity correlation was shown to be continuous (Sandage 1982, 1993a), the morphological parameter working to give a pseudo dichotomy.

The problem spawned much effort in the 1980's. A large literature grew concerning various explanations of the Oosterhoff dichotomy (Sweigart, Renzini, & Tornambe 1987; Caputo 1988; Buonanno, Corsi, & Fusi Pecci 1989; Lee, Demarque, & Zinn 1990; Catelan 1992, 1993a, b; Sandage 1990a, b; 1993a, b, c for reviews). A bare summary follows.

The correlation of individual RR Lyrae period distributions with metallicities in Galactic globular clusters is shown in Fig. 7.10. This is an expression

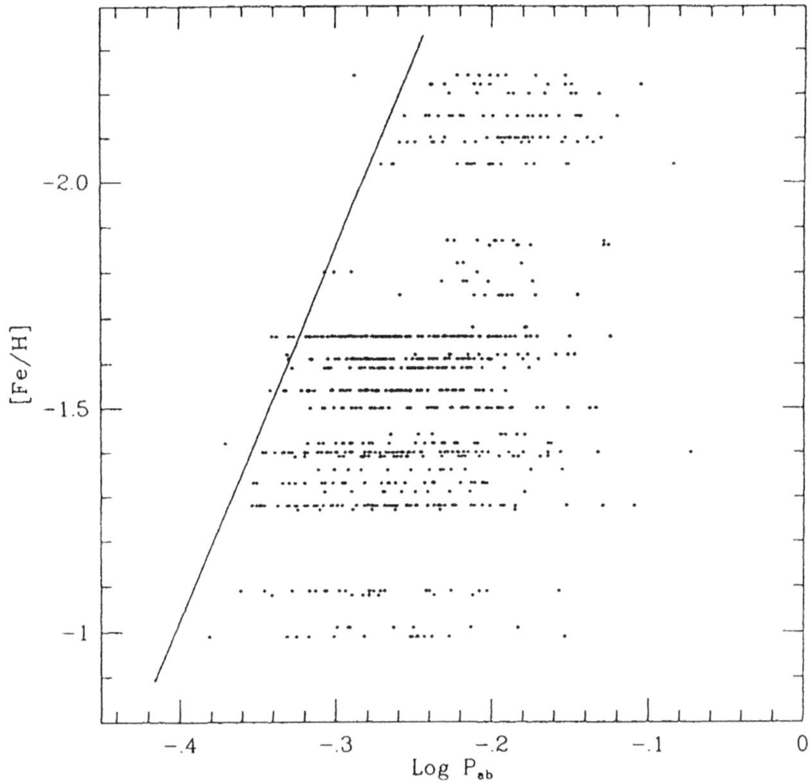

Fig. 7.10. Correlation of period distributions of the RR Lyrae type ab variables in 33 globular clusters with their [Fe/H] metallicities. The envelope line at the shortest periods in a given cluster defines the period-metallicity correlation at the fundamental blue edge of the instability strip. Diagram from Sandage (1993a).

of the Oosterhoff-Arp correlation, now not using mean periods but shown here to apply star-by-star at the fundamental blue edge of the instability strip.

An early explanation (the Vatican conference; O'Connell 1958) is that the HB luminosities differ between the two Oosterhoff groups, similar to the schematic in Fig. 7.11. Subsequent results have shown that not only must there be a luminosity difference between the clusters but also a slight temperature difference of the blue fundamental edge as a function of L (Iben 1971, Stellingwerf 1975, Caputo 1988, Simon & Clement 1993). To provide an explanation that is consistent with the available theoretical HB models such as those by Dorman (1992), following a long line of predecessors, when combined with evolution away from the zero-age HB such as by Lee, Demarque, & Zinn (1990), or Catelan (1993, with many references) requires the red and blue edges of the instability strip to be about 100 K cooler in Oosterhoff II clusters than in Oosterhoff I clusters (Sandage 1993b).

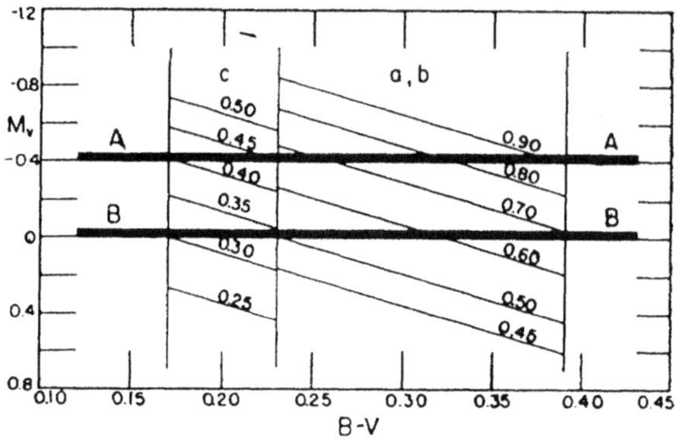

Fig. 7.11. Schematic of an early suggestion for the explanation of the period difference of the RR Lyrae variables between the two Oosterhoff groups. Lines of constant period thread the instability strip of a color-magnitude diagram. The HBs of the two schematic clusters A and B are put at different luminosities, showing a predicted period shift at constant color that will be a function of metallicity if the clusters have different [Fe/H] values. Diagram from Sandage (1958b).

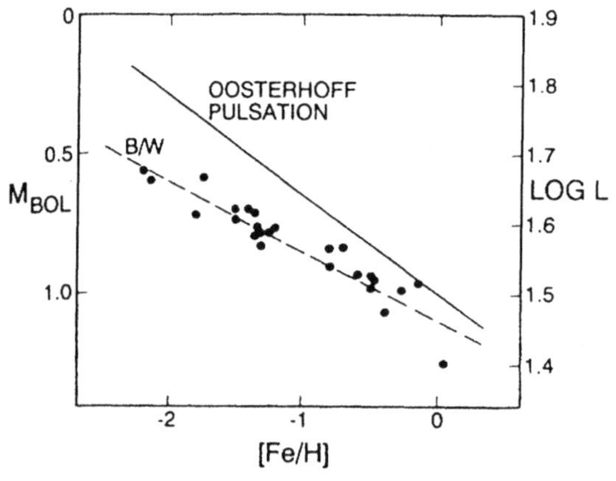

Fig. 7.12. The new calibration of the absolute magnitudes of RR Lyrae stars via the Oosterhoff-Arp period-metallicity correlation put in the pulsation equation, compared with an older calibration from data using the Baade-Wesselink method. Diagram from Sandage (1993b).

Combining this requirement with the size of the Oosterhoff-Arp period-metallicity correlation of $\mathrm{d}\log P/\mathrm{d}([\mathrm{Fe/H}]) = -0.122$ from Fig. 7.10 and from other data as well (Sandage 1993a), gives an absolute magnitude calibration via the pulsation condition $P\sqrt{\bar{\rho}} = \text{const.}$ as

$$M_{\mathrm{V}} = 0.30([\mathrm{Fe/H}]) + 0.94. \tag{7.5}$$

Equation (7.5) is between $0^{\mathrm{m}}_{\cdot}2$ and $0^{\mathrm{m}}_{\cdot}3$ brighter than the extant calibration via the Baade-Wesselink method in the relevant metallicity range. The pulsation calibration is $M_{\mathrm{bol}} = 0.36([\mathrm{Fe/H}]) + 1.00$ shown as a solid line in Fig. 7.12. The fainter Baade-Wesselink calibration, shown as the dashed line, has the equation $M_{\mathrm{bol}} = 0.25([\mathrm{Fe/H}]) + 1.09$.

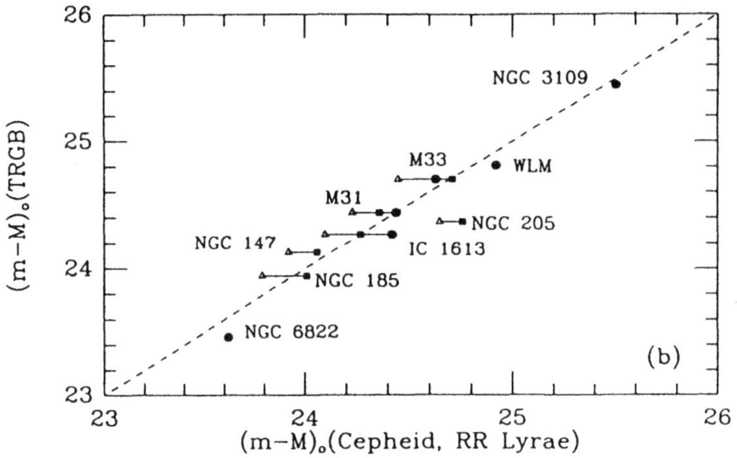

Fig. 7.13. (Abscissa) Comparison of distance moduli determined from classical Cepheids (closed circles) and RR Lyrae stars (triangles) with moduli determined from the top of the red giant branch (ordinate). Diagram from Lee *et al.* (1993).

It is this increase in L(RR Lyr) by an average of $0\overset{m}{.}25$ that lowers the average age of the globular clusters to $\langle T\rangle \sim 14 \times 10^9$ years. It remains to understand the difference between the two calibration methods. In passing, we note that the Baade-Wesselink method is highly sensitive to the systemic (gamma) velocity of each RR Lyrae variable. An error of only 1 km s^{-1} in γ produces an error of $0\overset{m}{.}1$ in $M_{\rm bol}$.

Such a drastic revision brightward in RR Lyrae luminosities has support from a number of independent results. The two most direct are (1) a comparison of the distance modulus of the Large Magellanic Cloud derived from classical Cepheids and from RR Lyrae stars requires a brightward revision of between $0\overset{m}{.}2$ and $0\overset{m}{.}3$ (Walker 1992), and (2) the same experiment performed in the Local Group galaxies of NGC 205, NGC 147, NGC 185, and IC 1613 by Saha *et al.* (1992) gives the same answer. A summary of this experiment is set out in Fig. 7.13 from Lee, Freedman, & Madore (1993) showing the stated offset in the derived distance moduli from Cepheids and RR Lyrae stars as abscissa compared with the moduli derived from the top of the red giant branch (ordinate). These (mostly) Saha *et al.* data again show the $0\overset{m}{.}3$ offset.

Corroborative evidence for the brighter calibration is had by (1) comparing the globular clusters in M31 (at the Cepheid distance) with the luminosity function (Secker 1992) of the globular clusters in the Galaxy at their RR Lyrae distances (Sandage & Tammann 1994), (2) analysis of the light curve shapes (Simon & Clement 1993), and (3) Eggen's (1994) intermediate-band photometry calibrated to give absolute magnitudes.

7.4 Age of the Galaxy and the Universe Using Globular Clusters

The result of using the absolute magnitude calibration of equation (7.5) with the age-dating equation (7.4) from the oxygen-enhanced models of Bergbush & VandenBerg (1992), applied to the photometric data of 24 clusters is shown in Fig. 7.14. The mean age of this system of halo clusters is $14.1 \pm 1.5 \times 10^9$ years. If globular clusters are the first to form in the initial density fluctuation that became the Galaxy (Peebles & Dicke 1968), then this can be taken to be the Galactic age.

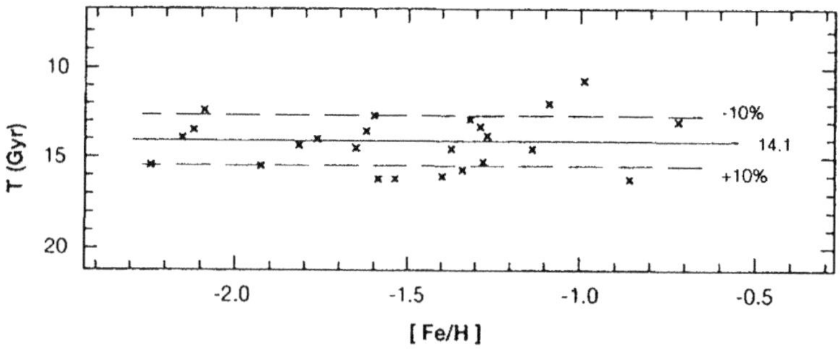

Fig. 7.14. Revised ages of 24 globular clusters whose main sequence turnoff luminosities have been determined by the luminosity of the HB via the RR Lyrae calibration of equation (7.5) using the oxygen-enhanced isochrones of Bergbush & VandenBerg (1992). Diagram from Sandage (1993c).

Evidence exists that the age of the universe is less than 1 Gyr older than the oldest stellar components of a galaxy. The most direct is the existence of quasars with redshifts as large as $z = 5$, corresponding to look-back times of less than $(5/6)H_0^{-1}$ if $q_0 > 0$ (see problem 7.1). The spectra of such quasars show lines of C, N, and O, proving that nucleosynthesis had already occurred in the quasars prior to this look-back time. The formation of these heavy elements presumably must have taken place in stars that themselves had to evolve to become supernovae, enriching the material that later became the central quasar in the parent galaxy. Hence, considerable stellar evolution must already have occurred before $z = 5$.

Similar evidence from spectrophotometry of high redshift radio galaxies (Chambers & Charlot 1990; McCarthy 1991; Persson 1991; McCarthy, Persson, & West 1992) shows an old stellar population even at these large look-back times. Furthermore, n-body simulations (e.g. Zurek, Quinn, & Salmon 1988) show collapse times for the initial aggregates (the denser globular clusters) in the collapsing envelope of a protogalaxy to be considerably less than 1 Gyr after the big bang singularity. Based on this we, therefore, add 1 Gyr to the mean age of the Galactic globular clusters in Fig. 7.14, to adopt thereby $T = 15 \pm 1$ Gyr as the age of the universe.

This is one of the numbers needed in Fig. 7.1 to make the timing test. The second required number, H_0, is the subject of the next two lectures.

References

Adams, W.S. 1916, Proc. Nat. Acad. Sci. **2**, 143
Adams, W.S. 1939, Pub. ASP **51**, 133
Adams, W.S., & Joy, A.H. 1917, ApJ **46**, 313
Adams, W.S., Joy, A.H., & Humason, M.L. 1926, ApJ **64**, 225
Adams, W.S., Joy, A.H., Humason, M.L., & Brayton, A.M. 1935, ApJ **81**, 187
Adams, W.S., Joy, A.H., Strömberg, G., & Burwell, C.G. 1921, ApJ **53**, 13
Adams, W.S., & Kohlschütter, A. 1914, ApJ **40**, 385
Arp, H.C. 1955, AJ **60**, 317
Arp, H.C., Baum, W.A., & Sandage, A. 1952, AJ **57**, 4
Arp, H.C., Baum, W.A., & Sandage, A. 1953, AJ **58**, 4
Baade, W. 1948, Pub. ASP **60**, 230
Baade, W. 1952, Trans. IAU **8** (Cambridge: Cambridge Univ. Press), p. 387
Baum, W.A. 1952, AJ **57**, 222
Bergbush, P.A., & VandenBerg, D.A. 1992, ApJS **81**, 163
Berry, M. 1976, *Principles of Cosmology and Gravitation*, (Cambridge: Cambridge University Press)
Blaauw, A. 1963, in *Basic Astronomical Data*, ed. K. Aa. Strand (Chicago: Univ. Chicago Press), p. 385
Bok, B.J. 1946, MNRAS **106**, 61
Buonanno, R. 1986, Mem. Soc. Astr. Ital. **57**, 333
Buonanno, R., Corsi, C.E., & Fusi Pecci, F. 1989, A&A **216**, 80
Caputo, F. 1988, A&A **189**, 70
Catelan, M. 1992, A&A **261**, 457
Catelan, M. 1993a, A&A **274**, 1013
Catelan, M. 1993b, A&A Suppl. **98** 547
Chambers, K.C., & Charlot, S. 1990, ApJ **348**, L1
Chandrasekhar, S. 1939, *Study of Stellar Structure* (Chicago: Univ. Chicago Press), Chapter 7
Ciardullo, R.B., & Demarque, P. 1977, Yale Obs. Trans. No. 33
Cowling, T.G. 1935, MNRAS **96**, 42
Deutsch, A.J. 1955, in *Principles Fondamentaux de Classification Stellaire*, Vol. **25** (Paris: CNRS)
Demarque, P.R. 1967, ApJ **150**, 943
Demarque, P.R. & Percy, J.R. 1964, ApJ **140**, 541
Demarque, P.R., Heasley, J.N. 1971, ApJ **163**, 547
Demarque, P., Mengel, J.G., & Aizenman, M.L. 1971, ApJ **163**, 37
Dorman, B. 1992, ApJS **81**, 221
Eggen, O.J. 1994, AJ in press
Epstein, I. 1950a, ApJ **112**, 207
Epstein, I. 1950b, ApJ **114**, 438
Gamow, G. 1939, Nature **144**; 575, 620
Gamow, G. 1940, *Birth and Death of the Sun* (New York: Viking Press)
Gamow, G. 1944, Phys. Rev. **65**, 20
Gamow, G. 1945, Phys. Rev. **67**, 120
Hartwick, F.D.A., & VandenBerg, D.A. 1973, ApJ **185**, 887
Henrich, L.R., & Chandrasekhar, S. 1941, ApJ **94**, 525

Hoyle, F., & Schwarzschild, M. 1955, ApJS **2**, 1
Iben, I. 1971, Pub. ASP **83**, 697
Iben, I., & Rood, R.T. 1970a, ApJ **159**, 605
Iben, I., & Rood, R.T. 1970b, ApJ **161**, 587
Jeans, J. 1929, *Astronomy and Cosmogony* (Cambridge: Cambridge Univ. Press), p. 381
Keller, G. 1948, ApJ **108**, 347
Kinman, T.D. 1959, MNRAS **119**, 559
Kuiper, G.P. 1936, Har. Bull. No. 903, p. 1
Kuiper, G.P. 1937, ApJ **86**, 176
Lee, M.G., Freedman, W.L., & Madore, B.F. 1993, ApJ **417**, 553
Lee, Y.-W., Demarque, P., & Zinn, R. 1990, ApJ **350**, 155
Mayall, N.U. 1946, ApJ **104**, 290
McCarthy, P. 1991, Ann. Rept. Carnegie Obs. (Carnegie Inst. Washington), Yearbook 90, p. 22
McCarthy, P., Persson, S.E., & West, S.C. 1992, ApJ **386**, 52
Morgan, W.W. 1956, Pub. ASP **68**, 509
Morgan, W.W. 1959, AJ **64**, 432
Oke, J.B. 1950, JRASC **44**, 135
Oke, J.B. 1955, JRASC **49**, 8
O'Connell, D.J.K. (ed.) 1958, Specola Astron. Vaticana, Vol. **5** (The Vatican Conference Report); (Vatican City: Specola Vaticana)
Oort, J.H. 1926, Groningen Pub. No. **40**
Oosterhoff, P. Th. 1939, Observatory **62**, 104
Oosterhoff, P. Th. 1944, BAN **10**, 55
Peebles, P.J.E., & Dicke, R.H. 1968, ApJ **154**, 891
Persson, S.E. 1991, Ann. Rept. Carnegie Obs. (Carnegie Inst. Washington), Yearbook 90, p. 27
Roman, N.G. 1954, AJ **59**, 307
Russell, H.N. 1910a, AJ **26**, 153
Russell, H.N. 1910b, Science **32**, 833
Russell, H.N. 1914, Popular Astronomy **22**, 342
Saha, A., Freedman, W.L., Hoessell, J., & Mossmann, A.E. 1992, AJ **104**, 1072
Saha, M.N. 1920, Phil. Mag. **40**, 472
Saio, H. 1977, Astrophys. Space Sci. **50**, 93
Sandage, A. 1953, AJ **58**, 61
Sandage, A. 1957, ApJ **126**, 326
Sandage, A. 1958a, ApJ **127**, 513
Sandage, A. 1958b, in *Specola Astron. Vaticana*, Vol. **5**, (The Vatican Conference Report, ed. D.J.K. O'Connell), p. 41
Sandage, A. 1961a, ApJ **133**, 355
Sandage, A. 1961b, ApJ **134**, 916
Sandage, A. 1982, ApJ **252**, 553
Sandage, A. 1986, Ann. Rev. A&A **24**, 421
Sandage, A. 1988, Pub. ASP **100**, 293
Sandage, A. 1990a, ApJ **350**, 603
Sandage, A. 1990b, ApJ **350**, 631
Sandage, A. 1993a, AJ **106**, 687
Sandage, A. 1993b, AJ **106**, 703
Sandage, A. 1993c, AJ **106**, 719
Sandage, A., & Cacciari, C. 1990, ApJ **350**, 645
Sandage, A., & Eggen, O.J. 1959, MNRAS **119**, 278
Sandage, A., Katem, B., & Sandage, M. 1981, ApJS **46**, 41

Sandage, A., & Schwarzschild, M. 1952, ApJ **116**, 463
Sandage, A., & Tammann, G.A. 1994, ApJ, in press
Sandage, A., & Walker, M.F. 1955, AJ **60**, 230A
Schönberg, M., & Chandrasekhar, S. 1942, ApJ **96**, 161
Schwarzschild, M. 1946, ApJ **104**, 203
Schwarzschild, M. 1958, *Stellar Structure and Evolution*, (Princeton: Princeton Univ. Press)
Seares, F.H. 1922, ApJ **56**, 97
Sears, R.L., & Brownlee, R.P. 1963, in *Stellar Structure*, ed. L.H. Aller, & D.B. McLaughlin (Chicago: Univ. Chicago Press), Chapter 11
Secker, J. 1992, AJ **104**, 1472
Shapley, H. 1930, *Star Clusters* (Harvard Obs. Monograph Series: Harv. Univ. Press)
Simoda, M. & Iben, I. 1968, ApJ **152**, 509
Simoda, M. & Iben, I. 1970, ApJS **22**, 81
Simon, N., & Clement, C. 1993, ApJ **410**, 526
Stellingwerf, R.F. 1975, ApJ **195**, 441
Strömgren, B. 1933, Zs. Astrophys. **7**, 222
Strömgren, B. 1952, AJ **57**, 65
Sweigart, A.V., & Renzini, A., & Tornambe, A. 1987, ApJ **312**, 762
Trumpler, R.J., 1925, Pub. ASP **37**, 307
van Agt, S., & Oosterhoff, P. Th. 1959, Ann. Stern. Leiden **21**, 253
VandenBerg, D.A. 1983, ApJS **51**, 29
Walker, A.R. 1992, ApJ **390**, L81
Zurek, W.H., Quinn, P.J., & Salmon, J.K. 1988, ApJ **330**, 519

8 Timing Test Continued (The Hubble Constant)

8.1 Historical Resume

Hubble's (1929) first calibration of the redshift-distance relation gave $H = 500$ km s^{-1} Mpc^{-1} based on (1) his estimates of distances to 22 galaxies using "brightest stars", calibrated as having a mean absolute magnitude of $\langle M_B \rangle = -6^{\rm m}.3$, and (2) an adopted mean absolute magnitude for galaxies as $\langle M_V \rangle = -15^{\rm m}.1$ and applied to Holetschek's estimated visual magnitudes for the same galaxies. These two adopted absolute magnitude scales were based on Hubble's calibrations from the few galaxies (M31, M33, LMC, SMC, NGC 6822) for which Cepheids had been studied, and the still fewer galaxies for which Hubble (1926) had used "other variables" to estimate distances (M81, NGC 2403, and M101, see footnote in Hubble & Humason, 1931).

Remarkably, Hubble's first value of the expansion rate is similar to $H = 627$ km s^{-1} Mpc^{-1} estimated by Lemaître (1927, translated and reprinted in 1931), and $H = 461$ km s^{-1} Mpc^{-1} by Robertson (1928) "before the redshift-distance relation had been discovered" (Lecture 5). But in fact, perhaps the agreement is not so remarkable because both Lemaître and Robertson based their estimates on applying Hubble's 1926 calibration of $\langle M \rangle_{\rm galaxies}$ to the extant data on apparent magnitudes and on Slipher's velocities as published by Eddington (1923) and by Strömberg (1925). These were the same data used by Hubble in 1929.

A second value of $H = 559$ km s^{-1} Mpc^{-1} was derived by Hubble and Humason (1931) using data on objects in their program galaxies, again identified as "brightest stars". The calibration method using the 10 local galaxies are set out in detail in their Table 1 and in the footnote to it. No major revision to this value was made in Hubble's (1936a, b, c, d; 1938, 1939) subsequent discussions and reviews throughout the two decades of the 1930s and 1940s.

As late as 1950 Hubble had given Holmberg (1950) his best estimates of the distance moduli to local galaxies and to the Virgo cluster. A comparison of Hubble's 1950 scale with a scale devised by Tammann and the writer up to 1982 is shown in Table 8.1. Note that Hubble's (1936c) distance modulus to the Virgo cluster is $m - M = 26^{\rm m}.8$ ($D = 2.3$ Mpc), which is a factor of ~ 10 smaller in distance than our current estimate of $m - M = 31^{\rm m}.7$.

The first successful challenge to the 1930's canonical scale was made with Baade's revision of the distance to M31 by a factor of about two, announced at the 1952 Rome IAU meeting (Lecture 5). This was widely misinterpreted in the semi-popular press and, because of that publicity, in the subsequent textbooks where it is generally written that all galaxy distances were to be increased by the same factor, and therefore that H was ~ 250 km s^{-1} Mpc^{-1}.

The conclusion was unwarranted. It was known even at that time that many factors connected with Hubble's calibration were uncertain and must

Table 8.1. Comparison of Hubble's 1950 scale with the scale prepared for the Vatican Conference on "Astrophysical Cosmology" [Sandage and Tammann (1982)]. $(m-M)_{\rm AB}$ is the apparent blue distance modulus.

Galaxy	$(m-M)_{\rm AB}$ Hubble (1950)	$D_{\rm Hubble}$ Mpc	$(m-M)_{\rm AB}$ here (1982)	$D_{\rm here}$ Mpc	$D(82)/D(50)$
LMC	$17^{\rm m}.1$	0.22	$18^{\rm m}.91$	0.52	2.4
SMC	$17^{\rm m}.3$	0.25	$19^{\rm m}.35$	0.71	2.8
M33	$22^{\rm m}.3$	0.24	$24^{\rm m}.68$	0.82	3.4
M31	$22^{\rm m}.4$	0.23	$24^{\rm m}.78$	0.67	2.9
NGC 6822	$21^{\rm m}.6$	0.16	$25^{\rm m}.03$	0.62	3.9
IC 1613	$22^{\rm m}.0$	0.23	$24^{\rm m}.55$	0.77	3.3
WLM	$22^{\rm m}.3$	0.25	$25^{\rm m}.54$	1.28	5.1
M81/2403 Group	$24^{\rm m}.0$	0.52	$27^{\rm m}.60$	3.25	6.2
M101	$24^{\rm m}.0$	0.52	$29^{\rm m}.3$	7.24	13.9
Virgo	$26^{\rm m}.8$	2.29	$31^{\rm m}.7$	21.9	9.6

be redetermined at each successively larger distance (Hubble 1951). The corrections for distances beyond M31 were suspected to be much larger than the factor of two caused "simply" by the change of zero-point of the Cepheid P-L relation.

The recalibration problems concerned (1) corrections to the apparent magnitude scales in the Selected Areas, begun by Baade (1944) and carried by photoelectric photometry into the more modern literature by Stebbins, Whitford, & Johnson (1950), (2) discovery of Cepheids in galaxies beyond the Local Group such as NGC 2403, M81, and M101 in which Hubble & Humason (1931) had only vague data on the brightest variables [later codified as HLVs meaning "high luminosity variables" whose prototypes had been followed by Hubble in the two decades of the 1930s and 1940s, (Hubble & Sandage 1953)], (3) the separation of HII regions from brightest stars using newly developed red sensitive photographic plates, use of which for the distance scale problem was to come only in 1956 (Humason *et al.* Appendix C; Sandage 1958), and (4) the development of new methods of distance determination such as supernovae, galaxy luminosity classes, globular clusters, linear diameters, galaxy rotation curves, etc., which are the subjects of this lecture.

By the late 1960's it had become evident that Hubble's scale had to be stretched, not merely re-zero-pointed (Table 8.1). The correction increased from a modulus difference of $1^{\rm m}.8$ for the Magellanic Clouds, to $2^{\rm m}.4$ for M31, M33, and NGC 6822, to $3^{\rm m}.6$ for M81 and NGC 2403, and to a finally stabilized value near $5^{\rm m}$ for M101 and beyond (i.e., to the Virgo cluster). A change of 5 magnitudes is a factor of 10 in distance. Hence, on that basis alone, Hubble's 1929 estimate of ~ 500 km s^{-1}Mpc^{-1} would have become $H \sim 50$, but the details of the actual program to correct the 1930 distance scale are, of course, more complicated. The purpose of this lecture is to set out the details.

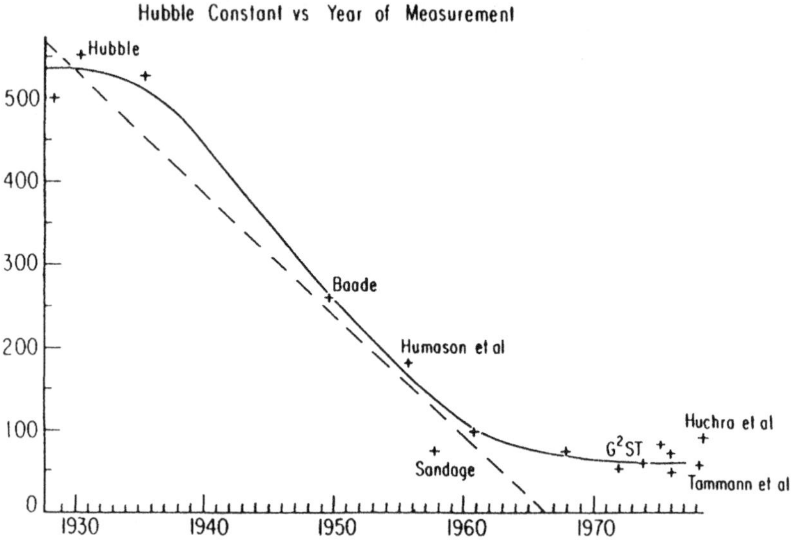

Fig. 8.1. Schematic from Schramm (1982) of the approach to the current range of H between 50 and 100 km s^{-1} Mpc^{-1} in the 60 years since Hubble's 1929 estimate. The earlier estimates of Lemaître (1927) of $H = 627$ and of Robertson (1928) of $H = 461$ are not shown.

An interesting summary of the chronology of H (but without the Lemaître 1927 nor the Robertson 1928 points) is in Fig. 8.1 from Schramm (1982). A realistic assessment of part of the recent effort to determine H is in Fig. 8.2 from Paturel (1982).

8.2 Crime, Chess, and the Philosophy of False Clues in Puzzles

The present literature on the distance scale is large and of different quality as to reality and prejudice. As in any detective story there are many false clues. The successful detective must devise means to ferret those parts of the total evidence that is germane, relevant, and correct. The rest which, although they seem to apply to the crime, in fact end in contradiction. These are eventually discarded as the puzzle approaches a correct solution.

As in all unsolved problems anywhere near the frontier of science, the dilemma is how to recognize the false clues. The methods vary amongst detectives. Sherlock Holmes always wins over Lestrade. Perry Mason always wins over Lt. Tragg of the police and Hamilton Burger of the prosecutor's office; Hercule Poirot over Inspector Japp, Nero Wolfe over Inspector Cramer, Sam Spade or the Saint over the local police, Richard Hannay over Scotland Yard, etc. Each works by intuition plus research.

Knowledge of how they spot the false clues and ignore them would of course be of enormous value in unsolved scientific puzzles. How can we know

Fig. 8.2. A not really so old way to H, devised by Paturel (1983).

what clues are relevant and which are not? Reading the distance-scale literature is like reading the crime literature.

But the problem is also similar (and herein lies its solution) to the strategy in a game of chess as practiced at five levels from a beginning student to a grand master[23]. Each knows the rules, but each approaches the game differently and with different results.

The novice has no intuition because an acquaintance with the game has just begun. There has been no opportunity to develop an expertise that may later become intuitive. The student has been taught, for example, to "always exchange if the total value of the captured pieces exceeds the value of the pieces lost", or "control the center when there are no advantageous exchanges".

The advanced beginner will have learned to recognize certain situations and will have stored these in memory, to be recalled when needed. He still remembers the rules, but now he has identified certain problems encountered before and also recalls the successes and the failures in his response.

For the competent performer, as Dreyfus writes "with more experience, the number of potentially relevant elements of a real-world situation that the learners are able to recognize becomes over-whelming". But the sense of what is important, or indeed relevant in any particular situation is missing.

[23] A remarkable discussion of the chess world is given by Dreyfus (1994) in his analysis of the problem of "teaching" a computer how to think. Is artificial intelligence as simple as Minsky (1967) once said, or is his later statement (1982, quoted by Gina Kolata) more accurate? We paraphrase part of Dreyfus's analysis of "learning" as an analogy to the way intuition works in the distance scale problem to identify its false clues.

The proficient performer, as a result of a large experience, although still guided by the rules and principles, will have his *modus operandi* nearly totally dominated by "situational discriminations and associated responses", i.e., he brings into play his intuition that has been developed by his half-vast, but not yet totally vast experience. Such a player "can recognize a large repertory of types of positions. Recognizing almost immediately and without conscious effort the sense of the position, the player sets about calculating [subconsciously] the move that best achieves the goal".

Finally, the grand master has a more subtle and refined ability. In most situations he/she "experiences a compelling sense of the issue and the best move". This player "knows" by intuition which clues are relevant. As in science, there never is a perfect set of data and methods. It is here that judgement-calls are crucial. "It is estimated that a master chess player can distinguish roughly 50,000 types of positions." In other words his or her intuition judges what is real in the game, what will or will not lead to contradiction, and what aspects of the data to ignore. Only in that way, where forks in the road require decision, can unimpeded travel proceed along the yellow brick road that leads to the emerald city. To a remarkable degree, Hubble and Baade were grandmasters of the game as it was then known and played. They began paving that yellow brick road and saw the emerald city some distance down the road even as it was far from their starting place.

None of this may sound much like objective science, but rather like opinion in philosophy. However, a case can be made that at some place in the path toward a solution of all really difficult scientific problems the scientist is faced with innumerable false clues. At that point progress comes only through intuition as to which to ignore at that moment, permitting the next series of decisions as to direction in the laboratory. Objective truth in science, as in detective stories, only comes near the end of the investigation.

For this lecture we discuss a sample of current astronomical methods to H_0, emphasizing those clues that seem to define the yellow brick road and also those which lead back into the woods on trails that will eventually narrow if they are flawed, or may eventually widen again down the way into a parallel avenue if their contradictions can be overcome.

Many reviews of the distance scale have been written since Baade began the revolution with his 1952 announcement. It would not be useful to summarize here all of the reviews, although a reading of the changing data and attitudes toward both the real and the false clues during the past 40 years (which by hindsight we can now sort out) would in fact be a history of practical cosmology during the period.

In this lecture we review the status (circa 1993) of methods that favor the long distance scale, mentioning more briefly those that do not. Concerning the short scale, two reviews, one by Jacoby *et al.* (1992) and the other by van den Bergh (1992) show prejudices and a choice of clues that differ from those espoused here.

Many reviews of the long scale exist. For this lecture, the single most valuable is that by Tammann (1992). It is one of the last of a series of reports setting out much of the evidence for the long scale (Tammann 1986, 1987, 1988; Sandage & Tammann 1982, 1984, 1986, 1990, 1994; Branch 1988; Sandage 1988a, b). The purpose of this lecture is to summarize these summaries.

8.3 Seven Astronomical Ways to H_0

(1) Distance to Virgo:

The observed redshift of the core of the Virgo cluster is not the cosmological redshift relative to the microwave background (CMB) although it is close to it. That the non-cosmological deviations of redshifts from a noiseless Hubble flow are small (certainly less than 200-300 km s^{-1} at the distance of the Virgo cluster) was known from the earliest days of the redshift-distance diagram. The Virgo cluster point in the cluster (m, z) Hubble diagram did not widely deviate from the more distant cluster points (see Fig. 5.5 of Lecture 5). The point was made explicitly at the 1961 Santa Barbara IAU Symposium 15 on "Problems in Extra Galactic Research" during the discussion of the report on the distance scale (Sandage 1961). That discussion shows the concerns even at that time of the problem of the Virgo "infall" velocity. The perturbation is actually the retarded expansion of the Local Group away from Virgo due to the cluster mass (Tammann & Sandage 1985 for a review and an entrance to the literature). Part of the 1961 discussion went as follows.

> *Abell:* To what extent do the values of the Hubble constant which you have listed depend on velocities in the Virgo Cluster or other objects in the Local Supercluster? Certainly one must base the Hubble constant upon velocities of remote objects, such as clusters that are at least as distant as the Coma Cluster. The distances of these objects must, of course, depend upon calibration with nearby galaxies.
>
> *Sandage:* If there was great difficulty in using the Virgo Cluster data to determine H, then this cluster would not fit onto the redshift-magnitude relation. But the Virgo cluster does fit this relation to within $0^m.3$, a divergence which amounts to only a factor of 1.15 in distance. At the present we do not know the Hubble constant this accurately for other reasons. [The value that had been derived from other considerations had already been determined to be $H = 75$ km $s^{-1}Mpc^{-1}$ (Sandage 1958). In the Symposium summary I had set out (Table 7 there) values of H determined by Sersic ($H = 113$), van den Bergh (106), Holmberg (112), and two by the writer (82 and 75)].
>
> *Hoyle:* I would like to comment on the applicability of the value of H to the more distant regions of the universe. If one doubts the applicability, on the grounds that the rate of expansion of the local supercluster might be different from that of the general field (say,

beyond 50 Mpc), then it seems reasonable that H for the general field should be taken larger than the local value. Otherwise, the local irregularity of the supercluster would have vanished into the general field. [Ed.: Not so if the rates are *identical* inside and outside the local supercluster (Lecture 5, Sec. 5.1.2).]

Sandage: I should again like to state that the observational data show that the Virgo cluster fits well on the $[m, z]$ relation defined by galaxies with redshifts greater than 5000 km s^{-1} and that, therefore, there cannot be a large peculiar expansion rate for this cluster.

Oort: I do not think that the Virgo cluster should be discredited for the determination of H. There is considerable dispersion of velocities in the Virgo cluster, but this can be averaged out.

After considering Hoyle's remarks, I agree that there should be some reserve in interpreting the result derived from the Virgo cluster, as the Galaxy might have a "random" motion. But it seems unlikely that this would be a large fraction of the expansion velocity. [Ed.: He simply restates here the (m, z) argument set out as the answer to Abell and Hoyle.]

Page: Your reply to Hoyle seemed to me the strongest argument *against* a local supercluster or metagalaxy. If the $[m, z]$ relation shows no lower slope for $z < 0.01$ than for higher values, then there can scarcely be a local supercluster. [Ed.: This of course is the modern (c. 1994) problem of why the random motions are so small in a $q_0 = 1/2$ universe. Of course, if $q_0 = 0$ (no mass) there would be no gravity and no random motions induced by gravity.]

Zwicky: I fail to see how the redshift constant can be accurately determined unless one ties the data in the Local Group directly to redshifts in clusters for which there is little dispersion of internal distance and of velocities. H from supernovae in the Cancer cluster, the Coma cluster, and the Local Group gives 175 km s^{-1} [sic].

These exchanges set out problems that were to occupy the field for the next 30 years such as the Virgo cluster infall, the motion of the Virgo complex relative to the CMB, the use of supernovae as distance indicators, the value of Ω, and the effect on the expansion field of the sheets and voids subsequently discovered (Chincarini & Rood 1972, 1975, 1976, 1979; Gregory & Thompson 1978; Tarenghi *et al.* 1979, 1980).

The problem of tying the Virgo cluster to the cosmic kinematic frame of the CMB, which presumably is the Machian proper frame of an ideal Hubble flow, has been solved decisively by Jerjen and Tammann (1993). Using a variety of methods (Tammann 1992, his Table V) they determined the distances of 17 clusters *relative* to Virgo (i.e., distance ratios), the absolute distances not being needed. From the known strength and direction of the CMB dipole ($l = 277° \pm 2°$, $b = 30° \pm 2°$ with amplitude 622 ± 20 km s^{-1} as the motion of the Local Group relative to the CMB), each of the measured

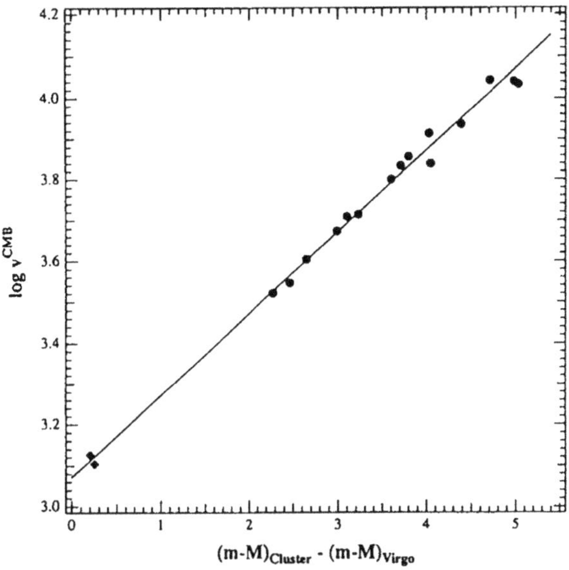

Fig. 8.3. The Hubble diagram of redshifts in the kinematic frame of the CMB vs. the differences in the distance moduli between each of 17 clusters and the Virgo cluster. The cosmic (Machian) redshift at the distance of the Virgo cluster core is read from the diagram as $v(\text{cosmic}) = 1179 \pm 17$ km s^{-1}. Diagram is from Jerjen & Tammann (1993).

redshifts (reduced to the centroid of the Local Group) of the 17 clusters were reduced to the kinematic frame of the CMB. The resulting Hubble diagram using redshifts in the CMB frame and the distance ratios (expressed as moduli differences relative to Virgo) is shown in Fig. 8.3. The line has the slope of 5 required from a linear redshift-distance relation.

Reading the intercept at zero modulus difference gives the cosmic redshift of the Virgo cluster core to be $v_{\text{CMB}} = 1179 \pm 17$ km s^{-1} *in the frame of the CMB.* This is the redshift the Virgo cluster would have had in the absence of the retarded redshift of the Local Group away from Virgo due to deceleration caused by the mass of the Virgo complex. Note that this cosmic redshift relative to the observed mean redshift of $v_0 = 932 \pm 38$ km s^{-1} implies an infall velocity of 247 ± 42 km s^{-1}, similar to the 220 km s^{-1} derived earlier (Tammann & Sandage 1985), and 233 ± 44 km s^{-1} by Jerjen & Tammann (1993).

Problem 8.1. (a) Using the same approximation as in problem 5.6 (section 5.6 of Lecture 5), find the approximate mass of the Virgo complex causing the deceleration using (1) the derived "Virgo infall" velocity of 247 ± 42 km s^{-1}, (2) a distance of $D = 22 \pm 2$ Mpc, and (3) the retarding force due to Virgo acting for the time $2/3H^{-1} = 15 \pm 1$ Gyr. Carry the errors with the calculation so as to estimate the certainty of the derived mass in this approximation to an exact proper integral calculation. (Note that any dark *smoothly* distributed mass beyond the Virgo complex does not cause any deceleration!)

(b) Estimate the total light of the galaxies in the Virgo Cluster Catalog as set out in the Virgo luminosity function (Sandage, Binggeli, & Tammann 1985), and derive therefrom the mass-to-light ratio of the Virgo core. This assumes that only the galaxies in the Virgo Catalog define the light corresponding to the retarding mass. If the total light is, say, four times greater than that defined by the Catalog galaxies (i.e. due to galaxies in the "envelope" of the complex surrounding the core), is the resulting M/L ratio more reasonable? Why?

Once the distance to the Virgo cluster core is known, H_0 follows by dividing the cosmic velocity of 1179 ± 17 km s^{-1} by the distance. Tammann (1992) argues from a variety of methods that the Virgo distance is $m - M = 31^{\mathrm{m}}.64 \pm 0^{\mathrm{m}}.08$ ($D = 21.3 \pm 0.8$ Mpc, giving $H_0 = 55 \pm 2$). The methods leading to the Virgo distance are discussed later in section 8.5.

(2) The Hubble Diagram for Sc I Galaxies

Sc I galaxies exhibit a well defined Hubble diagram of moderately small dispersion as shown in Fig. 8.4. The dots are the total sample of Sc I galaxies in the Revised Shapley Ames Catalog (RSA) whose catalog limit is about $m_{\mathrm{pg}} = 13^{\mathrm{m}}$. Open circles are fainter Sc I galaxies found in a search of the Palomar Sky Survey plates to the limit of the Zwicky *et al.* catalogs (m_{pg} limit $= 15^{\mathrm{m}}.7$). The redshifts were subsequently measured and listed in a catalog (Sandage & Tammann 1975). Similar diagrams made by combining these data with additional data of the same kind obtained in an independent program by Rubin *et al.* (1976) are discussed in Sandage (1988a, Fig. 5 there) where the bias properties of the sample are set out and corrected for using the method to be discussed in Lecture 10.

Knowledge of the mean absolute magnitude of Sc I galaxies combined with the equation of the unbiased ridge line in Fig. 8.4 would give the Hubble constant straightaway. And because Fig. 8.4 is well defined by galaxies at redshifts as large as $cz > 5000$ km s^{-1} (nearly all of which are in directions away from the direction of the dipole of the CMB) the redshifts (reduced here to the centroid of the Local Group) are in the kinematic frame of the CMB to better than 5% (Federspiel, Sandage, & Tammann 1994).

We presently know Cepheid distances to only one Sc I galaxy. Hence, the requirement in the future must be to determine such distances to a larger sample to properly calibrate Fig. 8.4. In the absence of such calibration we can only discuss the probability that the one calibrator (M101) is likely to be at the mean value of M_{pg}.

M101 is the nearest of the Sc I galaxies. Its distance modulus was stated to be $m - M = 24^{\mathrm{m}}$ by Hubble & Humason (1931) and also by Hubble 20 years later in his statement to Holmberg (1950). As part of the "Steps" program, Sandage & Tammann (1974b) derived a modulus of $m - M = 29^{\mathrm{m}}.3$. The large modulus near $29^{\mathrm{m}}.3$ was confirmed by Cook, Aaronson, & Illingworth

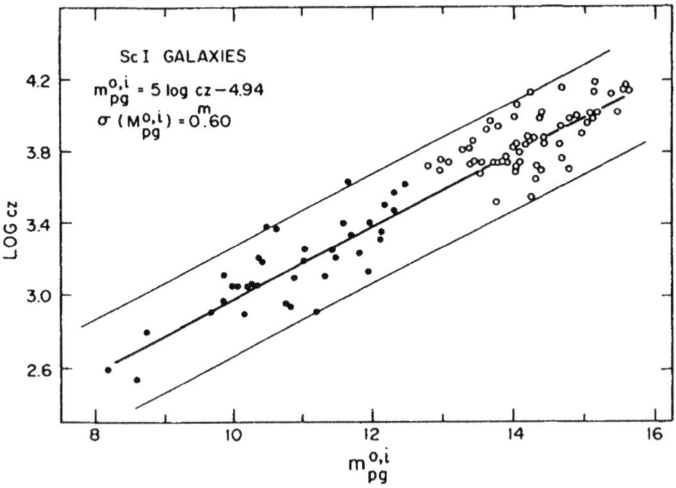

Fig. 8.4. Hubble diagram for Sc I galaxies from the RSA (dots) and from a fainter catalog. The lines have the slope of $dm/d\log z = 5$ required by a linear redshift-distance relation. Diagram from Sandage & Tammann (1975).

(1986), and may (or may not) have been confirmed at this reading by the results of an HST consortium (Freedman *et al.* 1994).

Because M101 is the nearest of the Sc I galaxies, it would be highly unlikely that it would be the largest and brightest Sc I in the local distance-limited sample. Indeed, as a sample of one, the most likely mean absolute magnitude $\langle M_B \rangle$ of a Gaussian luminosity function for Sc I galaxies is the observed absolute magnitude of M101. Using a distance modulus of $m - M = 29^{\mathrm{m}}\!.3$ and an apparent magnitude corrected for Galactic and for internal absorption of $B_{0,i} = 7^{\mathrm{m}}\!.89$ as listed in the RSA2, gives $M_B(\mathrm{M101}) = -21^{\mathrm{m}}\!.41$. The relation between the m_{pg} magnitude system in Fig. 8.4 and the B system is $B - m_{\mathrm{pg}} = 0^{\mathrm{m}}\!.1$ at the color of M101, (B magnitudes being fainter). Hence $M_{\mathrm{pg}}(\mathrm{M101}) = -21^{\mathrm{m}}\!.51$. Applied to the ridge-line equation in Fig. 8.4, the resulting best estimate for the Hubble constant is $H = 49 \pm 15$ where the error is put at $\pm 30\%$ based on $\sigma(M) = 0^{\mathrm{m}}\!.6$. If H were as large as 85, M101 would be the brightest Sc I galaxy in the distance-limited local (500 km s^{-1}) neighborhood, which is unlikely. Development of this argument is in the next section.

(3) Angular and Linear Sizes of Sb and Sc Galaxies Calibrated Using M31 and M101

In a way similar to the last method, a distance-limited sample of both M31 and M101 look-alikes, compiled from the classifications in the RSA2 and by a re-inspection of the original large-scale plates, provides the data from which a bias-free limit on H can be determined. Figure 8.5 shows the result for the M31 look-alike data.

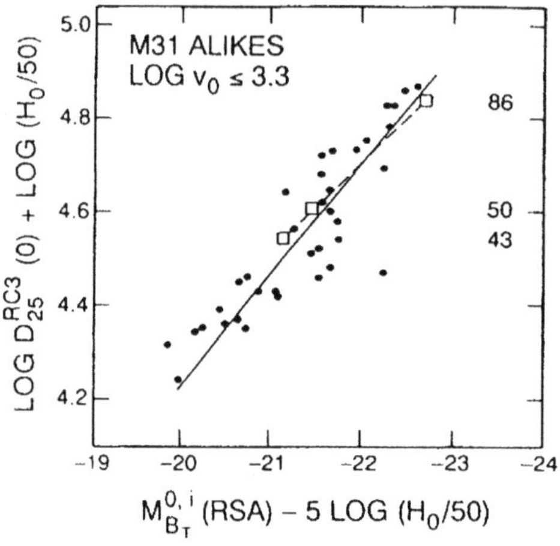

Fig. 8.5. Linear diameters (based on an arbitrary distance scale with $H = 50$) of M31 look-alikes in a distance-limited (and therefore unbiased) sample, compared with the known linear diameter of M31. The M31 calibrator will move relative to the sample galaxies for different actual values of H. Diagram from Sandage (1993b).

The conclusion (Sandage 1993a,b) is that the most probable value of the Hubble constant is $H = 43 \pm 11$ from the M101 data and $H = 45 \pm 12$ from the M31 data. A firm upper limit of $H < 85$ is derived from each sample on the precept that neither galaxy, each of which is the nearest galaxy of its type to us, is larger than the largest galaxy in the distance-limited sample. Restricting the method to a distance-limited sample removes all selection-bias problems, contrary to the discussion by van den Bergh (1992, 1993) in his criticism of the method.

Van den Bergh states that linear diameters (even of galaxies restricted to a given van den Bergh luminosity class such as Sc I) have too large a dispersion to be useful as distance indicators. Not only is this incorrect, but it contradicts his own precept (van den Bergh 1960) that galaxies in each luminosity class have a narrow enough dispersion in absolute magnitude to be useful for the problem. He initially derived (1960, his Table 4) $\sigma(M) = 0^m.3$ for Sc I galaxies, and further made a point of the distance scale capabilities of the "luminosity" classification.

Because linear diameters and absolute magnitudes are very tightly correlated (Sandage 1993a, Fig. 4 and 1993b, Fig. 1), if $\sigma(M)$ is small, then necessarily $\sigma(\log D)$ must also be small, the relation between the two being $\sigma(\log D) = 0.2\sigma(M)$. The data conform to this (Sandage 1993a,b). The observations show that $\sigma(\log D) \sim 0.10 \pm 0.02$, consistent within the errors with $\sigma(M) = 0^m.60$ from Fig. 8.4 where the Sc I sample from the RSA is increased by adding the fainter sample as indicated. [Note also in passing that a dispersion of $\sigma(M) = 0^m.6$ is smaller than for the Tully-Fisher relation as derived for field galaxies (Federspiel *et al.* 1994)].

Van den Bergh's statements concerning the linear-diameter method come from neglecting the difference between a distance-limited sample and one that is flux-limited. Because of this difference, his Fig. 4 (1992) is unportentous.

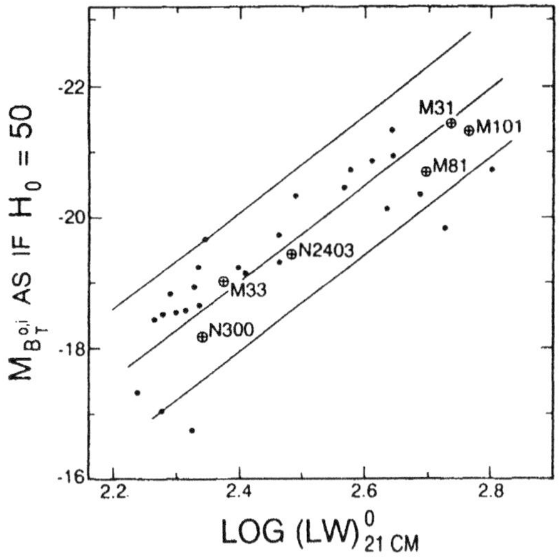

Fig. 8.6. Tully-Fisher correlation of log line-width (at the 50% level) with kinematic absolute magnitude ($H = 50$) for a distance-limited sample of field galaxies compared with the data on an absolute (Cepheid) calibration of five local calibrators. Evidence exists that the local calibrators define a distance scale that is $0^{\rm m}_{.}32$ (i.e. 16%) too short, leading finally to $H = 48 \pm 5$ km s^{-1} Mpc^{-1}. Diagram from Sandage (1994b).

NGC 309 has one of the largest redshifts in the RSA ($v_0 = 5786$ km s^{-1}) and cannot be used to compare with diameters in a local distance-limited sample. It is far out in the high-diameter wing of the diameter distribution function because of the enormous volume encompassed in its listing compared with M101 as the nearest Sc I to us.

The method of linear diameters is based on an argument due to Baade (unpublished but described by Sandage 1958). It has also been used by van der Kruit (1986) in his method for the distance to the Virgo cluster (see section 8.5).

(4) Tully-Fisher Method Applied to Unbiased Samples of Field Galaxies

The results of the Tully-Fisher (TF) method differ, depending on whether it is applied to incomplete (biased) data for groups and clusters or to distance-limited samples (Lecture 10) of field galaxies. Figure 8.6 shows the TF correlation of line width vs. relative (kinematic) absolute magnitude for galaxies in the strictly *distance-limited* 500 km s^{-1} sample (Kraan-Korteweg & Tammann 1979), shown as dots, compared with the calibrator galaxies with known Cepheid distances. The absolute magnitudes of the field galaxies are computed from the redshift (corrected for Virgo "infall") using $H = 50$. If this value of the Hubble constant is incorrect, then the field galaxies and the calibrators would disagree.

Evidence exists from the more secure 64-galaxies calibration by Richter and Huchtmeier (1984) that the 6 calibrating galaxies in Fig. 8.6 define a mean absolute magnitude that is $0^{\rm m}_{.}32$ too faint. Applying this more secure calibration to the unbiased data in Fig. 8.6 gives $H_0 = 48 \pm 5$ km s^{-1} Mpc^{-1}. The complications set out in more detail than would be interesting here are set out in Sandage (1994b), where the bias properties of non-distance-limited samples are discussed which nearly always lead to incorrectly larger values of H_0.

A case can be made that those estimates of H_0 which use incomplete TF data on "groups" and "clusters" leading to the short distance scale are also incorrect. Entrance to the extensive literature on the effects of selection bias using improper TF samples are in various archive papers with extensive references (Teerikorpi 1987; Bottinelli *et al.* 1987; Kraan-Korteweg, Cameron, and Tammann 1988; Sandage 1988a,b; 1994a,b; Federspiel, Sandage, and Tammann 1994).

(5) Luminosity Function of Planetary Nebulae

Hodge (1966) and then Ford (1978) early proposed use of PNe as distance indicators. Exploratory work began with the discovery by Ford, Jenner and Epps (1973) and by Ford and Jenner (1975) of PNe in NGC 185, NGC 205, and M31 (following Baade's much earlier initial discovery), and with Ford's (1978) discovery of the narrow range in the M(5007Å) absolute magnitude of the brightest PNe in seven Local Group galaxies. The work has continued into modern times with a series of discovery and calibration papers, as reviewed by Jacoby *et al.* (1992).

In a long series of calibration and application papers, the PNe method is said to support the short distance scale requiring $H \sim 85$. An encomium of the method is set out in Jacoby *et al.* (1992). A critique of the problems is given by Tammann (1992). Review of the physics and its effect on the use of PNe as distance indicators is made by Mendez (1993), showing that the PNe result of a high H cannot be dismissed lightly. Reconciliation of the PNe result with the many methods that require the long distance scale has not yet been achieved, but Bottinelli *et al.* (1991) and Tammann (1993) show the normalization effect of a non-sharp luminosity function which would solve the problem.

(6) Surface Brightness Fluctuations

Application of a novel method introduced by Tonry and Schneider (1988), based on a variant of the ingenious count-brightness technique of Baum and Schwarzschild (1955) appears to be the only method which now favors the short distance scale. Summaries of the method and its critique are in Tonry *et al.* (1988), Jacoby *et al.* (1992) and Tammann (1992). Reconciliation of the SB result favoring the short distance scale with the data favoring the long scale has not yet been achieved.

(7) Supernovae of Type Ia and of Type II

Beginning with Kowal's (1968) discovery that supernovae of type Ia have a tight Hubble $[m, z]$ diagram, a large literature on such Hubble diagrams has grown (reviews by Branch 1988; Tammann 1988; Branch and Tammann 1992; Sandage and Tammann 1993). If the intrinsic dispersion is as small as many (but not all) of these studies indicate, the Hubble constant follows from the ridge-line of the $[m, z]$ diagram once the absolute magnitude at maximum of SNe Ia has been calibrated. Lecture 9 is concerned with this problem where the extant calibration, circa 1994, from three supernovae is shown to be $\langle M_B(\max)\rangle = -19\overset{m}{.}55 \pm 0\overset{m}{.}08$, giving $H_0 = 52 \pm 8$ km s^{-1}Mpc^{-1}. The calibration in V from two SNe at $M_V(\max) = -19\overset{m}{.}58 \pm 0\overset{m}{.}09$ gives $H_0 = 55 \pm 8$. These observational calibrations relative to Cepheids are also consistent with theoretical expectations based on the putative physics of the explosions.

An early application of the "expanding atmosphere" method (a variation on the Baade-Wesselink method) to supernovae of type II gave $H_0 = 58 \pm 10$ (Schmidt *et al.* 1992). However, the method was later revised by these authors to permit them to favor the short distance scale of $H \sim 85$ (Schmidt *et al.* 1993), but their revision has been criticized on fundamental grounds (Branch 1994), and on the practical grounds that their revised method applied to the type II SN in M81 (1993J) fails to give the known distance of M81. At this writing, following Branch, we must discount the method using SNe II until these problems are understood.

8.4 Cepheid Distances to the Local Calibrators

Tammann (1987, 1992), in reviewing the stability over the past two decades of the distance determinations of galaxies in the Local Group and just beyond, concludes that "in spite of considerable progress in Cepheid research — the resulting distances have changed very little between 1974 and 1991". His table showing this is reproduced as Table 8.2 here, comparing distances adopted by Sandage and Tammann (1974a) at the beginning of the distance-scale series "Steps Toward the Hubble Constant" with modern distances compiled in a review by Madore and Freedman (1991, MF). Details of these various modern determinations are in the MF discussion and are similarly reviewed by Jacoby *et al.* (1992, their Table 1).

Table 8.2. Cepheid distance moduli of calibrating galaxies adopted by Sandage and Tammann (1974a) and Madore and Freedman (1991)

Galaxy	ST(1974) $(m-M)$	MF(1991) $(m-M)$
LMC	$18\overset{m}{.}59$	$18\overset{m}{.}50$
SMC	$19\overset{m}{.}27$	$18\overset{m}{.}87$
NGC 6822	$23\overset{m}{.}95$	$23\overset{m}{.}59$
IC 1613	$24\overset{m}{.}43$	$24\overset{m}{.}42$
M31	$24\overset{m}{.}12$	$24\overset{m}{.}63$
M33	$24\overset{m}{.}56$	$24\overset{m}{.}63$
NGC 2403	$27\overset{m}{.}56$	$27\overset{m}{.}51$
M81	$27\overset{m}{.}56$	$27\overset{m}{.}59$
M101	$29\overset{m}{.}3$	$29\overset{m}{.}38$

Table 8.2 is a highly condensed summary of much activity in the 17 years separating the studies. It is also devoid of some of the false conclusions which this writer made after the 1974 summary such as incorrectly revising the M81 modulus to $28^{\mathrm{m}}_{.}8$ (Sandage 1984) based on the difficulty to find Cepheids in M81 compared with NGC 2403.

Nevertheless, the 1974 scale upon which our "Steps" results depended is the same to within several hundredths in the moduli in the mean as the modern 1991 scale. Hence, the claim by Aaronson (1986) that "*all* (his italics) current estimates of the Hubble constant are plagued by the large uncertainties to the distances of nearby calibrating galaxies" is not correct.

Table 8.2 also corrects van den Bergh's (1992) statement and his Figs. 7 and 8 that our long scale becomes non-linear relative to modern accepted distances for $(m - M)_{\mathrm{Sandage}} > 26^{\mathrm{m}}_{.}2$. He neither plots nor discusses our value of $(m - M) = 27^{\mathrm{m}}_{.}6$ for NGC 2403 (Sandage and Tammann 1968b) nor $(m - M) = 29^{\mathrm{m}}_{.}3$ for M101 (Sandage and Tammann 1974b) which Table 8.2 shows to be the same as the 1991 modern scale.

In a most interesting commentary, Freedman and Madore (1993), although noting the agreement between the columns in Table 8.2 and also the comment by Tammann on the stability of the local calibration, are nevertheless scathing in their appraisal of the 1974 scale, and further on the validity of our 1968 calibration of the Cepheid P-L relation (Sandage and Tammann 1968a) upon which we based that scale. They set out their complaint even as their recalibration (Madore and Freedman 1991) of the P-L relation confirms the 1968 zero-points in both B and V to considerably better than $0^{\mathrm{m}}_{.}1$. They apparently confuse the galactic cluster data (leading to the 1968 Cepheid zero-point calibration) with the Cepheid campaign in the several galaxies in Table 8.2 (M33, M81) where I had obtained incorrect larger distances, maintained for only a finite time after the 1974 base. None of this affects the 1974 "Steps" distance scale, which is our present scale as well.

8.5 Six Methods to Obtain the Distance of the Virgo Cluster

This section follows the more detailed discussion of ways to the Virgo cluster distance given by Tammann (1987, 1988, 1992) in his series of reviews. Before introducing synopsis of the six methods, the results as set out by Tammann (1992) are shown in Table 8.3.

It should be remembered that advocates of the short scale require $(m - M)_{\mathrm{Virgo}} = 30^{\mathrm{m}}_{.}9$, considerably outside the range of this table.

(1) Globular Clusters

The method uses the luminosity function of globular clusters in E galaxies associated with the Virgo cluster "core" compared with the calibrated function for clusters in M31 and the Milky Way. The method and the data have been reviewed by Harris (1988, 1991) and applied by Harris *et al.* (1991) to obtain $(m - M)_{\mathrm{Virgo}} = 31^{\mathrm{m}}_{.}7$. Later changes in the precepts of how to apply the

Table 8.3. The distance of the Virgo cluster

Method	$(m-M)$	Galaxy Type	Calibrators
Globular clusters	$31^{m}_{.}62 \pm 0^{m}_{.}17$	E	RR Lyrae
Novae	$31^{m}_{.}57 \pm 0^{m}_{.}43$	S	M31, galactic novae
Supernovae	$31^{m}_{.}63 \pm 0^{m}_{.}25$	E,S	Cepheids, model
$D_n - \sigma$	$31^{m}_{.}85 \pm 0^{m}_{.}19$	E	Galaxy, M31, M81
21 cm line widths	$31^{m}_{.}60 \pm 0^{m}_{.}15$	S	13 nearby calib.
Size of the Galaxy	$31^{m}_{.}50 \pm 0^{m}_{.}20$	S	MW scale length
mean	$31^{m}_{.}64 \pm 0^{m}_{.}08$		$(D = 21.3 \pm 0.8$ Mpc)

method, forcing the answer to conform with $(m-M) = 30^{m}_{.}95$ from the PNe and SB fluctuation methods, caused Secker and Harris (1993) to compromise the basis of the method. A rediscussion using the new calibration of the RR Lyrae absolute magnitudes (Lecture 7), and using non-circular precepts for the basis of the globular cluster method recovers $(m-M) = 31^{m}_{.}7$ (Sandage and Tammann 1994).

(2) Normal Novae

In a tour de force, Pritchet and van den Bergh (1987) discovered nine normal novae in NGC 4472, the brightest E galaxy in the Virgo cluster (in subcluster B), for six of which they could determine the post-maximum decline rate. Using the relation between decline rate and absolute magnitude (Arp 1956; Rosino 1964, 1973) they obtained a modulus difference between M31 and NGC 4472 of $6^{m}_{.}8 \pm 0^{m}_{.}4$. This combined with a small correction to the M31 modulus as argued by Sandage and Tammann (1988) gives $(m-M)_{\mathrm{NGC\ 4472}} = 31^{m}_{.}57 \pm 0^{m}_{.}43$.

(3) Supernovae

From the discussion in the next lecture, following the detailed discussion by Tammann (1988, 1992) and by Branch in Jacoby *al.* (1992), the supernovae observed in Virgo cluster E galaxies give $(m-M) = 31^{m}_{.}54 \pm 0^{m}_{.}22$ [using a calibration of $M_B(\max) = -19^{m}_{.}6$]. Combining with data on two Virgo cluster SNe II from which Schmidt *et al.* (1992) determined $(m-M)_{\mathrm{Virgo}} = 31^{m}_{.}71 \pm 0^{m}_{.}26$ gives an average of $(m-M)_{\mathrm{Virgo}} = 31^{m}_{.}63 \pm 0^{m}_{.}25$ shown in Table 8.3.

(4) $D_n - \sigma$ relation

Combining the known relation between surface brightness and absolute magnitude for E galaxies (Lecture 6) with the Minkowski (1962) relation (later known as the Faber-Jackson relation) between absolute magnitude and central velocity dispersion gave a relation between surface brightness, velocity dispersion, and absolute magnitude as an expression of "the fundamental plane of E galaxies" by the seven authors discussed in Lecture 6 (Dressler *et al.* 1987). The relation was extended to the bulges of S0 and spiral galaxies

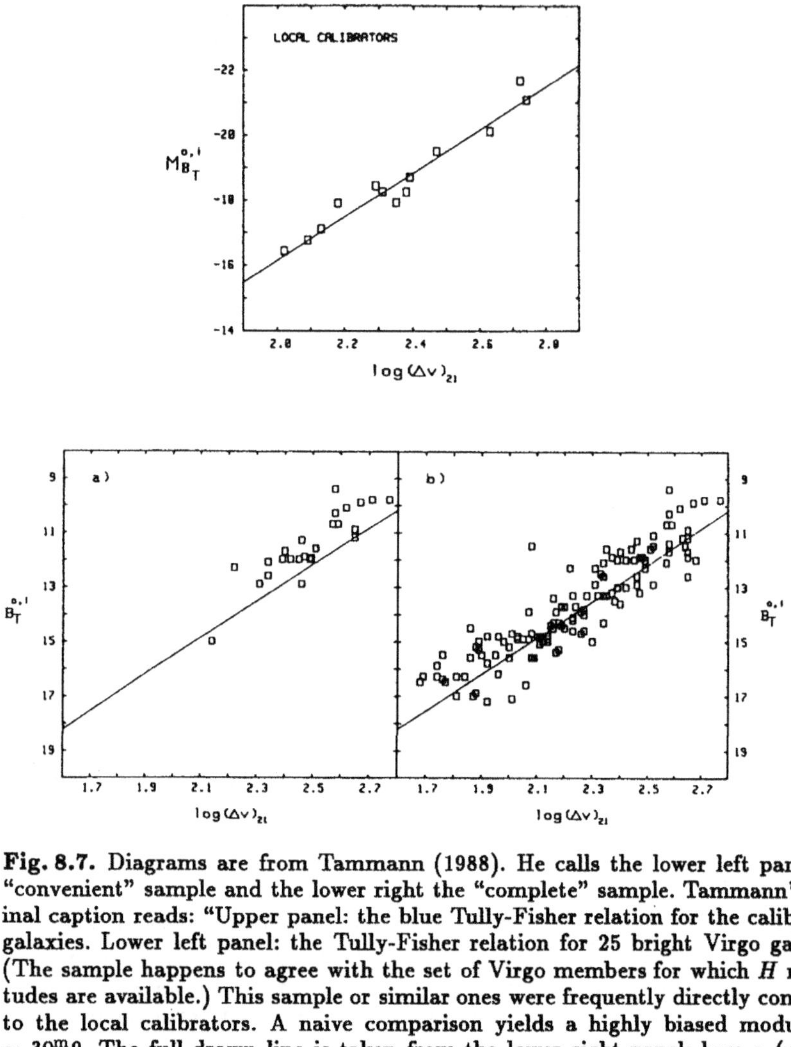

Fig. 8.7. Diagrams are from Tammann (1988). He calls the lower left panel the "convenient" sample and the lower right the "complete" sample. Tammann's original caption reads: "Upper panel: the blue Tully-Fisher relation for the calibrating galaxies. Lower left panel: the Tully-Fisher relation for 25 bright Virgo galaxies. (The sample happens to agree with the set of Virgo members for which H magnitudes are available.) This sample or similar ones were frequently directly compared to the local calibrators. A naive comparison yields a highly biased modulus of $\sim 30^{\rm m}.9$. The full drawn line is taken from the lower right panel: here a (nearly) *complete* sample of 87 Virgo Sa–Sm spirals with inclinations $> 45°$ is shown. The full drawn line is the best fit to the subset of 73 Sb–Sm galaxies and requires – if compared to the local calibrators – a modulus of $31^{\rm m}.60$."

in the Virgo cluster by Dressler (1987). Applying an absolute calibration by Tammann using the bulges of the Galaxy (see Terndrup 1988), M31 with $(m-M)_0 = 24^{\rm m}.26$, and M81 with $(m-M)_0 = 27^{\rm m}.65$, Tammann (1988) recalibrated the Dressler Virgo relation to obtain $(m-M)_{\rm Virgo} = 31^{\rm m}.85 \pm 0^{\rm m}.19$.

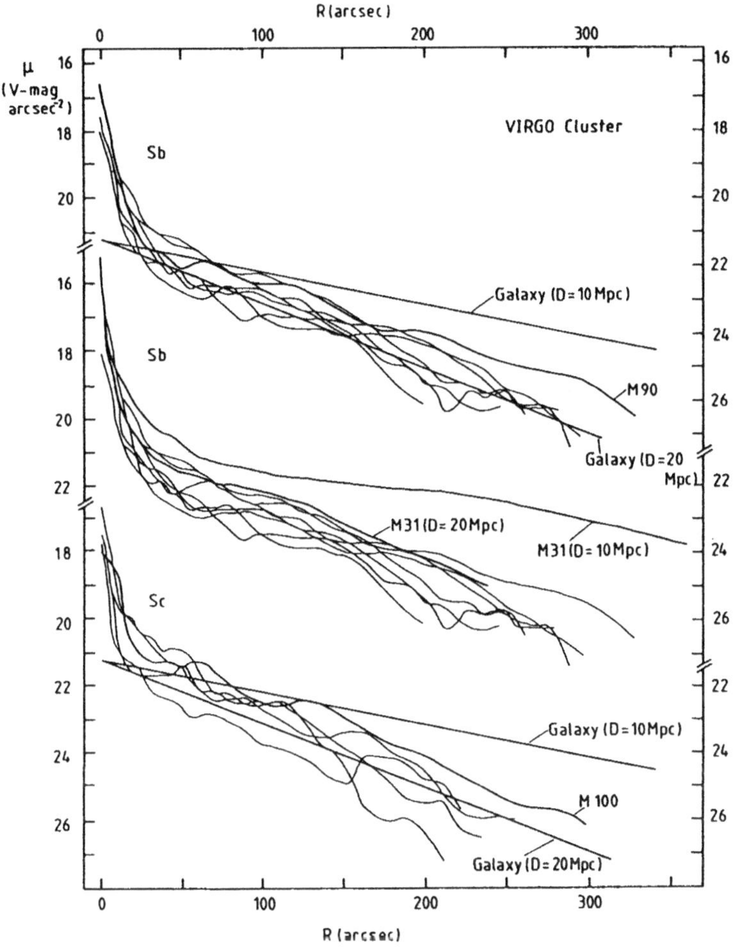

Fig. 8.8. Van der Kruit's (1986) diagram of scale lengths of Virgo Sb and Sc spirals placed at various distances compared with the absolute scale lengths of the Galaxy and M31.

(5) 21 cm line widths

In a most important study, important not only for its answer for the Virgo modulus but also because it demonstrates the effect of selection bias, Kraan-Korteweg, Cameron, and Tammann (1988) show the effect on values of $(m - M)$ using progressively more complete samples. They demonstrate that as the sample becomes more complete (progressively fainter apparent magnitude grasp) the derived modulus progressively becomes larger. The effect was also isolated by Teerikorpi (1987) and by Bottinelli *et al.* (1987) who named the problem "the cluster population incompleteness bias".

The effect is shown in Fig. 8.7 taken from Tammann (1988) where the caption, as he wrote it, is also shown as the explanation. The top panel shows the adopted calibration of the TF relation in M_B. The left panel is what Tammann calls the "convenient" Virgo sample. The right panel shows a virtually complete Virgo sample.

The complete sample, calibrated as shown, gives a Virgo modulus of $m - M = 31\overset{m}{.}60 \pm 0\overset{m}{.}15$, listed in the table.

(6) Size of the Milky Way and M31 Relative to Virgo Spirals

With arguments similar to those used in section 8.3.3 for Sc I galaxies, van der Kruit (1986) compares the scale lengths of Virgo cluster Sb and Sc galaxies with the absolute scale lengths of the exponential disks of the Milky Way and of M31 with the result shown in Fig. 8.8. From this comparison he concludes that the Virgo cluster must be at least at a distance of 20 Mpc, otherwise the Milky Way and M31 would be the largest galaxies of their class compared with Virgo spirals.

The average of the six methods in Table 8.3 is $m - M = 31\overset{m}{.}64 \pm 0\overset{m}{.}08$ or $D = 21.3 \pm 0.8$ Mpc. By the method of section 8.3.1 the resulting value of the Hubble constant, freed from all local velocity anomalies is

$$H_0 = 55 \pm 2 \text{ km s}^{-1}\text{Mpc}^{-1}$$

from the Virgo cluster method alone.

8.6 The Value of H_0

Table 8.4 summarizes the various results from the previous sections using each of the methods discussed there with the exception of that using PNe and the SB fluctuations. These are the only two that require $H_0 \sim 85$. These are too far outside the range to be included here.

Table 8.4. Summary of the various methods to H_0

Method	Section	H_0 [km s^{-1} Mpc^{-1}]
Virgo Distance	8.3.1; 8.5	55 ± 2
Sc I Hubble Diag.	8.3.2	49 ± 15
M101 Diameters	8.8.3	43 ± 11
M31 Diameters	8.8.3	45 ± 12
Tully-Fisher	8.3.4	48 ± 5
Supernovae (B)	8.3.7	52 ± 8
Supernovae (V)	8.3.7	55 ± 8
Unweighted mean		50 ± 2
Weighted mean		53 ± 2

8.7 The Timing Test

Adopting $H_0 = 50 \pm 5$ as the most reliable mean of these data, and using ± 5 km $s^{-1}Mpc^{-1}$ as a semi-realistic external error, gives the inverse as

$$H_0^{-1} = 19.5\ (+2.2, -1.8)\ \mathrm{Gyr}.$$

Comparing with the estimated age of the Universe at 15 ± 1 Gyr from Lecture 7 shows the possibility of sufficient deceleration so that $T_U = 2/3H_0^{-1}$ for $\Omega = 1$, but the errors are high. However, the conclusion is that for the first time the age of the globular clusters (and by inference the universe) and the inverse Hubble constant stand in the right relation with one another for the standard model to work at all (the inverse Hubble constant must be larger than the age of the universe, permitting some deceleration). From these data there is no need to introduce the cosmological constant to save the model.

References

Aaronson, M. 1986, in *Observational Cosmology*, IAU Symp. 124, ed. A. Hewitt, G. Burbidge, & L.Z. Fang (Dordrecht: Reidel), p. 187

Arp, H.C. 1956, AJ **61**, 15

Baade, W. 1944, ApJ **100**, 137

Baum, W.A., & Schwarzschild, M. 1955, AJ **60**, 247

Bottinelli, L., Fouqué, P., Gouguenheim, L., Paturel, G., & Teerikorpi, P. 1987, A&A **181**, 1

Bottinelli, L., Gouguenheim, L., Paturel, G., & Teerikorpi, P. 1991, A&A **252**, 550

Branch, D. 1988, in *The Extragalactic Distance Scale*, ed. S. van den Bergh, & C.J. Pritchet, ASP Conf. Ser. 4, p. 146

Branch, D. 1994, private communication

Branch, D., & Tammann, G.A. 1992, Ann. Rev. A&A **30**, 359

Chincarini, G., & Rood, H.J. 1972, AJ **77**, 448

Chincarini, G., & Rood, H.J. 1975, Nature **257**, 294

Chincarini, G., & Rood, H.J. 1976, ApJ **206**, 30

Chincarini, G., & Rood, H.J. 1979, ApJ **230**, 648

Cook, K.H., Aaronson, M., & Illingworth, G. 1986, ApJ **301**, L45

Dressler, A. 1987, ApJ **317**, 1

Dressler, A., Lyden-Bell, D., Burstein, D., Davies, R.L., Faber, S.M., Terlevich, R.J., & Wegner, G. 1987, ApJ **313**, 42

Dreyfus, H.L. 1994, in *The Key Reporter* (Phi Beta Kappa Journal), **59**, Winter 1993/94, p. 4

Eddington, A.S. 1923, *Mathematical Theory of Relativity* (Cambridge: Cambridge Univ. Press), p. 162

Federspiel, M., Sandage, A., & Tammann, G.A. 1994, ApJ **430**, 29

Ford, H. 1978, in IAU Symp. 76, *Planetary Nebulae*, ed. Y. Terzian (Dordrecht: Reidel), p. 19

Ford, H.C., & Jenner, D.C. 1975, ApJ **202**, 365

Ford, H.C., Jenner, D.C., & Epps, H.W. 1973, ApJ **183**, L73

Freedman, W.L. *et al.* 1994 (HST consortium on M101)

Freedman, W.L., & Madore, B.F. 1993, in *New Perspectives on Stellar Pulsation and Pulsating Variable Stars*, eds. J.M. Nemec and J.M. Matthews (Cambridge: Cambridge University Press), p. 61

Gregory, S.A., & Thompson, L.A. 1978, ApJ **222**, 784

Harris, W.E. 1988, in *The Extragalactic Distance Scale*, ASP Conf. Ser. No. 4, eds. S. van den Bergh & C.J. Pritchet, p. 231

Harris, W.E. 1991, Ann. Rev. A&A, **29**, 543

Harris, W.E., Allwright, J.W.B., Pritchet, C.J., & van den Bergh, S. 1991, ApJ Suppl. **76**, 115

Hodge, P.W. 1966, *Galaxies and Cosmology* (New York: McGraw Hill)

Holmberg, E. 1950, Lund Medd. Ser. 2, No. 128

Hubble, E. 1926, ApJ **64**, 321

Hubble, E. 1929, Proc. Nat. Acad. Sci. 15, 168

Hubble, E. 1936a, *Realm of the Nebulae* (New Haven: Yale Univ. Press)

Hubble, E. 1936b, ApJ **84**, 158

Hubble, E. 1936c, ApJ **84**, 270

Hubble, E. 1936d, ApJ **84**, 517

Hubble, E. 1938, *Observational Approach to Cosmology* (Oxford: Oxford Univ. Press)

Hubble, E. 1939, J. Franklin Inst. **228**, 131

Hubble, E. 1951 Proc. Amer. Phil. Soc. **95**, 461

Hubble, E. & Humason, M.L. 1931, ApJ **74**, 43

Hubble, E. & Sandage, A. 1953, ApJ **118**, 353

Humason, M.L., Mayall, N.U., & Sandage, A. 1956, AJ **61**, 97

Jacoby, G. *et al.* 1992, PASP **104**, 599

Jerjen, H. & Tammann, G.A. 1993, A&A **276**, 1

Kolata, G. 1982, Science **217**, 1237

Kowal, C.T. 1968, AJ **73**, 1021

Kraan-Korteweg, R.C., & Tammann, G.A. 1979, Astron. Nach. **300**, 181

Kraan-Korteweg, R.C., Cameron, L.M., & Tammann, G.A. 1988, ApJ **331**, 620

Lemaître, G. 1927, Ann. Soc. Sci. Bruxelles, **47**, 49

Lemaître, G. 1931, MNRAS **91**, 483

Madore, B.F. & Freedman, W.L. 1991, PASP **103**, 933

Mendez, R.H. 1993, Rev. Mex. A&A **26**, 35

Minkowski, R. 1962, in *Problems in Extragalactic Research*, IAU Symp. 15, ed. G.C. McVittie (Macmillan Co.), p. 112

Minsky, M. 1967, *Computation: Finite and Infinite Machines* (Englewood, N.J.: Prentice Hall)

Paturel, G. 1983, in *Highlights of Astronomy*, ed. R.M. West, IAU 18th Assemb. (Dordrecht: Reidel), p. 289

Pritchet, C.J. and van den Bergh, S. 1987, ApJ **318**, 507

Richter, O.-G. & Huchtmeier, W.K. 1984, A&A **132**, 253

Robertson, H.P. 1928, Phil. Mag. **5**, 835

Rosino, L. 1964, Ann. d'Ap. **27**, 498

Rosino, L. 1973, A&AS **9**, 347

Rubin, V.C., Ford, W.K., Thonnard, N., Roberts, M.S., & Graham, J.A. 1976, AJ **81**, 687

Sandage, A. 1958, ApJ **127**, 513

Sandage, A. 1961, in *Problems in Extragalactic Research*, IAU Sypm. 15, ed. G.C. McVittie, Macmillian Co. (New York), p. 359

Sandage, A. 1984, AJ **89**, 621

Sandage, A. 1988a, ApJ **331**, 583

Sandage, A. 1988b, ApJ **331**, 605

Sandage, A. 1993a, ApJ **402**, 3

Sandage, A. 1993b, ApJ **404**, 419

Sandage, A. 1994a, ApJ **430**, 1

Sandage, A. 1994b, ApJ **430**, 13

Sandage, A., Binggeli, B., and Tammann, G.A. 1985, AJ **90**, 1759

Sandage, A. & Tammann, G.A. 1968a, ApJ **151**, 531

Sandage, A. & Tammann, G.A. 1968b, ApJ **151**, 825

Sandage, A. & Tammann, G.A. 1974a, ApJ **190**, 525

Sandage, A. & Tammann, G.A. 1974b, ApJ **194**, 223

Sandage, A. & Tammann, G.A. 1975, ApJ **197**, 265

Sandage, A. & Tammann, G.A. 1982, in *Astrophysical Cosmology*, eds. H.A. Brück, G.V. Coyne, & M.S. Longair (Pont. Acad. Sci.: Vatican City), p. 23

Sandage, A. & Tammann, G.A. 1984, in *Large Scale Structure of the Universe, Cosmology, & Fundamental Physics*, First ESO-Cern Symp., ed. G. Setti, & L. van Hove (Geneva: Cern), p. 127

Sandage, A. & Tammann, G.A. 1986, in *Inner Space – Outer Space*, eds. E.W. Kolb *et al.* (Chicago: Univ. Chicago Press), p. 3

Sandage, A. & Tammann, G.A. 1988, ApJ **328**, 1

Sandage, A. & Tammann, G.A. 1990, ApJ **365**, 1

Sandage, A. & Tammann, G.A. 1993, ApJ **415**, 1

Sandage, A. & Tammann, G.A. 1994, ApJ submitted (Virgo globular clusters)

Schmidt, B.P., Kirshner, R.P., and Eastman, R.G. 1992, ApJ **395**, 366

Schmidt, B.P., Kirshner, R.P., and Eastman, R.G. 1993, in *Observational Cosmology*, eds. G. Chincarini, A. Iovino, T. Maccacaro, & D. Maccagni: ASP Conf. Ser. No. 51, p. 30

Schramm, D.N. 1983, in *Highlights of Astronomy*, ed. R.M. West, IAU 18th Assemb. (Dordrecht: Reidel), p. 241

Secker, J. & Harris, W.E. 1993, AJ **105**, 1358

Stebbins, J. Whitford, A.E., & Johnson, H.L. 1950, ApJ **112**, 469.

Strömberg, G. 1925, ApJ **61**, 353

Tammann, G.A. 1986, in *Observational Cosmology*, IAU Symp. 124, ed. A. Hewitt, G. Burbidge, & L.Z. Fang (Dordrecht: Reidel), p. 151

Tammann, G.A. 1987, in *Relativistic Astrophysics*, ed. M.P. Ulmer, 13th Texas Symp., World Scientific (Singapore), p. 8

Tammann, G.A. 1988, in *The Extragalactic Distance Scale*, ed. S. van den Bergh, & C.J. Pritchet, ASP Conf. Ser. No. 4, p. 282

Tammann, G.A. 1992, Physica Scripta T43, 31

Tammann, G.A. 1993, in *Planetary Nebulae*, IAU Symp. 155, eds. R. Weinberger & A. Acker (Dordrecht: Kluwer), p. 515

Tammann, G.A. & Sandage , A. 1985, ApJ **294**, 230

Tarenghi, M., Tifft, W.G., Chincarini, G., Rood, H.J., & Thompson, L.A. 1979, ApJ **234**, 793

Tarenghi, M., Chincarini, G., Rood, H.J., & Thompson, L.A. 1980, ApJ **235**, 724

Teerikorpi, P. 1987, A&A **173**, 39

Terndrup, D.M. 1988, in *The Extragalactic Distance Scale*, ASP Conf. Ser. 4, eds. S. van den Bergh & C.J. Pritchet, p. 211

Tonry, J.L., Luppino, G.A., & Schneider, D.P. 1988, in *The Extragalactic Distance Scale*, eds. S. van den Bergh, & C.J. Pritchet: ASP Conf. Ser. No. 4, p. 213

Tonry, J.L., & Schneider, D.P. 1988, AJ **96**, 807

van den Bergh, S. 1960, ApJ **131**, 215

van den Bergh, S. 1992, PASP **104**, 861

van den Bergh, S. 1993, Rev. Mex. A&A **26**, 73

van der Kruit, P. 1986, A&A **157**, 230

9 The Hubble Constant from Type Ia Supernovae

9.1 Introduction

Zwicky (1961) hoped that supernovae of type I would have the smallest dispersion in absolute magnitude of known distance indicators and that therefore "if true, — will enable us to establish a relative, as well as an absolute, cosmic distance scale which is more reliable than any of the scales currently in use". His supposition changed to a possibility when Kowal (1968) assembled a most remarkable Hubble diagram for SNe Ia that had "adequate" photometry to that date. Kowal's seminal Hubble diagram is reproduced as Fig. 9.1.

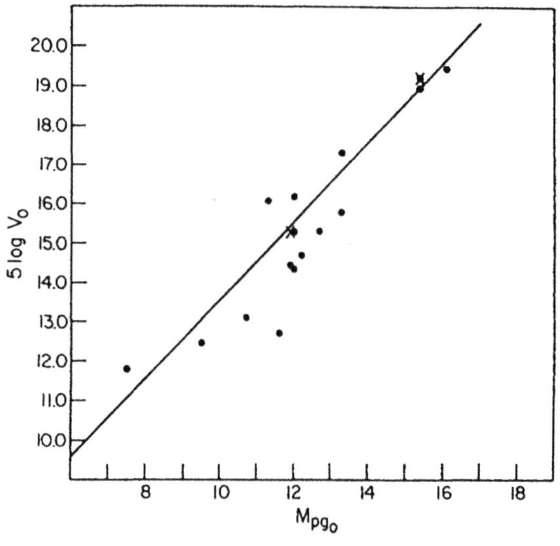

Fig. 9.1. Kowal's (1968) first Hubble diagram of SNe of type I. It was this diagram, especially with its two mean points (as crosses) for the many individual SNe in the Virgo and the Coma clusters, that constituted the first proof that such supernovae may have a small intrinsic spread in their absolute magnitude at maximum.

As the photometry improved over the years, not only had the knowledge stabilized, showing that the shape of the SNe Ia light curve represented a template, but also a technical improvement occurred concerning the values of the magnitudes reached at maximum light. Such data were compiled into photometric catalogs, and examples of light curves were given as templates. The most important of the catalogs and templates are (1) the Asiago Catalog (Barbon, Capellaro, & Turatto 1989), and (2) the Atlas of SNe I light curves (Leibundgut *et al.* 1991, hereafter called the Basel Atlas). An entrance to the extensive literature is available from the review by Branch and Tammann (1992).

It was early recognized by Pskovskii (1967) and by Kowal (1968) that the light curves of SNe Ia are phenomenally similar. A modern demonstration is shown in Fig. 9.2 from Cadonau's (1987) thesis (the diagram is more readily available in Cadonau *et al.* 1985) using the best photometry of individual light curves available to 1991, as set out in the Basel Atlas.

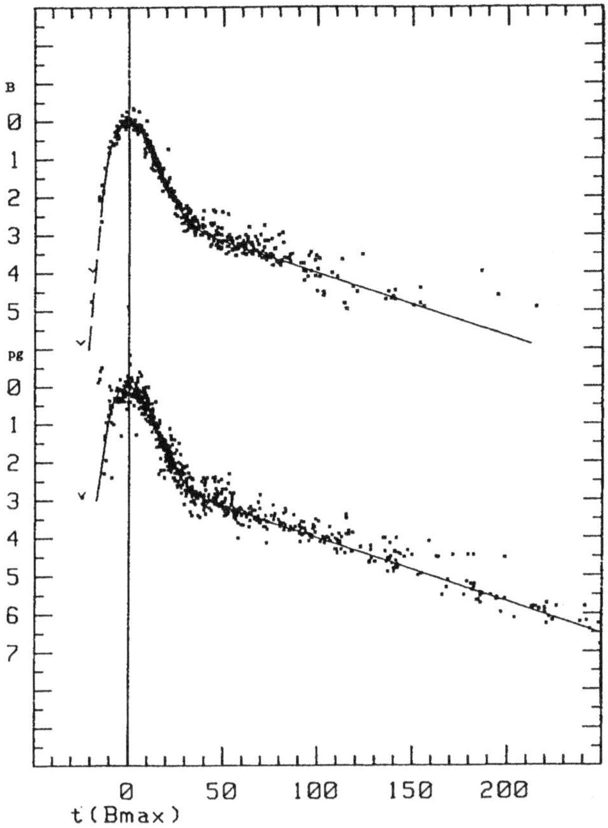

Fig. 9.2. (top) Composite light curve in the B photometric band made by superposing the individual light curves of 22 type Ia SNe. (bottom) Same as top but using 16 light curves in the m_{pg} photometric system. There is no evidence from these data for a gross difference in the decay rate after maximum light among the individual light curves. Diagram from Cadonau (1987) and Cadonau *et al.* (1985). The template is the same as is used in the Basel Atlas.

The top panel in Fig. 9.2 is a superposition of 22 light curves of SNe Ia that have B-band data. The lower curve is a superposition of 16 light curves where the data are on the international m_{pg} system, which was the near-B system used until the mid-1950's, generally determined from blue sensitive photographic plates. Because of technical advances since the 1950's, it is generally true that the photographic m_{pg} data are less accurate than B-band data, now almost universally obtained with photoelectric devices. Figure 9.3 shows similar data but for those supernovae where the observations were continued to longer times after maximum than in Fig. 9.2.

The nature of the decay as exponential is beyond doubt. However, on the basis of the data available even as late as the 1960's, Zwicky (1961) argued

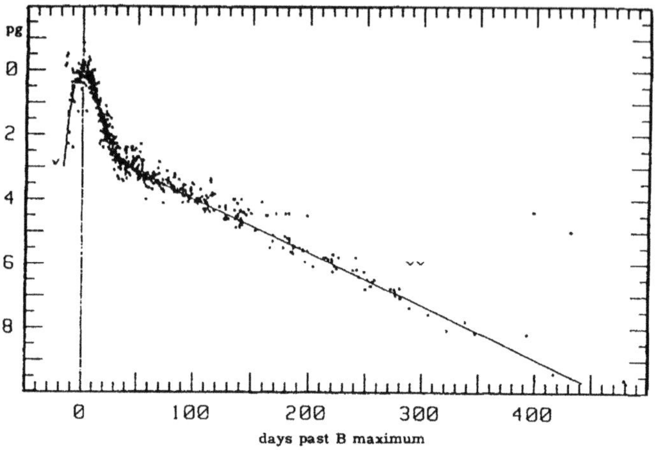

Fig. 9.3. Same as Fig. 9.2 but extended to 400 days past maximum, showing well the exponential decay due to the radioactivity of Ni^{56}.

against the exponential decay, shown as the straight line in the log flux-linear time representation in Fig. 9.3. This change of view since Zwicky's time shows the vast improvement in the data in the 40-year interval, and also the decay of prejudice with time. (As in all debates, as one set of critics leaves the field, another set appears to take their place. The critics will always be with you.)

9.2 The Hubble Diagram Circa 1994

9.2.1 Normal and Non-normal SNe I

Not all type I supernovae are identical in their minute spectral details or in all aspects of their light curves. Again the review by Branch and Tammann (1992) provides an entrance to the literature.

Two points of view exist in the current (circa 1994) literature concerning the differences that occur in a few SNe Ia compared with the prototypes of the class.

(A) Van den Bergh and Pazder (1992, vdBP) and van den Bergh (1993) claim the differences destroy SNe Ia as distance indicators. When the SNe that show differences from the prototype are put into a Hubble diagram in which "absorption" corrections are also applied to all the plotted SNe, the resulting Hubble diagram shows such a large scatter as to drive vdBP to their denial.

As a counter, Sandage and Tammann (1993) argue that such absorption corrections are spurious because vdBP (1) use too blue a $B-V$ color at maximum, and (2) adopt an unsupported premise that all SNe Ia have identical intrinsic colors at maximum. VdBP argue that their Hubble diagram, so corrected on the basis of identical color, prove non-similar absolute magnitude.

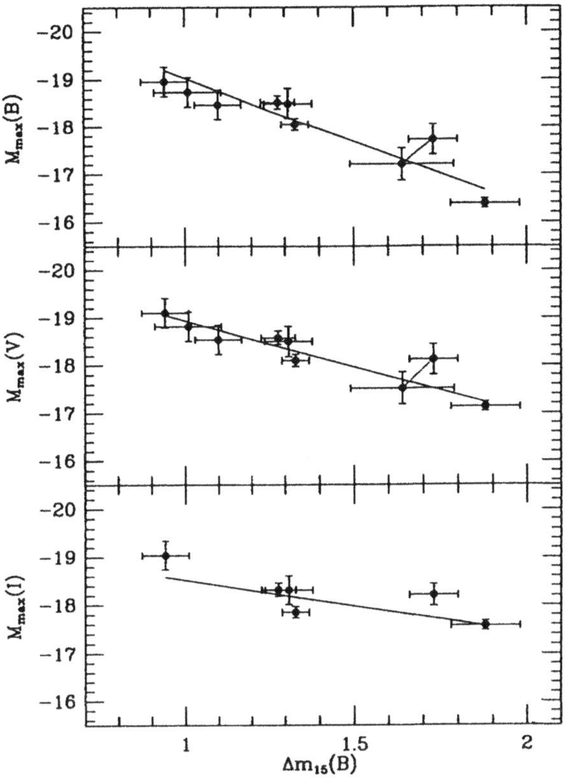

Fig. 9.4. Proposed relation by Phillips (1993) between the decay rate (days from maximum to decay by $1\overset{m}{.}5$) and the (relative) absolute magnitude at maximum. The absolute magnitudes are based on the short distance scale, obtained by PNe, SB fluctuations, and Tully-Fisher correlations uncorrected for bias (Lectures 8 and 10). Phillips has included three SNe Ia known to be peculiar in the sense of Branch *et al.* (1993). Exclusion of these give an intrinsic dispersion of $\sigma(M)_{max} \sim 0\overset{m}{.}2$ for the remaining five, consistent with Fig. 9.6. Use of the SNe Ia calibration with Cepheids gives absolute magnitudes $\sim 1\overset{m}{.}4$ brighter.

It is also to be noted that the SNe Hubble diagrams that are not corrected for "absorption" are tighter than that of vdBP, the only difference in procedure being the "absorption correction", making such corrections suspect for this reason alone. The case for no absorption corrections is set out by Branch and Tammann (1992), following Tammann (1982), Miller and Branch (1990), and others.

(B) Phillips (1993) takes a less extreme position, but is nevertheless anxious to include SNe Ia that are not normal (in the sense of Branch, Fisher, and Nugent 1993) despite the fact that such abnormal cases as SNe 1986G, 1991bg, and 1991T can be detected from spectra or photometry, and therefore can be excluded *a priori*, leaving only "normal" SNe to be discussed and calibrated.

As this most important point is to be addressed later, we set out Phillips' (1993) argument in more detail. Figure 9.4 from Phillips shows a putative correlation of relative absolute magnitude (on a scale with $H \sim 80$ km s^{-1} Mpc^{-1}) with the rate of decay, in magnitudes, in the first 15 days from maximum.

The three points in Fig. 9.4 with the fastest decays at the extreme right are the two peculiar supernovae, SN 1986G (plotted twice), and SN 1991bg. The data point farthest to the left is for SN 1991T. This light curve has the slowest decay. The SN is classified as peculiar.

What is the definition of "normal" and "peculiar"? How serious for Kowal's Hubble diagram, and for its modern representation is the existence of "non-normal" (whatever their definition) supernovae?

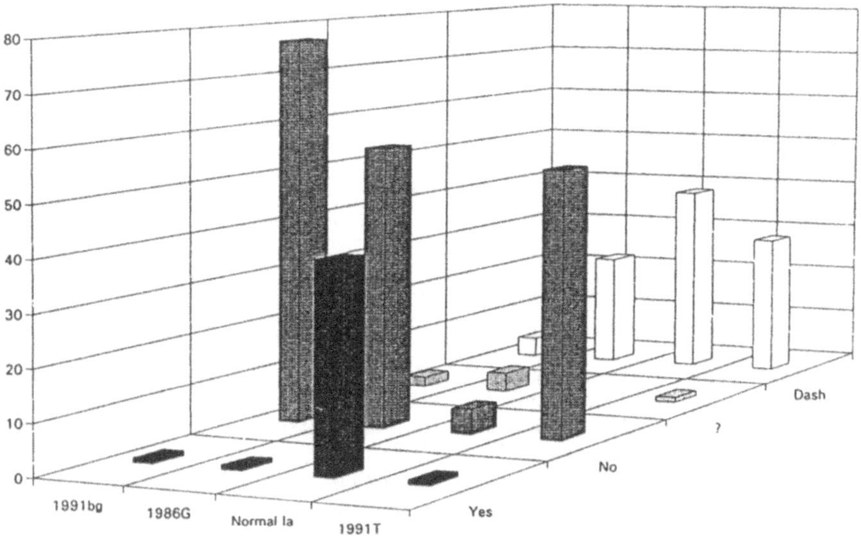

Fig. 9.5. Histogram by Branch, Fisher, and Nugent (1993) of the relative (%) frequencies of supernovae of type I that are classed either as "normal" by the spectroscopic criteria of these authors or that are similar to the three non-normal SNe 1991bg, 1986G and 1991T along the front axis. The side axis indicates if or not a particular SN is in the category indicated along the front axis.

Figure 9.5 shows the result of a survey of the available spectra of SNe of type I, reexamined in a study by Branch, Fisher, and Nugent (1993). Their aim was to determine the relative fraction of normal (as defined by them based on spectra) SNe in the available sample. Note that much of this sample has been found in discovery programs that are more *flux-limited* rather than distance-limited. Hence, intrinsically faint SNe, such as presumably are the

SNe Ib, are underrepresented compared with their true frequency per unit volume of space.

The result by Branch, Fisher, and Nugent (1993) is that the majority (89%) of the bona fide cataloged SNe Ia are normal SNe. On their criteria, the three non-normal SNe in Phillips' diagram (Fig. 9.4) would not appear if that diagram were restricted to "Branch normal" listings. They conclude "that the observational sample of SNe Ia is strongly peaked at spectroscopically normal. We further conclude that when arranged in the photometric sequence of Phillips, SNe Ia also form a spectroscopic sequence, and (therefore) that peculiar SNe Ia are over-represented in the Phillips sample".

The same conclusion was reached at the end of 1992 by Hamuy in an interim summary (conference report at the interregional meeting of the IAU at Viña del Mar, Chile), where he reported that of the 21 discoveries of type I supernovae in a flux-limited discovery program at Cerro Tololo over a two-year period, 20 were normal in the Branch spectroscopic sense. Only one was of type Ib. As said, it is generally believed that type Ib explosions are fainter than type Ia by 1 to 2 magnitudes. Even, then, if type Ib SNe are as frequent as type Ia in a volume-limited sample, they will be less frequent in any flux-limited discovery list.

If we restrict the sample in Fig. 9.4 to decay rates slower than $1\overset{m}{.}5/15$ days and also eliminate the peculiar cases such as SN 1991T, the remaining homogeneous (in the sense of spectra) sample has only a small range in $M(\max)$. But the ultimate proof of a small $\sigma(M)$ can only come from a small dispersion about the line of slope 5 in Hubble diagrams based on samples that have exquisite photometry.

9.2.2 Modern Hubble Diagrams

Kowal noticed that the several SNe Ia known at that time (1968) in the Virgo cluster had a remarkably small spread in the brightness at maximum. The same was true for the six SNe Ia in the Coma cluster. The mean points for the several SNe Ia in each cluster are shown in Kowal's diagram (Fig. 9.1) as crosses.

As the data accumulated, the Hubble diagram made from SNe Ia that had appeared in E galaxies (where the question of absorption was moot) continued to show a small scatter about a line of slope $dm/d\log v = 5$.

The data known in the early 1980's are shown in Fig. 9.6 using the maximum magnitudes from Barbon, Capaccioli, & Ciatti (1975), Pskovskii (1977), Kowal (1978), and Branch & Bettis (1978). It was the tightness of this diagram that supported the precept that $M(\max)$ for SNe Ia has a dispersion smaller than the extant measuring error. The data for the "normal" (in the sense of Branch) SNe Ia in the Virgo are in Table 2 of Branch and Tammann (1992).

Data of the same kind from E galaxies that have produced two or more SNe Ia show the same small dispersion in the apparent magnitudes at max-

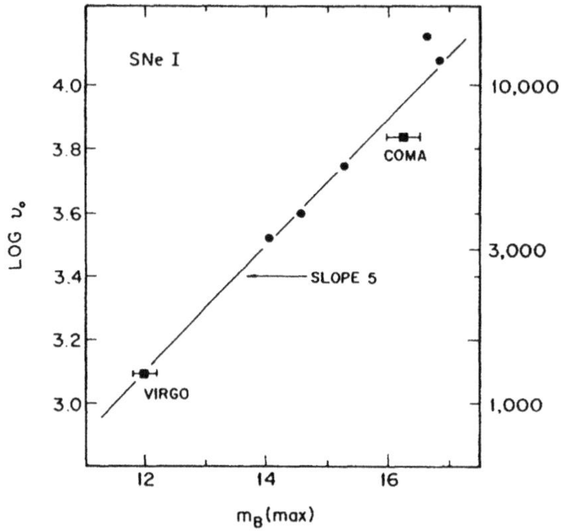

Fig. 9.6. Hubble diagram for 16 SNe Ia in E galaxies with known m_B(max) in 1982. Six supernovae make up the Virgo cluster point, and five are in the Coma cluster. Diagram is from Sandage and Tammann (1982).

imum. Branch and Tammann (1992) show three examples in their Table 1, giving a difference of less than $0\overset{m}{.}1$ between the maximum magnitudes of each of the separate SNe from the other in the same galaxy. Three more examples with the same result have been discussed by Rood (1994).

The premise of smaller $\sigma(M)_{\max}$ than can be currently measured continues to be supported by the wider data from the Basel Atlas. Figure 9.7 shows the Hubble diagrams that can be constructed from the data available to 1991, based on the archival photometry in the Basel Atlas.

It is expected that the m_{pg} data, obtained earlier without benefit of modern technology, are not as accurate as B and V photometry in the other two panels of Fig. 9.7. Nevertheless, the m_{pg} panel is similar to the diagram of Kowal (Fig. 9.1) but with more points (note the restriction in Fig. 9.7 to redshifts larger than 1000 km s^{-1}). Even with these relatively poor data, the dispersion about the line of slope 5 is small at $\sigma(M) = 0\overset{m}{.}65$. The true dispersion is undoubtedly smaller as shown by its decrease with increasingly precise data.

The middle panel of Fig. 9.7 shows B-band data where the dispersion has decreased to $0\overset{m}{.}51$, but again the value is an apparent one, determined by the considerable uncertainty in B(max) for any given SN.

The lower panel shows the V-band Hubble diagram where $\sigma(M) = 0\overset{m}{.}36$ which again is an upper limit, based on the known technical uncertainties of the data (poor coverage to determine the time and the value of the maximum brightness, etc.).

Intensive programs of discovery are now in place, such as that at Cerro Tololo (Hamuy 1992), and that of Evans (1994) by visual discovery. It is expected that within the next decade the improvement of Fig. 9.7 will be

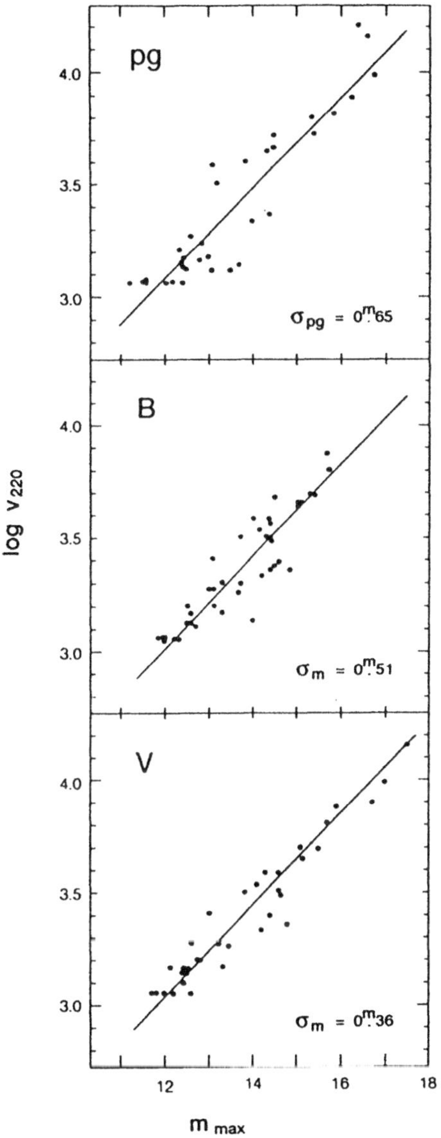

Fig. 9.7. The Hubble diagrams in the m_{pg}, B, and V pass bands of SNe Ia considered to be normal, based on the fit of the light curves to the standard template curve adopted in the Basel Atlas. Diagram is from Sandage and Tammann (1993). No corrections for absorption in the parent galaxy have been applied, based on the arguments by Branch and Tammann (1992).

substantial, either laying to rest the critics of it, or proving the case of small intrinsic dispersion directly.

Independent analysis by Miller and Branch (1990) concerning the effects of extinction for highly inclined parent galaxies has led them to conclude that the intrinsic dispersion in $\langle M(\max)\rangle$ is at most $0^{m}_{.}27$, and is likely to be less. If so, SNe Ia would be the most precise distance indicators known, as advocated in the first sentence of this lecture by the discoverer of the phenomenon.

We proceed on the premise that the ridge lines of the middle and lower panels of Fig. 9.7 will remain secure even as the technical data are improved, leading to the Hubble constant once $\langle M(\max)\rangle$ is known.

The equations of the two ridge lines are

$$\log v_{220} = 0.2B + (0.639 \pm 0.016), \qquad n = 41, \tag{9.1}$$

$$\log v_{220} = 0.2V + (0.653 \pm 0.012), \qquad n = 34, \tag{9.2}$$

which we use later.

9.3 Absolute Calibration of $M(\max)$ for SNe Ia

Once the mean absolute magnitude at maximum in B and in V are known from independent calibrations, the mean distance modulus, $m - M$, at any given redshift can be read from Fig. 9.7, or calculated from equations (9.1) and (9.2). The Hubble ratio of redshift-to-distance follows directly.

9.3.1 $\langle M(\max)\rangle$ From Methods Before 1992

The methods of calibration, both by brute force astronomical means before 1992, and/or model calculations based on premises concerning the physics of the explosion, are discussed by Branch and Tammann (1992). These include (1) the estimates (most uncertain to be sure) of the distances and apparent magnitudes of the four historical SNe in the Galaxy, (2) pre-1992 data for SN 1937C in IC 4182 and SN 1895B and SN 1972E in NGC 5253 with distances from brightest resolved stars in the two parent galaxies, (3) the SNe Ia in the Virgo cluster using an assumed cluster distance modulus of $m - M = 31^{m}_{\cdot}7$ from the other methods in Table 8.3 (neglecting the SNe method there), (4) model calculations based on the Nickel-Cobalt radioactivity explanation of the light curve (see Branch 1992), and (5) a thermal emission calculation based on a known temperature and radius at 25 days after maximum where the emergent flux is that of a blackbody, with then a subsequent reduction of the $M(25)$ absolute magnitude at that phase to maximum light (Branch 1990). These five methods give individual values of $\langle M(\max)\rangle$ shown in Table 9.1, taken from Branch and Tammann (1992). The average is $\langle M(\max)\rangle = -19^{m}_{\cdot}6 \pm 0^{m}_{\cdot}2$ using their precepts for the weights.

Impressive as the results of Table 9.1 are as to internal consistency, none of the methods is 100% secure. (1) The data for historical SNe are poor as regards apparent magnitudes at maximum (for example the SN of A.D. 1006 is said to have reached $m_V = -9$, that of A.D. 185 at $m_V = -8$, both of which are so bright as to wonder about the local photometric standards (but see Baade, 1945, on the use of historical records), (2) the distance to the

Table 9.1. Summary of five absolute magnitude calibrations (Branch and Tammann 1992)

Method	M_B
Historical Galactic supernovae	$-19\overset{m}{.}7 \pm 0\overset{m}{.}6$
SNe Ia in two nearby galaxies	$-19\overset{m}{.}8 \pm 0\overset{m}{.}3$
SNe Ia in the Virgo Cluster	$-19\overset{m}{.}6 \pm 0\overset{m}{.}4$
Nickel-Cobalt Radioactivity	$-19\overset{m}{.}4 \pm 0\overset{m}{.}4$
Thermal emission at 25 days	$-20\overset{m}{.}1 \pm 0\overset{m}{.}7$
Mean	$-19\overset{m}{.}6 \pm 0\overset{m}{.}2$

parent galaxies that produced SNe 1937C, 1895B, and 1972E had been determined, before 1992, by photometry of its brightest stars, a credible indicator (Sandage and Tammann 1974, Sandage and Carlson 1985) but not accurate enough for the present purposes, (3) the distance to the Virgo Cluster by other methods is the center of the current debate, and (4) the precepts for the two theoretical model calculations may be incorrect.

For these reasons, the advocates of the short distance scale, who need $\langle M(\max)\rangle = -18\overset{m}{.}2$, have not been convinced by Table 9.1. The method of calibration through Cepheids in IC 4182 and NGC 5253 became available with the commissioning of the Hubble Space Telescope. The remainder of this lecture describes the results of the first two experiments on this program that connects supernovae with Cepheids in galaxies where both have been observed, giving results that support the calibration in Table 9.1 and therefore the long distance scale ($H_0 \sim 50$).

9.3.2 $\langle M(\max)\rangle$ From Cepheids (1992–1994)

SN 1937C is a prototypical supernova of type Ia (it was once used to define the type). It was discovered by Zwicky at Palomar early in the campaign begun by Baade and Zwicky in the mid-1930's to discover and study such stars. The light curves in the m_{pg} and V photometric bands from the Basel Atlas are shown in Fig. 9.8 where all available photometric data are combined.

The supernova reached $m_{pg} = 8\overset{m}{.}57$ and $V = 8\overset{m}{.}64$ at maximum. Transforming to the modern photometric scale, after correcting Baade's 1937 scale (as originally set out in Baade and Zwicky 1938), gives maximum magnitudes of $B = 8\overset{m}{.}85 \pm 0\overset{m}{.}11$, $V = 8\overset{m}{.}64 \pm 0\overset{m}{.}10$ (Saha *et al.* 1994a; Schaefer 1994).

Before the discovery of Cepheids associated with SN 1937C, the distance modulus of the parent galaxy (IC 4182) had been determined to be $m - M = 28\overset{m}{.}21 \pm 0\overset{m}{.}2$ from a color-magnitude diagram of its 100 brightest resolved stars (Sandage and Tammann 1982), giving $M_B(\max) = -19\overset{m}{.}77$ for SN 1937C. This is one of the values that went into line 2 of Table 9.1. The number was not accepted by the advocates of the short distance scale.

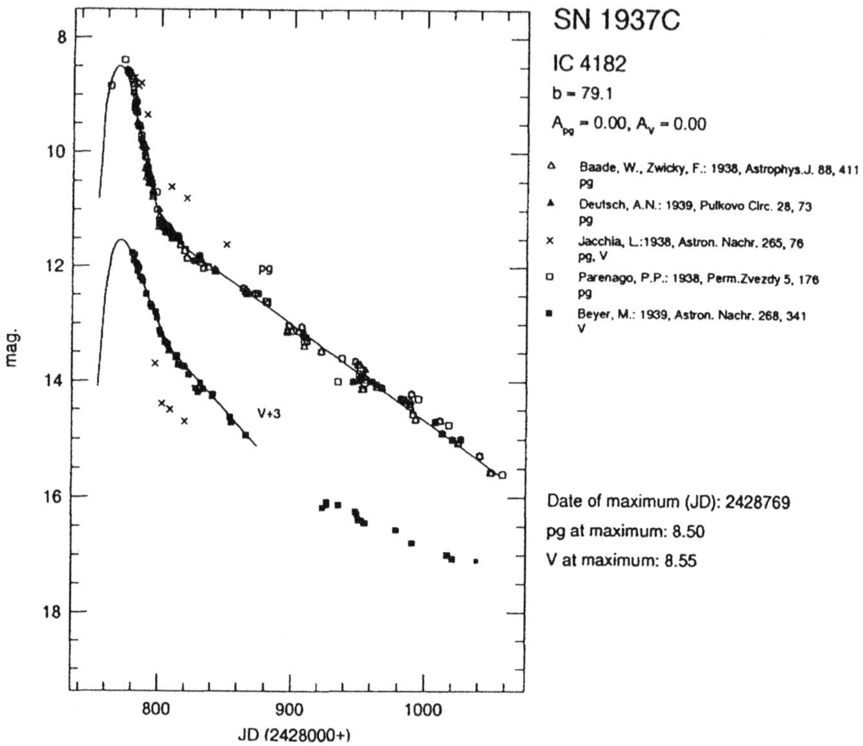

Fig. 9.8. Light curves of SN Ia 1937C in IC 4182 from the Basel Atlas.

The situation stabilized with the discovery of 27 Cepheids in IC 4182 during the first two years of operation of the Hubble Space Telescope. The galaxy is so nearby that the data could be obtained and the photometry made accurate enough with the telescope even in its initial aberrated condition.

Before setting out the Cepheid data in IC 4182, it is useful to recall the form of the period-luminosity relation and its absolute calibration as it existed after the major correction of $\sim 1^{\rm m}\!.5$ made by Baade (1952) (Lecture 7).

The relation in the B and the V pass band is shown in Fig. 9.9. The scatter is real, due to the finite width of the Cepheid instability strip in the H-R diagram, causing the sloping lines of constant period in that diagram to cut the blue and red edges of the strip at different absolute magnitudes (Sandage 1958; Sandage and Tammann 1968).

The shape of the PL relation is determined by the individual PL relations for the four Local Group galaxies identified in the code. The absolute magnitude zero point was set by several Cepheids in Galactic clusters (black dots) and by Cepheids suggested at the time to be associated with the Perseus stellar complex with h + χ Per as its nucleus.

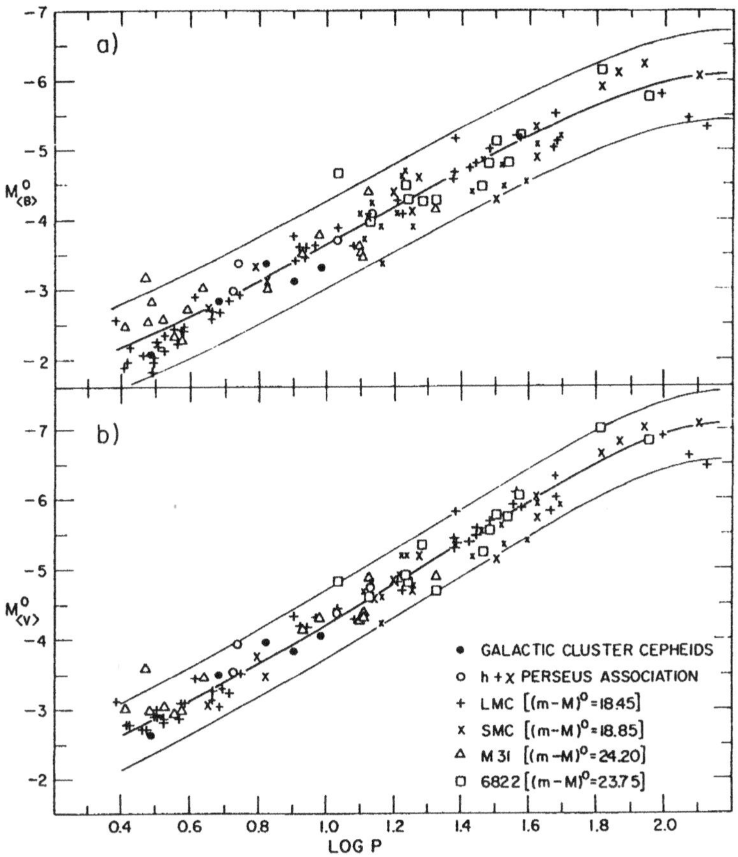

Fig. 9.9. The period-luminosity relation for Cepheids based on Local Group galaxies to determine the shape, and on star clusters in the Galaxy for zero-point determination. Diagram from Sandage and Tammann (1968).

The shape and the zero point differ by less than $0\overset{\mathrm{m}}{.}1$ over the relevant period range from the more modern calibrations (Feast and Walker 1987; Madore & Freedman 1991), calibrations based on much photometry in the 25 years between the epoch of Fig. 9.9 and these two modern reviews.

A linear approximation for the PL relation in V in Fig. 9.9 is

$$M_V = -2.83 \log P - 1.37, \tag{9.3}$$

the slope of which we shall use shortly in another context. The linearized relation reviewed by Feast and Walker (1987) has the equation

$$M_V = -2.78 \log P - 1.35. \tag{9.4}$$

Madore and Freedman (1991), giving PL relations in B and V, also list the equations in other photometric bands as well. Their equations relevant to this discussion are

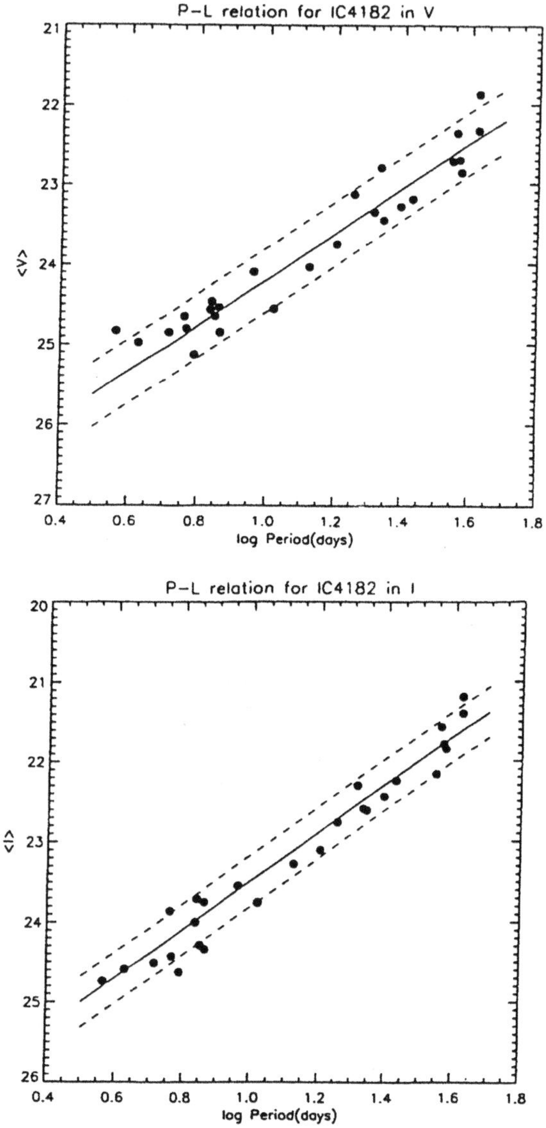

Fig. 9.10. The PL relation in V and I for the 27 Cepheids found in IC 4182 with the Hubble Space Telescope. The slopes of $\mathrm{d}V/\mathrm{d}\log P = -2.83$, and $\mathrm{d}I/\mathrm{d}\log P = -3.06$ have been drawn, based on equations (9.3) and (9.7). Diagram from Saha *et al.* (1994a).

$$M_{\mathrm{V}} = -2.76 \log P - 1.40, \tag{9.5}$$

or

$$M_{\mathrm{V}} = -2.88 \log P - 1.24, \tag{9.6}$$

depending on the sample. Both equations agree with equations (9.3) and (9.4) to within $0\overset{\mathrm{m}}{.}1$ over the relevant period range.

The PL relation of Madore and Freedman in the I photometric band is

$$M_{\rm I} = -3.06 \log P - 1.81, \tag{9.7}$$

or

$$M_{\rm I} = -3.14 \log P - 1.70, \tag{9.8}$$

again depending on the sample.

The PL relation for the 27 Cepheids in IC 4182, from Saha *et al.* (1994a), is shown in Fig. 9.10 in the V and the I photometric bands. The slopes of the solid ridge lines in the panels have not been determined by the IC 4182 data themselves but have been adopted as -2.83 in V from equation (9.3) and -3.06 in I from equation (9.7).

Envelope lines have been drawn parallel to the ridge line with a scatter of $0\overset{\rm m}{.}4$ and $0\overset{\rm m}{.}3$ respectively, expected as the intrinsic scatter due to the finite width of the instability strip. The agreement of the data with the expectations concerning the intrinsic scatter show that there is no detectable differential absorption of one Cepheid compared with another within the Cepheid domain.

Adopting apparent magnitude zero points in V and I so as to minimize the residuals in Fig. 9.10, and using the absolute calibrations from equations (9.3) and (9.7) give apparent distance moduli of IC 4182 to be $(m - M)_{\rm V} = 28\overset{\rm m}{.}31 \pm 0\overset{\rm m}{.}12$ and $(m - M)_{\rm I} = 28\overset{\rm m}{.}41 \pm 0\overset{\rm m}{.}12$, for an adopted mean of

$$(m - M)_0 = 28\overset{\rm m}{.}36 \pm 0\overset{\rm m}{.}09. \tag{9.9}$$

Evidence from colors of the brightest resolved stars and from the agreement within the errors of the modulus itself show that the internal reddening is negligible. But, in fact, we need only consider the *differential* absorption between the Cepheids and the supernova to determine $M(\rm max)$ of SN 1937C, which clearly is negligible if the general absorption itself is negligible.

The absolute magnitudes at maximum for 1937C, using $B(\rm max) = 8\overset{\rm m}{.}85 \pm 0\overset{\rm m}{.}11$ and $V(\rm max) = 8\overset{\rm m}{.}64 \pm 0\overset{\rm m}{.}10$, are then

$$M_{\rm B}(\rm max) = -19\overset{\rm m}{.}51 \pm 0\overset{\rm m}{.}14, \tag{9.10}$$

and

$$M_{\rm V}(\rm max) = -19\overset{\rm m}{.}72 \pm 0\overset{\rm m}{.}15, \tag{9.11}$$

confirming line two of Table 9.1 and thereby giving a degree of credence to the other entries as well.

To determine the intrinsic dispersion of $M(\rm max)$ of SNe Ia it is necessary to increase the number of such calibrations beyond simply the one afforded by SN 1937C. A second experiment could be made with the aberrated Hubble Space Telescope using the two SNe produced by NGC 5253, a member of the Centaurus Group whose dominant member is NGC 5128.

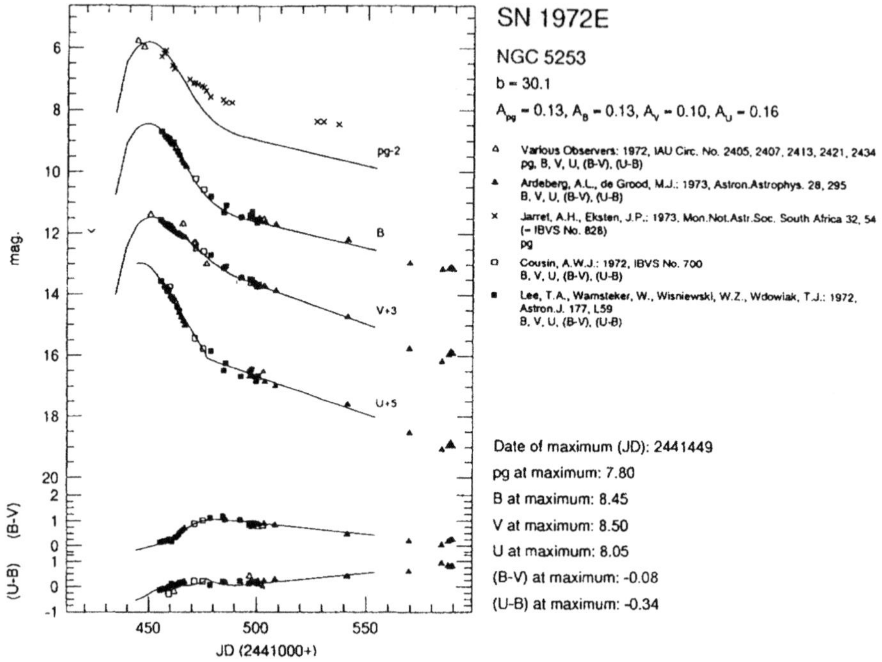

Fig. 9.11. Light curves in m_{pg}, B, V, and U for SN 1972E in NGC 5253 from the Basel Atlas.

NGC 5253 produced SN 1895B (Z Cen) and SN 1972E which, as with 1937C, has been defined at various times to be a prototype of the Ia class. The light curves of 1972E, with the Basel template curves superposed, are in Fig. 9.11. The maximum observed magnitudes of 1972E are $B(\max) = 8\overset{\mathrm{m}}{.}58$ and $V(\max) = 8\overset{\mathrm{m}}{.}60$. These are the same as listed at the right in Fig. 9.11 when the Basel listings are corrected back to the observed frame by subtracting $0\overset{\mathrm{m}}{.}13$ from their listed B and $0\overset{\mathrm{m}}{.}10$ from their V, which had been added for Galactic extinction.

Eleven Cepheid variables were discovered in NGC 5253 with the HST (Sandage *et al.* 1994). The PL relation in V is shown in Fig. 9.12. The PL relation was also obtained in I. These and other data show that, as in IC 4182, the differential absorption between the Cepheids and the supernovae is smaller than we could measure at a level of $A_{\mathrm{V}} < 0\overset{\mathrm{m}}{.}1$ (Sandage *et al.* 1994; Saha *et al.* 1994b).

The resulting distance modulus of NGC 5253 is

$$(m - M)_{\mathrm{V}} = 28\overset{\mathrm{m}}{.}06 \pm 0\overset{\mathrm{m}}{.}06, \tag{9.12}$$

which, when combined with the apparent magnitudes at maximum in B for SN 1895B and in B and V for SN 1972E, gives the calibrations of $M_{\mathrm{B}}(\max) = -19\overset{\mathrm{m}}{.}69 \pm 0\overset{\mathrm{m}}{.}21$ for SN 1895B, $M_{\mathrm{B}}(\max) = -19\overset{\mathrm{m}}{.}51 \pm 0\overset{\mathrm{m}}{.}14$, and $M_{\mathrm{V}}(\max) = -19\overset{\mathrm{m}}{.}72 \pm 0\overset{\mathrm{m}}{.}15$ for SN 1972E.

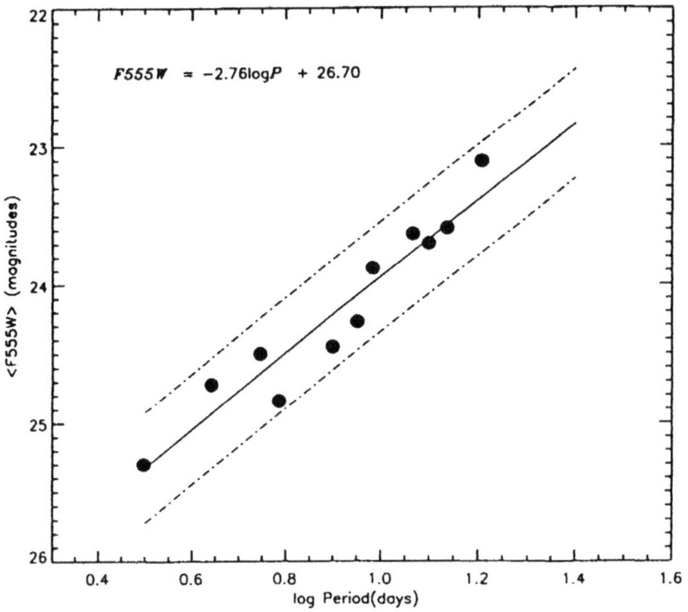

Fig. 9.12. PL relation in the F555W (close to standard V) photometric band of HST for eleven Cepheids in NGC 5253. The slope of the ridge line has been adopted to be $\mathrm{d}V/\mathrm{d}\log P = -2.83$ from equation (9.3). Diagram from Sandage *et al.* (1994).

Combining the data for the three SNe calibrated in this way gives

$$\langle M_{\mathrm{B}}(\max)\rangle = -19^{\mathrm{m}}.55 \pm 0^{\mathrm{m}}.08, \tag{9.13}$$

from three determinations, and

$$\langle M_{\mathrm{V}}(\max)\rangle = -19^{\mathrm{m}}.58 \pm 0^{\mathrm{m}}.09, \tag{9.14}$$

from two determinations.

9.4 H_0 and the Cosmological Timing Test

Combining equations (9.13) and (9.14) with the ridge-line solutions of the SNe Hubble diagrams in Fig. 9.7 [equations (9.1) and (9.2)], gives Hubble constants of

$$H_0(B) = 52 \pm 8 \text{ km s}^{-1}, \tag{9.15}$$

and

$$H_0(V) = 55 \pm 8 \text{ km s}^{-1}. \tag{9.16}$$

The quoted errors are external, based on an assumed true intrinsic dispersion of SNe about the Hubble diagram ridge line of $\sigma(M) = 0^{\mathrm{m}}.3$, which may actually be less. But to determine the true dispersion will require many more SNe-Cepheid calibrations of this kind.

The timing test is now complete. We saw from Lecture 7 that the age of the Universe can be put at $T_U = 15 \pm 2$ Gyr based on the age of the Galactic globular clusters at 14 Gyr plus 1 Gyr. If $H_0 = 50 \pm 2$ as in Table 8.4 of Lecture 8, then $H_0^{-1} = 19.5$ Gyr. Hence, for the first time since the time-scale dilemma arose over Hubble's short age of 2 Gyr compared with Holmes' age of the Earth of 3.5 Gyr in the 1930's, there is no discrepancy between the expansion age of the universe and the astrophysical ages of the objects in it.

It remains to be seen, however, if the ages can be made to stand in the ratio of 2 to 3 as is required in the standard model of cosmology if $\Omega = 1$, with all its consequences for dark matter.

References

Baade, W. 1945, ApJ **102**, 309

Baade, W. 1952, Trans. IAU **8**, (Cambridge: Cambridge University Press) p. 387

Baade, W., & Zwicky, F. 1938, ApJ **88**, 411

Barbon, R., Capaccioli, M., & Ciatti, F. 1975, A&A **44**, 267

Barbon, R., Capellaro, E., & Turatto, M. 1989, A&A Suppl. **81**, 421 (The Asiago Supernova Catalog)

Branch, D. 1990, in *Supernovae*, ed. A.G. Petschek (New York: Springer), p. 30

Branch, D. 1992, ApJ **392**, 35

Branch, D., & Bettis, C. 1978, AJ **83**, 224

Branch, D., Fisher, A., & Nugent, P. 1993, AJ **106**, 2383

Branch, D., & Tammann, G.A. 1992, Ann. Rev. A&A **30**, 359

Cadonau, R. 1987, PhD thesis, Univ. Basel

Cadonau, R., Sandage, A., & Tammann, G.A. 1985, in *Supernovae as Distance Indicators*, ed. N. Bartel, Lecture Notes in Physics No. 224 (New York: Springer), p. 151

Evans, R. 1994, Proc. Aust. AS **11**, 7

Feast, M., & Walker, A. R. 1987, Ann. Rev. A&A **25**, 345

Hamuy, M. 1992, paper at 7th IAU regional reunion (Viña del Mar, Chile)

Kowal, C.T. 1968, AJ **73**, 1021

Kowal, C.T. 1978, private communication

Leibundgut, B., Tammann, G.A., Cadonau, R., & Cerrito, D. 1991, A&A Suppl. **89**, 537 (The Basel SNe Ia Atlas)

Madore, B.F., & Freedman, W.L. 1991, PASP **103**, 933

Miller, D.L., & Branch, D. 1990, AJ **100**, 530

Phillips, M.M. 1993, ApJ **413**, L105

Pskovskii, Yu. P. 1969, Sov. Astron. AJ **12**, 750

Pskovskii, Yu. P. 1977, Astron. Zh. **54**, 1188

Rood, H. 1994, PASP **106**, 170

Saha, A., Labhardt, L., Schwengeler, H., Macchetto, F.D., Panagia, N., Sandage, A., & Tammann, G.A. 1994a, ApJ **425**, 14

Saha, A., Sandage, A., Tammann, G. A., Labhardt, L., Schwengeler, H., Panagia, N., & Macchetto, F.D. 1994b, ApJ in press

Sandage, A. 1957, ApJ **125**, 435

Sandage, A. 1958, ApJ **127**, 513
Sandage, A., Saha, A., Tammann, G.A., Labhardt, L., Schwengeler, H., Panagia, N., & Macchetto, F.D. 1994, ApJ **423**, L13
Sandage, A., & Carlson, G. 1985, AJ **90**, 1464
Sandage, A., & Tammann, G.A. 1968, ApJ **151**, 531
Sandage, A., & Tammann, G.A. 1974, ApJ **191**, 603
Sandage, A., & Tammann, G.A. 1982, ApJ **256**, 339
Sandage, A., & Tammann, G.A. 1993, ApJ **415**, 1
Schaefer, B.E. 1994, ApJ **426**, 493
Tammann, G.A. 1982, in *Supernovae: A Survey of Current Research*, eds. M.J. Rees, & R.J Stoneham (Dordrecht: Reidel), p. 371
van den Bergh, S. 1993, Revs. Mex. A&A **26**, 73
van den Bergh, S., & Pazder, J. 1992, ApJ **390**, 34
Zwicky, F. 1961 in *Problems in Extragalactic Research*, ed. G.C. McVittie (New York: Macmillan Co.), p. 356

10 Observational Selection Bias

In Lecture 4 (section 4.5.1) we saw an example of observational selection bias in the sample of the brightest radio sources from the 3C, 4C, and the Parkes catalogs whose optical identification and redshifts had been determined by 1970. The correlation of absolute radio power with distance for this sample (Fig. 4.7) shows a z^2-dependence of radio power on redshift. However, it was clear from the nature of the sample that the correlation is not real but is due to the pernicious effect of a selection bias caused by using a flux-limited sample confined to the brightest known objects of the class when the total class has a very large range of absolute power.

Proof that the apparent correlation is due to selection bias is made by adding a fainter sample in apparent flux and seeing the lower limit line of the correlation move faintward in absolute power in lock-step with the level of the fainter apparent flux. The method of detecting bias in this way is proposed using examples of galaxy redshift, apparent magnitude, and linewidth with particular galaxian distance estimators.

Demonstrations of the method using observational data have now begun in the archive literature (Kraan-Korteweg, Cameron, and Tammann 1988; Sandage 1988a, b, 1994a, b; Bottinelli, *et al.* 1988; Fouqué, *et al.* 1990; Federspiel, Sandage, & Tammann 1994, 1995). An early shadow of how selection bias affects the slope of the Hubble diagram for field E galaxies had been given (Sandage, Tammann, & Yahil 1979, STY) as a precursor of this general method and as an explanation of the effect seen, but not explained, in the early Hubble diagrams for field galaxies by Humason *et al.* (1956, HMS).

The necessity to identify observational selection bias and to devise practical methods to correct for it are the subjects of this lecture.

10.1 Observational Selection Bias in the Hubble Diagrams of Sb I and Sc I Galaxies

An illustration of the effects of selection bias on the slope of the Hubble $(m, \log v)$ diagram is seen in Figs. 10.1 and 10.2. The sample consists of all field Sb I, Sbc I, and Sc I galaxies in the flux-limited Revised Shapley-Ames Catalog (the RSA), shown as filled dots, to which fainter field Sc I galaxies are added. These are taken from two special catalogs of such objects compiled from inspection of the Palomar Survey material (Sandage and Tammann 1975; Rubin *et al.* 1976), shown as open circles and triangles in the two diagrams.

The envelope lines and the center ridge line in each diagram are drawn with the slope of $\mathrm{d}B/\mathrm{d}\log v = 5$ required if the redshift-distance relation is linear (Lecture 5). The data have not been corrected for internal absorption (i.e., no CIA).

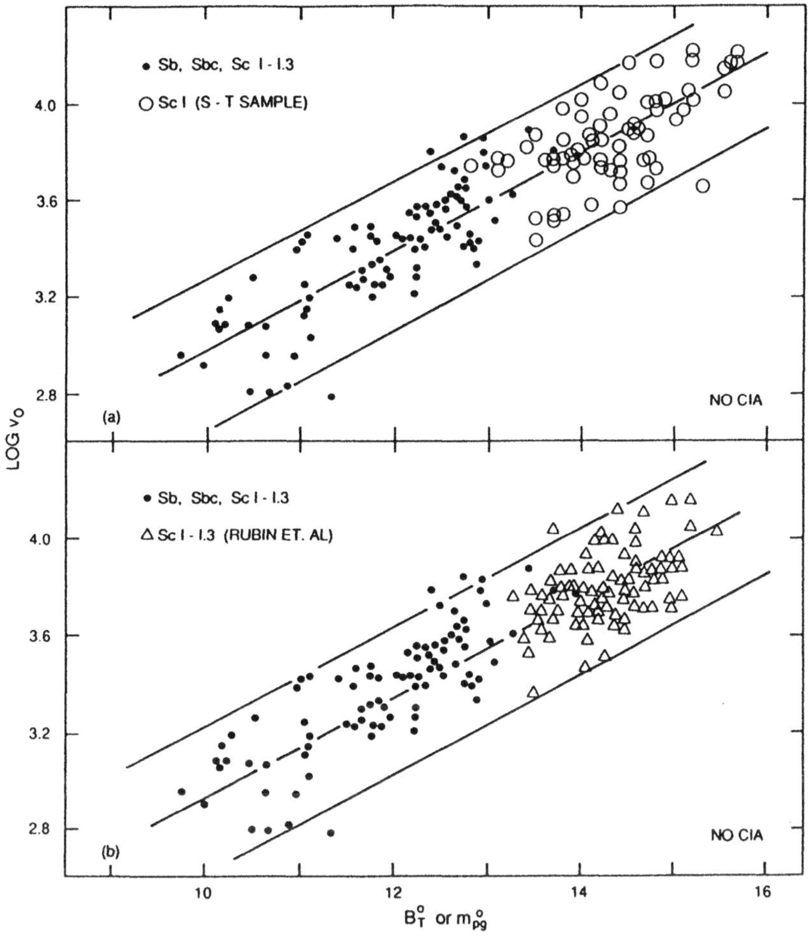

Fig. 10.1. Hubble diagram for Sc I galaxies to two different flux limits. Dots are galaxies from the RSA whose limiting magnitude is $m_{pg} = 13$. Open circles are from a fainter catalog to $m_{pg} = 15.7$ (Sandage and Tammann 1975). Triangles are from a similar but independent catalog from Rubin *et al.* (1976). Diagram from S88a.

Cursory inspection shows that the data follow the linear expansion expectations moderately well, but more detailed inspection shows the same effect seen in the Hubble diagrams in HMS and in Figs. 1-4 of STY (1979). The data points near the magnitude limits of both catalogs have brighter magnitudes at a given redshift than a uniform distribution about the ridge line would imply. The effect is shown in the formal least-squares solutions for the slope of the correlation of v on m for each subsample separately shown in Fig. 10.2. Our main task is to show that the effect is due to selection bias caused by the flux limitation of each of the catalogs.

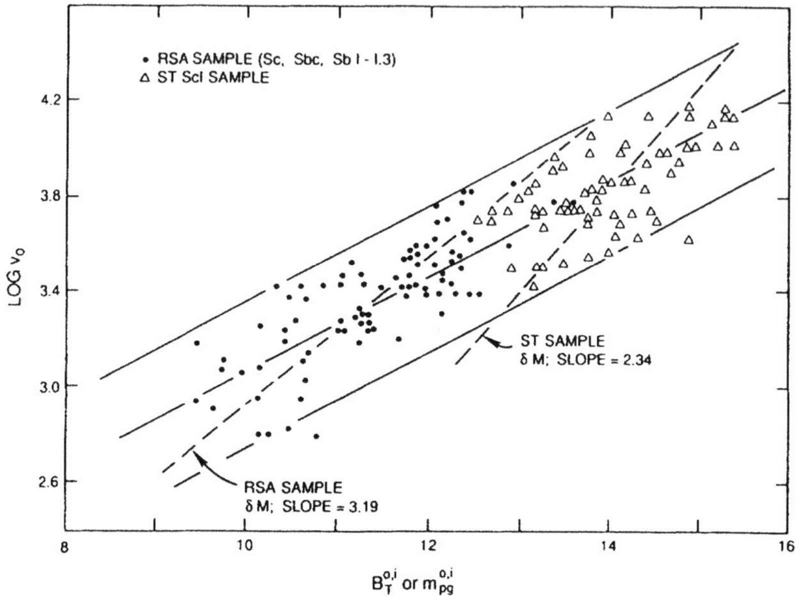

Fig. 10.2. Least-squares correlation lines for the data in Fig. 10.1 separately for the bright and faint samples showing the effects of selection bias near each of the limits of the data. Diagram from S88a.

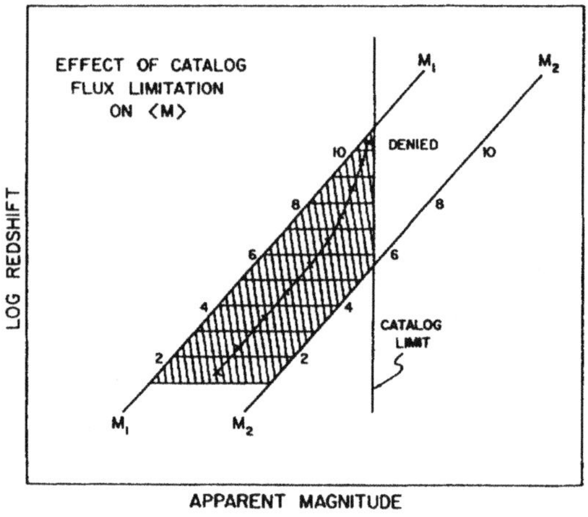

Fig. 10.3. The bias effect on the mean m, $\log v$ ridge line at a given $\log v$ due to flux limitation. The mean $\log v$ at a given m is unaffected. Diagram from S88c.

Figure 10.3 shows schematically the sense of the deviation of the "ridge line" from the mid points of the envelope lines near the flux limitation line marked "catalog limit". We now approach the problem systematically, leading to an analytical treatment.

10.2 The Spaenhauer Diagram

Experience has shown that a simple but powerful way to detect observational selection bias in particular samples is with a diagram similar to Fig. 4.7 where the absolute magnitude of each object in the sample is plotted against the log of its distance. In such a diagram, loci of apparent log flux (or, apparent magnitude) are straight lines. In what follows, we give such a plot the eponymous name of a Spaenhauer (1978) diagram (SD) after the astronomer who used it to great advantage in problems of galactic structure.

The use of SDs in detecting observational selection bias is illustrated in Fig. 10.4, taken from a simulation by Spaenhauer. The top panel shows the distribution of absolute magnitude as a function of distance for a sample of 500 galaxies whose mean absolute magnitude is $\langle M \rangle = -18^{\mathrm{m}}$, distributed with a Gaussian luminosity function of dispersion $\sigma(M) = 2^{\mathrm{m}}$. The top panel shows the distribution of a total distance-limited sample. The upper and lower envelope lines open toward the right due to the increased volumes surveyed at larger distances. Hence, the luminosity function is sampled further into the bright and faint wings of the distribution as the volume normalization increases.

If a subsample is drawn from a portion of the total sample by considering only those galaxies brighter than a particular apparent magnitude, i.e., $m = 13$ in Fig. 10.4, part of the total sample is denied the observer, being fainter than the assigned (catalog) limit. The relation between absolute magnitude, apparent magnitude, and distance (in parsecs) is $m = M + 5 \log D(\mathrm{pc}) - 5$, based on the definition of M as that magnitude which equals the observed apparent magnitude at a distance of 10 parsecs.

The point of Fig. 10.4 is that the mean absolute magnitude of the partial sample in the bottom panel becomes progressively brighter with increasing distance. The determination of the difference between the true mean absolute magnitude, M_0, for the distance-limited sample (at $M_0 = -18^{\mathrm{m}}$ in the top panel) and the progressive mean magnitudes at various distances, i.e., the function $M_0 - M(m) = f(D)$, is the analytical problem we need to solve. Note that this function is *not* the Malmquist bias correction which applies to the total sample but rather is the function that defines the individual corrections required *at each distance*, the sum of which, weighted by the volume elements, is what Malmquist calculated as the (ensemble) mean absolute magnitude of the total sample.

As a preliminary, consider the upper and lower envelope loci in Fig. 10.4 such that one galaxy is statistically expected on those loci. The problem reduces to renormalizing the individual Gaussian distributions in total number depending on the volume contained in an annulus at distance D of thickness $\mathrm{d}D$ (or in redshift space at z in $\mathrm{d}z$). This volume increases as $z^2 \mathrm{d}z$, which is the factor by which to increase the total number of objects under each Gaussian at each distance. The upper and lower limit absolute magnitudes of the envelopes, such that one galaxy at each of these luminosities are found,

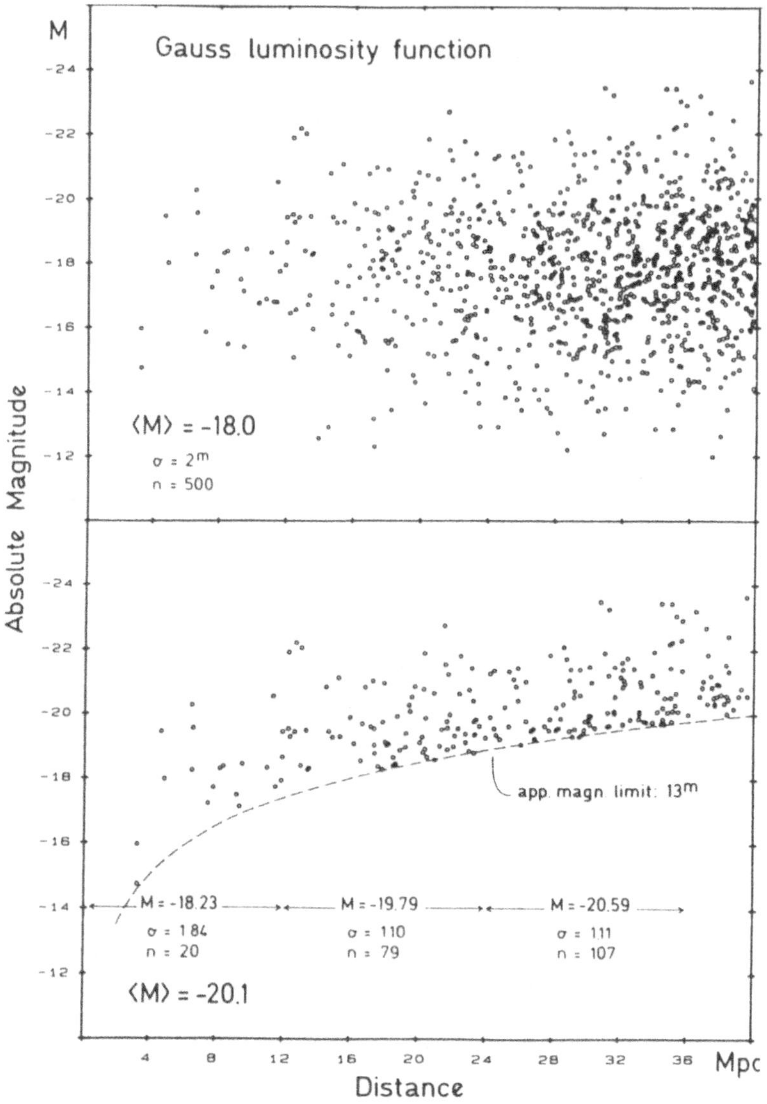

Fig. 10.4. Spaenhauer diagram from a simulation by Spaenhauer using a sample with an intrinsic Gaussian dispersion of $\sigma(M) = 2^{\rm m}$. If the abscissa had been log distance, the lower flux limit line would be a straight line. Diagram from Tammann and Sandage (1983).

is obtained by reading the renormalized Gaussians at appropriate places in the bright and faint wings where $n = 1$. The result is shown in Fig. 10.5 (Sandage 1994a, S94a, Fig. 3 there).

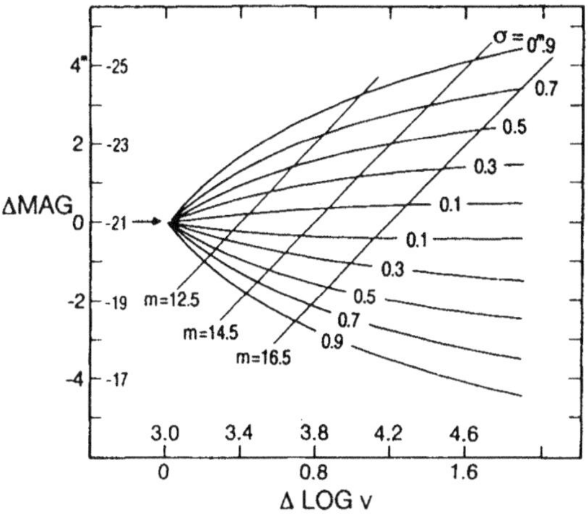

Fig. 10.5. Envelope lines in a log redshift, absolute magnitude Spaenhauer diagram for five values of the intrinsic dispersion of an assumed Gaussian luminosity function. The envelope lines are the loci of the distribution such that one galaxy would be expected at that limit line. The lines open outward due to the increased volume for increasing redshift.

10.3 The SD Applied to Real Data

The method can be illustrated using real data. The discussion here summarizes the details given elsewhere (S88a).

Figure 10.6 shows the Spaenhauer diagram for the sample of bright Sb and Sc galaxies from the RSA1 (Sandage & Tammann 1981). The top panel shows absolute magnitude of the sample galaxies as ordinate, plotted against log redshift reduced to the kinematic frame of the Virgocentric system using the model by Kraan-Korteweg (1986a,b). The absolute magnitudes are computed from the redshifts using $H_0 = 50$. The assumption is that the kinematic luminosities so determined are more accurate than any obtained by other methods, namely that the redshift-distance relation is linear and that deviations from the Hubble flow are negligible. Justification for these precepts are in Lecture 5 and in the discussion on circularity in Federspiel *et al.* (1994). The adopted magnitude corrections for the various morphological types that reduce the data to the Sc I zero point are shown in the inside border of the top panel.

That the apparent increase in the luminosity of the sample galaxies with increasing distance can be explained by the selection effect of Figs. 10.3 and 10.4 is seen in the bottom panel of Fig. 10.6 by placing an appropriate template curve from Fig. 10.5 over the data. A lower apparent magnitude limit line of $m = 12.5$ fits the lower boundary well.

The good fit of the template to the data in Fig. 10.6 does not prove that the apparent increase in luminosity of this sample *is* due to selection bias, only that it is a possible explanation. With only these data we could have just as logically concluded that the actual Hubble constant increases outward (Tully 1988), and therefore that the calculation of the absolute magnitude, M_B, in the manner of Fig. 10.6 is faulty. Proof that this is not the case is made by adding the fainter sample and watching the properties of the distribution

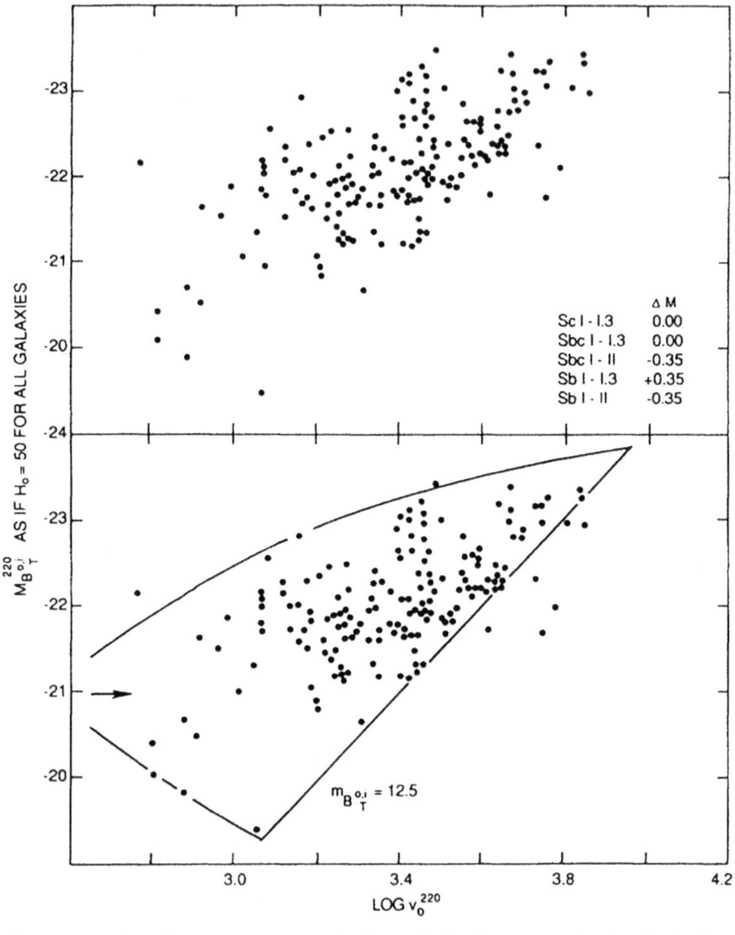

Fig. 10.6. (Top) Apparent correlation of absolute magnitude (calculated from redshifts) with distance (redshift) for the Sc I galaxies from the RSA set out in Fig. 10.1. The apparent increase of brightness with redshift is a selection effect caused by the flux limitation of the RSA. The effect is identical to that in Fig. 4.7. (Bottom) Envelope lines from Fig. 10.5 superposed on the data using a dispersion of $\sigma(M) = 0\overset{m}{.}7$. Diagram from S88a.

move to larger redshifts at the rate of 0.2 dex (due to the increased volume element) in the redshift for every magnitude increase in the catalog limit.

The bottom panel of Fig. 10.7 shows the result of adding a fainter sample to the RSA1 sample, shown again in the top panel. A new limit line at $m = 15$ is drawn. Fig. 10.8 shows the superposition of template curves from Fig. 10.7 for dispersions of $\sigma(M) = 0\overset{m}{.}5$ and $0\overset{m}{.}7$. The fit is excellent. The effect is due to selection bias rather than to H_0 increasing outward.

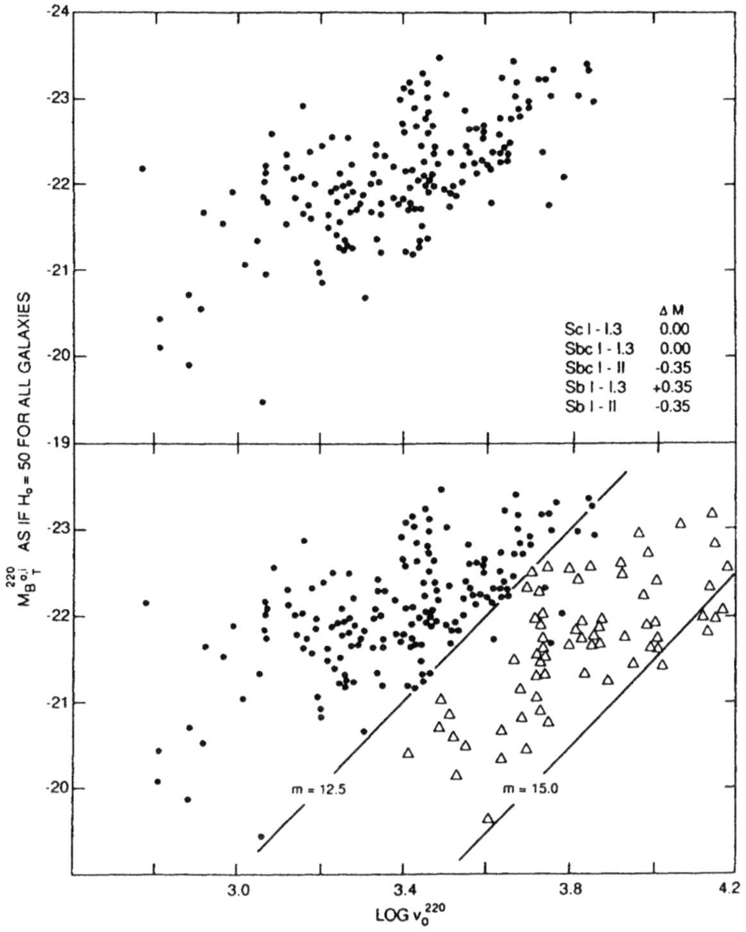

Fig. 10.7. Same as Fig. 10.6 but with the faint Sc I sample of ST (1975) superposed showing the effect of adding a fainter sample. Diagram from S88a.

10.4 The Malmquist $M_0 - \langle M(m) \rangle$ Calculation

As a preliminary to the calculation of the required correction to absolute magnitudes at every redshift and for any catalog limit (section 10.5), it is useful to show the calculation of the "global" Malmquist correction. This correction is global in the sense that it is the average absolute magnitude of a total sample. It is the integration over all redshifts of the individual corrections *that are needed at each redshift*, the individual corrections thereby becoming lost.

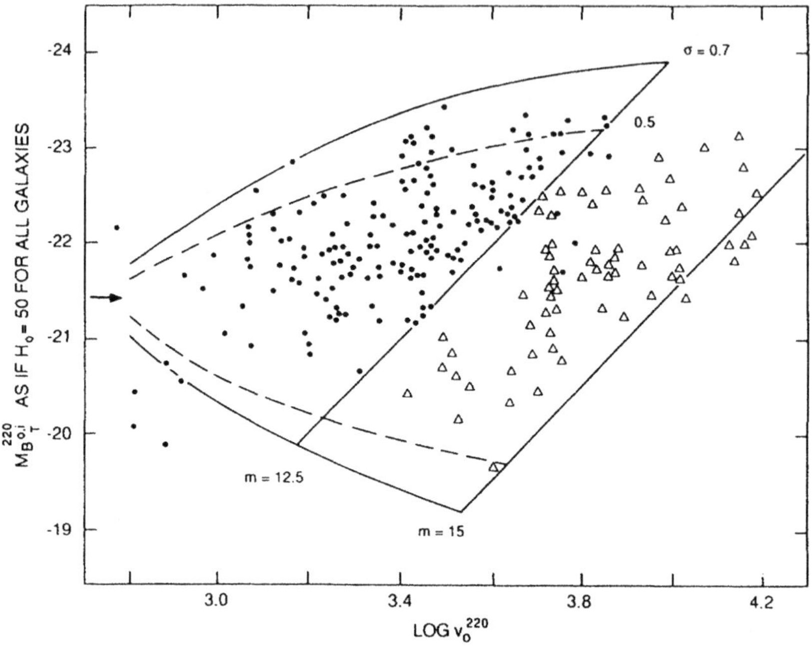

Fig. 10.8. Envelope lines from Fig. 10.5 superposed on the data in Fig. 10.7. Diagram from S88a.

10.4.1 The Malmquist Global Bias Equation (First Pass)

We first set out the derivation of the bias as Malmquist (1920) initially derived it.

The derivation follows from the spread-sheet method of solving the fundamental equation of stellar statistics discussed in section 3.3.1 of Lecture 3 from an $(m, \log r)$ table. Recall from Fig. 3.1 that each box of the spread sheet contains an entry, $a(m, r)$, that is the number of stars in the absolute magnitude interval $M + 1/2$ and $M - 1/2$ in the apparent magnitude interval $m + 1/2$ and $m - 1/2$ in the distance interval $\Delta \log r$ at the appropriate mean distance corresponding to these apparent and absolute magnitude intervals. Clearly

$$a(m, r)\Delta m = \Delta \mathrm{Vol} \times \Phi(M)\Delta M = 4\pi r^2 \Phi(M)\Delta r, \tag{10.1}$$

where m, r, and M are connected by

$$m = M - 5 + 5\log r. \tag{10.2}$$

Summing any particular column of the spread sheet (i.e., at a particular apparent magnitude, m, and summing over all distances) gives the total number of stars, $A(m)\Delta m$, at m in the interval $m + \Delta m/2$ and $m - \Delta m/2$ contributed

Fig. 10.9. Gunnar Malmquist (1893-1982).

by all stars of all absolute magnitudes at the various appropriate distances. The equation for $A(m)$ is

$$A(m) \equiv \int_{r=0}^{\infty} a(m,r)\mathrm{d}r = \omega \int_{r=0}^{\infty} r^2 \exp\left\{-\frac{(m+5-5\log r - M_0)^2}{2\sigma^2}\right\}\mathrm{d}r, \quad (10.3)$$

where $\omega = 4\pi/41,253$.

From the definition of the mean absolute magnitude of the ensemble of objects in a flux-limited sample that is complete to a given apparent magnitude, m, denoted by $M(m)$, it follows that

$$M(m) \equiv \frac{\int_0^{\infty} Ma(m,r)\mathrm{d}r}{\int_0^{\infty} a(m,r)\mathrm{d}r} \equiv \frac{1}{A(m)}\int_0^{\infty} Ma(m,r)\mathrm{d}r. \quad (10.4)$$

It also follows from equation (10.3) that

$$\frac{\mathrm{d}A(m)}{\mathrm{d}m} = -\frac{\omega}{\sigma^2}\int_0^{\infty} r^2(\ )\exp\left\{-\frac{(\ )^2}{2\sigma^2}\right\}\mathrm{d}r, \quad (10.5)$$

where the symbol ( ), not always written out in the following, is, by definition,

$$(\ ) \equiv m + 5 - 5\log r - M_0 \equiv M - M_0. \quad (10.6)$$

Hence, (10.5) with (10.6) gives

$$\frac{\mathrm{d}A(m)}{\mathrm{d}m} = -\frac{\omega}{\sigma^2}\left[\int_0^\infty Mr^2\exp\left\{-\frac{(\;)^2}{2\sigma^2}\right\}\mathrm{d}r \right.$$

$$\left. - \int_0^\infty M_0 r^2\exp\left\{-\frac{(\;)^2}{2\sigma^2}\right\}\mathrm{d}r\right], \tag{10.7}$$

which reduces by (10.3) and (10.4) to

$$\frac{\mathrm{d}A(m)}{\mathrm{d}m} = -\frac{\omega}{\sigma^2}\left[\frac{M(m)A(m)}{\omega} - \frac{M_0 A(m)}{\omega}\right], \tag{10.8}$$

or

$$M(m) = M_0 - \frac{\sigma^2\mathrm{d}A(m)}{A(m)\mathrm{d}m} \equiv M_0 - \frac{\sigma^2}{0.4343}\frac{\mathrm{d}\log A(m)}{\mathrm{d}m}. \tag{10.9}$$

For a homogeneous distribution of the objects

$$\log A(m) = 0.6m + \mathrm{const.}, \tag{10.10}$$

hence,

$$M(m) = M_0 - 1.38\sigma^2, \tag{10.11}$$

which is Malmquist's famous equation.

log v

M	2.9-3.1 3.0	3.1-3.3 3.2	3.3-3.5 3.4	3.5-3.7 3.6	3.7-3.9 3.8
−24					
−23					
−22					
−21					
−20	11.5	12.5	13.5	14.5	15.5
−19	12.5				
−18	13.5				
−17	14.5				
−16					
m-M	31.5	32.5	33.5	34.5	35.5

Fig. 10.10. Spread sheet for the calculation of the generalized $M(m, v)$ bias corrections at every redshift and for any flux-limitation line.

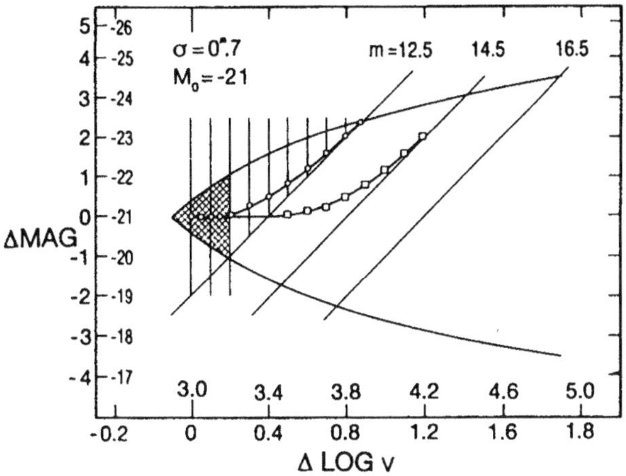

Fig. 10.11. Same as Fig. 10.10 but displayed as a Spaenhauer diagram. Diagram from S94a.

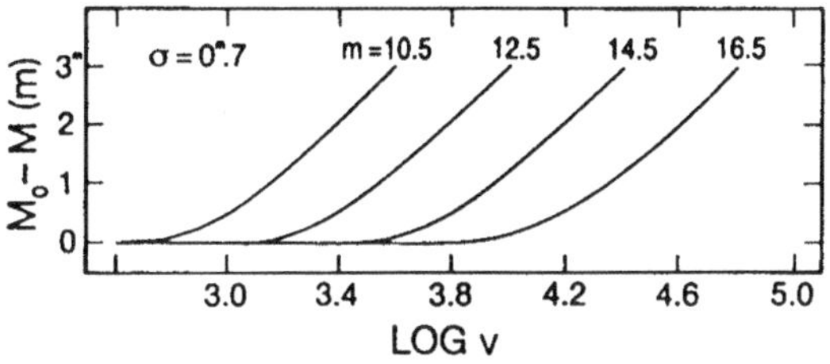

Fig. 10.12. The $M_0 - M(m)$ correction at every redshift as a function of m(limit) and redshift. Diagram from S94a.

10.4.2 A Different Method Not Used by Malmquist

A different, and sometimes easier way to visualize the calculation is via the Spaenhauer diagram, again constructed as a spread sheet as in Fig. 10.10. Intervals of log redshift are along the top. Absolute magnitude is along the ordinate as in Fig. 10.5. Figure 10.11 is the same as 10.10 but with the envelope lines drawn to show the extent of the vertical integration now described.

Lines of constant apparent magnitude thread the arrangement, calculated from $m = M - 5 + 5\log v/H$ where H is the Hubble constant, giving $m = M + 5\log v(\text{km s}^{-1}) + 16.5$ if $H = 50$ km s^{-1} Mpc^{-1}.

The entry in each box of Fig. 10.10 is the volume contained in the redshift interval along the top, times the number of galaxies at M in the interval $M + 1/2$ and $M - 1/2$ given by the adopted luminosity function. Summing the columns from the upper envelope line (see Fig. 10.11) to the apparent magnitude limit line of the catalog at the faint end, and dividing by the total number of objects in the sample at that redshift gives $M(m, v)$ for objects at that redshift. Note that $M(m, v)$ changes with redshift. It is the correction needed *at each redshift* to account for selection bias.

Summing each column horizontally, appropriately weighted, gives the grand mean $M(m)$ that was calculated by Malmquist, given by equation (10.9).

Figure 10.11 shows this procedure for three samples of limiting magnitude at $12^{\rm m}\!.5$, $14^{\rm m}\!.5$, and $16^{\rm m}\!.5$. The hatched area encloses that part of the $m = 12.5$ sample that is distance limited, defined by the volume whose outer redshift is where the magnitude limit line meets the lower envelope of the Spaenhauer configuration. Figure 10.11 shows why adding a fainter sample increases the volume-limited sample, and why the bias properties of the fainter sample repeat the same properties of the brighter sample but move faintward by $1^{\rm m}$ for every 0.2 dex increase in the redshift.

Figure 10.12 illustrates this property in a schematic calculation using the spread-sheet method of Fig. 10.10 for an assumed Gaussian luminosity function with dispersion of $\sigma(M) = 0^{\rm m}\!.7$. Note that the individual corrections $M_0 - M(m, v)$ are functions of both m and v, and show the isomorphism just mentioned that $2^{\rm m}$ in the limit moves the corrections by 0.4 dex in redshift.

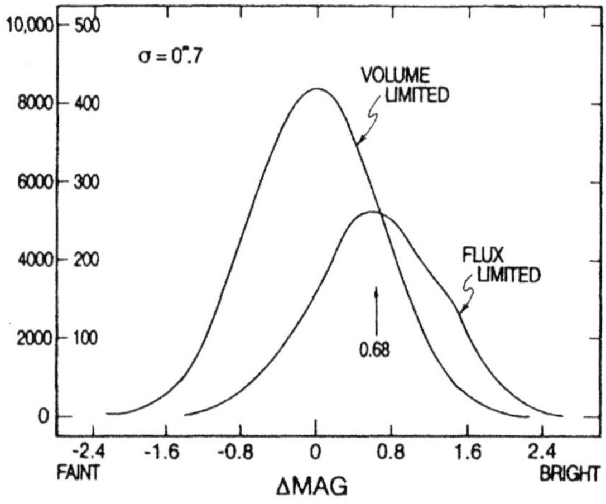

Fig. 10.13. Comparison of the distribution of absolute magnitudes for a volume-limited sample whose intrinsic dispersion is $\sigma(M) = 0^{\rm m}\!.7$ with the distribution of a subset of that sample that is flux-limited. Note that the dispersion of the two distributions are nearly the same but that the mean absolute magnitudes of the samples differ. This *is* the global Malmquist correction. It is composed of the individual $f(m, v)$ corrections when they are integrated over all redshifts, weighted by the volume elements. But then these needed individual corrections have been lost.

The result of a numerical integration using $\sigma(M) = 0^{\rm m}\!.7$, is shown in Fig. 10.13 giving the change in the distribution of absolute magnitudes between the volume-limited sample and the flux-limited sample. This is the Malmquist bias for the total sample, comprised of the sum of the individual biased distributions at each relevant redshift weighted by the relevant volumes. Note that the displacement of the grand mean of the flux-limited distribution is $0^{m}_{.}68$ brightward from the proper volume-limited mean. This, in fact, is given by equation (10.11), as it must if our procedure via the numerical integration using Fig. 10.10 (or 10.11) is correct.

10.5 The Hubble Constant Does Not Increase Outward

If the most probable absolute magnitude, M_0, valid for a distance-limited sample is used for galaxies beyond the distance limit shown as the hatched region of Fig. 10.11, photometric distances become progressively more incorrect at progressively larger redshifts. It is to avoid this progressive error that the corrections similar to those in Fig. 10.12 are required.

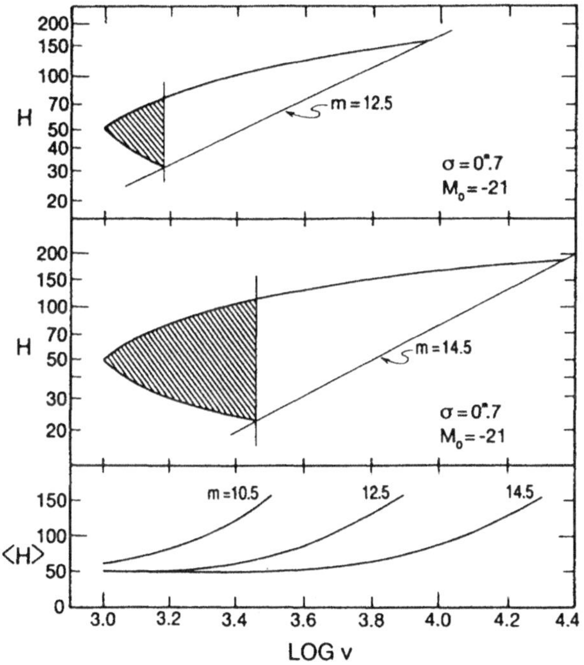

Fig. 10.14. The (log) Hubble constant vs. redshift (distance) for flux-limited samples to two flux levels. An unbiased value of log H can be obtained only within the hatched area. Diagram from S94a.

However, use of only the M_0 calibration, which was the normal practice in the literature of the mid-1980's to mid-1990's, will result in a false signal that the Hubble constant increases outward (e.g. Tully 1988). The reason is seen in Fig. 10.14 which follows in an obvious way from the Spaenhauer diagram formalism in the previous sections.

The characteristic Spaenhauer configuration, but now in H, generated using Fig. 10.5 and shown in Fig. 10.14, will be symmetrical in its upper and

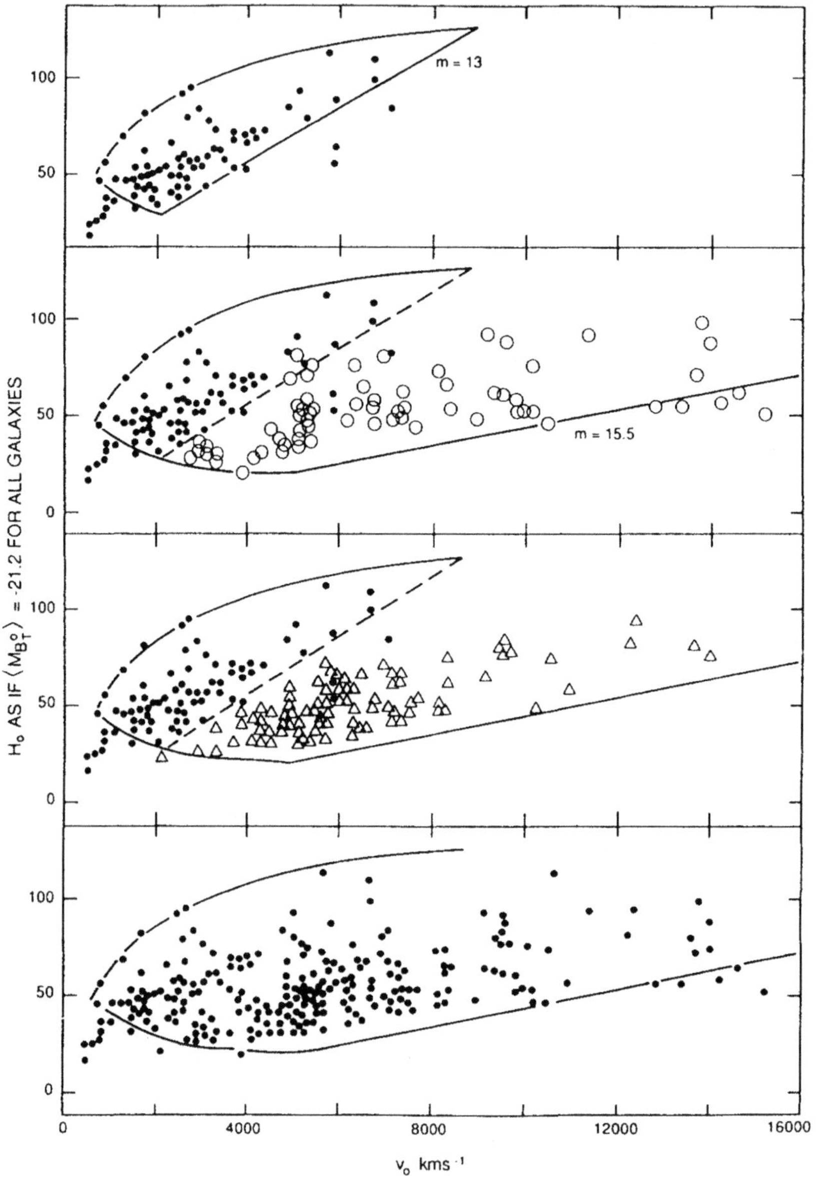

Fig. 10.15. Same as Fig. 10.14 but with the data of Fig. 10.1 superposed. All three catalogs are displayed in the various panels. Diagram from S88a.

lower envelope if the ordinate of the Hubble ratios are plotted as logarithms. The effect of the apparent magnitude limit, progressively cutting away the luminosity function at larger distances, is evident. The galaxies remaining in the sample for redshifts beyond the hatched area give mean Hubble ratios that increase with distance. The bottom panel in Fig. 10.14 shows that without correction for this bias effect, the mean Hubble constant obtained by averaging the data in various redshift intervals appears to be multivalued at any given redshift, depending on the apparent magnitude cutoff. The effect is clearly unphysical and is a result of the bias.

Figure 10.15 shows the result for the mean Hubble ratios using the actual data for the Sb I and Sc I galaxies from Fig. 10.1. The top panel shows that galaxies from the RSA with redshifts larger than 2000 km s^{-1} (magnitude limit of $m = 13^m$) give increasingly larger $\langle H \rangle$ values at larger redshifts. The remaining panels demonstrate the filling of the Spaenhauer configurations by adding the two fainter samples.

Note that the ordinate is in linear measure, not logarithmic. Hence the upper and lower envelopes are not symmetric. Note also that the range over which the sample is distance-limited has increased to 5000 km s^{-1} in the second, third, and fourth panels, compared with 2000 km s^{-1} for the brighter sample in the top panel. This is close to the increased range expected for the $\sim 2^m$ change in the catalog limits (see Fig. 10.8), based again on the isomorphic condition that a change of 0.4 dex in the distance corresponds to a 2^m change in the limits. (The actual change in catalog limit is not precisely known for these data because of the uncertainty of perhaps $0\overset{m}{.}4$ in the faint limit of the two fainter samples. We have used $2\overset{m}{.}5$ in the diagrams but, from the known corrections to the Zwicky magnitudes upon which the estimates for the faint sample are based, the difference could be as small as 2^m.)

10.6 Selection Bias in the Tully-Fisher Method

We have set out in the foregoing sections a method of detecting and correcting the effects of observational selection bias for a single set of objects with a well defined mean absolute magnitude, M_0, of its (true, i.e., distance-limited) luminosity function. We generalize these results in this section for the more complicated case of many sets of objects, each with different mean absolute magnitudes.

Such a case applies to the data for galaxies analyzed for distance by the line width-absolute magnitude correlation discovered by Roberts (1962, 1969) and by Balkowski *et al.* (1974) and developed by Tully and Fisher (1977). Here, line width separates the total sample into discrete sets (binned by line-width interval). We seek a method to detect and correct for bias effects in each set (i.e., at each line width, at each redshift, and for any given catalog flux cutoff). The discussion follows the precepts in the archive literature (S88b; S94b; FST94) to which the reader is referred for details. Only an outline is given here.

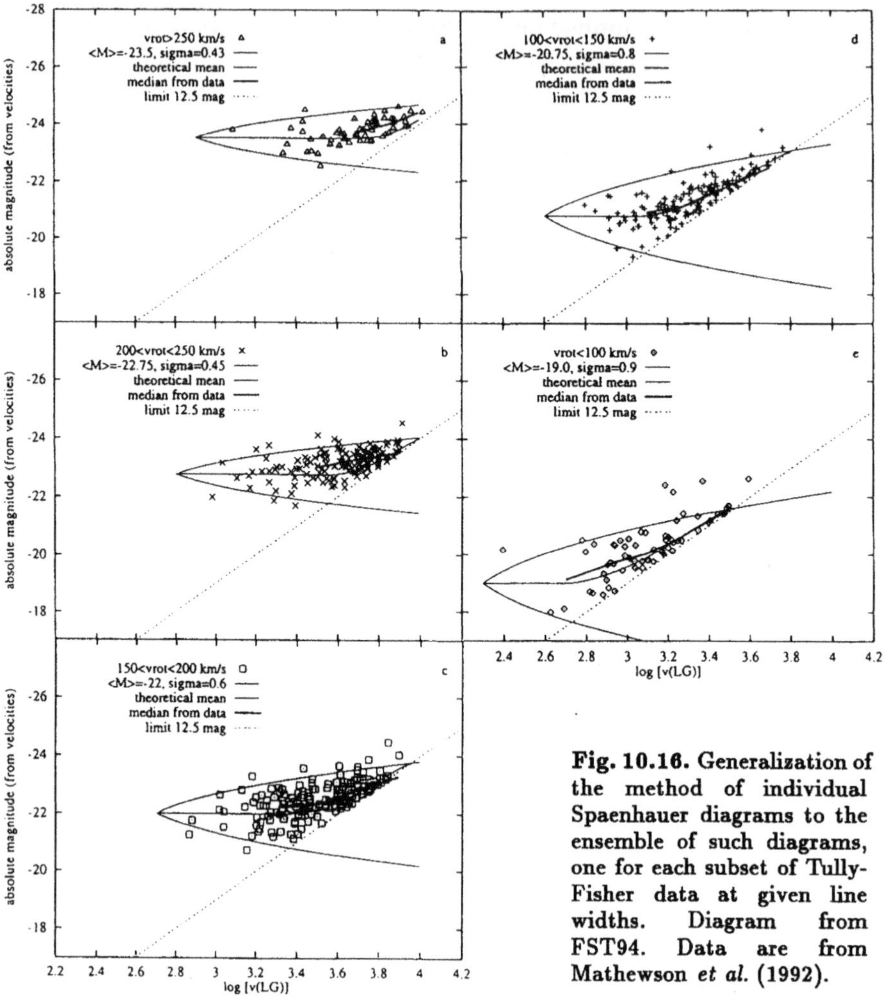

Fig. 10.16. Generalization of the method of individual Spaenhauer diagrams to the ensemble of such diagrams, one for each subset of Tully-Fisher data at given line widths. Diagram from FST94. Data are from Mathewson *et al.* (1992).

Figure 10.16 shows the generalization of the Fig. 10.5, 10.6, and 10.11 Spaenhauer diagrams as applied to the Tully-Fisher case. Individual Spaenhauer configurations are shown for five discrete intervals of rotation velocity (i.e., LW/2). The data are from the large sample of Mathewson, Ford, and Buchhorn (1992; MFB). The diagram and discussion are from Federspiel, Sandage, and Tammann (1994). The "bright", not the "total", sample of MFB is used here, cut at an I magnitude of $I = 12^{\rm m}\!.5$.

The absolute magnitudes along the ordinate are computed from the redshifts ($H_0 = 50$) on the precept that any perturbations on the Hubble flow are smaller than errors in other distance indicators. Justification using available data are given in S94b and FST94. The redshifts along the abscissa are in the kinematic frame of the centroid of the Local Group.

The two items to note in Fig. 10.16 are (1) the dispersions of the distributions increase as the rotational velocities decrease (i.e., reading down the columns and then from left to right), and (2) the intersection of the sloping limit line with the lower envelopes of the individual SDs define the separation of the data into the distance-limited (to the left of the intersection) and the flux-limited parts. An unbiased TF relation can only be obtained from the distance-limited subset of the data.

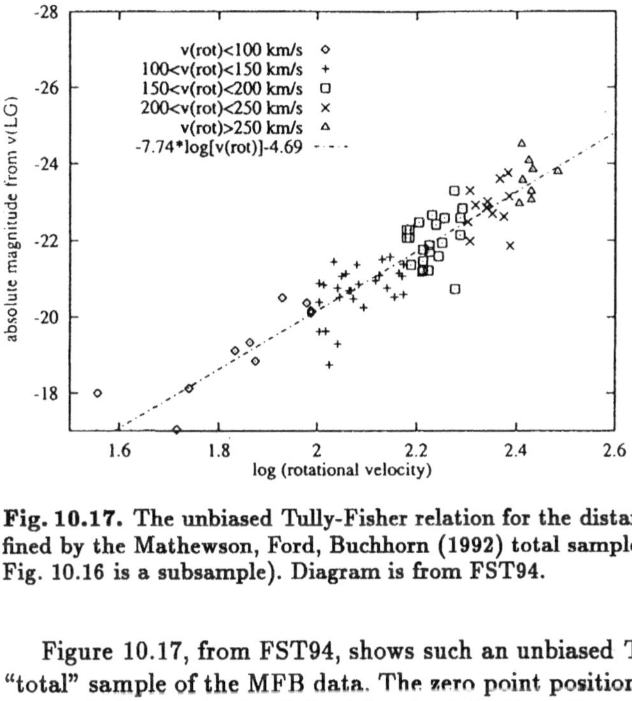

Fig. 10.17. The unbiased Tully-Fisher relation for the distance-limited sample defined by the Mathewson, Ford, Buchhorn (1992) total sample (of which the data in Fig. 10.16 is a subsample). Diagram is from FST94.

Figure 10.17, from FST94, shows such an unbiased TF relation using the "total" sample of the MFB data. The zero point positions in absolute magnitude, which can also be defined by the individual Spaenhauer configurations by the "apex" absolute magnitudes in Fig. 10.16, are seen by inspection to vary from $M_{\mathrm{I}} = -23^{\mathrm{m}}$ for the highest rotational velocities (upper left panel in Fig. 10.16) to $M_{\mathrm{I}} = -19^{\mathrm{m}}$ for the slowest rotators (lower right panel).

The most important bias property of the method is shown in Fig. 10.18 for the "total" MFB sample, showing again the TF correlation. The top panel shows the entire data set. This is broken into seven discrete redshift intervals in the bottom panel where the ridge line in each interval is shown. The progression with redshift, discovered empirically by Kraan-Korteweg, Cameron, and Tammann (1988, their Fig. 1), is predicted by the bias model (S94b; FST94), based on the changing dispersions and the effects of the different limit lines relative to the individual SDs in Fig. 10.17. The multivaluedness, depending on redshift in Fig. 10.18, is clearly unphysical. It is the result of observational selection bias.

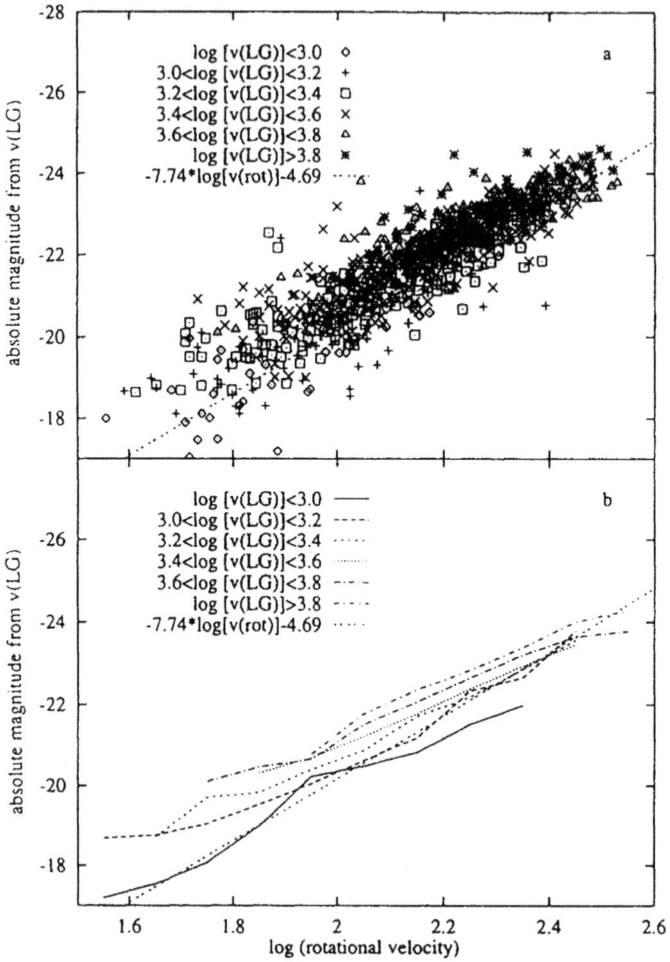

Fig. 10.18. (Top) The Tully-Fisher diagram from the Mathewson *et al.* (1992) data, using kinematic absolute magnitudes ($H_0 = 50$). The distance-limited calibration of Fig. 10.17 is shown near the lower envelope of the data points in the top panel. (Bottom) Mean ridge lines for the data in the top panel for different redshift intervals. The separation is the expected result from the bias model of the text. Correction for observational selection bias removes the separation and the resulting apparent, non-physical, multi-valuedness of the Hubble constant with distance, line width, and catalog apparent magnitude limits.

FST 94 show that correcting for the bias removes the multivaluedness and also removes the concomitant non-physical multivalued Hubble constant that results from a naive (i.e., non-bias-corrected) application of the TF data. The consequence of Fig. 10.18 is the manifestation of a non-physical Hubble constant increasing outward and having a different value for each line-width interval, and each redshift interval. The neglect of the effect and its correction in all the extant discussions (e.g. Aaronson *et al.* 1982; Huchra 1986, 1992; Pierce & Tully 1988; Jacoby *et al.* 1992; van den Bergh 1992) is the reason that each of these authors have derived the incorrectly high value of H_0 using the Tully-Fisher method.

The same effects seen here in the field galaxy sample have been shown to apply also to flux-limited cluster data (Teerikorpi 1987; Kraan-Korteweg, Cameron, and Tammann 1988, KKCT; Fouqué *et al.* 1990; Federspiel, Sandage, & Tammann 1995). The model and the argument will not be repeated here, but the reader can gain an appreciation of the problem and the necessity to address it by study of these four papers.

The one conclusion from the cluster study which we do emphasize is that this "cluster incompleteness bias" has the effect not only of obtaining too large a Hubble constant by as much as 1^{m} (KKCT, their Fig. 6) but, as importantly, gives a dispersion of the TF relation that falsely appears to be too small (Pierce and Tully 1988) by a factor of about two, thereby misleading the practitioners into asserting that the bias problems are four times less severe than they actually are.

We close this lecture with a demonstration that $H_0 = 50$ is required by even the earliest field galaxy data (Aaronson *et al.* 1982) when they are corrected for selection bias. The fact that the cluster data also require the long distance scale (KKCT88; Fouqué *et al.* 1990; FST 1995), also when corrected for bias, should again be noted.

10.7 The Hubble Constant from the TF Method Using a Distance-Limited Field Galaxy Sample

Figure 10.19 shows the TF relation using kinematic absolute magnitudes for the Aaronson *et al.* (1982) sample in the *B* photometric band. The least-squares ridge-line solution using the 249 galaxies in the line-width interval between log LW of 2.35 and 2.75 is drawn with the equation shown in the upper left corner. The envelope lines are put symmetrically about the ridge line.

The dispersion in *M* about this line, read at constant LW, is $\sigma(M) = 0\overset{m}{.}64$. We assert this large dispersion to be real (S94b, FST94), not artificially increased by putative streaming motions as sometimes claimed. The argument is based on the nature of the residuals (photometric distance minus redshift distance) at different redshifts. The *increase* of these "distance" residuals with redshift (FST94, their Fig. 19c) shows the opposite characteristic (the distribution plotted vs. redshift opens outward) from what would obtain if the large dispersion in Fig. 10.19 was caused by streaming motions.

This argument concerning the dispersion is important in view of derived dispersions of half as large in TF relations based on incomplete cluster data (Pierce & Tully 1988). It has been shown that this is due to the "cluster incompleteness selection bias" (Teerikorpi 1987; FST 1995), the true disper-

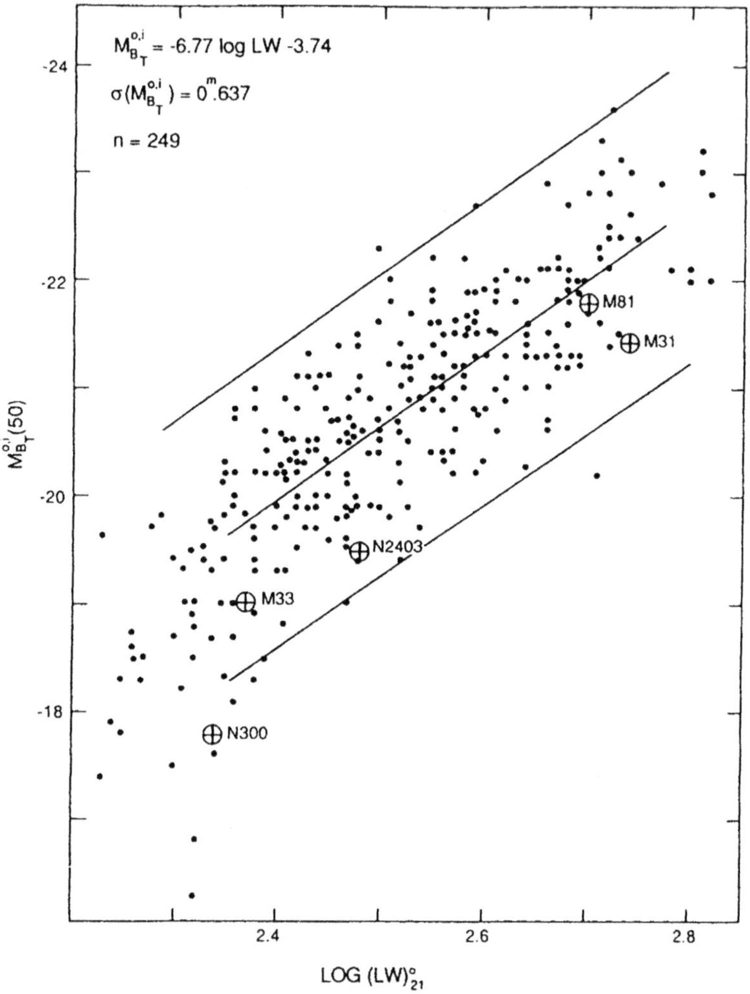

Fig. 10.19. The Tully-Fisher correlation for the Aaronson *et al.* (1982) sample using kinematic absolute magnitudes ($H_0 = 50$). Diagram from S88b. Position of the calibrators are as their absolute magnitudes had been adopted in 1988. More modern calibration values are in Fig. 10.20.

sion being approximately twice the apparent dispersion in such biased cluster samples.

The effect of the global bias in the flux-limited field galaxy data is seen from the faint position of the five calibrators in Fig. 10.19. The absolute magnitude data adopted in 1988 for these calibrators (S88b) have changed in detail, but not in substance. The more modern calibration data (S94) are set out in Fig. 10.20 below. The point to be made from Fig. 10.19 is that the position of the calibrators is confined to the lower half of the spread of

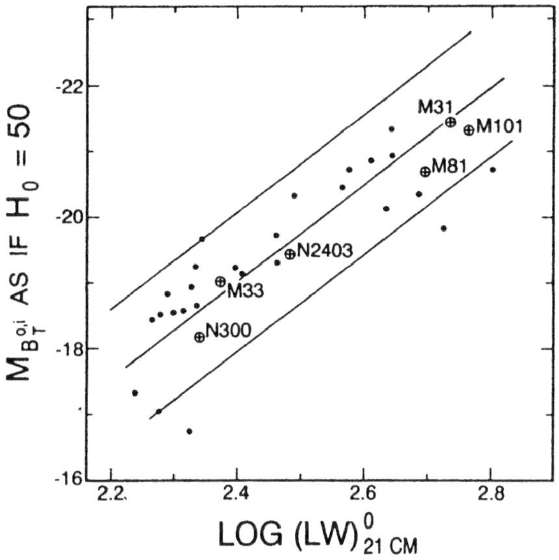

Fig. 10.20. The Tully-Fisher correlation using the distance-limited KKT 500 km s^{-1} sample and the local calibrators. Diagram from S94b. New positions for the calibrators compared with Fig. 10.19 are based on the precepts adopted in 1994.

the total field galaxy sample in the diagram. This is due to selection bias, not due to a real difference, as shown earlier using the model of the bias via the Spaenhauer configurations. The naive interpretation of Fig. 10.19 would have been that the scale of absolute magnitudes along the ordinate, based on $H_0 = 50$, is wrong. However, the ridge line is brighter than the mean of the calibrators because of the selection bias.

Astronomers interpreting Fig. 10.19 in this naive way have derived a magnitude difference of $\sim 0^{\rm m}.9$ which translates to a correction to the assumed Hubble constant of a factor of 1.5, giving $H(\text{naive}) \sim 75$. Note, that with $\sigma(M) = 0^{\rm m}.64$, the full Malmquist bias, which in fact is the effect shown, predicts via eq. (10.11), that the difference between the ridge line and the calibrators should be $0^{\rm m}.6$ if the sample were ideal, similar to what is observed.

Proof that the effect is due to bias is shown in Fig. 10.20. The data are for galaxies in the *distance-limited* sample of Kraan-Korteweg and Tammann (1979) using radio line-width data from Richter and Huchtmeier (1984). The absolute magnitudes along the ordinate are again based on $H_0 = 50$. The data for the six calibrators are based on distances from Cepheids. The agreement is satisfactory between the $H_0 = 50$ field galaxy data and the calibrators, differing from Fig. 10.19 based on the *flux-limited* Aaronson *et al.* sample.

More detailed discussion of the experiment in Fig. 10.20, set out elsewhere (S94b), shows that the six calibrators define a mean absolute magnitude system that is probably $0^{\rm m}.32$ too faint compared with a 64 galaxy calibrating sample from Huchtmeier & Richter (1986). If so, the available field galaxy data, at this writing, give

$$H_0 = 48 \pm 5 \text{ km s}^{-1} \text{ Mpc}^{-1}, \tag{10.12}$$

using the Tully-Fisher method.

References

Aaronson, M. *et al.* 1982, ApJ Suppl. **50**, 241
Balkowski, C., Bottinelli, L., Chamaraux, P., Gouguenheim, L., & Heidmann, J. 1974, A&A **34**, 43
Bottinelli, L., Gouguenheim, L., Paturel, G., & Teerikorpi, P. 1988, ApJ **328**, 4
Federspiel, M., Sandage, A., & Tammann, G.A. 1994, ApJ **430**, 29 (FST94)
Federspiel, M., Sandage, A., & Tammann, G.A. 1995, ApJ submitted
Fouqué, P., Bottinelli, L., Gouguenheim, L., & Paturel, G. 1990, ApJ **349**, 1
Huchtmeier, W.K., & Richter, O.-G. 1986, A&A Suppl. **63**, 325
Huchra, J. 1986 in *13th Texas Symposium on Relativistic Astrophysics*, ed. M.P. Ulmer (Singapore: World Scientific), p. 1
Huchra, J. 1992, Science **256**, 321
Humason, M.L., Mayall, N.U., & Sandage, A. 1956, AJ **61**, 97 (HMS56)
Jacoby, G. *et al.* 1992, PASP **104**, 599
Kraan-Korteweg, R.C. 1986a, A&A Suppl. **66**, 255
Kraan-Korteweg, R.C. 1986b, *A Catalog of 2810 Nearby Galaxies* (Basel Pub. Ser. No. 18)
Kraan-Korteweg, R.C., Cameron, L.M., & Tammann, G.A. 1988, APJ **331**, 620 (KKCT88)
Kraan-Korteweg, R.C., & Tammann, G.A. 1979, Astron. Nach. **300**, 181 (KKT79)
Malmquist, G. 1920, Lund Medd., Ser. 2, No. 22
Mathewson, D.S., Ford, V.L., & Buchhorn, M., 1992, ApJ Suppl. **81**, 413
Pierce, M., & Tully, R.B. 1988, ApJ **330**, 579
Richter, O.-G., & Huchtmeier, W.K. 1984, A&A **132**, 253
Roberts, M.S. 1962, AJ **67**, 437
Roberts, M.S. 1969, AJ **74**, 859
Rubin, V.C., Ford, W.K., Thonnard, N., Roberts, M.S., & Graham, J.A. 1976, AJ **81**, 687
Sandage, A. 1988a, ApJ **331**, 583, (S88a)
Sandage, A. 1988b, ApJ **331**, 605, (S88b)
Sandage, A. 1988c, Ann. Rev. Astron. Astrophys. **26**, 561 (S88c)
Sandage, A. 1994a, ApJ **430**, 1 (S94a)
Sandage, A. 1994b, ApJ **430**, 13 (S94b)
Sandage, A., & Tammann, G.A. 1975, ApJ **197**, 265, (ST)
Sandage, A., Tammann, G.A., & Yahil, A. 1979, ApJ **232**, 352 (STY79)
Sandage, A., and Tammann, G.A. 1981, *A Revised Shaples-Ames Catalog of Bright Galaxies*, Carnegie Institution of Washington Pub. 635 (RSA1)
Spaenhauer, A. 1978, A&A **65**, 313
Tammann, G.A., & Sandage, A. 1983, in *Highlights Astron.* Vol. **6**, ed. R.M. West (Dordrecht: Reidel), p. 301
Teerikorpi, P. 1987, A&A **173**, 39
Tully, R.B. 1988, Nature **334**, 209
Tully, R.B., & Fisher, J.R. 1977, A&A **54**, 661
van den Bergh, S. 1992, PASP **104**, 861

Evolution in the Galaxy Population

Richard G. Kron

University of Chicago, Yerkes Observatory, Box 0258, Williams Bay, WI 53191, USA

1 Foundations of Galaxy Evolution Models

1.1 Introduction

This series of lectures will describe some of what is known about normal galaxies at high redshift, namely those galaxies common in flux-limited surveys. The motivation is to search for the systematic differences in the properties of these galaxies with respect to nearby samples, and to use these observations as constraints on ideas for how galaxies may have changed with cosmic epoch. (The term "galaxy evolution" is used universally in this context, but "galaxy aging" might better describe the phenomenon we are looking for.)

There are several good reasons why normal field galaxies are useful probes of the deep universe. In the first place, by definition they shine by star light, and since we know something about stars and stellar populations, we have a basis for interpreting what we see. Second, the extended images of galaxies provide a wealth of information that can be exploited if one has sufficient angular resolution. Third, galaxies are abundant, which means that large statistical samples can be amassed and appropriate statistical tests devised. Finally, nature has provided us with a sky that is dark and transparent in the visible band, where much of the radiation is to be found.

1.2 Cosmology and Evolution

The appearance of distant galaxies depends on cosmological parameters. Here we will consider the Friedman-Robertson-Walker models with zero cosmological constant, as discussed in detail in the parallel lectures by A. Sandage and M. Longair. For reasons related to the ages of star clusters, the Hubble constant is adopted to be 50 km $\sec^{-1}$ Mpc^{-1}. The cosmological model is then specified by the value of the deceleration parameter q_0. The value $q_0 = 0.5$ will be assumed in numerical examples.

Galaxies may appear systematically brighter at high redshift either because their stellar content was intrinsically more luminous at a younger age, or because of the properties of the cosmological model. If one of these two effects were known, the other could be determined. (Other things that are not directly related to either of these can also affect the appearance of distant galaxies. For example, the intergalactic medium may absorb light from distant sources, and gravitational lensing may be important in some instances.) In practice, neither galaxy evolution nor the cosmological model is generally considered to be known well enough to enable a clean test for the other; hence, the use of galaxies as cosmological probes and their study for evolutionary effects continue to be intertwined.

Nevertheless there are some ways to separate evolution from cosmology. The apparent brightness of a standard candle as a function of redshift can only have a particular form that depends on the cosmological model, whereas evolutionary trends are constrained only by our relatively poor understanding of the astrophysical processes. As described by Sandage, there are a variety of types of cosmological tests: in addition to the brightness - redshift test, there is also the number - redshift test, for example. The effects of galaxy evolution and cosmology have different dependencies in these different tests, thus allowing them to be distinguished in principle. A clean test for galaxy evolution is provided by the dependence of the dimming of surface brightness on redshift, since this effect is independent of the cosmological model (see Sandage, Lecture 6).

To keep track of these possibilities, it is helpful to realize that cosmological effects may change the *amplitude* of some observable, such as how big or bright something looks, or how many things are counted in a certain volume, but cosmological effects cannot change the *shape* of something, such as the shape of a galaxy image profile, the shape of a spectrum, or the shape of a statistical distribution function (like the luminosity function). This aspect of the problem provides one of the most powerful tools available for separating cosmological from evolutionary effects.

For example, one can compare the spectrum of a distant galaxy with that of a nearby galaxy which is deemed to be of the same type. Figure 1.1 shows an example of this procedure, where the data refer to three distant first-ranked cluster galaxies, and the solid line refers to a composite of nearby cluster ellipticals. If all of these galaxies are coeval, then the ones at higher redshift will have a stellar population with a main-sequence turnoff at hotter temperature, and hence the galaxies would be expected to have a bit more flux at shorter wavelengths. If indeed these galaxies were observed to be systematically bluer in their rest-frame, the amount of luminosity change corresponding to the color change could be deduced from a model for the stellar population, and applied as a correction to the apparent brightness to enable a cosmological test.

Regardless of whether one is trying to conduct a cosmological test or trying to find evidence for galaxy evolution, one proceeds by comparing a statistical sample of galaxies at high redshift with a comparable sample at

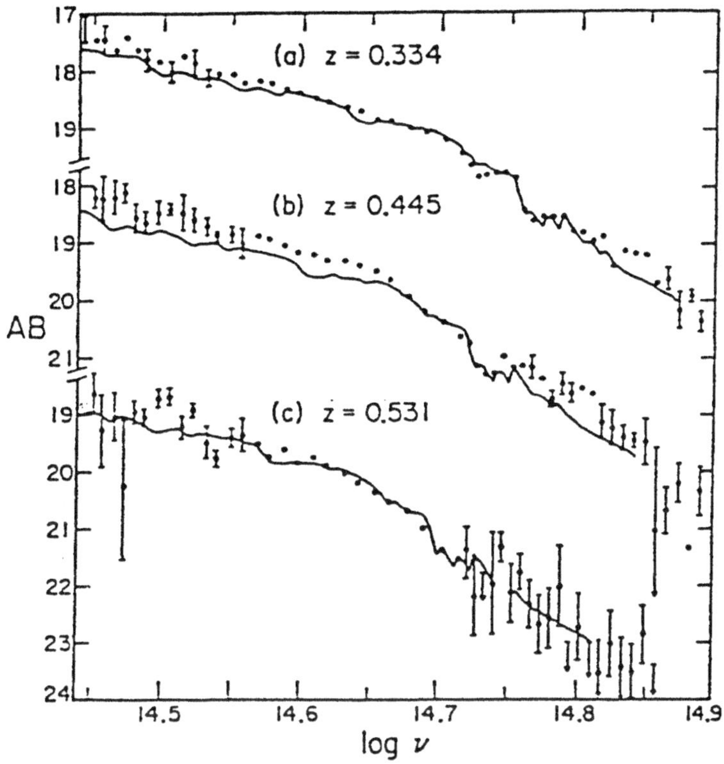

Fig. 1.1. Spectra of high-redshift luminous cluster galaxies from Oke (1984). "AB" is defined as AB = $-2.5 \log f_\nu - 48.60$, where f_ν is in erg sec^{-1} cm^{-2} Hz^{-1}. The solid line is a template spectrum derived from local elliptical galaxies, shifted to the appropriate redshift. The galaxies are (top to bottom) 2154.2+0508; 2158.0+0429; 2142.6+0348.

low redshift. Unfortunately, one does not normally observe volume-limited samples, but rather flux-limited or diameter-limited samples. One aspect of the problem is illustrated in Figure 1.2, which shows schematically the distribution function of some observable property (for example, the rest-frame colors) for a nearby sample and for a distant sample. The shape of the distribution has evidently changed, but there is no information on the nature of the mapping from one distribution to the other. Two galaxies with the same spectral shapes at the present epoch may have had different colors in the past, and *vice versa.* The galaxies at the peak of one distribution do not necessarily correspond to the galaxies at the peak of the other distribution. Moreover, if galaxies are selected to be within some range of the rest-frame color at all redshifts, it clearly does not guarantee that they are really drawn from the same population of objects – indeed they cannot have been if their absolute number densities differ at different redshifts, as indicated by the shaded area

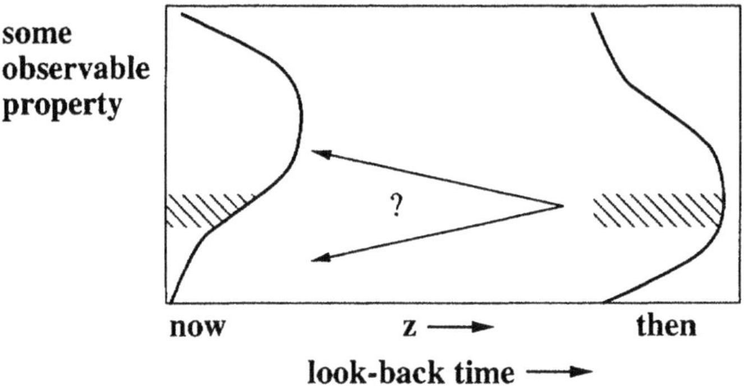

Fig. 1.2. A generic distribution of an observable property of galaxies is plotted against cosmic time or redshift. While the overall distribution can be traced as a function of redshift, the trajectories of individual galaxies are ambiguous.

in Figure 1.2. There are two lessons here: first, in general it is safer to define samples as inclusively as possible and look at whole distribution functions as opposed to a narrow range of properties; and second, in addition to the properties of direct interest, one must also measure other independent properties that serve to control for selection effects.

1.3 Star Light

Since the luminosity density in the local Universe in the optical and near-infrared bands can be shown to be completely dominated by light from stars (as opposed to active nuclei), these lectures will concentrate on star light. Some star light is of course absorbed by dust and re-emitted in the far-infrared spectral range, but only in spectacular cases like IRAS F10214+4724 is the detection of this far-infrared emission currently possible for galaxies at high redshift.

It was recognized many years ago that from the colors of galaxies one might be able to deduce something about their stellar constituents (Whipple 1935). The color temperature of spirals was known to be hotter than that of ellipticals, implying a greater proportion of hotter stars. The radiation field in the solar neighborhood is dominated by A dwarf stars and K giant stars, with some smaller contribution from F dwarf stars. Since the F stars have intermediate temperature, we can anticipate that their exact fractional contribution might be hard to determine from the integrated spectrum. Indeed, Aaronson (1978) showed that the near-infrared colors of galaxies could be remarkably well modeled by the sum of different proportions of A dwarf and M giant stars, and a similar approach to spectral classification of galaxies has more recently been pursued by Bershady (1994). In a color-color diagram, normal galaxies fall along a well-defined sequence of colors, suggesting that there is a single dominant parameter, presumably to be identified with the ratio of numbers of hot dwarf stars to cool giant stars. Models such as those of Struck-Marcell and Tinsley (1978) have shown that the effect of reddening internal to the galaxies moves their colors approximately parallel to the prevailing locus, as does the effect of blanketing from a different assumed metal abundance. It appears that colors provide a useful index of the basic mix of stellar temperatures, but that a detailed dissection of the stellar population using colors alone would be limited by uncertainties in these other effects.

The baseline in wavelength from which these conclusions are derived is less than one decade, from the U band (0.36 μm) to the K band (2.2 μm). In principle, a handle on a second parameter could be obtained by extending the baseline to shorter and longer wavelengths. In practice, shorter wavelengths are severely affected by absorption by dust, and longer wavelengths sample emission from dust. An alternative to extending the baseline in wavelength is to expand the spectral resolution: then, the model for the stellar population must not only reproduce the general flux distribution, but also line strengths. A pioneering attempt along these lines was the work of Morgan and Mayall (1957). The crucial point about galaxies that was emphasized by these authors (Humason had earlier called attention to this) was the correlation between the integrated spectrum and the appearance of a galaxy – whether extended or more centrally concentrated (disk-like or bulge-like). The sense was that galaxies with a more diffuse distribution of light tended to have spectra in the blue with relatively strong Balmer lines, i.e., A-type spectra, and galaxies with more condensed profiles generally had spectra similar to K giants. These considerations lead to the Yerkes classification system for galaxies, which is a *morphological* system based on the degree of central concentration of the light, but uses the nomenclature *a, f, g,* and *k* for the sequence. The connection between the observable properties of image profile shape, surface brightness, and spectral type, must rank as one of the most important features of the optical appearance of galaxies.

Nevertheless, the Yerkes system is still basically a sequence from hot to cool, and it remains to see what spectroscopy can reveal about the details of the stellar population of a galaxy. Examples of things that might be derivable from a spectrum are the age of the oldest generation of stars that are making a detectable contribution to the light; whether the light is dwarf-dominated or giant-dominated; whether there is any constraint on the mass-to-light ratio for the stellar population; the metallicity; and whether it is possible to construct a picture for the build-up of the population, i.e., the history of star formation in the galaxy.

Photoelectric scanner measurements were first used to disentangle the contributing stellar types (Spinrad and Taylor 1971). The basic idea is that one constructs a "library" of spectra of stars that are candidates for the types of stars contributing to the light from other galaxies, and one then tries to find what combination of these stars best synthesizes the integrated spectrum of different types of galaxy. Such efforts have continued with better spectral resolution, better signal-to-noise ratio, and libraries that include a wider variety of stars, especially evolved stars with high abundances. As it turns out, the uniqueness of the spectral synthesis is hard to demonstrate for basically the same reasons it is hard to extract more than a certain amount of information from broad-band colors: the light from minority populations of stars is veiled by light from the majority populations. While the spectrum of a galaxy may appear to be rich in information in the sense that there are numerous well-detected lines, in fact much of this information is effectively degenerate because of the strong correlation of stellar spectra at similar effective temperature.

Populations with very different age and metallicity can still have nearly identical giant branches, and their spectra would be correspondingly similar (e.g., Worthey 1992). The similarity of the integrated spectra of galaxies that are dominated by K giants is due to the insensitivity of the colors of K giants to their respective masses. Another problem that is repeatedly encountered is that the libraries of stellar spectra that are used in these synthesis models are deficient in some types of stars that are known to be important in galaxies, for example stars with high [Mg/Fe] (Worthey 1992).

Another example of why we need to be cautious in the detailed interpretation of spectra is the comparison by O'Connell (1976) of the spectra of M31 and the spectra of giant ellipticals in the Virgo cluster. The spectra are remarkably similar over the whole optical range, regarding both the broad-band colors and the line strengths. Still, it is hard to imagine that the stellar populations are really closely similar, considering that the M31 observations refer to a region of 100 parsecs around the nucleus of a disk galaxy, and the Virgo observations refer to the inner 800 pc of massive spheroidal systems that presumably had a different formation history.

In spite of these various concerns, the spectral synthesis results of, for example, Turnrose (1976) for late-type spirals (which includes a determination of the internal reddening), and the spectral synthesis results of Pickles (1985) for a number of giant ellipticals in the Fornax cluster, are illuminating and worth careful study.

Many of the spectral synthesis efforts have derived conclusions that were not strictly empirical, but have relied on sensible constraints for the stellar population. For example, it is reasonable to suppose that the number of stars on the main sequence increases monotonically and smoothly towards lower masses. Also, the number of giants can be supposed to be related to the number of stars now evolving off the main sequence.

1.4 The Star-Formation Rate and the Initial Mass Function

Instead of applying more and more such astrophysical constraints to a stellar synthesis, one can adopt a different approach called evolutionary synthesis. Here, one constructs artificial galaxies that contain stellar populations characterized by different assumed forms for the initial stellar mass function and the history of star formation. At each time step, the evolution of stars in each interval of mass is computed in the Hertzsprung-Russell diagram, and the integrated luminosity and spectrum is calculated. It can then be established what region of this model parameter space (mass function, star formation history, age) matches real galaxy spectra. This approach is described in the next lecture; here, we introduce the subject by reviewing what is known about the rate of star formation and the initial stellar mass function, at least for the solar neighborhood. Because of the central role that these two key functions play in astronomy, they have been repeatedly studied, good examples being the papers of Rana and Wilkinson (1986) for the history of star formation in the Milky Way, and Rana (1987) for the mass function in the solar neighborhood. In fact, the determination of these functions are connected, since in order to derive the initial mass function from the present-day observed mass function of field stars, one needs to know the history of star formation.

No disastrous inconsistencies result from the assumption that the rate of star formation in the solar neighborhood was about the same in the past several Gyr as it is now. More quantitatively, the model of Rana and Wilkinson (1986) has the star-formation rate changing by no more than about 50% over the past 15 Gyr. If star formation occurs episodically, the effects of such fluctuations can be considered separately.

In the nearest galaxies for which the brightest stars can be resolved, we can determine directly the current rate of formation of massive stars by counting them and adjusting for their lifetimes. For more distant populations, an observation of the ultraviolet energy distribution yields essentially equivalent information: for a given stellar mass function, the ultraviolet luminosity is directly proportional to the star-formation rate. (Unfortunately, it is difficult to determine accurately the slope of the mass function at the high-mass end from the observed energy distribution because good knowledge of the effect of dust is needed.) The emission in the Lyman continuum is partially reprocessed as Hα emission in the gas; hence the flux of Hα can be used as a diagnostic of the number of massive stars currently alive. A similar measure is provided by the thermal radio continuum flux from H II regions. The flux from massive stars is also partly absorbed by circumstellar and interstellar dust associated with sites of star formation, and the re-processed continuum radiation is detectable at far-infrared wavelengths.

A somewhat different technique for quantifying the current rate of massive star formation takes advantage of the "flocculent" appearance of regions that are actively forming stars. By separating the fraction of light that appears at high spatial frequency from that which appears at low spatial frequency, one can isolate the contribution to the total light from OB associations (Isserstedt and Schindler 1986).

All of these techniques yield only partial information about the past history of the rate of star formation, in the sense that what is measured is the current (or at least very recent) rate of formation of massive stars. Even in the solar neighborhood, where we have an opportunity to explore the age distribution of lower-mass stars, the picture is surprisingly sketchy. If the measured current rates of formation of massive stars are assumed to be representative of the past, and extrapolated to give the total consumption rate of gas due to formation of stars of all masses, then a large fraction of giant disk galaxies would deplete their supply of gas in less than 5 Gyr. Some interpretations of this result are reviewed by Kennicutt (1983).

The initial mass function in the solar neighborhood may be simply described as a power-law in the mass. According to Rana (1987), this description is valid for $M > 1.5\ M_\odot$; at lower masses, the mass function is much flatter and may display bumps and dips. The detailed shape of the initial mass function may depend on time if the prevailing conditions for star formation depend on time. For example, the supply of gas and the metallicity in the protostellar gas change with time, and the metallicity is expected to influence the fragmentation process in protostellar clouds. The nature of the dust also depends on the metallicity and the intensity of the ultraviolet radiation field. The ability of gas to condense into clouds is determined in part by the energetics of bulk motion in the interstellar medium, which is governed by hot stellar winds, supernovae, and galactic dynamics.

The slope of the initial mass function at the main-sequence turnoff governs the rate at which stars enter the red giant phase, and therefore has a profound effect on the evolution of the light of a single generation of stars. The characteristic age of the Galactic disk corresponds to a main-sequence turn-off mass of about $1\ M_\odot$, and 5 Gyr ago the turnoff mass for the oldest stars was not much greater. Therefore, for old stellar systems generally, the slope of the mass function needs to be known accurately only in a restricted range of stellar masses to calculate the evolution of the spectrum. As just mentioned, there may be a change in the power-law slope for masses below $1.5\ M_\odot$, and one must be appropriately cautious.

Upper and lower limits need to be specified for the initial mass function. The lower mass limit does not directly affect the light (as long as it is moderately low compared to $1\ M_\odot$), but it does affect the amount of mass locked up in stars (most of which do not return any gas, processed or otherwise, over the age of that generation). Besides being important for computing the amount of gaseous material left over for subsequent generations of stars, the lower mass limit is important for computing the overall galaxy mass-to-light ratio (M/L) as a constraint. In contrast, the upper mass limit affects the bolometric luminosity very much, but the details are sensitive to the instantaneous birth rate and the fraction of the total luminous mass of the galaxy that is in the form of young stars. The heating, ionization, and bulk motion of the interstellar medium also depend on the upper mass limit.

References

Aaronson, M. 1978, ApJ **221**, L103
Bershady, M.A. 1994, AJ in press
Isserstedt, J. and Schindler, R. 1986, A&A **167**, 11
Kennicutt, R.C. 1983, ApJ **272**, 54
Morgan, W.W., and Mayall, N.U. 1957, PASP **69**, 291
Oke, J.B. 1984, in *Clusters and Groups of Galaxies*, eds. F. Mardirossian, G. Giuricin, and M. Mezzetti (Dordrecht: Reidel), p. 99
O'Connell, R.W. 1976, ApJ **206**, 370
Pickles, A. 1985, ApJ **296**, 340
Rana, N.C. 1987, A&A **184**, 104
Rana, N.C. and Wilkinson, D.A. 1986, MNRAS **218**, 497
Spinrad, H. and Taylor, B.J. 1971, ApJS **22**, 445
Struck-Marcell, C. and Tinsley, B.M. 1978, ApJ **221**, 562
Turnrose, B.E. 1976, ApJ **210**, 33
Whipple, F. 1935, Harvard College Observatory Circ. No. 404
Worthey, G. 1992, PhD thesis, University of California, Santa Cruz

2 Evolutionary Synthesis Models

The goal of galaxy evolution models is to predict the time behavior of the rest-frame spectra for different types of galaxy. From such information, the broad-band colors, line strengths, and other spectral indices can be calculated. In order to compute the net flux at each wavelength at each time step, it is necessary to specify the mix of luminous stellar types, the gaseous emission, and the effect of absorption by dust.

2.1 Evolutionary Synthesis Models

As described in Lecture 1, the mix of stellar types depends on the mass function and the rate of star formation. To keep the number of adjustable parameters to a minimum, most models assume a shape for the mass function (and high- and low-mass limits) that is temporally and spatially invariant. The shape is almost always taken to be a power-law in the mass, at least over some particular range in mass (Salpeter 1955).

Once the past history of the rate of star formation has been specified, each star can be evolved in the Hertzsprung-Russell diagram, and the net galaxy spectrum at each time step is then the integral over all luminous stars. Therefore, one must also specify the stellar evolutionary isochrones, plus the emergent spectrum to be associated with each part of the Hertzsprung-Russell diagram. Both depend on metallicity, but the metallicity can be calculated because the same stellar evolutionary models also provide the yield of heavy elements and the fraction of the mass of a star that is eventually returned to the interstellar medium.

The actual modeling procedure can perhaps be made clearer by following a single stellar generation. Suppose we start by turning instantaneously some fraction of the primordial gas reservoir of a galaxy into stars. The initial abundances are the primordial cosmic mix, and we can also allow for some pre-enrichment. We specify the initial mass function including its upper and lower limits; since the total mass in stars is known, this yields the absolute number of stars born within a given range of mass. If we have model atmospheres for zero-metallicity stars, we can calculate the spectrum of the ensemble. As time goes by, the most massive stars die, the total luminosity decreases, and the color temperature of the galaxy cools. The rate of these changes rapidly slows as the turnoff mass on the main sequence decreases, such that at ages of 10 to 15 Gyr the evolution in both spectral shape and in bolometric luminosity are very small indeed.

The gas released from this first generation becomes available for subsequent generations of stars, and qualitatively the same time dependencies apply to these chemically enriched generations. One important difference between the first generation and subsequent generations is the appearance of dust. A galaxy evolution model thus consists of the specification of the critical parameters to allow the spectrum and luminosity of a single generation to be tracked, plus the ability to compute the chemical evolution and a formula for adding different generations together.

The initial mass function $\phi(m)$ for star formation is clearly of critical importance, and it is worthwhile to present in greater detail the particular ways that the mass function appears in these calculations. In the first place, the *evolutionary flux* of stars through different regions of the Hertzsprung-Russell diagram depends on the mass function (see, for example, the discussion in Renzini 1993). For simplicity, let us consider a coeval stellar population of uniform metallicity with an age such that the mass of a star at the main sequence turnoff is $m_{\rm TO}$. If N_j is the number of stars in a particular post-main-sequence evolutionary phase j, and t_j is the time spent by stars in that phase, then

$$N_j/t_j \approx \phi(m)_{\rm TO}|({\rm d}m/{\rm d}t)_{\rm TO}|,$$

since for masses below $m_{\rm TO}$ the evolution is comparatively weak and for masses above $m_{\rm TO}$ the evolution is both strong and rapid with respect to main-sequence lifetimes.

The mass function $\phi(m)$ also governs the *returned mass fraction* R of gas that goes back into the interstellar medium:

$$R = \int_{m_{\rm TO}}^{m_{\rm u}} (m - w_{\rm m})\ \phi(m){\rm d}m,$$

where $w_{\rm m}$ is the mass of stellar remnants and $m_{\rm u}$ is the upper limit of the mass function.

Finally, the *element production yield* y is defined as

$$y = \frac{1}{(1 \mid R)} \int_{m_{\rm TO}}^{m_{\rm u}} m\ p_{\rm zm}\ \phi(m){\rm d}m,$$

where $p_{\rm zm}$ is the mass fraction that is converted to metals and subsequently ejected from the star. [The notation used here follows Tinsley (1980).] Stellar interior models are required to specify the values for $({\rm d}m/{\rm d}t)_{\rm TO}$, $w_{\rm m}$, and $p_{\rm zm}$ as well as of course the dependence of the turnoff mass $m_{\rm TO}$ on the age of the population.

Since in the solar neighborhood we can perform a census of the population and determine the distribution of ages and chemical abundances, it is natural that the solar neighborhood has taken on a special significance in terms of providing boundary conditions that models which are supposed to represent spiral disks are required to match. Some of the specific constraints at the present epoch are:

- the age - metallicity relation (e.g., Carlberg *et al.* 1985);
- the abundance in the interstellar gas (e.g., Talbot and Arnett 1975);
- the rate of star formation (e.g., Rana and Wilkinson 1986);

- the broad-band colors;
- the fraction of mass in the form of gas;
- the mass-to-light ratio.

With the exception of the age - metallicity relation for dwarf stars, these properties can also be measured in other galaxies. In fact, it is necessary to appeal to other galaxies for some data. For example, the rate of supernovae is of critical importance in several applications, but the rate would be very poorly determined from observations in the solar neighborhood alone.

2.2 Evolution of the Solar Neighborhood

A good example of an evolutionary synthesis model based on properties of the Milky Way is that of Mazzei, Xu, and De Zotti (1992). Specifically, the age-metallicity relation, the present metallicity of the interstellar medium, the present rate of star formation, and the stellar mass function, are all based directly on observations in the solar neighborhood. The predicted spectrum at an age of 10 to 15 Gyr is similar to the actual integrated spectrum in the vicinity of the Sun, which of course is a necessary consistency check. The model accounts for the effect of chemical evolution on the integrated light and incorporates absorption and emission from dust. Another important feature of this model is the use of the isochrones of Bertelli *et al.* (1990), which provide a consistent grid of models extending up to and beyond the asymptotic giant branch phase. (Other efforts have required patching together different interiors models in different mass ranges, and using a semi-empirical approach to the specification of later stages of stellar evolution.)

More specifically, the Mazzei, Xu, and De Zotti (1992) model has the following ingredients. The initial mass function is simply that of Salpeter (1955),

$$\phi(m)\, \mathrm{d}m = A\, m^{-2.35}\, \mathrm{d}m,$$

where A is a normalization factor. The lower mass limit is taken to be 0.01 $M_\odot$, and the upper mass limit is taken to be 100 $M_\odot$.

The rate of star formation is assumed to depend linearly on the mass fraction of gas, f_g:

$$\Psi(t) = \Psi_0\, f_g\, M_0\, \mathrm{yr}^{-1}$$

where $f_g = m_{gas}/m_{disk}$ and $\Psi_0 = 12.3(m_{disk}/10^{11}\mathrm{M}_\odot)\mathrm{M}_\odot\mathrm{yr}^{-1}$. At an age of 10 to 15 Gyr, this formulation leads to a current rate of star formation that is close to what is observed (Rana and Wilkinson 1986). No infall of gas is assumed. To reproduce the age-metallicity relation, it is assumed that the initial metallicity of the disk gas $Z_g(0) = 0.005$. The fraction of mass ejected by stars during the course of their evolution – both enriched and unenriched – has been adopted from Renzini and Voli (1981) and from Maeder (1981).

According to Mazzei, Xu, and De Zotti (1992), the general results of the evolutionary synthesis model (change of luminosity of the stellar population with time, color evolution, etc.) are not sensitive to the power of the gas density assumed in the expression for the rate of star formation, as long as other parameters are adjusted so that the current rate of star formation and other key functions agree with what is observed.

This model stellar population is thus especially simple: the mass function is assumed to be a single power-law, the star formation depends on just the gas fraction, and the stellar interiors are derived from a single set of models. It is designed to be consistent with the overall picture for the characteristics of the solar neighborhood reviewed by Rana and Wilkinson (1986) and Rana (1987). The simplicity of this model stellar population makes it convenient for comparison with other models.

The luminosity of the model is calculated as a function of time and as a function of wavelength. Results are given for the individual contributions of phases of stellar evolution (e.g., main sequence, red giants, asymptotic giants), and for two assumptions regarding the internal extinction. Infrared emission from both circumstellar envelopes and from diffusely distributed dust is also calculated; these evolve with time, according to the absorbed star light.

The luminosity and color evolution are plotted in Figure 2.1. The luminosity evolves very slowly: the model is only $0\overset{m}{.}26$ brighter in rest-frame V at $t = 10$ Gyr than at $t = 15$ Gyr. This behavior reflects the slow change in the rate of star formation, which itself is limited by the model gas supply and is consistent with the metallicity data in the Milky Way. The color evolution is also slow, for basically the same reason. Moreover, the actual colors are close to what is observed in real galaxies. The best example is the $V - K$ color shown in Figure 2.1d, which shows that the model is close to the observed average colors of Sb and Sc galaxies, unlike the other models shown, which are consistently too blue in this color index. This success may be due to the treatment of the asymptotic giant branch in the Mazzei, Xu, and De Zotti model. The evolution of both rest-frame $V - K$ and $U - V$ is slow, with changes of only tenths of a magnitude in many Gyr.

There are a few observables that are not well matched by the Mazzei, Xu, and De Zotti (1992) model. One of these is the present-day gas fraction, which is observed to be about 0.10, and predicted by the model to be about 0.3. Another is the mass-to-light ratio, which according to the model is 13 in the B band, too high for normal disk populations. Both of these parameters depend on the lower stellar mass limit, but in different ways: if the lower mass limit is increased, then the M/L will be smaller, as desired, but the residual gas fraction and the metallicity will not match. These problems can be dealt with by adding extra free parameters (such as infall of gas).

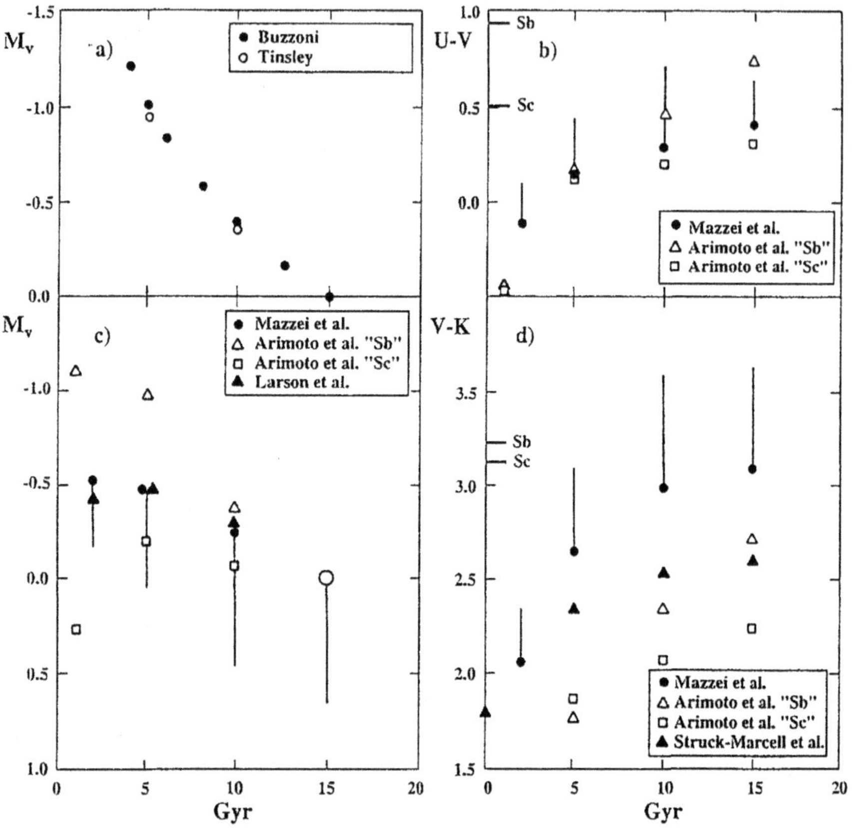

Fig. 2.1. a) Evolution of the V-band light from an elliptical galaxy (filled dots). The model is Buzzoni's (1989) $Y = 0.25$, $Z = 0.017$, and $\eta = 0.3$. The open circles are the one-parameter Tinsley model, $\mathrm{d}M_V/\mathrm{d}\ln t = 1.3 - 0.3x$, where x is the slope of the mass function (assumed here equal to 1.0).

b) Models for the evolution of the V-band light for disk galaxies. The normalization of the luminosity scale of each model is at 15 Gyr, and is indicated by the common large circle. The bars connect the dust-free model model of Mazzei, Xu, and De Zotti (1992) (top of bar) with the same model including absorption by dust (bottom of bar). The other models are from Arimoto, Yoshii, and Takahara (1992) and Larson, Tinsley, and Caldwell (1980).

c) Predicted change of the rest-frame $U - V$ color of disk galaxies, from the same models as shown in Figure 2.1b. The present-day mean colors of Sb and Sc galaxies are indicated.

d) Predicted change of the rest-frame $V - K$ color of disk galaxies, from some of the same models as shown in Figure 2.1b, plus the model of Struck-Marcell and Tinsley (1978). The present-day mean colors of Sb and Sc galaxies are indicated.

2.3 Effect of the Interstellar Medium

Emission from gas and absorption by dust may affect the broad-band colors of galaxies, a possibility explored in the evolutionary synthesis models of Guiderdoni and Rocca-Volmerange (1987). They adopt nine distinct prescriptions for the star formation rate as a function of time to represent that number of galaxy types. Gas emission is included by scaling from the observed emission from H II regions in the Milky Way and other nearby galaxies. Essentially one counts the number of Lyman continuum photons generated by the OB stars, and corrects for the photons absorbed by dust or which escape from the galaxy. The remainder are converted into Balmer lines, and other important lines can be added based on their relative empirical strengths. Adjustment of the fraction of photons that escape guarantees that the net line strengths match those observed in nearby galaxies. (A self-consistency check is also available in principle, since the current star formation rates and information on the upper end of the mass function is derived from these same observations of Balmer equivalent widths.)

The treatment of internal extinction due to dust is idealized by assuming that the dust, gas, and stars are mixed everywhere in the same proportions, and that there are no effects due to chemical evolution. The key parameter is the column density of gas (inclination effects are easily incorporated by assuming that the disk is thin with respect to its diameter). The column density of gas evolves according to the astration rate. More specifically, the assumed optical depth due to dust depends on the column density of metals in the gas, where empirically $\tau_\lambda \sim Z_{\rm g}(t)^{3/2}$. Since $Z_{\rm g}(t)$ is determined by the evolutionary model, the absorption can be computed with a few other parameters derived from observation: the average extinction for a given column density of gas, and the extinction as a function of wavelength. Mazzei, Xu, and De Zotti (1992) adopted this same model for the effects of the interstellar medium, except that they allowed also for chemical evolution in the gas and followed the consequences through for the dust absorption.

A feature of the Guiderdoni and Rocca-Volmerange prescription for the internal extinction is that the effect is maximized for a gas fraction of around 0.2. The B-band absorption can be quite large at high inclination angles, but more typically ($i = 30°$) $A_{\rm B} = 0\overset{\rm m}{.}4$. The effect on the $B - V$ color is mild, being at $i = 30°$ less than $0\overset{\rm m}{.}1$ (and very insensitive to the gas fraction).

The effect of the emission lines on the broad-band fluxes are not more than a few tens of percent. For example, blue galaxies ($B - V \sim 0.5$) are predicted to have an equivalent width in Hα + [N II] of about 40 Å, which would not greatly perturb the R-band magnitude which is measured over a band of $\sim$ 1500 Å.

The net conclusions of the Guiderdoni and Rocca-Volmerange (1987) models are at least qualitatively similar to what we deduced earlier: the evolution in optical luminosity is expected to be small. Quantitatively, for their intermediate galaxy classes, which are supposed to span the range of giant spiral galaxies, the evolutionary brightening is about $0^{\mathrm{m}}.1$/Gyr for the past 10 Gyr or so. Five Gyr ago, these same galaxy types were on average only $0^{\mathrm{m}}.2$ bluer in rest-frame $B - V$.

2.4 Other Disk Galaxy Models

Bruzual and Charlot (1993) have developed a numerical technique called *isochrone synthesis*, which provides a way to interpolate smoothly between a coarse grid of stellar masses. They have included an extensive library of stellar spectra, which includes for the first time spectrophotometry of stars in the near-infrared bands. These models are based on stars with solar metallicity, but they nevertheless match well the observed spectra of giant ellipticals, normal spirals, hot disk systems, and star clusters in the Galaxy and in the Magellanic Clouds. They confirm that the light in elliptical galaxies at $\lambda < 2500$ Å may arise from planetary nebula nuclei as opposed to residual formation of massive stars. Another result confirms the slow evolution in the $V - K$ color for galaxies at late times previously found by Arimoto and Yoshii (1986) and others. Thus the K-band and V-band fluxes vary at about the same rate.

It is worth commenting that most of the galaxy evolution models have no dependence on the physical scale of the galaxy, that is, in these models small and large galaxies evolve as scaled versions of each other. Arimoto and Yoshii (1986) attempted to include explicitly the connection between depth of the potential well and the ability to retain processed gas. In the "self-propagating star formation" picture, an essential ingredient is the statistical fluctuations due to patches of star formation, which would naturally be greater in galaxies of lower luminosity (Gerola, Seiden, and Schulman 1980). If star formation is enhanced in mergers of small systems with larger systems, and if galaxies grow substantially by such accretion, then larger galaxies may have had more merger events and a higher specific amount of metal enrichment, as reviewed by Zaritsky (1993).

Arimoto, Yoshii, and Takahara (1992), like Mazzei, Xu, and De Zotti (1992), have considered a model for disk galaxies that explicitly accounts for chemical evolution. (They also adopt the Salpeter mass function, albeit with different upper and lower limits.) Arimoto, Yoshii, and Takahara (1992) explore the consequences of assuming an appreciable infall of gas. The rate of infall is adjusted so that the age-metallicity relation is reproduced. In practice, the quantity $\beta = a/\alpha\Psi$ is fixed for a given model, where a is the rate of infall, α is the fraction of mass in stars that is not recycled over the age of the galaxy, and Ψ is the star formation rate. Arimoto, Yoshii, and Takahara (1992) also explore what happens if the rate of star formation depends on the square of the gas fraction (with a different value of β to maintain agreement with the age-metallicity relation). They conclude that galaxies evolve in the past several Gyr in much the same way, regardless of which formulation is chosen – infall or closed system, and which power of the gas fraction determines the rate of star formation. This result breaks down at early epochs, where the particular form of the rate of star formation is paramount.

Some of the earliest detailed evolutionary synthesis models were by Tinsley in the 1960's and 1970's, who developed most of the conceptual foundations of this field. It is interesting to see how closely these older models agree with the newer generation which include spectral energy distributions for each contributing stellar type. One example is the series of disk models of Larson, Tinsley, and Caldwell (1980), where the evolution of M_V and $B - V$ is computed as a function of the characteristic time scale for gas depletion due to astration. The special case of constant star formation was presented earlier by Larson and Tinsley (1978). The particular case of $\tau = 10$ Gyr, the characteristic time scale for star formation, is compared in Figure 2.1b with the more recent models; at least for this choice of τ, the agreement is excellent, which suggests that the conclusions are generally robust.

2.5 Single-Generation Models

The evolution of a single generation of stars of uniform metallicity has been considered by many authors, and exemplified by the detailed treatment of Buzzoni (1989). The principal purpose of single-generation (also called "single-burst" and "initial-burst") models is to serve as templates for the evolution of elliptical galaxies and perhaps the bulges of spirals, since the present-day properties of such systems can be understood on this basis (Renzini 1993). Buzzoni (1989) gives tables for the time evolution of such single-generation, single-metallicity models for a wide range of input parameters. These include the metallicity, the helium abundance, the power-law slope of the mass function s, the mass-loss parameter η, and the morphology of the adopted horizontal branch. Buzzoni's tables give several descriptive data aside from the bolometric luminosity and the rest-frame broad-band colors, such as the mass at the main-sequence turnoff, the fractional contributions to the total light from various evolutionary phases, and the mass-to-light ratio in various bands. The particular model $Z = 0.017$, $Y = 0.25$, $s = 2.35$, and $\eta = 0.3$ is shown in Figure 2.1a.

In summary, a variety of models for the evolution of stellar populations have been computed. In many instances, taste governs the particular choices – for example, one can choose to use the Salpeter mass function as do Mazzei, Xu, and De Zotti (1992) on the grounds of limiting the number of adjustable parameters, or one can use a mass function that is more complex (e.g., Guiderdoni and Rocca-Volmerange 1987) on the grounds that it is more realistic. The same could be said of the choice of whether to include the effects of infall of gas (e.g. Arimoto, Yoshii, and Takahara 1992). Different models assume different forms for the past history of star formation in galaxies of different types, but the constraints of smoothness in the dependence of amount of mass in gas on time and the present-day residual gas fraction guarantee that these assumed forms are generically similar. One can choose to include the effects of the interstellar medium, but it may be only at the level of tenths of a magnitude and the inclusion of absorption and emission does add the complication of specifying the evolution of this component as well. The atmospheres to be associated with the interiors models can be chosen to be either theoretical or empirical. The advantage of using theoretical atmospheres is that a wide range of wavelengths may be available, the grid in effective temperature and surface gravity may be relatively fine, and the bolometric corrections are known precisely. The advantage of using an empirical library of stellar spectra is that all sources of opacity are automatically included. Finally, different interiors models may be chosen, from which one derives not only the energy released during different phases of evolution and the durations of these phases, but also the mass of enriched and unenriched material that is expelled from the star. The Bertelli *et al.* (1990) models address specifically the effect of convective overshoot. The overshoot parameter can be adjusted so that theoretical color-magnitude diagrams match those of the globular clusters in the Magellanic Clouds, for example. Besides the overshoot parameter, other important parameters like the helium abundance and the mass-loss parameter need to be specified. There are still some features that need to be put in by hand, such as the morphology of the horizontal branch. Even though there may appear to be a large number of parameters that can be adjusted, there are also a lot of constraints, particularly the constraint of matching the relative numbers of stars in each region of globular cluster color-magnitude diagrams.

2.6 Summary

In spite of this wealth of choices, galaxy evolution models are in qualitative and even quantitative agreement on the important general consequences, such as the evolution in luminosity and in color. The evolution in luminosity is expected to be very mild, amounting to about $0^{\rm m}_{.}1$/Gyr for typical galaxies (Figure 2.1). Elliptical galaxies tend to brighten somewhat faster, but late-type galaxies are expected to be *fainter* in the past. Perhaps one-third of the present-day luminosity density is due to bulge light (Lecture 3). However, as detailed in later lectures, flux-limited samples strongly favor intrinsically bluer galaxies at high redshifts and faint fluxes. Thus, the evolutionary brightening expected for bulges may not dominate the overall result expected for faint-galaxy samples, and the luminosity evolution will be correspondingly even milder. The same is true for colors because at all cosmic times except the earliest, each generation of young stars that contribute to the blue light also generates K giants that contribute to the red light. Hence, the mean rest-frame color of galaxies selected in a flux-limited sample may also change very slowly with time (Figure 2.1). This is the basic reason why the "no-evolution" case is a convenient – and even realistic – null hypothesis against which to test observations. These general conclusions could be changed if the rate of star formation in galaxies were much higher at some observable recent epoch, perhaps because of a greater rate of galaxy-galaxy merging. However, the extent of this greater rate of processing is constrained by the present-day metallicity, star-formation rate, and residual gas fraction.

Notwithstanding the agreement of the models in such general predictions as the luminosity and color evolution of galaxies, we should not lose sight of some critical assumptions that could be in error in important ways. Chief among these is the stellar initial mass function, which is almost always assumed to be not only the same for all galaxy types (and within different regions of the same galaxy), but also the same at all times. The simplification universally adopted merely reflects both our ignorance of the details of star formation under different conditions, and the absence of strong theoretical reason to do otherwise.

References

Arimoto, N. and Yoshii, Y. 1986, A&A **164**, 260
Arimoto, N., Yoshii, Y., and Takahara, F. 1992, A&A **253**, 21
Bertelli, G., Betto, R., Bressan, A., Chiosi, C., Nasi, F., Vallenari, A. 1990, A&AS **85**, 845
Bruzual A., G. and Charlot, S. 1993, ApJ **405**, 538
Buzzoni, A. 1989, ApJS **71**, 817
Carlberg, R.G., Dawson, P.C., Hsu, T., and Vanden Berg, D.A. 1985, ApJ **294**, 674
Gerola, H., Seiden, P.E., and Schulman, L.S. 1980, ApJ **242**, 517
Guiderdoni, B. and Rocca-Volmerange, B. 1987, A&A **186**, 1
Larson, R.B. and Tinsley, B.M. 1978, ApJ **219**, 46
Larson, R.B., Tinsley, B.M., and Caldwell, C.N. 1980, ApJ **237**, 692
Mazzei, P., Xu, C., and De Zotti, G. 1992, A&A **256**, 45
Maeder, A. 1981, in *The Most Massive Stars*, eds. S. D'Odorico, D. Baade, K. Kjär (Garching: European Southern Observatory), p. 173
Rana, N.C. 1987, A&A **184**, 104
Rana, N.C. and Wilkinson, D.A. 1986, MNRAS **218**, 497
Renzini, A. 1993, in *Galaxy Formation*, eds. J. Silk and N. Vittorio (in press)
Renzini, A. and Voli, M. 1981, A&A **94**, 175
Salpeter, E.E. 1955, ApJ **121**, 161
Struck-Marcell, C. and Tinsley, B.M. 1978, ApJ **221**, 445
Talbot, R.J. and Arnett, W.D. 1975, ApJ **197**, 551
Tinsley, B.M. 1980, Fundamentals of Cosmic Physics, **5**, 287
Zaritsky, D. 1993, PASP **105**, 1006

3 Basic Statistics of Galaxies

In this lecture we will review what is known about the relative frequencies of galaxies with different star formation rates at the present epoch. There are also several other relevant statistical distribution functions, such as the luminosity function, parameters that describe the distribution of light in an image, and how these relate to the rate of star formation. Before discussing these, we first outline an elementary framework that may help interpret some of the observations of galaxies at visible and near-infrared wavelengths.

Galaxies can be usefully classified into the groups *giants* and *dwarfs*. Giant galaxies can usually be characterized by a *bulge* component and a *disk* component, where the bulge population is generally dominated by light from metal-rich red giant stars that are arguably old. The disk population is generally like the solar neighborhood, and can be characterized by the fraction of its light from a young generation – that is, some disk populations look almost like bulge populations, and some are dominated by light from young massive stars. On this basis, one can classify a giant galaxy on the basis of two dominant parameters: the ratio of bulge light to disk light, and a measure of the relative youth of the disk light. Hubble's scheme implicitly takes account of these parameters, combining them into a one-parameter sequence. Since the detection of a distant galaxy depends both on its image structure and on its spectral energy distribution, it is worthwhile to maintain a two-dimensional classification.

3.1 Giant Galaxies

The radial distribution of light in the central regions of a typical giant galaxy can be described well by four parameters: a surface brightness and size scale for the bulge component, and a surface brightness and size scale for the disk component. Examples of approximate functional forms are, for bulges (Schweizer 1979),

$$I_{\mathrm{B}} = \frac{I_{0,\mathrm{B}}}{(r/r_c)^2 + 1}, \tag{3.1}$$

and for disks (Freeman 1970),

$$I_{\mathrm{D}} = I_{0,\mathrm{D}} \exp(-r/h). \tag{3.2}$$

In practice, it is often acceptable to regard disks as having similar central surface brightness $I_{0,\mathrm{B}}$, differing mainly in their size scale h. For bulges, the surface brightness scale (which is more often characterized by the surface brightness at the radius that contains half the total light, called the effective surface brightness I_{e} and the effective radius r_{e}, respectively) depends on luminosity (Kormendy and Djorgovski 1989).

Naturally, there are a number of practical complications. The "decomposition" of the image into disk and bulge components may be sensitive to the inclination angle of the galaxy – for example, it may be impossible to detect a weak disk if it is face-on in a bulge-dominated system. Moreover, one needs to consider the effect of internal absorption by dust. The disk-to-bulge ratio (D/B) will depend on wavelength since the color of the disk is generally bluer than the color of the bulge, and if there are color gradients in either component, the interpretation of the derived parameters is dependent on the wavelength of observation in an even more complicated way.

An important question is the relative contribution to the luminosity density in the Universe from "bulge-type" light and from "disk-type" light. This was considered by Schechter and Dressler (1987), who found that the disk-to-bulge ratio was correlated with total luminosity, in the sense that more luminous galaxies have lower D/B. When integrated over all giant galaxies, the ratio of disk light to bulge light was about 2:1 in the V band. (If one were to account for the different mass-to-light ratios for these different populations, it turns out that approximately equal amounts of *mass* are represented by the two populations.)

Some systems may have turned gas into stars more efficiently because they satisfy a threshold surface density of gas in the disk (Kennicutt 1989). That is, if the gas surface density is below a threshold (which depends on dynamical parameters for the disk, specifically the Toomre stability parameter Q), star formation will proceed at a much lower rate. This might explain why the visible disks of galaxies tend to have a cutoff at around 4.5 scale lengths (van der Kruit 1987), and why dwarf galaxies often have a low rate of star formation for their current gas content.

Bothun (1990) and colleagues (e.g., van der Hulst *et al.* 1993) have called attention to the existence of giant galaxies that may be overlooked because their disks have low surface brightness. Since some such galaxies are hydrogen-rich, the low surface brightness may be related to a relatively feeble rate of turning gas into stars because of not satisfying the threshold gas density criterion. The total luminosities of some of these galaxies may be high because of the very long scale lengths of their disks. It is not yet clear what fraction of the local luminosity density is due to these galaxies, and whether the selection effects against their inclusion into catalogs are strong enough to bias extrapolations of properties of the local Universe to high redshift.

Radial gradients in the stellar populations of galaxies will in general result in color gradients. At high redshift, the shape of the profile of the galaxy will appear to change because of the different spectral qualities of the center and the outer regions, and the detectability of the galaxy becomes more complicated to compute. As a zeroth approximation, we expect spiral galaxies to be redder at their centers because of the bulge population. As a first approximation, we expect gradients in the disk itself, since from a variety of arguments it appears that the rate of star formation has been highest in the center: more gas has been consumed in the inner regions, there is less current star formation, and the colors are correspondingly redder.

In practice, this expected trend from redder colors in the inner regions to bluer colors in the outer regions is not always obvious, as in the sample of spiral disks studied by Wevers, van der Kruit, and Allen (1986). Kent (1986) has plotted several $B-I$ color profiles, with the same conclusion: while mean colors may get somewhat bluer with increasing radius (decreasing surface brightness), the trend is mild and is masked by the considerable scatter within individual galaxies and between different galaxies. De Jong and van der Kruit (1994) have established a larger sample which includes near-infrared surface photometry, which does show a mean trend to bluer colors at larger radii.

3.2 Dwarf Galaxies

The correlation between D/B and luminosity is consistent with the notion that bulges – at least high-surface brightness bulges – appear only at luminosities higher than a certain threshold. Low-luminosity galaxies are either describable as a disk without an appreciable bulge component (but possibly with a bright nucleus), or they are describable by a bulge-type profile, but where the parameters place the galaxies into a qualitatively different category – the dwarf spheroidals. A recent review of the morphological characteristics of dwarf galaxies has been given by Binggeli (1994). Besides morphology as measured either by dimensionless shape parameters or by physical scales, the dwarf galaxies are also distinct in their stellar population, which tends to have lower metallicity and often to have a higher gas mass to total mass ratio than giant galaxies. There is typically a wide dispersion in the ratio of gas mass to total mass and the ratio of total mass to light, which may also reflect the statistics of a small number of star-forming sites. These features suggest that dwarf galaxies are less evolved in the sense that there have been fewer cycles of gas through stars.

Besides being bluer and having lower surface brightness and lower metallicity, dwarf galaxies often have irregular shapes. If star formation occurs episodically in OB associations with a characteristic size and luminosity, then the smaller the galaxy, the greater the relative importance of a single site of new star formation. Thus the patchy appearance of many low-luminosity galaxies may be less apparent at long wavelengths and when averaged over time. When the image profile looks sufficiently well concentrated and symmetrical to permit fitting a radial profile, the profile is often like that of an exponential disk.

It is important to determine the statistical properties of dwarf galaxies because there are a number of reasons to expect their contribution to flux-limited samples to increase with decreasing flux of a survey, as described later.

3.3 Distribution Functions

The distributions per unit volume that matter most for calculating the appearance of faint-galaxy populations are those of luminosity, surface brightness, central concentration of light, and spectral energy distribution. It is also of interest to know how the distributions of these properties may depend on local density or other environmental influences. In principle, we would like to know how galaxies populate the multi-dimensional space defined by these properties, but unfortunately no completely adequate catalogue exists for this purpose. Hence, we will concern ourselves with the traditional marginal distributions.

The Luminosity Function

The luminosity function $\phi(M)$ specifies the number of galaxies per cubic Mpc with luminosity in the range $M \pm 0\overset{\text{m}}{.}5$. The functional form is often parameterized by a characteristic space density ϕ^* and a characteristic luminosity M^*, but as we will see below this simplification may obscure important information.

If galaxies were uniformly distributed in transparent space, then the number of galaxies per square degree, $A(m)$, seen in the apparent magnitude interval $m \pm 0\overset{\text{m}}{.}5$ would be related to $\phi(M)$ by the fundamental equation of stellar statistics:

$$A(m) = 800\pi \ln 10 \; 10^{0.6m} \int \phi(M) \; 10^{-0.6M} \mathrm{d}M. \tag{3.3}$$

The integrand is a bell-shaped curve which has an important interpretation – it gives directly the frequency distribution of absolute magnitudes that appear in a flux-limited sample of sources, where, as might be expected, the peak is near M^* (Figure 3.1). The half-width of this curve at half-maximum is about $1\overset{\text{m}}{.}25$. This is why the notion of "galaxies of ordinary luminosity" is often abbreviated to "M^* galaxies". The value of the integral evidently provides the important datum of the normalization for the counts of galaxies. Figure 3.1 shows the effect of varying the slope parameter α for the faint end of the luminosity function, using Schechter's (1976) formulation. The point is that a small uncertainty in α can make a large difference in the galaxies sampled.

However, the cosmological volume element is not Euclidean, and the apparent magnitude m samples progressively shorter wavelengths at higher redshifts. Therefore, the integral is not actually independent of distance (or m or z). The effect is that as the depth of a survey increases, the peak of the bell curve shifts to values of M that are progressively fainter than M^*, and the slope of the counts,

$$\mathrm{d} \log A(m)/\mathrm{d}m = \gamma,$$

shifts to values that are smaller than 0.6.

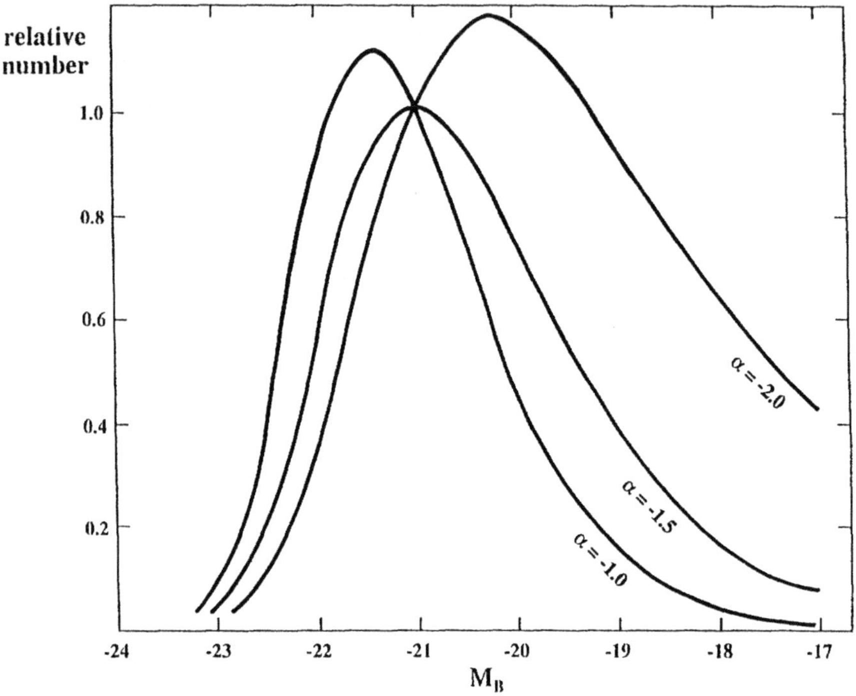

Fig. 3.1. The function $\phi(m) \times 10^{-0.6M}$, where $\phi(M)$ is the Schechter (1976) galaxy luminosity function with $M^* = -21^{\mathrm{m}}.0$ and $\alpha = -1.0, -1.5$, and -2.0. The function gives the distribution of absolute magnitudes in a flux-limited sample; there is a close connection between this function and the distribution of redshifts in a flux-limited sample.

(In many applications of faint-galaxy counts, the slope γ happens to be close to $0.43 = (\ln 10)^{-1}$. If γ were constant at this value, the integral counts $N(m)$, namely the number of galaxies brighter than m, would be equal to the differential counts $A(m)$. This is why it is convenient to express the differential counts $A(m)$ in an interval $\Delta m = 1^{\mathrm{m}}.0$.)

In rich clusters of galaxies, the luminosity function can be readily determined if the cluster surface density overwhelms the projected field, or if membership can be reliably assigned on morphological grounds. The luminosity function so determined in the Virgo cluster has been studied extensively by Sandage, Binggeli, and Tammann (1985), and a similar study has been conducted in the Fornax cluster (Ferguson and Sandage 1988). These papers show that the shape of the luminosity function (and M^* and the relative ϕ^*) is strongly dependent on the type of galaxy. Giant galaxies have bell-shaped luminosity functions (and different Hubble-types have different M^* and relative ϕ^*), whereas the number of dwarf galaxies increases exponentially at fainter luminosities. See Binggeli, Sandage, and Tammann (1988) for a comprehensive review of the B-band luminosity function when considered by morphological type.

The luminosity function broken down by Hubble morphological class by Sandage, Binggeli, and Tammann (1985) shows that the number of types S0–Sc declines at lower luminosity. These types have a bulge component, and part of this behavior reflects the absence of low-luminosity bulges. In addition, it is evident that the existence of a spiral pattern in the disk requires a particular gravitational potential, such that low-luminosity galaxies cannot support a spiral pattern.

The determination of the luminosity function for field galaxies naturally requires having redshifts for a complete sample, and a volume must be surveyed that is large enough to average out effects of large-scale structure. (The obvious strong density fluctuations from one place to another should alert us to the possibility that the shape and other features of the luminosity function could also fluctuate over large regions - see Santiago and Strauss 1992.) Determinations of the field-galaxy luminosity function have recently been made by Efstathiou, Ellis, and Peterson (1988), de Lapparent, Geller, and Huchra (1989), and Loveday *et al.* (1992). In all these cases, the luminosity function is presented in the two categories of "spirals" and "ellipticals". As will be emphasized in the following, the very poor connection between morphological type and intrinsic color means that the luminosity function may look quite different when computed according to groups of *color*. (Also, it would be of great importance to know what the luminosity function looks like when divided into classes of surface brightness.)

Color Distribution and Relation to Morphological Type

The distribution of colors of galaxies is important because the galaxy-evolution models are constrained by the present-day colors. In particular, the recent rate of star formation governs the observed colors more than any other parameter. Hence, we would like to know the distribution of galaxy colors in order to specify the "mix" of the individual galaxy-type models. At least at bright magnitudes, there are however remarkably few adequate determinations of the field-galaxy color distribution. The distribution was examined by Tinsley and Danly (1980) in a similar context; since then, the field-galaxy color distribution has been presented by Butcher and Oemler (1985) and only a few others. The distribution has a peak somewhat blueward of $B - V = 1\overset{\text{m}}{.}0$ (due to bulge-dominated galaxies), with a broad shoulder to bluer colors, declining rapidly around $B - V = 0\overset{\text{m}}{.}5$. The ratio of number of galaxies in the shoulder to galaxies in the peak is consistent with Schechter and Dressler's (1987) ratio of 2:1 of disk light to bulge light.

Aside from constraining the evolution models, the color distribution is needed because any calculation of the appearance of high-redshift field galaxies depends on knowing the relative importance of different kinds of spectral energy distributions. The "color distribution" could be specified per unit volume, but it may be more convenient to have the distribution of a flux-limited sample. As the flux limit is decreased and the redshift increased, the observed-frame color distribution naturally changes, but that effect can be modeled in a straightforward way, as described in the next lecture.

It is tempting to derive the color distribution from the morphological type distribution, since the luminosity functions are presented in terms of morphological type (and so are the k-corrections, unfortunately). There are two problems with this, however. The less fundamental problem is that different determinations of the field-galaxy morphological type distribution are radically different. For example, in the Peterson *et al.* (1986) sample, 43% of the galaxies are classified as "E" or "S0", and only 5% of the galaxies are classified as "Sc", whereas in the *Revised Shapley-Ames Catalog* (Sandage and Tammann 1981) these same fractions are 28% and 25%, respectively.

The more fundamental problem is that there is in fact *a very poor relationship between Hubble class and color.* The very broad range of color at a given Hubble type for field galaxies can be seen in $U-B$, $B-V$ plots of de Vaucouleurs and de Vaucouleurs (1972) and Huchra (1977). This same poor correlation also applies in clusters, as can be seen for example in the plot of $(B-V)_0$ vs. Hubble type for galaxies in the Pegasus cluster by Bothun, Schommer, and Sullivan (1982).

Color–Luminosity Relation

Galaxies show a trend between color and luminosity in the sense that more luminous galaxies are redder. The effect is present in both elliptical and spiral types, but may be due to different origins. More luminous ellipticals have stronger lines, indicating that the effect is driven by metallicity. More massive ellipticals have deeper potential wells, and therefore their ability to retain processed gas is enhanced. In spirals, the mixture of different generations and the prevalence of dust make the interpretation more convoluted. More massive spirals may have a disproportional amount of internal reddening (Peletier and Willner 1992).

The color-luminosity relation for elliptical galaxies is dramatically evident in color-magnitude diagrams for rich clusters of galaxies, where each interval of absolute magnitude is well populated and thus the red envelope is well defined. The effect is also clear in looser clusters that contain spirals. For example, Pierce and Tully (1992) show a diagram of $B-I$ versus $\log W$, where W is the 21-cm line width (and for purposes of this discussion is equivalent to luminosity), for spirals in Virgo and Ursa Major, as well as for a complete sample of field galaxies. The mean colors change by 1^{m} over a luminosity range of 8^{m}. A similar strong correlation between the $B-H$ color and $\log W$ was found by Peletier and Willner (1992). Still another example is a survey of field galaxies by Mobasher, Ellis, and Sharples (1986) which includes near-infrared (JHK) band photometry; here the slope $\Delta(B_J-K)/\Delta M_K$ is about 0.2. Giant galaxies may be characterized by one color-luminosity relation and dwarfs by another. Dwarfs as a class are bluer than giants, and display among themselves a color-luminosity relation that is less steep but of the same sign.

Regardless of the details of the form of the color- luminosity relation, the existence of such a relation means that *the shape of the luminosity function must depend on both the waveband of observation and on the spectral type of the galaxy*. The sense is that the luminosity function is expected to have a steeper slope $\Delta \log \phi(M)/\Delta M$ at the faint end for bluer galaxies. This is just what was observed by Shanks (1990), who constructed luminosity functions for field galaxies according to rest-frame $B - V$ color.

Disk Scale Lengths

While the exponential-disk formulation of Eq. 3.2 may be only approximate in many cases, it does provide a useful format for discussion. A galaxy so describable has an integrated luminosity $L = 2\pi\, h^2\, I_0$. Instead of the luminosity function $\phi(L)$, we can consider the bi-dimensional distribution of the two parameters h and I_0, $\phi(h, I_0)$.

One sample that provides a useful look at the distribution of these parameters is Watanabe's (1983) photometry of disk galaxies in the Virgo and Ursa Major clusters. The parameters I_0 and h were derived and plotted by van der Kruit (1987), and shown in Figure 3.2. I have added a locus labelled "V = 10" that corresponds to M^* galaxies. Note the well-known result that the dispersion in I_0 seems to be relatively small for luminous disks, and at least approximately independent of the value of h. (The plotted quantity is the *apparent* central surface brightness, which will of course be affected by the bulge. In the usual discussion of the variation of I_0, the bulge light is already subtracted from the profile.) It does however appear that low-luminosity galaxies may have somewhat fainter central surface brightnesses, which is confirmed by other studies (e.g. Cornell *et al.* 1987).

The Diameter Function

The distribution of intrinsic sizes of galaxies has not been studied nearly as extensively as the distribution of luminosities, despite the fact that isophotal diameters are readily measured. The nature of the diameter function can be expected to differ from the luminosity function on account of variations in the surface brightness of galaxies, and on account of variations in the distribution of light as measured by a dimensionless shape parameter. Van der Kruit (1987) concluded from a modest-sized sample of disk galaxies that the distribution of scale lengths h in a volume-limited sample declines by a factor of e for each increase in h of 1 kpc. Bardelli *et al.* (1991) used the much larger ESO galaxy profile survey of Lauberts and Valentijn (1989) to derive the diameter function for field galaxies (not just disk systems) at a B-isophote of 25^{m} arcsec^{-2}. Clearly, this is operationally very different from van der Kruit's procedure of fitting an exponential disk and deriving the scale length, and it yields diameters that are about 5 times larger that depend on the type of the galaxy. (In fact, one of the main points of the Bardelli *et*

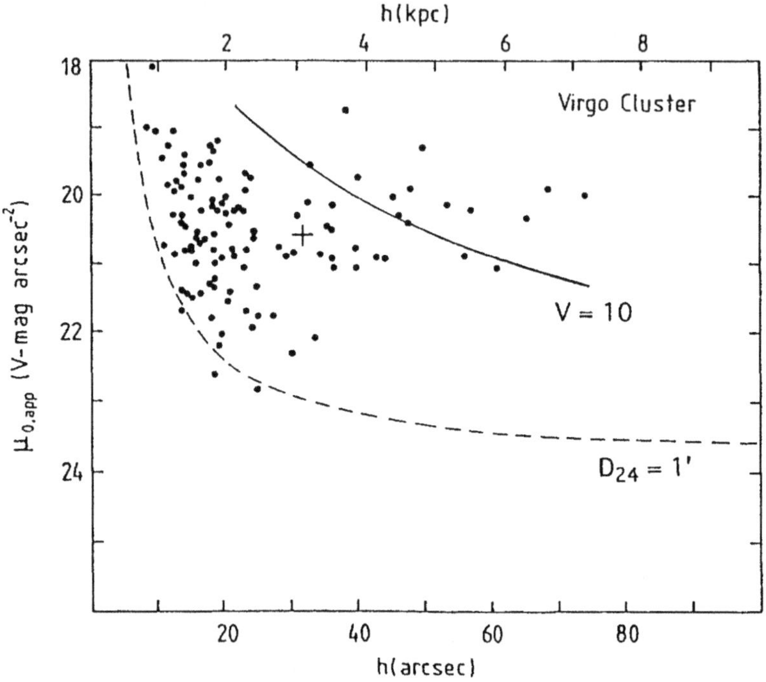

Fig. 3.2. Galaxy profile parameters I_0 and h derived from Watanabe's (1983) photometry of galaxies in the Virgo cluster, as plotted by van der Kruit (1987). The central surface brightness I_0 has not been corrected for inclination or other effects. The locus labelled "V = 10" corresponds to M^* galaxies assuming their total luminosity is $L = 2\pi\ h^2\ I_0$ (a distance to Virgo of 20 Mpc is assumed here). The cross indicates the position of a "standard" galaxy adopted in Lecture 6.

al. analysis was to investigate what biases are encountered when galaxies are selected by diameter.) Their derived diameter function, the log of the number of galaxies per unit volume per log interval of diameter, has a flat distribution for isophotal diameters smaller than about 30 kpc (converting to $H_0 = 50$ km sec^{-1} Mpc^{-1}), and declines smoothly at larger diameters.

Bivariate Distribution Functions

Instead of considering surface brightness and size, one can equivalently consider luminosity and size, as discussed by Sodré and Lahav (1993). As was the case for the Bardelli *et al.* work, Sodré and Lahav used a subsample of the ESO-Lauberts-Valentijn catalogue with complete redshift information, and worked with isophotal ($\mu_B = 25$) diameters. They derive an analytic expression for the number - luminosity - size relation, where the parameters are given for the whole sample, the sample of E and S0 galaxies, and the sample of S and Irr galaxies separately.

As remarked already, in general the surface brightnesses of galaxies are not strongly correlated with luminosity. However, for elliptical galaxies, the most luminous and least luminous ones have lower surface brightness (Sandage and Perelmuter 1990). Pierce and Tully (1992) and Mathewson, Ford, and Buchhorn (1992) have shown that for disk galaxies there is a correlation between a characteristic surface brightness and the width of the 21 cm line, in the sense that lower-mass disks have fainter surface brightness, consistent with van der Kruit (1987).

As mentioned earlier, Pierce and Tully (1992) also showed that for the same sample there was a correlation between color and line strength, in the sense that the lower-mass galaxies have bluer colors; presumably this implies that there is a correlation between color and surface brightness. In particular, van der Hulst *et al.* (1993) find that galaxies with very low surface brightness can be exceptionally blue. One should keep in mind that such generalizations may depend on the definition of diameter and surface brightness.

References

Bardelli, S., Zucca, E., Vettolani, G., and Zamorani, G. 1991, A&A **248**, 354

Binggeli, B. 1994, in: *Panchromatic View of Galaxies – their Evolutionary Puzzle*, eds. G. Hensler *et al.* (Gif-sur-Yvette: Editions Frontières), p. 173

Binggeli, B., Sandage, A., and Tammann, G.A. 1988, ARA&A **26**, 509

Bothun, G.D. 1990, in *Evolution of the Universe of Galaxies*, ed. R.G. Kron (San Francisco: ASP), p. 54

Bothun, G.D., Schommer, R.A., and Sullivan, W.T. 1982, AJ **87**, 725

Butcher, H.R. and Oemler, A. 1985, ApJS **57**, 665

Cornell, M.E., Aaronson, M., Bothun, G., and Mould, J. 1987, ApJS **64**, 507

de Jong, R.S. and van der Kruit, P.C. 1994, in *Infrared Astronomy with Arrays: The Next Generation*, ed. I.S. McLean (Dordrecht: Kluwer)

de Lapparent, V., Geller, M.J., and Huchra, J.P. 1989, ApJ **343**, 1

Efstathiou, G., Ellis, R.S., and Peterson, B.A. 1988, MNRAS **232**, 431

Ferguson, H.C. and Sandage, A. 1988, AJ **96**, 1520

Freeman, K.C. 1970, ApJ **160**, 811

Huchra, J.P. 1977, ApJS **35**, 171

Kennicutt, R.C. 1989, ApJ **344**, 685

Kent, S.M. 1986, AJ **91**, 1301

Kormendy, J. and Djorgovski, S. 1989, ARA&A **27**, 235

Lauberts, A. and Valentijn, E.A. 1989, *The Surface Photometry Catalogue of the ESO-Uppsala Galaxies* (Garching: European Southern Observatory)

Loveday, J., Peterson, B.A., Efstathiou, G., and Maddox, S.J. 1992, ApJ **390**, 338

Mathewson, D.S., Ford, V.L., and Buchhorn, B. 1992, ApJS **81**, 413

Mobasher, B., Ellis, R.S., and Sharples, R.M. 1986, MNRAS **223**, 1

Peletier, R.F. and Willner, S.P. 1992, AJ **103**, 1761

Peterson, B.A., Ellis, R.S., Efstathiou, G., Shanks, T., Bean, A.J., Fong, R., and Zen-Long, Z. 1986, MNRAS **221**, 233

Pierce, M.J. and Tully, R.B. 1992, ApJ **387**, 47

Sandage, A., Binggeli, B., and Tammann, G.A. 1985, AJ **90**, 1759

Sandage, A. and Perelmuter, J.-M. 1990, ApJ **361**, 1

Sandage, A. and Tammann, G.A. 1981, *A Revised Shapley-Ames Catalog of Bright Galaxies* (Washington, D.C.: Carnegie Institution of Washington)
Santiago, B.X. and Strauss, M.A. 1992, ApJ **387**, 9
Schechter, P.L. 1976, ApJ **203**, 297
Schechter, P.L. and Dressler, A. 1987, AJ **94**, 563
Schweizer, F. 1979, ApJ **233**, 23
Shanks, T. 1990, in *Galactic and Extragalactic Background Radiation: Optical, Ultraviolet, and Infrared Components*, IAU Symp. 139, eds. S. Bowyer and Ch. Leinert (Dordrecht: Kluwer), p. 269
de Vaucouleurs, G. and de Vaucouleurs, A. 1972, Mem. R. Astron. Soc. **77**, 1
Sodré, L. and Lahav, O. 1993, MNRAS **260**, 285
Tinsley, B.M. and Danly, L. 1980, ApJ **242**, 435
van der Hulst, J.M, Skillman, E.D., Smith, T.R., Bothun, G.D., McGaugh, S.S., and de Blok, W.J.G. 1993, AJ **106**, 548
van der Kruit, P.C. 1987, A&A **173**, 59
Watanabe, M. 1983 Annals of the Tokyo Astronomical Observatory, 2nd Ser. **19**, No. 2, 121
Wevers, B.M.H.R., van der Kruit, P.C., and Allen, R.J. 1986, A&AS **66**, 505

4 Computing Models for Faint-Galaxy Samples

Given evolutionary models for galaxies with various histories of star formation, it remains to compute the number of each type as a function of apparent magnitude. This computation automatically yields the number as a function of redshift and the number as a function of apparent color, both also observable distribution functions. (One can think of other observables, of course: the distribution of surface brightness, image compactness parameters, rotation curve amplitudes and shapes, etc.) The goal is to compare the computation directly with the observed properties of magnitude-limited samples of field galaxies to see how good the match is, and what parameters the model is most sensitive to.

4.1 Effect of Redshift

In the nearby Universe, namely the region close enough that the effects of redshift are for practical purposes negligible, the number of galaxies seen per square degree is given by Eq. 3.3. In the more distant Universe, the redshift has a strong effect on what is seen in magnitude-limited samples, thus making faint samples appear to be quite different, even though the actual galaxy population may be identical to the local census. Hence, these effects must be understood very accurately before we can properly interpret apparent differences as real differences.

The redshift has two basic effects that change the appearance of magnitude-limited samples: the so-called k-correction, and the dimming of surface brightness. Galaxies naturally radiate different amounts of flux at different rest-frame frequencies; hence, a fixed band in the observed frame will intercept different fractions of the bolometric flux as a function of redshift. This is called the k-correction. From the perspective of optical observations at low to moderate redshifts, in general galaxy spectral energy distributions are "concave downwards", meaning that as the redshift increases, there is progressively less flux in the observed band (by convention, the k-correction is numerically positive in this case: $m_{\rm cor} = m_{\rm obs} - k(z)$. This spectral curvature means that observed colors generally become redder with redshift, up to the redshift where an inflection point in the spectrum is sampled by the two bands that define the color. The dependence of apparent color on redshift is sometimes called a "differential k-correction".

The dimming with redshift will be discussed more fully in Lecture 6, but the essence of the effect (first discussed by Tolman) is that the surface brightness

$$\frac{s_\nu}{\pi\Theta^2} \propto (1+z)^{-3},$$

where Θ is a measure of the apparent size of a galaxy or region in a galaxy and s_ν is the corresponding monochromatic flux. [The *bolometric* relation has a $(1+z)^{-4}$ dependence.]

Detection techniques are often advertised to have a magnitude limit, but images of faint galaxies are actually most often catalogued according to the area of the image above a certain contour of surface brightness. A typical practical algorithm demands that the image have some minimum average apparent surface brightness within some minimum area. As the redshift increases, the dimming of surface brightness means that a smaller fraction of the image of the galaxy will appear above the threshold. To model the selection of faint galaxies in practice, we would need to know for each galaxy both the intrinsic surface brightness scale of the galaxy image, called here I_0, and the degree of central concentration of the light, a dimensionless shape parameter called here C.

Aside from the kinematical effects of the redshift (the k-correction and the surface-brightness dimming), there are also cosmological effects: the bolometric apparent flux m depends on the cosmological model, as does the volume element $\mathrm{d}V/\mathrm{d}z$ and hence the number of galaxies $A(m,z) \sim (\mathrm{d}V/\mathrm{d}z)\Delta z$ that populate a particular shell of redshift. The transformation between cosmic time and redshift also depends on the cosmological model. The models for galaxy evolution are tied to star clusters and the theory of stellar evolution; the predicted evolutionary state of a galaxy at a given redshift thus depends on the "look-back time" (given by the *cosmic* time scale) associated with that redshift. These various cosmological factors can also substantially modulate the appearance of faint-galaxy samples.

4.2 Model Input Parameters

To compute the properties of a statistical sample of faint galaxies, we need to know the statistics of galaxies in the local volume (Lecture 3) and then do the scaling as outlined above. We need to specify the following ingredients:

- $\Phi(M, SED, I_0, C)$: This is the joint spatial frequency of galaxies according to the key parameters that govern their visibility – intrinsic luminosity, intrinsic spectral energy distribution, intrinsic surface brightness scale, and image compactness. (M, I_0, and C together constrain the intrinsic size R; one might as well substitute R for I_0 or M.)
- $M(t), SED(t), C(t)$: This is the evolution model, allowing for changes in the bolometric flux, the distribution of the flux over wavelength, and the distribution of the flux over the image of the galaxy.
- q_0, Λ, H_0 : These parameters specify the cosmological model. H_0 provides the link between look-back time and redshift; otherwise, its value cancels out of the scaling between near and distant samples. The numerical value used here, $H_0 = 50$ km s^{-1} Mpc^{-1}, enables consistency of the galaxy models with the ages of star clusters; this is a requirement since in effect the galaxy models use star clusters as building blocks.

- *Environment(t)*: It is known that galaxy properties depend on their environment, at least for environments containing a high density of galaxies. Since environment changes with time as galaxy clustering evolves, this aspect should also appear in our model.
- *Detection model*: Real samples of faint galaxies are affected by details of the imposed limits on size and on threshold surface brightness. These practical aspects need to be included in the model before a meaningful comparison can be made; some of these issues will be described in the following two lectures. (Sometimes it is attempted to correct the data to the models, rather than the other way around. To do so requires *a priori* knowledge of critical properties of faint galaxies.)

For a variety of reasons, models fall short of including all of the ingredients listed here. For example, the important joint distribution function Φ is not known! The catalogues of nearby galaxies from which Φ might be determined are either deficient in listing all of the needed parameters M, SED, I_0, and C, or one cannot be sure that the sample with these properties is statistically complete. The best we can do is work with the usual marginal distributions such as $\Phi(M)$, $\Phi(SED)$, etc. As another example, the galaxy evolution models usually do not specify $C(t)$, the evolution of the shape of the profile. Finally, the environmental time dependence is not well enough understood at present to be included except on an *ad hoc* basis.

4.3 Time Scales

There are still more parameters that need to be specified, namely the epoch of galaxy formation and the spread in this quantity. The galaxy-evolution models are given to us as a series of spectral energy disributions at successive ages, and there is some freedom to choose which part of the time-line we are observing over some interval of redshift. In practice this is done by choosing the age of the galaxy now, t_G. If we define the "epoch of galaxy formation", t_F, as the time when the first generation of stars formed that currently matter in the context of our evolution model, then $t_G + t_F = t_0$, where t_0 is the age of the Universe. Different galaxies may naturally "form" at different times, even if they are apparently the same type and have the same colors at the present epoch; we can allow for this by introducing still other parameters. For example, we can assume that galaxies formed uniformly over some interval t_Δ centered on the epoch t_F.

The mapping of these times into the observable quantity z depends sensitively on the cosmological model, especially at high redshifts. To see this explicitly and to introduce some simple forms, consider the cases $q_0 = 0$ and $q_0 = 0.5$, both with $\Lambda = 0$. The respective relations between the look-back time τ to a given redshift z are:

$$\tau/t_0 = 1 - (1+z)^{-1}, \qquad t_0 = H_0^{-1}$$

and

$$\tau/t_0 = 1 - (1+z)^{-3/2}, \qquad t_0 = \frac{2}{3}H_0^{-1}.$$

Table 4.1 gives the look-back time in Gyr at $z = 0.5$ for $H_0 = 50$ km s^{-1} Mpc^{-1}.

Table 4.1. Cosmological Time Scales (Gyr)

	$q_0 = 0$	$q_0 = 0.5$
t_0	19.6	13.0
$\tau(z = 0.5)$	6.5	5.9
τ/t_0	0.333	0.46

The point of this exercise is to show that the look-back time to a given redshift is larger for $q_0 = 0$ when reckoned in Gyr, but larger for $q_0 = 0.5$ when reckoned as a fraction of the cosmic time t_0. Thus, which cosmological model yields the larger evolutionary effects depends on how the physical changes are modeled, for example the choice for the parameters t_F and t_Δ.

4.4 Modeling the Count Distribution

Given all of the ingredients described above, how does one proceed to construct a model for, say, the number of galaxies as a function of magnitude and redshift, $A(m, z)$? One starting procedure is to

- choose a particular type of galaxy as specified by a particular spectral energy distribution at the present day;
- assume this spectrum corresponds to a unique evolutionary path, in particular characterized by a unique t_F, profile type (I_0, C), and free of dependences on the environment and mass or luminosity scale;
- attach a particular luminosity function shape $\phi(M)$ to this spectral class, normalized in the luminosity scale M^* but not necessarily in the spatial density ϕ^*.

This procedure allows the (unnormalized) $A(m, z)$ to be computed for that particular galaxy type. Other types, characterized by their spectral energy distributions and luminosity functions $\phi(M)$ are computed in turn. The remaining task is to add the separate galaxy types together with the right relative weighting.

A way to do this is to demand that the color distribution at some bright magnitude is reproduced, where cosmological and evolutionary effects are negligible and where the redshifts are low and have small range. One can then normalize the ensemble to reproduce the known galaxy counts per square degree within this same bright magnitude interval. Hence, by construction the observed distributions $A(m)$ and $A(B-V)$ are matched at bright magnitudes

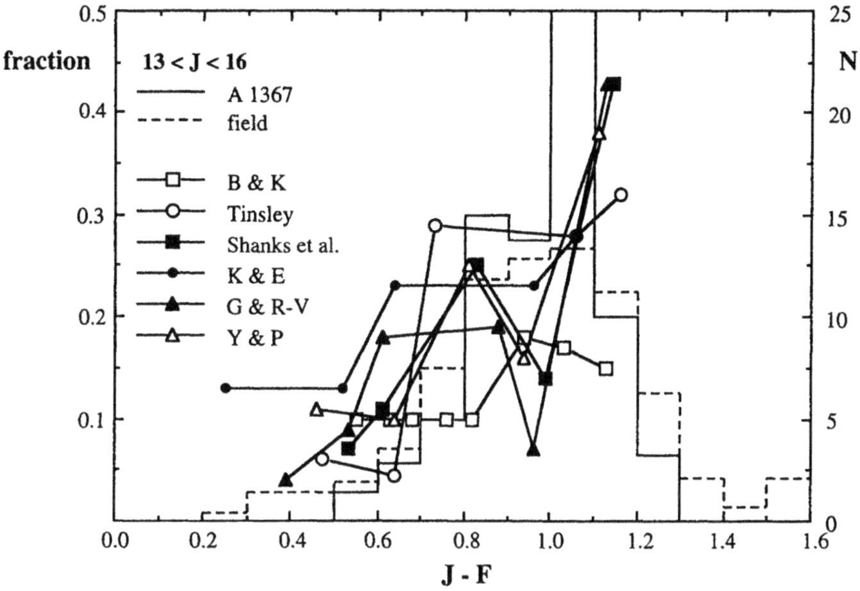

Fig. 4.1. The color distribution from Butcher and Oemler (1984), shown as histograms for the cluster Abell 1367 and for the field, both on the observed $(J - F)$ system. For various models the fraction of the total at the color of a galaxy class is plotted. "B&K" = Bruzual and Kron (1980); Tinsley (1980); Shanks *et al.* (1984); "K&E " = King and Ellis (1985); "G&R-V" = Guiderdoni and Rocca-Volmerange (1990); "Y&P" = Yoshii and Peterson (1991).

and low redshifts; $A(z)$ must also be matched if the luminosity scale M^* is correct. Then, the machinery of the model can be used to predict the various observable distribution functions at faint magnitudes and high redshifts. The model parameters that are often varied are t_G, q_0, and the formulation for the past history of the star formation rate.

Many such models have been constructed over the past 15 years. It is constructive to compare how the modelers have chosen their "mix" of different galaxy spectral classes. Typically, only five or six classes are computed, and the models are characterized by tabulating the fraction of each class that contribute to the counts at bright magnitudes. The color-redshift curves tend to diverge as the redshift increases, which means that at a given redshift, the small number of available spectral classes may only sparsely sample the physically plausible range of observed colors.

Figure 4.1 shows by how much the models disagree on the galaxy mix. The histogram is the galaxy color distribution in the interval $15^m < B_J < 16^m$ from Butcher and Oemler (1984); this is given on the observed $B_J - R_F$ color system [also called $(J - F)$], the system on which many of the modelers have based their calculations. The mixes are indicated by the fraction of the total at the color of a galaxy class. The colors and fractions are evaluated at $B_J = 15^m_{\cdot}5$ or close to it (not all of the models are given in exactly this way). Even allowing for errors in the color measurements to generate a smooth distribution from this coarse sampling, it is evident that these models do not, in fact, reproduce the Butcher and Oemler color distribution very well.

In view of these differences in the input ingredients – namely the spectra as characterized by the low-redshift color and their relative frequency – it is easy to imagine that the model predictions at high redshift may diverge. The point of Figure 4.1 is to raise questions about how accurate these models really are, apart from differences in the evolution model and even apart from uncertainties in our understanding of the properties of local galaxies.

Different forms for the luminosity function $\phi(M)$ may apply to different spectral classes. The existence of a color-luminosity relation in the sense that faint galaxies are bluer (Lecture 3) can be modeled by specifying a fainter M^* for bluer spectral classes, or a steeper faint-end slope, or both.

4.5 Selection Effects

These models allow the selection effects with redshift mentioned earlier to be quantified. We expect the most-frequently-seen absolute magnitude in a flux-limited sample to shift progressively to fainter absolute magnitudes because of selection against the distant, luminous galaxies. This bias occurs partly because of the k-correction and color-luminosity relation, and partly because of the "volume squeeze".

As mentioned earlier, most galaxies when redshifted appear fainter because a smaller fraction of their bolometric light is seen in a fixed band at visible wavelengths. This effect is smaller for intrinsically bluer galaxies than for intrinsically red galaxies, and the result is an effect that favors bluer galaxies at small redshifts with respect to red galaxies at large redshift as the depth of a flux-limited sample is increased. This is amplified by the existence of a correlation between luminosity and intrinsic color, since the most luminous galaxies in a flux-limited sample, i.e., the ones at the greatest redshift, will tend to be intrinsically redder.

The "volume squeeze" refers to the property that the cosmological volume element $\mathrm{d}V(z, q_0)$ is – at least for non-pathological cases – smaller than the Euclidean $z^2\mathrm{d}z$, and the difference increases with increasing redshift. This means that Eq. 3.3 is invalid; the effect is that the nearer volumes are weighted disproportionately more with respect to the distant volumes, and again the lower-redshift, lower-luminosity, bluer galaxies are progressively favored as the depth of a survey is increased.

The combined amplitude of these effects depends of course on all of the assumed parameters. I calculated (Kron 1978) some representative cases for luminosity functions of different shapes, different spectra, and different cosmological models, and found that the amplitude was between one and two

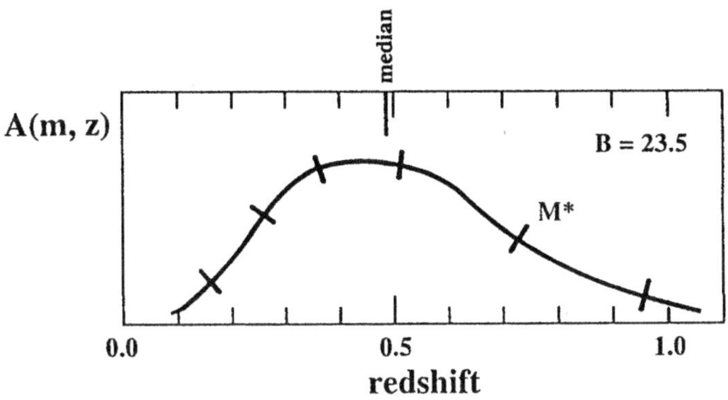

Fig. 4.2. Schematic expectation for the redshift distribution at $B \sim 23^{\mathrm{m}}_{.}5$ and the assumption of no luminosity evolution. The most-frequently-seen type of galaxy is substantially fainter than M^*, due to geometrical effects and the fact that k-corrections are generally positive. Each tick mark indicates an interval of $\Delta M = 1^{\mathrm{m}}$.

magnitudes of dimming by $B_{\mathrm{J}} = 23^{\mathrm{m}}_{.}5$. Conservatively, the most-frequently-seen absolute magnitude is expected to be about $M^* + 1^{\mathrm{m}}_{.}2$ at this depth, shown schematically in Figure 4.2. Hence, the models predict that apparently faint samples should be relatively rich in lower luminosity galaxies in the absence of evolutionary effects.

Another prediction is that the colors should shift blueward at faint magnitudes, even in the absence of evolutionary effects, and even though each individual galaxy class becomes redder (at least initially) as the redshift increases. This was also shown in Kron (1978; see Bruzual and Kron 1980), and happens because the selection against the redder, more luminous galaxies at larger redshift is so strong.

All of the models already cited, and many others since then (e.g. Franceschini *et al.* 1991) have yielded $A(m)$, the differential galaxy counts in selected wave bands. When integrated over m, this function in turn yields the integrated surface brightness of the night sky from galaxies. The general concave-down curvature of galaxy spectra, and the volume squeeze at high redshift, result in half of the detected photons in the B_{J} band having been emitted from redshifts less than about $0^{\mathrm{m}}_{.}22$, assuming no evolution. Alternatively, half of the integrated flux originates from galaxies brighter than $B_{\mathrm{J}} = 21^{\mathrm{m}}_{.}5$. [This is why models for the amount of background light due to distant galaxies are sometimes computed excluding bright galaxies, e.g. Tinsley (1980).]

A magnitude of $B_{\mathrm{J}} = 21^{\mathrm{m}}_{.}5$ is well within the range of spectroscopic investigation, so the model expectations cited above can be checked directly. This will be done in Lecture 7, but for now it is worth noting that since the redshifts are small, we should not *expect* the counts $A(m)$ to be very sensitive to evolution; a difference between the predicted counts and the data might more reasonably indicate deficiencies in the model.

The same models predict the redshift distribution $A(z)$ for a fixed interval of apparent magnitude. The $A(z)$ distribution is usually bell-shaped with a tail to high redshifts. (In some cases the predicted $A(z)$ is bimodal: the high-redshift bump comes from an early luminous phase in ellipticals and spiral bulges, the visibility of which depends sensitively on the choices for t_F and t_Δ, as well as the particulars of the initial burst.)

4.6 Approximate Analytic Form for the *k*-Correction

One of the intermediate steps of the model calculation is calculation of the flux from a galaxy with specified spectrum as seen in a given band at a given redshift – in other words, the *k*-correction. Empirical spectra of representative galaxies can be used to make this calculation, and they have the appeal that complications like internal reddening and emission lines are automatically taken into account. Nevertheless, evolutionary synthesis model galaxy spectra have two important advantages: an arbitrarily dense set of present-day spectral shapes can be generated by adjusting the model input parameters, and the effects of evolution in the spectral shape with look-back time can be calculated.

Empirical *k*-corrections require accurate spectrophotometry over a wide range of wavelengths to enable *k*-corrections to be computed over a range of wavebands and a range of redshifts. In addition, large-aperture spectrophotometry is desirable since the application is almost always for the integrated light from distant galaxies. Lastly, the menu of galaxies chosen for such spectrophotometry needs to be truly representative of the galaxies appearing in the faint counts (but one cannot determine this without knowing their *k*-corrections!). The standard reference for empirical *k*-corrections is Coleman, Wu, and Weedman (1980), which gives tabular and graphical *k*-corrections in the $UBVR$ bands for five different morphological classes. This paper is an extension of the work of Pence (1976) in that it added newer ultraviolet data and so extended the range of the *k*-corrections to higher redshift. Pence's (1976) work was in turn based directly on data obtained by Wells (1972). In view of the poor correlation of morphological type and color that has been repeatedly stressed, it is unfortunate that this essentially color-related issue has always been couched in terms of morphological type. These *k*-corrections should be used with care; see Koo and Kron (1992) for a critique.

It is sometimes convenient to adopt an analytic form for the *k*-correction. The inspiration for the form of the spectrum comes from Figure 4.3, which shows a series of spectra of the central regions ($\sim$ 800 pc diameter) of eight Sc I galaxies obtained by Turnrose (1976). Redward of roughly $\lambda5200$ the spectra are power laws of the form $f_\nu \sim \nu^\alpha$. The value of α is about -2; very blue galaxies have $\alpha \sim -1$, and the "flat-spectrum" galaxies with $\alpha = 0$ seem to be very rare at any redshift. [The convention and rationale for the sign of the index α comes from Moffet (1975).] For redshifts such that this

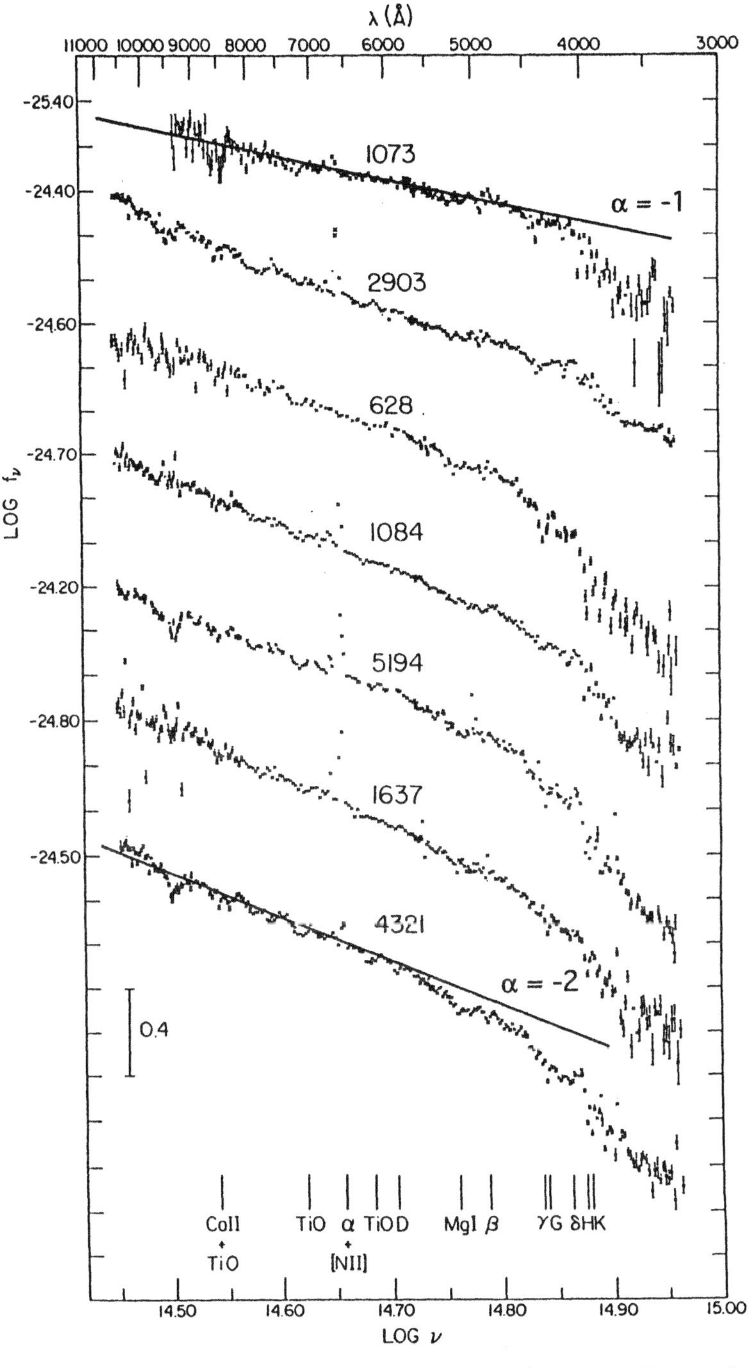

Fig. 4.3. Series of spectra of the central regions (~ 800 pc diameter) of seven ScI galaxies by Turnrose (1976). In the red the spectra are power laws of the form $f_\nu \sim \nu^\alpha$; examples are indicated for $\alpha = -1$ and $\alpha = -2$.

part of the spectrum is included in the band, it is easy to derive that the k-correction, including the bandwidth term, is

$$k(z) = -2.5\ (1 + \alpha)\ \log(1 + z).$$

This formalism reproduces quite well the R-band k-corrections for the various galaxy types of Coleman, Wu, and Weedman (1980) out to $z = 0.4$, beyond which it fails in the expected way because of the curvature of the actual spectra at shorter rest-frame wavelengths.

At near-infrared wavelengths, the spectral index changes sign but there is again a section of the spectrum that can be usefully approximated with a power-law. In the K-band, the value of α is roughly $+1.2$ for all galaxy types, an approximation that holds only for redshifts less than about 0.3 (Bershady 1994). This states quantitatively the well-known property that K-band k-corrections are *negative*, i.e. proportionately *more* of the bolometric flux is redshifted into the band with increasing redshift.

References

Bershady, M.A. 1994, AJ, in press
Bruzual A., G. and Kron, R.G. 1980, ApJ **241**, 25
Butcher, H. and Oemler, A. 1984, ApJ **285**, 426
Coleman, G.D., Wu, C.-C., and Weedman, D.W. 1980, ApJS **43**, 393
Franceschini, A., Toffolatti, L., Mazzei, P., Danese, L., and De Zotti, G. 1991, A&AS **89**, 285
Guiderdoni, B. and Rocca-Volmerange, B. 1990, A&A **227**, 362
King, C.R. and Ellis, R.S. 1985, ApJ **288**, 456
Koo, D.C. and Kron, R.G. 1992, ARA&A **30**, 613
Kron, R.G. 1978, Ph.D. dissertation, University of California, Berkeley
Moffet, A.T. 1975, in *Galaxies and the Universe*, vol **IX**, eds. A. Sandage, M. Sandage, and J. Kristian (Chicago: University of Chicago Press), p. 211
Pence, W. 1976, ApJ **203**, 39
Sandage, A., Binggeli, B., and Tammann, G.A. 1985, AJ **90**, 1759
Shanks, T., Stevenson, P.R.F., Fong, R., and MacGillivray, H.T. 1984, MNRAS **206**, 767
Tinsley, B.M. 1980, ApJ **241**, 41
Turnrose, B.E. 1976, ApJ **210**, 33
Wells, D.C. 1972, Ph.D. dissertation, University of Texas, Austin
Yoshii, Y. and Peterson, B.A. 1991, ApJ **372**, 8

5 Distant Galaxy Observations in the Real World

The flux from a particular high-redshift galaxy, or the combined flux from some sample of galaxies, can be considered to be a stream of information about the distant Universe. The stream can be so diluted by the expansion of the Universe that random errors due to photon counting statistics become important and need to be accounted for in the comparison of models with data. Computing the photon statistics is elementary, but its elaboration is still worthwhile to support a number of arguments made in these lectures.

5.1 Detection Model

The first point to establish is that faint-galaxy observations are *fundamentally* difficult, partly because of the low surface brightness of typical regions in normal galaxies, and partly because of the bright background of the night sky. To demonstrate this, we will adopt a "standard" patch of a galaxy, such as would be seen by an observer outside the Milky Way looking back at the position of the Sun, and explore what this patch looks like as a function of redshift.

For didactic purposes we can consider the radiation field to be characterized by a dilute blackbody. The projected surface brightness at the effective radius (roughly at the solar circle) is about $\mu_V = 22\overset{\mathrm{m}}{.}5$ arcsec^{-2} (de Vaucouleurs and Pence 1978). Since the $B - V$ color of the nearby Galactic disk is probably not far from that of the Sun (Mattila 1980), we can adopt $T_{\mathrm{eff}} = 5800$ K, which results in a dilution of the intensity of the radiation field by about 6×10^{-14} with respect to a blackbody at that temperature. [Modeling the radiation field in a spiral disk by a dilute thermal spectrum is reminiscent of Hubble's (1936) approach.]

Similarly, the background radiation from the night sky can be modeled as a dilute blackbody if we consider the continuum as being dominated by zodiacal light and by diffuse stellar light. The scattered zodiacal light of course has $T_{\mathrm{eff}} = 5800$ K, and to match the intensity near the ecliptic poles requires a dilution factor of about 6×10^{-14}, the same as for the adopted galaxy surface brightness. (It can be anticipated that the solar neighborhood is unrepresentatively low in surface brightness since the Sun is located a few scale lengths away from the center of the Galaxy; however, compensating for this, the background has also been underestimated since no account has been made for airglow and other contributions to it.)

The photon flux from an extended source with a dilute blackbody spectrum is:

$$\left[\frac{p}{(\lambda/2D)}\right]^2 \frac{\pi\eta\beta}{8R}\,\nu\,\frac{1}{e^{(h\nu/kT)}-1}, \tag{5.1}$$

where

p = pixel size in radians;

D = diameter of telescope entrance pupil;
β = dilution factor;
η = instrumental efficiency;
$R = \lambda/\Delta\lambda$ = spectral resolution.

The formula gives the photon flux per angular resolution element (given by the first factor) and per spectral resolution element. Note that if the pixel size is scaled to the size of the Airy diffraction pattern, the photon flux per pixel is independent of the size of the telescope, D.

The instrumental efficiency η is the ratio of the number of detected photons to the original number of photons incident at the top of the atmosphere. The table below gives, in order of the light path, values for the individual components of the calculation for η for a typical broad-band imaging application. The losses can be due to any combination of scattering, undesired reflection from surfaces, and actual absorption. This table is meant only to be illustrative; a detailed calculation would provide for the variation of η with wavelength, which is especially important for the atmospheric factor and for the detector sensitivity.

Table 5.1. Instrumental Throughput

atmosphere	0.85
primary mirror	0.80
secondary mirror	0.85
central obscuration	0.80
field correcting lenses	0.91
filter	0.92
dewar window	0.97
CCD quantum efficiency	0.60
net for imaging	$\eta = 0.23$
diffraction grating	0.65
re-imaging optics	0.80
net for spectroscopy	$\eta = 0.12$

The case for spectroscopic work is similar in most respects, but the efficiency is lower by the two factors listed above. The actual spectroscopic efficiency is usually even lower because atmospheric dispersion spreads each image into a small spectrum depending on the zenith angle. Hence, if a narrow slit is centered at the light from λ5000, then the light at λ3500 and λ8500 will not be centered and some of this light will be lost.

We can proceed to calculate the actual count rates. The adopted surface brightness of $\mu_V = 22\overset{\mathrm{m}}{.}5$ arcsec^{-2} is equivalent to

$35 L_\odot$ pc^{-2} or

1.55×10^5 Jy steradian^{-1} or

$$3.6\ \mu\text{Jy arcsec}^{-2}.$$

(1 Jy $= 10^{-26}$ W m^{-2} Hz^{-1} and $V = 0$ corresponds to 3640 Jy.) A useful conversion factor is

$$1\ \mu\text{Jy} = 15.1 \text{ photons sec}^{-1}\ \text{m}^{-2}\ (\Delta\lambda/\lambda)^{-1}.$$

Let's adopt a net efficiency of 0.23 as derived above and typical values for the other parameters: a collecting aperture of 12.5 m^2 (i.e., a 4 m-telescope), a bandwidth of $\Delta\lambda/\lambda = 0.2$, and a pixel size of 0.30 arcsec. Then, multiplying all of these factors together, the standard patch has a photon flux of only

$$2.8 \text{ detected photons sec}^{-1}\text{pixel}^{-1}.$$

5.2 Signal-to-Noise Ratio as a Function of Redshift

This calculation is for a patch of a galaxy at $z = 0$. To compute what happens at higher redshift, we take advantage of the formalism developed by the other lecturers (e.g., Longair, Chapter 2). For a spectrum described by $f_\nu \sim \nu^\alpha$, the apparent flux

$$\begin{aligned} s_{\nu_o} &= \frac{L_{\nu_e}}{4\pi\ D^2\ (1+z)} \\ &= \frac{L_{\nu_o}\ (1+z)^\alpha}{4\pi\ D^2\ (1+z)} \end{aligned}$$

and

$$\Theta = R\ (1+z)/D,$$

where subscript "o" indicates the observed frame, subscript "e" indicates the emitted frame, R is a physical length scale, and D is the quantity

$$D = \frac{c}{H_0\ (1+z)} z(1+z/2) \qquad (q_0 = 0)$$

$$D = \frac{c}{H_0 q_0^2\ (1+z)} \left[q_0 z + (q_0 - 1)(\sqrt{1+2q_0 z} - 1)\right]. \qquad (q_0 \neq 0)$$

(The quantity $D\ (1+z)$ is called the *luminosity distance.*) Therefore, the *surface brightness* is equal to

$$\begin{aligned} \frac{s_{\nu_o}}{\pi\Theta^2} &= \frac{1}{4\pi} \frac{L_{\nu_o}}{\pi R^2}\ (1+z)^\alpha\ (1+z)^{-3} \\ &= I_{\nu_o}(0)\ (1+z)^\alpha\ (1+z)^{-3} \\ &= 2.8 \text{ photons sec}^{-1}\ \text{pixel}^{-1}\ (1+z)^{-5} \quad (\text{for } \alpha = -2). \end{aligned}$$

The signal-to-noise ratio, S/N, can be computed very simply with a few approximations. First, we assume that the noise is entirely due to photon-counting statistics (as opposed to detector noise, for example). Second, we assume that the background flux is so much larger than the source flux that the photon noise in the source itself is negligible - this is the "background-limited regime" and is justified on the basis of the $(1+z)^{-5}$ factor. The signal is just the source flux times the integration time $t_{\rm exp}$, and the noise is approximated by the square root of the background flux times $t_{\rm exp}$, in this case both evaluated for a single pixel. That is,

$$S/N = \sqrt{2.8}\,(1+z)^{-5}\ \ t_{\rm exp}^{1/2}.$$

According to this formula, to get $S/N = 20$ per 0.3-arcsec pixel at $z = 0.5$ requires $t_{\rm exp} = 2.25$ hours with a 4 m-telescope. This expectation for the required exposure time is consistent with common practice for good photometry of distant galaxies. However, note the steep dependence on redshift: to do as well for $z > 0.5$ requires much longer exposures. More specifically, the exposure time needed to reach a fixed value of the S/N per pixel depends on redshift like $(1+z)^{10}$!

This looks very unpromising, to say the least. If we look back to identify where the various factors of $(1+z)$ came from, we find that the k-correction is an important part (we assumed $\alpha = -2$, which corresponds to $(1+z)^{-2}$ in the S/N and $(1+z)^4$ in the time). These factors can be avoided by shifting the central wavelength of our band redward and stretching the bandwidth to correspond to a chosen part of the rest-frame spectrum for a galaxy with known redshift. Unfortunately, this does not solve the problem, since the noise from the sky background is then a function of redshift (= observed wavelength): the sky background is so much more intense at longer wavelengths (Fig. 5.1) that the S/N turns out to be more-or-less the same, regardless of which procedure is followed.

The discussion so far has concerned direct imaging; we now consider what is involved with a spectroscopic measure of the patch of the galaxy at high redshift. Aside from the lower net efficiency, the following instrumental differences are representative: the spectral resolution is much higher, say $\Delta\lambda/\lambda = 0.001$ as opposed to 0.2, and the angular sampling is coarser, say 0.55 arcsec per pixel as opposed to 0.3 arcsec per pixel. One wants a sensible match between the projected width of the slit on the detector (say, 3 pixels) and the size of the galaxy (3×0.55 arcsec is about right for a typical distant galaxy). Since we are not interested in the spatial dimension in the spectrum, and since in the wavelength direction the resolution has been blurred to some extent by the finite width of the slit, we consider the photon counts along the spectrum in 3 pixel $\times$ 3 pixel bins. After collecting all of these factors, we find that at $z = 0$ the photon flux is

$$I_{\nu_\circ}(0) = 0.22 \text{ photons sec}^{-1}(3\ \times 3 \text{ pixels})^{-1},$$

and the dependence on redshift is the same as for the imaging case.

From this result, we derive that at $z = 0$, $S/N = 20$ requires $t_{\rm exp} = 30$ minutes, and at $z = 0.5$, $t_{\rm exp} = 30$ hours.

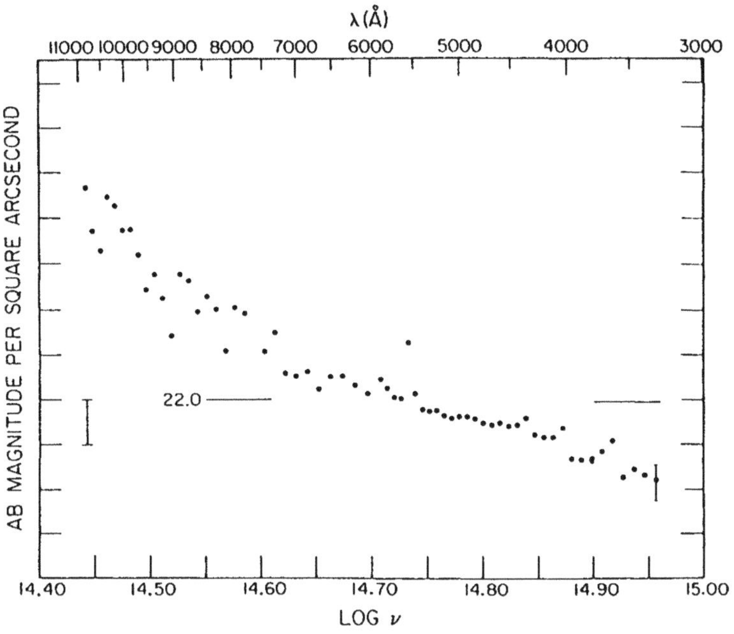

Fig. 5.1. Spectrum of the night sky at Palomar Observatory, from Turnrose (1974). Each vertical division is 1^{m} in the flux per unit frequency, and $m(\mathrm{AB}) = 22\overset{\mathrm{m}}{.}0$ is indicated.

This exercise has revealed an important lesson. Since observations of $z > 0.5$ galaxies reported in the literature have not, in fact, required many nights of integration with a 10 m-telescope, one of the following must be true:

1) the S/N per angular and spectral resolution element is much lower than 20;
2) the angular and/or spectral resolution is much coarser at high redshift than is typically employed at lower redshift;
3) the galaxy is unusual in some important way: it either has unusually high intrinsic surface brightness, or the observations were made in the light of an intense emission line.

Regarding item 2), note that since galaxy images are small, there is a limit to how coarse the angular sampling can reasonably be. To maintain the signal-to-noise ratio, the sampling cannot be much worse than about 1 arcsec. The limit on how coarse the spectral resolution can reasonably be is determined by the ability to recognize the strongest spectral features, such as Ca II K and H, $\lambda\lambda 3933$, 3970. By this argument, $R_{\min} \sim \lambda/\Delta\lambda = 100$. Point 3) suggests a great danger from selection effects – the galaxies for which we have measured a high redshift comprise a highly biased version of the underlying population.

It is sometimes remarked that galaxy luminosity evolution compensates for the cosmological dimming of surface brightness. Indeed, the models discussed in Lecture 2 suggest that for ellipticals and bulges of spirals, the V-band magnitude is brighter by an amount that turns out to be roughly equal to the redshift (that is, $0\overset{m}{.}4$ brightening at $z = 0.4$, see Fig. 2.1a). Since the models assume that the intrinsic sizes of galaxies are essentially stable, the luminosity increase is equivalent to a surface brightness increase. This picture cannot continue indefinitely of course; if such evolution were actually very strong, we would have clear manifestations, such as a high value for the extragalactic background radiation at optical wavelengths.

5.3 Other Aspects of Faint-Object Detection

When other observational aspects are considered, the situation regarding photon detection rates goes from bad to much worse. Ideally we want to model *all* aspects of the detection process, say as represented by the distribution of photons on a two-dimensional detector. Here is a partial list of the aspects that should be considered quantitatively:

Atmospheric seeing: Regions of galaxies with intrinsically high surface brightness tend to occur at high spatial frequency, for example the nuclear regions or regions of active star formation. Therefore any degradation of the sharpness of the image degrades the information that is most valuable in the sense of the previous discussion.

Scattered light: A surprising fraction of the total light from an image is scattered to large angles [e.g. Capaccioli and de Vaucouleurs (1983) found 32% is scattered to radii greater than 30 arcsec] by a combination of the atmosphere and the telescope/instrument optics. This can inject low-spatial-frequency modulation of the detected photons, depending on the nature of the sources in the field and just outside of the field.

Image crowding: Recent imaging at optical and near-infrared wavelengths has reached flux levels that are so faint that galaxy images are close to each other with respect to their diameters (a surface density of 10^6 galaxies per square degree corresponds to a mean separation of only 3.6 arcsec). The "sea" of galaxies that are too faint to be individually detected will combine to form a lumpy background, affecting the measurement of the brighter images. (These phenomena are well-known in other astronomical applications where the signal is confusion-limited - see Chapter 4 of M. Longair.

Statistical bias: A histogram of pixel values will show more pixels to have lower observed flux, down to the level of the night-sky background. This means that at a given measured level, it is more likely that the pixel actually sampled a fainter flux and was erroneously measured too high, as opposed to sampling a brighter flux and being erroneously measured to be too low.

This bias also applies to the integrated flux from individual galaxies, since there are more faint galaxies - where the errors are larger - than bright galaxies. In this context, the distortion of the shape of the actual counts is a classic problem in statistical astronomy. Trumpler and Weaver (1953) showed that, to first order, the observed counts $A_o(m)$ are related to the true counts $A_{true}(m)$ by the relation

$$A_o(m) = [1 + 1/2(\sigma\gamma \ln 10)^2] \, A_{true}(m),$$

where σ is the *r.m.s.* photometric error in magnitudes and γ is the underlying slope $\gamma = d\log A(m)/dm$. For example, for $\gamma = 0.43$ and $\sigma = 1^m$, the observed counts are 50% too high, clearly a very strong bias. (This formula also serves to provide a working definition for the expression "limiting magnitude". The *survey limit* has been reached when the second term in the above expression is important.)

Isophote determination: This is a combination of the statistical bias mentioned above, aggravated by image crowding. Many detection and measurement algorithms depend on a determination of a particular level of surface brightness that acts as a threshold, i.e., an isophote. For example, a threshold that is one-thousandth of the zodiacal light would be equivalent to $\mu_V = 30^m$ arcsec^{-2}. Returning to our earlier example, at a photon flux of 2.8 photons per second due to the night sky background, a ten-hour exposure would yield a statistical error of 0.3% in each pixel, and when averaged over 1 arcsec^2, one could indeed obtain a precision of 0.1%. However, these are formal errors in the sense that they presume a strictly smooth background. In reality, the background is modulated on all relevant angular scales with significant amplitudes, and in a given field the modulation due to the source distribution is *a priori* unknown. This modulation must make the identification of the proper position of an isophote far more uncertain because of the bias introduced whenever there is a slope in the intensity distribution. A more meaningful number to quote would be the number of sky photons detected per pixel, rather than the "limiting surface brightness" in mag arcsec^{-2}.

The various observational difficulties just mentioned refer both to source detection and to the measurement of the properties of the sources once they have been detected. Photometric procedures in crowded fields might best combine these two aspects into a single algorithm. At this point, it makes little sense to try to generalize, since each technique will have its own features that need to be understood. One obvious approach is to make simulations of the appearance of the sky, and then test directly the performance of the algorithms on the simulated data (e.g. Yee 1991). The simulator provides a first guess of the properties of faint images – the distribution over magnitude, surface brightness, and angular size – and includes the important instrumental characteristics of photon statistics, sky brightness, seeing, scattered light, and perhaps detector fixed-pattern noise. Then, various detection/measurement algorithms can be checked for their efficiency (fraction of sources that were properly recovered), their reliability (fraction of detected sources that were actually noise), accuracy with which the known source image parameters were measured, and success of deblending close images. Once the performance of the analysis procedures has been determined in this way, the input parameters of the simulation can be varied to check the robustness of the conclusions.

Proceeding in this way implies a substantial sophistication in the modeling, since a lot must be specified about both the properties of faint images and the properties of the detection process. The usual separation of the activities of predicting the counts and reducing faint imaging data is likely to be inappropriate because of the possibility that subtle biases strongly influence the apparent results.

References

Capaccioli, M. and de Vaucouleurs, G. 1983, ApJS **52**, 456
Hubble, E. 1936, ApJ **84**, 517
Mattila, K. 1980, A&A **82**, 373
Turnrose, B.E. 1974, PASP **86**, 545
Trumpler, R.J. and Weaver, H.F. 1953, *Statistical Astronomy* (Berkeley: University of California Press)
de Vaucouleurs, G. and Pence, W.D. 1978, AJ **83**, 1163
Yee, H.K.C. 1991, PASP **103**, 396

6 Galaxy Profiles at High Redshift

As detailed in the last lecture, galaxies are intrinsically low-surface brightness objects that appear with even lower surface brightness as the redshift is increased, an effect that is usually aggravated by a positive k-correction. Since the angular size diminishes with increasing redshift (at least for $z < 1.25$, $q_0 < 0.5$), the solid angle at any particular intrinsic surface brightness also decreases. Thus, fewer pixels sample the corresponding isophote, and the statistical precision of its determination is therefore diminished. In addition, a smaller fraction of the galaxy luminosity appears above a fixed apparent isophote as the redshift increases. This is a complicated way of saying that as a galaxy is redshifted, its image fades away into the noise of the sky background.

6.1 Calculating Apparent Sizes and Surface Brightnesses

Various aspects of the detection problem as a function of redshift have been considered. A clear example of the dimming below an isophotal threshold is given in Figure 6.1. This shows that the radius of the detection isophote shrinks to zero at sufficiently high redhift, depending on the intrinsic galaxy profile.

Models for galaxy profiles commonly refer to azimuthally symmetric cases. Bohlin *et al.* (1991) have taken the next step by using images of nearby galaxies observed in the ultraviolet to simulate their two-dimensional appearance at high redshift at the corresponding observed wavelengths, including realistic noise for different types of observation and the effect of blurring by a point-spread function. The ultraviolet images of spiral galaxies are usually rich in high-spatial-frequency information – spiral arms and OB associations – but the appearance at high redshift is strongly moderated by blurring by seeing.

Since we are interested in computing the appearance of statistical samples of galaxies, we need to know the distribution functions of different types of profile shapes and surface brightness scales in a volume-limited, flux-limited, or diameter-limited sample. Figure 3.2 shows that the luminous spirals have central surface brightnesses that cluster around a value of $\mu_V = 20\overset{m}{.}5$ arcsec^{-2}. Since most faint galaxies can be anticipated to be spirals, and since many spirals have a surface brightness profile that can be usefully approximated by the exponential form of Eq. 3.2, we will proceed for purposes of illustration to use this representation. We can adopt $I_0 = 20\overset{m}{.}5$ arcsec^{-2} in V on the strength of Figure 3.2. The scale length h is determined by ensuring that the luminosity $L = 2\pi\, h^2 I_0$ is equal to the expected most-frequently-seen luminosity in faint samples. Recall from Lecture 4 that the most-frequently-seen luminosity at faint magnitudes can be, say, $1\overset{m}{.}2$ dimmer than M^*; on this basis, we adopt $M_V = -20\overset{m}{.}45$, which yields $h = 3$ kpc.

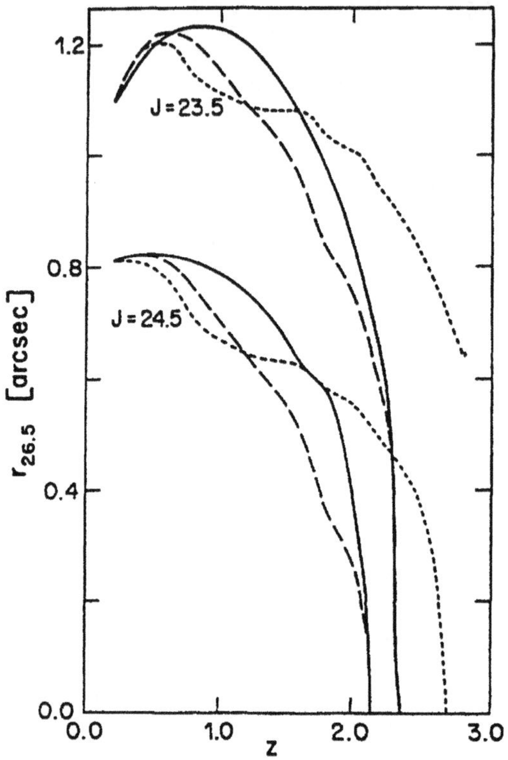

Fig. 6.1. The angular size as a function of redshift of the isophote at $\mu_J = 26\overset{m}{.}5$ arcsec^{-2} for disk galaxies observed to have fixed apparent magnitudes $J = 23\overset{m}{.}5$ and $J = 24\overset{m}{.}5$, for three assumed histories of star formation (from Pritchet and Kline 1981).

The model galaxy with these values for I_0 and h is marked with a cross in Figure 3.2.

A variety of different characteristic radii and surface brightnesses can be derived from this analytic model, as summarized in Table 6.1.

Table 6.1. Characteristic Radii and Surface Brightnesses

radius (units of h)	μ_V mag arcsec^{-2}	description
0	20.5	central
1	21.6	one scale-length
< 1.68	21.6	average SB within effective radius
1.68	22.3	1/2-light or effective radius
1.79	22.4	optimum radius

In Lecture 5 we used a characteristic surface brightness of $\mu_V = 22\overset{m}{.}5$ arcsec^{-2} which was partially motivated by this model galaxy. The "optimum radius" listed above is the formal result for an exponential profile in the limit where the noise is dominated by the brightness of the sky background (see

Lecture 5); this is the radius within which the S/N is highest (Phillipps 1984). This radius also corresponds to the maximum gradient in the light growth curve $l(\theta)$,

$$l(\theta) = 2\pi \int_0^\theta \theta' I(\theta') \mathrm{d}\theta',$$

where θ is the angular radius of measurement on the detector ("diaphragm"). Hence, a small error in θ will result in a large difference in $l(\theta)$. From the point of view of photometry, it is thus better to measure the light within a larger radius.

We now compute how big the model galaxy looks as a function of redshift, and what the apparent surface brightness is at the effective radius. The relation between redshift, cosmological model, physical size and apparent size has been derived by both M. Longair (Chapter 2) and A. Sandage (Lecture 4). For $q_0 = 0.5$, $\Lambda = 0$ the relation reduces to

$$\theta = (RH_0/2c)\ (1+z)^2\ \left[(1+z) - (1+z)^{1/2}\right]^{-1}.$$

The effective radius $R = 1.68 \times h = 1.68 \times 3$ kpc for $H_0 = 50$ km sec^{-1} Mpc^{-1}, and the leading factor is then equal to 0.173 when θ is expressed in seconds of arc. Table 6.2 gives the apparent sizes for the effective radius and the apparent V-band surface brightness at the effective radius.

Table 6.2. Apparent Properties of Model Disk

redshift	θ	μ_V mag arcsec^{-2}
0	(5 kpc)	22.30
0.25	1.03 arcsec	23.49
0.53	0.69 arcsec	25.09

The example redshifts were so chosen because at $z = 0.25$, the B-band flux is shifted into the V band, and at $z = 0.53$, the U-band flux is shifted into the V band. Our standard galaxy has $B - V = 0.65$ and $U - B = 0$; on the AB system, the rest-frame surface brightnesses at the effective radius are then $\mu_V(\mathrm{AB}) = 22\overset{\mathrm{m}}{.}30$ arcsec^{-2}, $\mu_B(\mathrm{AB}) = 22\overset{\mathrm{m}}{.}76$ arcsec^{-2}, and $\mu_U(\mathrm{AB}) = 23\overset{\mathrm{m}}{.}71$ arcsec^{-2}. The entries in Table 6.2 are these values, plus the cosmological dimming term $-2.5\log(1+z)^{-3}$. By choosing the example redshifts in this way, we do not need to explicitly calculate the k-corrections as long as the rest-frame broad-band colors are known. The conclusion from Table 6.2 is that a normal spiral galaxy will look both very small and have very low surface brightness at $z = 0.53$ (the surface brightness of the night sky including airglow is typically brighter than $\mu_V = 22\overset{\mathrm{m}}{.}0$ arcsec^{-2}).

We should consider measures of image morphology that use few parameters, do not make assumptions about the detailed nature of the galaxies at high redshift, and which take into account the practicalities of real measurements of faint images (low signal-to-noise ratio and the fact that the point-spread function is normally not small in comparison with a characteristic angular size for the galaxy). One such approach is to derive a dimensionless measure of image compactness, such as the ratio of radii containing 25% of the total light and 75% of the total light. An example of this is given by Kent (1985), who showed that, for large disks, the image compactness parameter is positively correlated with μ_e. Doi, Fukugita, and Okamura (1993) considered classification of faint galaxy images based on a compactness parameter versus surface brightness plot. In this diagram, the principal distinguishing parameter between different types of galaxy is the image compactness, and the surface brightness plays a secondary role because at the effective radius, spirals and bulges have similar μ_e.

6.2 Evolution and Cosmology

There are two common applications that exploit faint-galaxy surface brightness profiles:

1) if we think we know how the surface brightness scale I_0 may have evolved with time, then we can make a direct test of Tolman's prediction for surface-brightness dimming with redshift, independent of q_0;
2) if we understand how the structure of the galaxy may have evolved, then its apparent size can be used as a test for the value of q_0 in the same way that apparent brightness is often used.

A major problem in both cases is the effect of image blur on the apparent profile, which will affect the central regions more because the gradients tend to be steeper there. Hence, such measurements are normally done at relatively large values of θ, despite the very low values of surface brightness so encountered.

A power-law galaxy profile such as that of Eq. 3.1 is shown in Figure 6.2 to illustrate the limitations of using the profile at large radii for either of these tests. The high-redshift profile is shifted to a smaller apparent size by an amount that depends on the cosmological model, and it is shifted to lower intensities by an amount that depends only on the redshift (and on the evolution of the intrinsic surface brightness). Evidently, different combinations of the shifts $\Delta \log I$ and $\Delta \log \theta$ will allow the nearby galaxy profile to be scaled to the distant-galaxy profile, and the two tests are mixed up. The way out of this difficulty is to use the information not shown in Figure 6.2, namely the constraint from the amount of light in the center of the galaxy: while the *distribution* of this light is affected by image blur, its *amount* is of course preserved.

It should be clear that an isophotal diameter is a very crude measure of the size of an image of a galaxy. Two galaxies that are identical in all respects

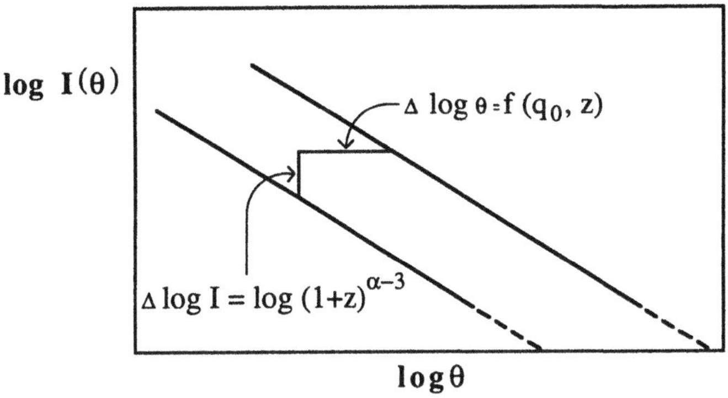

Fig. 6.2. The schematic effect of redshifting a power-law galaxy profile. There is a scaling in $\log \theta$ which depends on the redshift and the cosmological model. There is a scaling in $\log I(\theta)$ due to evolution, the cosmological dimming of surface brightness, and the k-correction.

except a scale factor in column density of stars will be measured to have different sizes. Moreover, measures of isophotal diameters of faint galaxies whose redshifts are unknown will correspond in each case to a physical diameter that depends on the unknown redshift (and the unknown k-correction), thus making such data hard to interpret.

6.3 Petrosian's Formulation

An elegant and powerful approach to measuring galaxy sizes was suggested by Petrosian (1976) [see also Gunn and Oke (1975)]. The profile $I(\theta)$ is assumed to be azimuthally symmetric and generally declining from a peak central value. Petrosian considered the logarithmic derivative of the growth curve $l(\theta)$ with respect to the logarithmic derivative of θ; that is, the function of interest is how fast the light is growing with respect to the radius. (For example, this is what is wanted to compare to the growth of the noise from the background as the measurement radius is increased.) Specifically,

$$\eta = \frac{1}{2}\frac{d \ln l(\theta)}{d \ln \theta}.$$

[I am at risk here of muddling conventions: Petrosian actually defined η to be the inverse of this function, but I find the version given above to be more convenient because it tends to 1 for $\theta \to 0$ and it tends to 0 for $\theta \to \infty$. Djorgovski and Spinrad (1982) expressed Petrosian's function in terms of magnitudes, and this convention was adopted by Sandage and Perelmuter (1990).]

The function η as defined above has the nice property that

$$\eta = I(\theta)/\langle I \rangle_\theta,$$

that is, it is the ratio of the surface brightness at radius θ to the average surface brightness within θ. This relation also helps to appreciate what η is actually measuring.

From the formula above, η is clearly independent of I_0 since the surface-brightness scale cancels out; hence a plot of η *versus* $\log\theta$ depends only on the shape of the profile and how big it really looks. Similarly, a plot of η *versus* $\log I$ is independent of the apparent size, depending only on the shape of the profile and the scale of surface brightness. Splitting the size information from the surface brightness information in this way is equivalent to splitting cosmological information from information about galaxy evolution (provided the profile shape does not evolve).

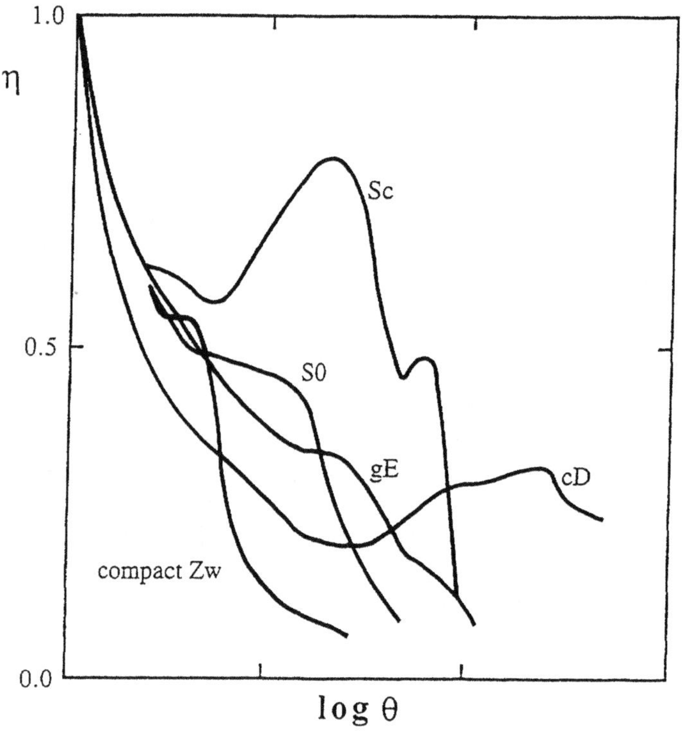

Fig. 6.3. Examples of the η function for various of galaxy profile types. Data from miscellaneous sources.

The η function is also good at displaying differences in shape between galaxy profiles. Figure 6.3 gives some examples of what the η function looks like for a variety of galaxy types. The η function is not always monotonic, which means that for a particular value of η, there is an ambiguity in which value of $\log \theta$ to associate with it. However, this is no worse in principle than the conventional approach, for which more than one radius may be associated with a particular isophote.

The η function measures a *metric* size for galaxies because when η has some particular value, it points to a specific part of the galaxy, independently of redshift or cosmological model. [Gunn and Oke (1975) and Hoessel (1980) considered things the other way around – what the value of η was for a particular radius in kpc. However, the projection of that radius into arcsec depends on q_0.] The use of the η function obviates the need for an aperture correction of the kind discussed by Sandage (1972), which as he pointed out lessens the sensitivity of the angular-size test to the cosmological model.

Since the η function when plotted *versus* $\log \theta$ does not depend on the scale of surface brightness, it can be generated without the need to calibrate the surface photometry. The amount of extinction between us and the galaxy also does not matter, at least to first order, and if color gradients are known to be small, it does not even matter which bandpass is used for the measurement.

6.4 Selection by Signal-to-Noise Ratio

Having elaborated in some detail on the measurement of galaxy profiles, we can return briefly to the theme of the previous lecture concerning the detectability of galaxies at high redshift. Suppose for sake of illustration that a galaxy has an image that is a disk with uniform surface brightness, and beyond some radius the surface brightness falls rapidly to zero. For such disks characterized by different values of angular diameters θ and surface brightness I, in the background-limited case the S/N for a measurement of the integrated flux is simply proportional to the quantity $\theta \times I$ (as opposed to the signal itself, which depends on $\theta^2 \times I$). A detection algorithm based on a minimum size θ and a minimum surface brightness I within this size is thus limited by S/N in an especially simple way, and other algorithms would behave similarly. Nevertheless, since models for faint-galaxy counts (Lecture 4) rarely take account of the angular extent of galaxy images (nor their variation), it is virtually universal to consider samples as if they were, in fact, magnitude-limited. Since high-redshift galaxies have lower surface brightness, as is well known they will be preferentially missed in such surveys. I showed (Kron 1988) schematically the strong effect on the predicted distribution of redshifts if one were to select by S/N as opposed to magnitude (Figure 6.4). The model was computed for a specific profile type, but it should still capture the main qualitative features of the variation of surface brightness with luminosity, and the behavior of the luminosity function as a function of both surface brightness and physical size. The solid curves and dashed curves in Figure 6.4 were normalized with respect to each other such that at bright magnitudes, they would select analogous samples of galaxies. Similar arguments about the strong selection because of dimming by surface brightness were made by Yoshii and Fukugita (1991) and by Phillipps, Davies, and Disney (1990), and we will see the importance of this bias in the next Lecture.

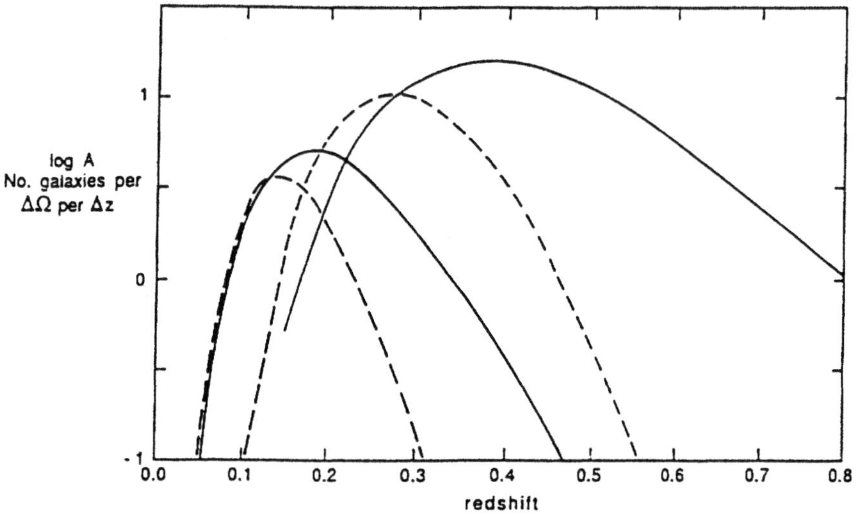

Fig. 6.4. The predicted distribution of redshifts if one were to select by $(S/N)^2$ (dashed curves) as opposed to magnitude (solid curves), from Kron (1988). The solid curves correspond roughly to one-magnitude intervals at $B = 20^{\mathrm{m}}$ and $B = 22\overset{\mathrm{m}}{.}5$.

References

Bohlin, R.C., Cornett, R.H., Hill, J.K., Hill, R.S., Landsman, W.B., O'Connell, R.W., Neff, S.G., Smith, A.M., and Stecher, T.P. 1991, ApJ **368**, 12
Doi, M., Fukugita, M., and Okamura, S. 1993, MNRAS **264**, 832
Djorgovski, S. and Spinrad, H. 1982, ApJ **251**, 417
Kent, S.M. 1985, ApJS **59**, 115
Kron, R.G. 1988 in *Astronomy, Cosmology, and Fundamental Physics*, eds. M. Caffo, R. Fanti, G. Giacomelli, and A. Renzini (Kluwer: Dordrecht), p. 383
Gunn, J.E. and Oke, J.B. 1975, ApJ **195**, 255
Hoessel, J.G. 1980, ApJ **241**, 493
Petrosian, V. 1976, ApJ **210**, L53
Phillipps, S. 1984, A&A **136**, 371
Phillipps, S., Davies, J.I., and Disney, M.J. 1990, MNRAS **242**, 235
Pritchet, C. and Kline, M.I. 1981, AJ **86**, 1859
Sandage, A. 1972, ApJ **173**, 485
Sandage, A. and Perelmuter, J.-M. 1990, ApJ **350**, 481
Yoshii, Y. and Fukugita, M. 1991, in *NATO Workshop on Observational Tests of Cosmological Inflation*, ed. T. Shanks *et al.*, p. 267

7 Reconciling Counts with the Redshift Distribution

This lecture reviews some of what is currently known about the statistical properties of faint field galaxies. To provide a framework for discussion, I review the topical issue of how to reconcile the steep counts of faint galaxies with the paucity of high-redshift galaxies in the faintest spectroscopic samples.

7.1 Statement of the Problem

A reasonable presumption is that since less gas in galaxies was locked up in the form of low-mass stars at earlier epochs, this greater amount of gas would have generated a greater rate of star formation. Galaxies at high redshift would then be expected to be generally more luminous (e.g. Quirk and Tinsley 1973). In the event of a well-orchestrated early burst of star formation (say, associated with the formation of galactic bulges), galaxies might have been very luminous and thus visible to quite high redshift, the so-called "primeval galaxy" phase. Furthermore, the active star formation would naturally be expected to be accompanied by blue colors, and since galaxies in distant volumes would be counted, the numbers of galaxies at a given magnitude would be relatively high.

This picture was partly built on the premise put forward by Partridge and Peebles (1967) that elliptical galaxies and spiral bulges may have formed the bulk of their stars within one dynamical free-fall time, yielding a high luminosity during this brief phase. Early metal enrichment was one solution to the "G dwarf problem" posed by van den Bergh and by Schmidt in the early 1960's. Without specific limits to the redshift of formation and the length of the interval of intense star formation, models predicted a wide range of observational signatures: while the original Partridge-Peebles idea predicted low-surface-brightness images with red colors visible in the near-infrared ($z \sim 10$), Meier (1977) advocated Larson's picture where star formation was more distributed in time but more concentrated spatially, with the result that the images were predicted to be blue, compact, and at lower redshift ($z \sim 2$).

An early report by Mattila (1976) of a detection of the extragalactic background light at an unexpectedly high level inspired Tinsley (1977) to propose that high-redshift galaxies should be visible in spectroscopic samples even at $B = 19^{\mathrm{m}}$ (a limit that was considered reasonably faint at the time). Tinsley's suggestion in turn inspired E. Turner, W. Sargent, and J. Gunn to make an exploratory investigation of the $B = 19^{\mathrm{m}}$ redshift distribution, using a target list that I prepared. Since no one knew what a high-redshift galaxy with intense star formation looked like, I deliberately chose galaxies with a variety of morphological types, colors, and surface brightnesses. The results of this early survey were reported by Turner (1980; see also Koo 1985), and needless to say no spectacularly high redshifts were found. (The distribution of redshifts was in fact about what was expected for the case of nothing dramatic happening.) As it turned out, the report of a high level of the extragalactic background light was not confirmed by subsequent efforts (see specifically Dube, Wickes, and Wilkinson 1979).

While the possibility of very high redshifts at bright magnitudes was not confirmed, these ideas nevertheless left a legacy that favored the notion that galaxies were indeed brighter in the past, perhaps by a substantial amount, and that evidence for evolution should be detectable even at modest redshifts. This expectation is also a natural outcome of models which are predicated on a star-formation rate that depends exponentially on the gas supply and is basically in the spirit of the old picture of Searle, Sargent, and Bagnuolo (1973). The good correspondence between the exponential star-formation rate picture and the color distribution of galaxies at the present epoch does not actually provide a strong constraint on the *history* of the star-formation rate; a wide variety of alternative forms could equally well reproduce the distribution of present-day colors. From the results described in Lecture 2, we should not be surprised by quite mild luminosity evolution in important classes of galaxies when an average is taken over a volume-limited sample.

The early work on counts (see review by Kron 1980a) and colors (Kron 1980b) of faint field galaxies showed that the counts are steep and high at faint magnitudes ($B \sim 23^{\rm m}$), and that there is a trend to bluer colors with increasing magnitude. This is just what is expected if galaxies are seen in a state of higher luminosity and higher rate of star formation, and thus these findings supported the prevailing picture. However, I also showed (Kron 1980b) that even a no-evolution model yielded a bluing trend of the right amplitude (but not quite the right shape), and that the steep counts could be properly interpreted only with good knowledge of the random errors and a detailed model of the detection process.

The crucial test of the models of galaxy counts and colors is of course the redshift distribution. David Koo and I set out to acquire these data upon the availability of an efficient multiple-object spectrograph at Kitt Peak called the Cryogenic Camera. This effort was discussed in Koo and Kron (1987, 1988), who showed that the effect of galaxy clustering in the small spectroscopic fields was so strong that the initial intent of determining the shape of the high-redshift tail of the redshift distribution was at best highly problematic. In effect, one of the main thrusts of that program has become the study of large-scale structure (e.g., Broadhurst *et al.* 1990).

In the meantime, other groups had also been conducting redshift surveys of field galaxies. One of the most influential papers was that of Broadhurst *et al.* (1988), who presented 187 redshifts in five fields in the interval $20^{\rm m} < B_{\rm J} < 21\overset{\rm m}{.}5$. Each of the five fields has a very different-looking redshift distribution due to large-scale structure. Their combined redshift distribution was compared to models, and the argument was made that the shape of the distribution was like that of their no-evolution model, and unlike that of their evolution model which predicted more high-redshift galaxies than were observed. But, the *counts* were still steep and high, implying active evolution of some kind. One way out of this puzzle was to propose that the *shape* of the luminosity function evolved, such that at $z \sim 0.2$ the galaxies now populating the faint end of the luminosity function were at that time close to M^* in luminosity. This enhanced luminosity would naturally have been driven by enhanced star formation, and this was supported by the higher fraction of strong-emission-line galaxies in the faint sample than in other brighter samples of field galaxies.

This picture has not gone unchallenged: both Eales (1993) and Lonsdale and Chokshi (1993) have used the Broadhurst *et al.* (1988) data to derive the luminosity function *in situ*, without being able to confirm the proposed form of evolution; Boroson *et al.* (1993) have questioned whether the space density of strong-emission-line galaxies at $z > 0.2$ is in fact greater than expected; and Koo and Kron (1992) and Smetanka (1992) have shown how a selection effect may account for the apparent trend of emission-line strengths with apparent magnitude.

7.2 The Count Distribution

It is remarkably difficult to extract model-independent results from galaxy counts alone, without benefit of other constraints such as colors or redshifts. Such constraints of course do exist and are routinely imposed, but the spectroscopic constraints invariably refer to substantially brighter magnitudes than the detection limit of broad-band imaging, and there is often disagreement on the numerical values of the colors and the change of the mean color with magnitude at faint limits (cf. Tyson 1988, Lilly, Cowie, and Gardner 1991; Steidel and Hamilton 1993). The available constraints are not sufficient to provide a clear picture – for example, the form of the redshift distribution at $B = 26^{\mathrm{m}}$ is quite unknown – and the interpretation of the features in the faint counts are still debated.

Koo and Kron (1992) reviewed the counts in three observed bands, B_{J}, r, and K, collected from the major efforts to date. The agreement between different observers is quite good. The key result is that a no-evolution model provides a good fit to the K-band counts everywhere, falls a bit lower than the r-band counts at the faint end, and lower still (but only by a factor of 2 or 3) than the B_{J} band counts. In other words, the "steeper and higher" phenomenon is wavelength-dependent, being most pronounced at the shortest wavelengths, confirming Koo's (1986) earlier result. Two other points are noteworthy: the difference between the data and the no-evolution model in the B_{J} band does not appear to become progressively greater with increasing magnitude. Rather, the excess appears between $B_{\mathrm{J}} = 21^{\mathrm{m}}$ and $B_{\mathrm{J}} = 23^{\mathrm{m}}$, and then stays at about the same factor to the faint end. Second, the K-band counts do not seem to be in any way remarkable: the shape and normalization are consistent with entirely conventional expectations, as indeed is the

$r - K$ color distribution. [However, Gardner, Cowie, and Wainscoat (1993) attribute a change in median color at $K \sim 17^{\mathrm{m}}$ to an evolutionary effect.]

In summary, the strongest evidence for some kind of evolution comes from the faint B_{J} counts, but even there the amount is relatively modest and is not progressive with magnitude. The rest-wavelengths that appear in the B_{J} band are in the ultraviolet, for which we have only sketchy empirical constraints on the luminosity function (e.g. Donas *et al.* 1987). For $\lambda_{\mathrm{rest}} < 4000$ Å, the spectral energy distributions of galaxies are steep and have large curvature, meaning that relatively small errors in either the individual or ensemble spectral energy distributions will be amplified at high redshift. We can also expect that the effects of reddening will be more pronounced at these short wavelengths, and model spectral energy distributions may be unreliable on that account. If at $B_{\mathrm{J}} = 24^{\mathrm{m}}$ the typical redshift is $z \sim 0.5$ (Allington-Smith 1993), then the observed rest-wavelength is λ3100 and a typical galaxy in the sample is about $1\overset{\mathrm{m}}{.}5$ fainter than M^* (about what is expected – see Lecture 4). Finally, we can reasonably anticipate that the typical galaxy is intrinsically blue, if only because of the color-luminosity relation (Lecture 3). We do not have good constraints on the space density of such galaxies viewed at this wavelength – in fact, the *data* at $z = 0.5$ might be superior to the extrapolations from samples at $z \sim 0$! All of this is to promote skepticism that our models are sufficiently reliable to enable definitive conclusions regarding evolution based on the current B_{J} counts and the current spectroscopic samples selected in the B_{J} band. At longer wavelengths the differences with respect to the no-evolution model are smaller, and so too are the uncertainties in the models, since the rest-wavelengths correspond to spectral regions actually observed in local samples.

7.3 The Redshift Distribution

It has become standard practice to quantify the distribution of redshifts in a complete magnitude-defined sample by the *median redshift*, and the claim has been made that for $B > 20^{\mathrm{m}}$ the median redshift is too low to be consistent with conventional models that match the galaxy counts. Koo and Kron (1992) collected all of the available redshift surveys and produced a cumulative redshift distribution, broken down by B_{J} magnitude. The assumption is that the galaxies populating each bin of apparent magnitude – the ones with measured redshifts – are representative of all galaxies at that magnitude. Koo and Kron (1992) also presented a no-evolution model and discussed an evolution model. Figure 7.1 summarizes the results for the median redshift. The agreement of the data and the model at bright magnitudes was built in as a constraint, equivalent to specifying the value of M^*. Otherwise, for $B_{\mathrm{J}} < 22^{\mathrm{m}}$ the data scatter around by an amount that is larger than the difference between the no-evolution and the evolution model, despite the large number of galaxies in these bins (typically 300). The scatter is due to effects of large-scale clustering that has not been averaged out. On this basis, for

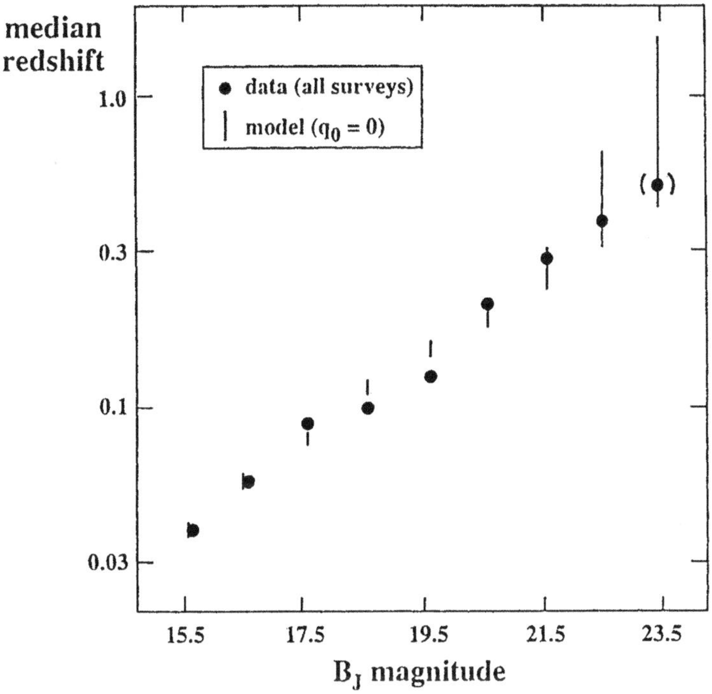

Fig. 7.1. The median redshift in bins of B_J magnitude, from the compilation of redshift surveys by Koo and Kron (1992). The faintest point, which was quite uncertain from that work, has now been substantially confirmed by Allington-Smith (1993). The bars give the model results discussed by Koo and Kron (1992), where the lower end of each bar assumes no evolution in luminosity for any of the galaxies, and the upper end is the prediction according to Bruzual's formulation for evolution.

$B_J < 22^m$, there is in fact *no* evidence that the observed median redshift can yet discriminate between the non-evolution model and one with conventional amounts of evolution.

The discussion has naturally shifted to the faintest available survey, where the evolution model of Figure 7.1 clearly predicts redshifts that are too high. There are however many available parameters (epoch of formation, mix of types of galaxy, details of the luminosity function for each type, chosen star formation history for each type) and Koo and I made no special effort to explore the range of possibilities. Had an excellent fit to the data including the faintest point been obtained, it would have been interesting but still not definitive. (Later, I will mention new results by Koo, Gronwall, and Bruzual on a no-evolution model that has its parameters adjusted automatically, rather than by hand.)

From Figure 7.1 one cannot immediately decide if the low observed median redshift at the faintest magnitudes is due to a deficiency in the expected number of high-redshift galaxies or to an excess in the expected number of low-redshift galaxies. This failure illustrates the limitation of using the median redshift as an index for galaxy evolution, since the galaxies at low redshift just add noise from the point of view of searches for evolutionary effects. (The shape of the redshift distribution at low redshift has only to do with the shape of the faint end of the luminosity function and, because of the small volume sampled, fluctuations due to large-scale structure.) As another example of the limitations of the median redshift as a tool for interpreting the redshift distribution, suppose that in a sample of 20 faint galaxies, one galaxy were found to lie at $z > 3$. The presence of such a galaxy might completely change the conclusions of the survey, yet it would have little effect on the value of the median redshift.

One of the recent new faint redshift surveys is by Lilly (1993), who selected galaxies at long wavelengths to $I_{\rm AB} = 22\overset{\rm m}{.}5$ in a single 10 arcmin^2 field. Of the 37 objects in the interval $21^{\rm m} < I_{\rm AB} < 22\overset{\rm m}{.}5$, 27 are galaxies with known redshift. This sample can be used to determine the space density of luminous galaxies at high redshift, which can then be compared to the local density of galaxies of similar luminosity. The following gives a modified version of the argument developed by Lilly. Lilly's $I_{\rm AB}$ band is at $\lambda 8500$, and at a redshift of $z = 0.83$ the $I_{\rm AB}$ band samples the same rest-wavelengths that appear in the local determinations of the space density of galaxies ($B_{\rm J} \sim \lambda 4650$). As it turns out, a galaxy at $z = 0.83$ and $I_{\rm AB} = 22^{\rm m}$ is almost exactly M^* (for $q_0 = 0.5$). We can compute, using the machinery developed by A. Sandage, the expected number of galaxies within some magnitude and redshift bin. I choose the faintest whole magnitude bin, $21\overset{\rm m}{.}5 < I_{\rm AB} < 22\overset{\rm m}{.}5$, and the redshift range $0.74 < z < 0.92$ (corresponding to $\pm 5\%$ in the quantity $(1 + z)$ with respect to $z = 0.83$). Within Lilly's area and these limits in magnitude and redshift, we would expect to see 2.2 galaxies based on the normalization ϕ^* of Loveday *et al.* (1992), whereas he actually measured 4 (and some of the unidentified objects in this same magnitude range might also have $0.74 < z < 0.92$). The statistics are obviously still quite poor, but the point is that this technique should be robust, since no model of the mix of different galaxy types is needed, no k-corrections are needed, and since the normalization ϕ^* in the $B_{\rm J}$ band is well determined. At least there is no indication that there are *fewer "M^* galaxies"* at high redshift than expected. Hence, the resolution of why the median redshift is unexpectedly low may more likely involve the faint end of the luminosity function (Tresse *et al.* 1993).

Progress has been made in ensuring completeness in the faint samples selected in the $B_{\rm J}$ band. For example, the survey of Colless *et al.* (1990) was 80% complete in the faintest magnitude bin, and it could be argued that the missed galaxies were preferentially at high redshifts (say, if they had large sizes and low surface brightness at a given magnitude) according to the arguments given in Lecture 6. The Colless *et al.* (1993) extension has now raised the completeness level to 95% for the original sample. The most recent faint B_J sample is given by Allington-Smith (1993), which extends to $B_J = 24^m$ and at that point is 80% complete. This new survey does show a tail to high redshift: fainter than $B_J = 23^m$, 9 galaxies, or more than 20%, have redshifts greater than $z = 0.7$, and two of these have $z > 1$. These are clearly very luminous galaxies, but since for these we are observing rest-wavelengths of roughly $\lambda 2500$, it is hard to evaluate whether their space density is consistent with galaxies in the local Universe.

A contrarian interpretation of the faint field-galaxy data can be summarized as follows: we know that there must exist strong biases against the detection of high-redshift galaxies (Lecture 6). In the absence of luminosity evolution, half of the photons that we see from galaxies are expected to originate at $z < 0.22$. Detailed models for luminosity evolution (Lecture 2) support the notion that evolutionary effects should be mild at these redshifts. Moreover, there is no reason to expect galaxies to be accurately co-phased in their evolution at $z \sim 0.2$, meaning that the evolution of individual galaxies will tend to be washed out in the ensemble. Therefore, if the blue counts are seen to be higher than expected, it may be more economical to suspect systematic modeling errors at low redshifts and low luminosities, rather than unexpectedly dramatic evolutionary phenomena. The models for the counts and the redshift distribution cannot be taken too seriously for several reasons:

1) in general, they do not account for the angular extension of real galaxy images nor do they include an explicit model of the detection process;
2) we do not have an adequate knowledge of the relation between profile shape, surface brightness scale, and intrinsic color for field galaxies;
3) the spectral energy distributions for field galaxies are not well specified, especially in the near ultraviolet.

7.4 The Color Distribution

Because most galaxies have spectral energy distributions in the visible band that are concave downwards with increasing slope at shorter wavelengths, the colors of individual galaxies tend to become redder with increasing redshift. Depending on the details of which bands are used and the type of galaxy spectrum, the color peaks at some redshift and thereafter becomes bluer. The dependence of the distribution of colors of an ensemble of galaxies with increasing apparent magnitude is more complicated, as it depends not only on the relative frequency of different intrinsic spectral energy distributions, but also on the distribution of luminosities (i.e., the luminosity function) of each spectral type. Because the most-frequently-seen type of galaxy changes with increasing apparent magnitude, the mean color of an ensemble can become bluer with increasing magnitude, even though each galaxy is becoming redder with increasing redshift.

If the metallicity of a model stellar population is increased, or the foreground or internal reddening is increased, or the fraction of hot stars is decreased, the spectral energy distribution gets steeper in such a way that the model galaxy shifts on a color-color diagram parallel to the main galaxy locus (which itself arises from the range of these same parameters). Thus metallicity, reddening, and age mix are approximately degenerate as far as broad-band colors in the visible spectral range are concerned. On the contrary, the effect of redshift is to change the *curvature* of the apparent spectrum, and the model galaxy shifts perpendicular to the main galaxy locus. Hence, the distribution of colors for faint galaxies has encoded in it the distribution of redshifts; in principle, one can even contemplate using colors of individual galaxies to derive "photometric redshifts" which may be statistically useful (Koo 1985; Loh and Spillar 1986). This possibility is especially attractive when applied at $B > 24^{\rm m}$ where there is essentially no other diagnostic for the redshifts (but, for the same reason, there is no way to check the conclusions).

The color distribution as a function of apparent magnitude also depends on the mix of galaxy types and on the luminosity function, as described above. A goal that is less ambitious than trying to determine individual redshifts is to model the overall color distribution as a global check on the assumed parameters. This was demonstrated by Koo (1986) for a complete sample of faint field galaxies, using a four-band photometric system. The key result was that the Bruzual μ-models, normalized by the statistical properties of nearby galaxies, were indeed able to describe the observed color distributions – the absolute value of the color centroid, the spread in colors, and the asymmetry in the color distribution – with no *ad hoc* extra features. This agreement provides strong encouragement that the model spectral energy distributions must be generally matching real galaxies (see also Bruzual 1988 and 1990). The agreement between the model spectral energy distributions and the data is a necessary but not sufficient condition for the viability of Bruzual's picture for the past history of the rate of star formation.

A good example of a prediction based on colors from these models was given by Koo (1990), who proposed that galaxies with a measured $B_{\rm J} - I$ color less than $1^{\rm m}\!.0$ should have redshifts greater than 1.0 and should be prevalent at $B_{\rm J} > 23^{\rm m}$. This suggestion motivated Colless *et al.* (1993) to make a color-selected faint redshift survey, specifically aimed at these relatively blue galaxies. They found that, on the contrary, none of the galaxies meeting Koo's criteria had redshifts greater than 1.0 (they obtained reliable redshifts for 6 galaxies at $B_{\rm J} \sim 23^{\rm m}$ with $B_{\rm J} - I < 1^{\rm m}\!.0$); instead, the typical redshift was about 0.5. This finding might suggest that the Bruzual μ-models evolve unrealistically strongly.

Nevertheless, as mentioned before it is known that the colors derived from Bruzual's models do reproduce the data (except for the bluest galaxies) and are therefore useful in situations where the amount of luminosity evolution is not the key concern. One example of such a study is that of Koo, Gronwall, and Bruzual (1993), who produced a new no-evolution model that attempts to account for all of the principal observed marginal distributions: counts, colors, and redshifts. They used an automated procedure to determine what local mix of galaxies can account for the faint-galaxy observations; that is, they left the form of the local luminosity function and the local intrinsic color distribution unconstrained. Then, these functions derived directly from the faint-galaxy data can be compared with the local volume. They find that, indeed, a match can be made simultaneously to the B_J, r, and K counts; the $B_J - R_F$ colors; and the faint-galaxy redshift distribution that is as good as any other model, and which yields a reasonable local luminosity function. With this particular no-evolution model, the difference with respect to the faint B_J counts is even lower than Koo and Kron's (1992) model. They are also able to reproduce the number and slope of the counts at $B_J \sim 19^m$, thus challenging the conclusion of Maddox *et al.* (1990) that the counts at these magnitudes indicated a surprising amount of evolution at surprisingly bright magnitudes.

References

Allington-Smith, J.R. 1993, in Gemini Project Newsletter No. 4

Boroson, T.A., Salzer, J.J., and Trotter, A. 1993, ApJ **412**, 524

Broadhurst, T.J., Ellis, R.S., and Peterson, B.A. 1988, MNRAS **235**, 827

Broadhurst, T.J., Ellis, R.S., Koo, D.C., and Szalay, A.S. 1990, Nature **343**, 726

Bruzual A., G. 1988, in *Towards Understanding Galaxies at Large Redshift*, eds. R.G. Kron and A. Renzini (Dordrecht: Kluwer), p. 161

Bruzual A., G. 1990, in *Evolution of the Universe of Galaxies*, ed. R.G. Kron (San Francisco: A.S.P.), p. 185

Colless, M., Ellis, R.S., Taylor, K., and Hook, R.N. 1990, MNRAS **244**, 408

Colless, M., Ellis, R.S., Broadhurst, T.J., Taylor, K., and Peterson, B.A. 1993, MNRAS **261**, 19

Donas, J., Deharveng, J.M., Laget, M., Milliard, B., and Huguenin, D. 1987, A&A **180**, 12

Dube, R.R., Wickes, W.C., and Wilkinson, D.T. 1979, ApJ **232**, 333

Eales, S. 1993, ApJ **404**, 51

Gardner, J.P., Cowie, L.L., and Wainscoat, R.J. 1993, ApJ **415**, L9

Koo, D.C. 1985, AJ **90**, 418

Koo, D.C. 1986, ApJ **311**, 651

Koo, D.C. 1990, in *Evolution of the Universe of Galaxies*, ed. R.G. Kron (San Francisco: A.S.P.), p. 268

Koo, D.C. and Kron, R.G. 1987, in *Observational Cosmology*, IAU Symp. 124, eds. A. Hewitt, G. Burbidge, and L.Z. Fang (Dordrecht: Reidel), p. 383

Koo, D.C. and Kron, R.G. 1988, in *Towards Understanding Galaxies at Large Redshift*, eds. R.G. Kron and A. Renzini (Dordrecht: Kluwer), p. 209

Koo, D.C. and Kron, R.G. 1992, ARA&A **30**, 613

Koo, D.C., Gronwall, C., and Bruzual A., G. 1993, ApJ **415**, L21

Kron, R.G. 1980a, Phys. Scripta **21**, 652

Kron, R.G. 1980b, ApJS **43**, 305

Lilly, S.J. 1993, ApJ **411**, 501
Lilly, S.J., Cowie, L.L., and Gardner, J.P. 1991, ApJ **369**, 79
Loh, E.D. and Spillar, E.J. 1986, ApJ **303**, 154
Lonsdale, C. and Chokshi, A. 1993, AJ **105**, 1333
Loveday, J., Peterson, B.A., Efstathiou, G., and Maddox, S.J. 1992, ApJ **390**, 338
Maddox, S.J., Sutherland, W.J., Efstathiou, G., Loveday, J., and Peterson, B.A. 1990, MNRAS **247**, 1P
Mattila, K. 1976, A&A **47**, 77
Meier, D.L. 1977, ApJ **207**, 343
Partridge, R.B. and Peebles, P.J.E. 1967, ApJ **147**, 868
Quirk, W.J. and Tinsley, B.M. 1973, ApJ **179**, 69
Searle, L., Sargent, W.L.W., and Bagnuolo, W.G. 1973, ApJ **179**, 427
Smetanka, J.J. 1992, in *The Evolution of Galaxies and Their Environment*, NASA Conf. Publ. 3190, eds. D. Hollenbach, H. Thronson, and J.M. Shull (Ames Research Center: NASA) p. 27
Steidel, C.C. and Hamilton, D. 1993, AJ **105**, 2017
Tinsley, B.M. 1977, ApJ **211**, 621
Tresse, L., Hammer, F., Le Fèvre, O., and Proust, D. 1993, A&A **277**, 53
Turner, E.L. 1980, in *Objects of High Redshift*, IAU Symp. 92, eds. G.O. Abell and P.J.E. Peebles (Dordrecht: Reidel), p. 71
Tyson, J.A. 1988, AJ **96**, 1

8 The Butcher-Oemler Effect – A Case Study

So far we have concentrated on the evolution of galaxies in the field, without special regard to their environment. It is well known that the properties of galaxies do depend on their neighborhoods, at least for high-density environments, where S0 and E morphological types are common and spirals and dwarf irregulars are rare. The study of the evolution of galaxies in high-density regions, namely the cores of rich clusters of galaxies, therefore has special interest.

Indeed, the case that evolutionary effects have been detected has been made more forcefully for distant cluster galaxies than for distant field galaxies, and the changes seen are commonly quoted as primary evidence for evolution of any type of galaxy in the Universe. Evolutionary changes associated with rich clusters are often referred to loosely as the "Butcher-Oemler effect", but it is important to distinguish between different types of evolution, for example evolution of colors as opposed to evolution of spectral lines or evolution of morphology.

The actual "Butcher-Oemler effect" not surprisingly is best defined by the authors themselves (Butcher and Oemler 1984): the colors of cluster core galaxies at high redshift are systematically bluer with respect to samples selected in a similar way at low redshift. The specific selection rules are:

1) the measured color is $B - V$ in the rest-frame of the cluster;
2) the absolute magnitudes are brighter than $M_V = -20^m$ for $H_0 = 50$ km sec^{-1} Mpc^{-1};
3) the colors are corrected for the color-luminosity relation for elliptical and S0 galaxies;
4) the galaxies are selected within a radius that contains 30% of the total galaxies in the cluster (typically about 1 Mpc);
5) the clusters have to be compact, as measured by the ratio of radii containing 60% of the galaxies and 20% of the galaxies;
6) the color distribution is corrected statistically for the background of projected field galaxies.

Then, the claim is that the fraction of the galaxies with colors bluer than $0\overset{m}{.}2$ of the peak of the color distribution (a parameter called f_b) grows with increasing redshift, being noticeable already at $z = 0.2$. Specifically, the upper envelope for the distribution of f_b grows from a negligible value at $z = 0$ to $f_b = 0.4$ at $z = 0.5$, about the value for field galaxies. Verification that rich clusters at $z = 0.2$ contained blue galaxies was forthcoming from the work of Couch and Newell (1984) and others.

The surprising thing about this result is that such low redshifts correspond to a look-back time of only a few cluster crossing-times, and it is hard to see why such strong evolution should be happening so recently, and for so many clusters all in rough synchronization. The clusters are expected to be dynamically old (that is, having experienced many crossing times) because the ones chosen for such studies tend to be centrally condensed; for these clusters, one might have expected the critical processes that depend on cluster dynamics to have reached some kind of equilibrium long ago.

The intriguing state of affairs to which Butcher and Oemler called attention naturally inspired a number of pictures of the evolutionary process. The original Butcher-Oemler conjecture (1978) was based on the similarity of the number of blue galaxies in the high-redshift clusters to the number of S0 galaxies seen in local clusters. This suggested that at high redshift we are seeing spiral disks in a state of active star formation; subsequently, they either exhaust the supply of gas or the gas was stripped out, and they now appear as disk galaxies with little or no star formation, i.e., S0 galaxies. Since the frequency of morphological types is very sensitive to galaxy density (at least at high densities), perhaps this relationship is strongly time-dependent (Oemler 1992). Another idea is that the blue galaxies have enhanced star formation because they are falling into the cluster for the first time: ram pressure from the ambient intracluster medium enhances the star formation rate in the affected galaxies by inducing cloud collapse (Gunn and Dressler 1988). This view is based on the observation that the blue galaxies tend to have a wider angular distribution and a wider distribution in velocity space. Still another possibility is that the blue galaxies are mergers or milder interactions (the events having induced a greater rate of star formation, as in many specific instances seen locally) which supposedly were more frequent in more youthful clusters (Lavery and Henry 1988).

8.1 A Collection of Concerns

The question of the redshift at which an evolutionary effect is seen raises an interesting issue that applies generally to galaxy-evolution studies. At very small distances, the Universe is sufficiently inhomogeneous that a representative sample of some type of source cannot be reliably constructed, no matter what the average space density is. For rare objects such as powerful radio galaxies or especially massive clusters, it may be necessary to make a census out to substantial redshifts to yield a useful statistical sample. As the volume that defines the local sample becomes larger, the random fluctuations should of course decline, but the possibility that real evolutionary effects are encountered increases. If a sample at $z = 0.2$ appears to differ from another sample at $z = 0.05$, how is it to be determined whether the nearby sample is non-representative, or whether the more distant sample is actually showing the effects of evolution?

No matter how carefully the nearby and distant clusters are compared, the result ultimately depends on the integrity of the respective samples. Since the number density of field galaxies increases with magnitude, and the number of galaxies in a cluster brighter than a fixed magnitude decreases with increasing redshift, the contrast of the cluster decreases with redshift. This effect has been modeled by Cappi *et al.* (1989) and Couch *et al.* (1991), among others, with the result that the decrease in contrast with redshift is indeed quite steep and depends on the bandpass of observation. Hence, there is expected to be a strong preference for selection in favor of the richest, most condensed clusters at high redshift.

As just mentioned, if one selects clusters in the optical band and subsequently looks for evolution also in the optical band, there is an obvious danger of building in a bias. One way around this problem is to select the clusters based on the presence of a radio source; a good example of this approach is the work of Allington-Smith *et al.* (1993), who studied the properties of groups of galaxies surrounding strong radio galaxies. (In principle, one still needs to ensure that there is no connection between the existence of the radio source and the optical properties of the galaxies in the group. A test of this nature was made by Allington-Smith *et al.*, who showed that the luminosity function in the groups was indistinguishable from the field.) They argued that whereas the nearby groups showed the usual tendency for redder galaxies to predominate in richer environments, this relation is not present at $z \sim 0.4$, where the fraction of blue galaxies is independent of the richness of the group surrounding the radio galaxy.

Since clusters that have a condensed morphology in the X-ray band tend to coincide with rich, high-velocity-dispersion, dynamically well-mixed systems, X-ray surveys may provide a powerful means of specifying a homogeneous sample of rich clusters. While the number of clusters with a particular X-ray luminosity may be expected to evolve with cosmic epoch, clusters with similar X-ray properties may still be arguably similar. The sample of clusters seen by *Einstein* has been discussed by Edge *et al.* (1990) and Henry *et al.* (1992). The sample of clusters found in uniform surveys by ROSAT are eagerly anticipated (see Shanks *et al.* 1991 for an early effort).

8.2 A Rogue's Gallery

The most fundamental diagram showing the Butcher-Oemler effect is Figure 3 of Butcher and Oemler (1984). In this diagram, the five clusters at highest redshift form a natural grouping. All of these clusters were in the literature and had prior spectroscopic measurements. [Because of the great difficulty of spectroscopic studies of faint galaxies, these existing clusters by default were the clusters measured for f_b by Butcher and Oemler (1984).] Each one of them has some associated tale: for example, A370 at $z = 0.37$ is one of the most distant clusters in the Abell (1958) catalogue that has a spectroscopic redshift based on a number of galaxies. [Its extreme distance is evident from the deep survey of Abell clusters by Huchra *et al.* (1990), in which no cluster was found as far as this one.] 3C 295 at $z = 0.46$ had its redshift determined by Minkowski (1960) and had already been the subject of intense study (Baum 1962) since the apparent characteristics of the cluster suggested an exceptionally high redshift. The two clusters 1446+2619 and 0024+16 were both found by Allan Sandage (1957) and were included in the distant-cluster study of Gunn and Oke (1975); their redshifts are 0.37 and 0.39, repectively. It seems reasonable to suppose that the clusters selected by Sandage are also extreme cases, since they too were originally found on Palomar Schmidt plates, and the procedure outlined by Humason was to find the most distant clusters, not necessarily to establish a representative sample of distant clusters. Indeed, 0024+16 has proved itself to be exceptional in many ways – it is exceptionally rich, contains one of the first gravitational arcs discovered (like A370), and has peculiarities in its spatial and velocity fields. Finally, 0016+16 at $z = 0.54$ is the most distant cluster in the Butcher-Oemler sample, and as it turns out it has a *low* value for f_b (Koo 1981). I found it on Mayall 4-m prime focus plates, and suggested to H. Spinrad that it deserved spectroscopic attention as a likely high-redshift cluster because of its red color and the faintness of the brightest candidate members. The cluster is also exceptionally rich (none of the other 4-m plates that I have obtained – covering roughly 5 deg^2 – show anything like this cluster), and it is an exceptionally luminous X-ray source (White, Silk and Henry 1981). It seems that 0016+16 was discovered in much the same way that 1446+2619 and 0024+16 were except that the color was available and used to reinforce the claim for high redshift. The existence of a spectroscopic redshift then promoted subsequent work such as the long-exposure, pointed *Einstein* observations, and the study by Koo (1981) of the fraction of blue galaxies.

Since 1984 there have of course been a number of other high-redshift clusters that have been added to the sample (e.g., Dressler and Gunn 1992), but these also tend to be drawn from older sources (e.g. Gunn and Oke 1975) and from visual searches of photographic plates (e.g. Gunn, Hoessel, and Oke 1986, Couch *et al.* 1991). In the survey by Gunn, Hoessel, and Oke (1986), the waveband of detection was shifted to the red to attempt to tune to the same rest-frame wavelengths. It was also shown that the number of clusters in the catalogue was consistent with a selection that picked the same type of cluster at all redshifts, at least up to $z = 0.5$. Automated searches for faint clusters is definitely an idea whose time has come (Postman 1993), and there has been encouraging progress in developing techniques at brighter magnitudes (Lumsden *et al.* 1992; Dalton *et al.* 1992). It is hoped that in the not-too-distant future the size and quality of major redshift surveys will enable selection of clusters in *redshift* space.

In summary, the clusters that formed the foundations of the Butcher-Oemler effect in 1984 are evidently very exceptional objects – otherwise, objects discovered in the mid-1950's would not continue to appear so prominently in current research efforts. The discovery technique was subjective (with the exception of the cluster surrounding 3C 295, a powerful radio source) and hard to quantify, and the procession from cluster discovery to published redshift to selection for color study is also not controlled. The kinds of checks that one might want to make to ensure that a sample of distant clusters was in fact representative might well be the same tests used to search for evolutionary effects! Clearly, more information is needed: first to establish the comparability of the near and distant cluster samples, and then, using independent information, to show that there are real differences in their properties.

8.3 Galaxy Morphological Types

The apparent presence of a higher fraction of blue galaxies in distant clusters could just be a selection effect, since whatever blue galaxies there are in the cluster will have smaller k-corrections and therefore appear relatively bright when seen at shorter rest-wavelengths. In a prescient paper, Greenstein (1938) said as much:

> "A possible observational test of the extreme difference in behavior of late-type spirals and ellipsoidal nebulae would be afforded by the relative frequency of spirals and ellipsoidals in distant clusters of extragalactic nebulae. The higher [color] temperatures would result in smaller magnitude corrections arising from the red shifts for spirals than for ellipsoidals. The relative frequency of spirals should, therefore, increase among the fainter clusters of nebulae."

The most effective and straightforward technique for dealing with this problem is to measure the color distributions of galaxies in their respective rest-frames (Butcher and Oemler 1984). It is also important to extend this methodology to the selection of the clusters (Gunn, Hoessel, and Oke 1986) since the change in appearance of the cluster could affect its inclusion in a sample.

The association of elliptical galaxies with the cores of rich clusters naturally suggests that the color distribution should be red and relatively narrow. Until recently, good color distributions for nearby clusters were not available, and so the distribution of morphological types in clusters like Coma were used to infer the color distribution (Butcher and Oemler 1978). Measuring different quantities at different redshifts is obviously not desirable; to address this, more recent work (e.g. Butcher and Oemler 1984) has produced color distributions in more nearby clusters that provide the needed frame of reference.

In Lecture 3 I commented on the poor connection between color and morphological type for field galaxies, and the same is true for cluster galaxies. This is worth re-emphasizing, since many of the arguments about the interpretation of the blue galaxies in distant clusters relate to morphology (e.g., whether interactions are responsible for the bluer stellar population, and whether the blue light is in a disk).

Another example of the lack of a good connection between colors and morphological type is provided by the Coma cluster itself. Abell (1977) plotted the distribution of all spiral galaxies recognizable on a V-band Palomar Schmidt plate (36 deg^2) centered on Coma, and showed that the cluster

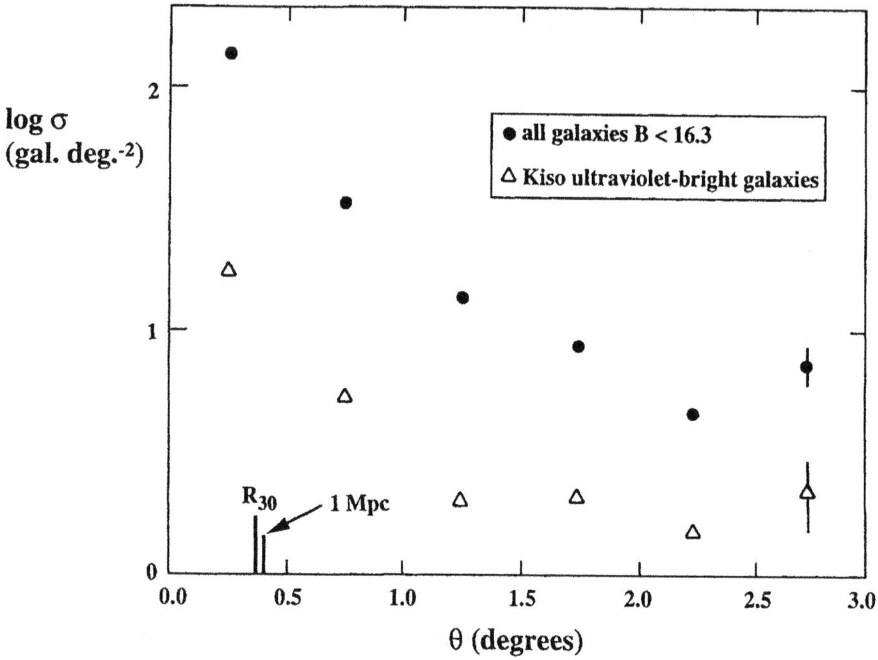

Fig. 8.1. Number of galaxies per square degree, corrected statistically for the field background, as a function of radial zone in the Coma cluster, from the Kiso survey by Takase (1980). The dots represent all galaxies brighter than about $B_J = 16\overset{m}{.}3$, and the open triangles represent the distribution of ultraviolet-bright galaxies. $\sqrt{N}$ error bars are given for the outermost radial bin. The fraction of ultraviolet-bright galaxies in the field using the same color criterion is about 25%.

was not even visible! However, Takase (1980) in a study of ultraviolet-bright galaxies in a number of fields showed that there is a clear concentration of these galaxies to the center of the Coma cluster (although the *fraction* of ultraviolet galaxies was not as high in the center). His central bin corresponds roughly to the radius containing 30% of the total galaxies (1 Mpc), and there the fraction of ultraviolet-bright galaxies is 13% (Figure 8.1), to be compared to the value for the field similarly defined, namely 25%. The magnitude limit can be derived from the mean field surface density of 5.25 galaxies per square degree; his band (G) is centered at λ4900 Å, not too far from the photographic B_J band. On this basis, the magnitude limit is about $B_J = 16\overset{m}{.}3$, which at the distance of the Coma cluster corresponds to $M_V = -19\overset{m}{.}5$, not much fainter than Butcher and Oemler's recipe. Nevertheless the point is that there *are* blue galaxies in the core of even the Coma cluster, and these are evidently not the spirals of Abell's 1977 study. It is intriguing in this regard that Caldwell *et al.* (1993) found that 1/3 of the early-type galaxies in a field 40 arcmin away from the center of Coma show spectroscopic signatures of recent star formation, but this effect was not present for the galaxies in the core.

Much of what is known about the properties of galaxies in nearby clusters, specifically their morphological type distribution and the distribution of strengths of spectral features (such as [OII] 3727 and the 4000 Å spectral "break") comes from studies by Dressler and his colleagues (Dressler, Thompson, and Shectman 1985; Dressler and Shectman 1987; Dressler and Shectman 1988). The last two studies used an entrance aperture measuring 2×4 arcsec for most galaxies; since the redshifts are typically $z \sim 0.04$, the spatial scale sampled is about 3×6 kpc ($H_0 = 50$ km sec^{-1} Mpc^{-1}). One might wonder whether a bias could be introduced because galaxies in distant clusters are measured with an aperture that projects to a substantially larger physical size. To avoid this problem, Dressler and Gunn (1982) were careful to measure the nearby comparison sample with apertures that projected to about 15 kpc.

The first results from *Hubble Space Telescope* imaging of distant, rich clusters are now becoming available. These images are in principle capable of falsifying models for the origin of the Butcher-Oemler effect. One such study by Dressler *et al.* (1994) of 0939+4713 at $z = 0.41$ showed that among the bluest galaxies, there were indeed many images with disturbed morphology suggestive of interactions, as argued already for other clusters from high-resolution ground-based imaging by Lavery, Pierce, and McClure (1992). At the same time, many galaxies with less extreme colors look like "normal ... late-type spirals". In another study, Couch *et al.* (1993) correlated their spectroscopic classification in AC 114 ($z = 0.31$) with the morphology derived from *Hubble Space Telescope* imaging, and also found a high incidence of systems that they interpreted as merging and having tidal interactions.

In conclusion, the present situation regarding the Butcher-Oemler effect is that there is abundant spectroscopic evidence that interesting phenomena are at work in the cores of rich clusters at high redshift (and other interesting phenomena at low redshift). It is still difficult to place this result in proper context because of the lack of a way to evaluate the *particular* distant clusters, since they were selected under non-controlled conditions. It may be that a sample of X-ray-selected clusters will provide the needed independent check on the sample selection for both near and distant clusters.

References

Abell, G.O. 1958, ApJS **3**, 211
Abell, G.O. 1977, ApJ **213**, 327
Allington-Smith, J.R., Ellis, R.S., Zirbel, E.L., and Oemler, A. 1993, ApJ **404**, 521
Baum, W.A. 1962, in *Problems of Extra-Galactic Research*, IAU Symp. No. 15, ed. G.C. McVittie (New York: MacMillan), p. 390
Butcher, H. and Oemler, A. 1978, ApJ **219**, 18
Butcher, H. and Oemler, A. 1984, ApJ **285**, 426

Cappi, A., Chincarini, G., Conconi, P., and Vettolani, G. 1989, A&A **223**, 1
Caldwell, N., Rose, J.A., Sharples, R.M., Ellis, R.S., and Bower, R.G. 1993, AJ **106**, 473
Couch, W.J., Ellis, R.S., Malin, D.F., and MacLaren, I. 1991, MNRAS **249**, 606
Couch, W.J., Ellis, R.S., Sharples, R.M., and Smail, I.R. 1993, in *Observational Cosmology*, A.S.P. Conf. Ser. vol. 51, eds. G. Chincarini, A. Iovino, T. Maccacaro, and D. Maccagni (San Francisco: A.S.P.), p. 240
Couch, W. and Newell, E.B. 1984, ApJS **56**, 153
Dalton, G.B., Efstathiou, G., Maddox, S.J., and Sutherland, W.J. 1992, ApJ **390**, L1
Dressler, A. and Gunn, J.E. 1982, ApJ **263**, 533
Dressler, A. and Gunn, J.E. 1992, ApJS **78**, 1
Dressler, A., Oemler, A., Butcher, H.R., and Gunn, J.E. 1994, ApJ **430**, 107
Dressler, A. and Shectman, S.A. 1987, AJ **94**, 899
Dressler, A. and Shectman, S.A. 1988, AJ **95**, 284
Dressler, A., Thompson, I.B., and Shectman, S.A. 1985, ApJ **288**, 481
Edge, A.C., Stewart, G.C., Fabian, A.C., and Arnaud, K.A. 1990, MNRAS **245**, 559
Greenstein, J.L. 1938, ApJ **88**, 605
Gunn, J.E. and Dressler, A. 1988, in *Towards Understanding Galaxies at Large Redshift*, eds. R.G. Kron and A. Renzini (Dordrecht: Kluwer), p. 227
Gunn, J.E., Hoessel, J.G., and Oke, J.B. 1986, ApJ **306**, 30
Gunn, J.E. and Oke, J.B. 1975, ApJ **195**, 255
Henry, J.P., Gioia, I.M., Maccacaro, T., Morris, S.L., Stocke, J.T., and Wolter, A. 1992, ApJ **386**, 408
Huchra, J.P., Henry, J.P., Postman, M., and Geller, M.J. 1990, ApJ **365**, 66
Koo, D.C. 1981, ApJ **251**, L75
Lavery, R.J. and Henry, J.P. 1988, ApJ **330**, 596
Lavery, R.J., Pierce, M.J., and McClure, R.D. 1992, AJ **104**, 2067
Lumsden, S.L., Nichol, R.C., Collins, C.A., and Guzzo, L. 1992, MNRAS **258**, 1
Minkowski, R. 1960, ApJ **132**, 908
Oemler, A. 1992 in *NATO Advanced Study Institute on Clusters and Superclusters of Galaxies*, ed. A.C. Fabian (Dordrecht: Kluwer), p. 29
Postman, M. 1993, in *Observational Cosmology*, A.S.P. Conf. Ser. vol. 51, eds. G. Chincarini, A. Iovino, T. Maccacaro, and D. Maccagni (San Francisco: A.S.P.), p. 260
Sandage, A.R., in *1956-1957 Annual Report of the Director, Mt. Wilson and Palomar Observatories*, p. 62
Shanks, T., Georgantopoulos, I., Stewart, G.C., Pounds, K.A., Boyle, B.J., and Griffiths, R.E. 1991, Nature **353**, 315
White, S.D.M., Silk, J., and Henry, J.P. 1981, ApJ **251**, L65
Takase, B. 1980, PASJ **32**, 605

9 Deep-Universe Programs for the Future

The next generation of astronomical instrumentation will provide new observational opportunities for studies of distant galaxies. In this lecture I describe a few examples of directions that might be followed. The common theme is that with sufficient perseverance, we can learn as much about distant galaxies as about nearby galaxies. This lecture is a *potpourri* of four experiments: measuring galaxy rotation curves at high redshift; using adaptive optics to study galaxy morphology at high redshift; detection of distant supernovae; and plans for taking advantage of a low-background "window" in the near-infrared spectral range.

9.1 Galaxy Rotation Curves at High Redshift

As Allan Sandage has stressed in his lectures, the key element of a cosmological test is the association of a proper distance to an independently measured redshift. If we can establish the physical scale of a galaxy – its intrinsic luminosity, size, or mass – then its distance can be deduced from the corresponding apparent properties. The Tully-Fisher method is an example of how a measurement that is distance-independent (the amplitude of the rotation curve) can be used to deduce something about the physical scale of the galaxy, from which its distance can be determined.

There is no reason why something like the Tully-Fisher (or Faber-Jackson, or line width - luminosity) relation cannot be used at high redshift, thereby providing a route to q_0 as well as H_0. For example, the 21-cm line width used in the Tully-Fisher technique is integrated over the galaxy, that is, essentially no spatial resolution is needed. Hence, the fact that distant galaxies are only marginally resolved is not necessarily a fundamental problem. However, one cannot use the 21-cm line at large redshifts ($z > 0.07$ or so) for a variety of reasons, not the least of which is that the redshifted line may suffer interference from the broadcast television band.

Disk galaxies usually have flat rotation curves and have most of their H I outside the central regions, i.e., the H I is mostly located on the flat part of the rotation curve. The result is a characterisitc bi-cusped line profile that has a well-defined width. The integrated spectra of optical emission in disk galaxies will show the same effect, as long as the bulk of the emission comes from regions a few kiloparsecs away from the nucleus. Van der Kruit and Pickles (1988) calculated what the integrated profile of Hα should look like under plausible conditions, and concluded that indeed the bi-cusped shape should be apparent. Hence, a measurement of the line profile (with a Fabry-Perot interferometer or with a slit spectrograph) will yield the amplitude of the rotation curve at high redshift, without requiring high spatial resolution. (Hα itself gets redshifted into a region of the spectrum with intense emission from the Earth's atmosphere. However, other strong emission lines at shorter wavelengths would serve the same purpose.)

We also need the inclination angle of the disk so that the measured width of the line can be corrected for the projection angle. Moreover, even though the classical application of the Tully-Fisher technique does not do so directly, if a reliable measure of the characteristic size of the galaxy were available, then the mass itself can be obtained from $GM \sim R\ (V/\sin i)^2$. Structural parameters can be determined from a high-resolution image of the galaxy (from the refurbished *Hubble Space Telescope* cameras). More specifically, one would use the image to determine:

1) the inclination angle i;
2) the degree of concentration of light, which is related to the bulge-to-disk ratio and therefore to the properties of the rotation curve;
3) the characteristic size R;
4) assurance that the galaxy is a normal disk system.

The point is that while the spectroscopy itself does not require the high spatial resolution, a full analysis does.

In the above application the parameter R is derived from the light distribution but applied to the mass distribution. Presumably variations in the mass-to-light ratio M/L are an important part of the scatter in Tully-Fisher distances. In the central regions of galaxies, the M/L is roughly what is expected from the stellar population alone, and some high mass-to-light ratio material is needed in the outer parts.

This rather involved state of affairs was worked out long ago in a physically transparent way by Öpik (1922), who showed how the distance to M 31 could be deduced (this was at a time when there was not yet a consensus on even the nature of the spiral nebulae). Öpik was the first to devise the dynamical technique of determining distances to galaxies. Actually, his method is more general and more powerful than the Tully-Fisher/Faber-Jackson technique, since full advantage is taken of the angular information.

The idea is simple: assume for sake of argument a galaxy at distance d that is circularly symmetric and supported by rotation, and consider quantities as a function of angle θ from the center, e.g. $l(\theta)$ is the apparent flux integrated within θ, $M(\theta)$ is the mass within θ, and so on. Then by combining the inverse-square law with $\alpha GM = d\theta\ (V/\sin i)^2$ (the dimensionless parameter α describes the geometrical distribution of the mass), one arrives at the distance

$$d = \frac{1}{4\pi\alpha G}\ \frac{L}{M}\ \frac{\theta}{l}\left(\frac{V}{\sin i}\right)^2,$$

or, in terms of the surface brightness I,

$$d = \frac{1}{4\pi^2\alpha G}\ \frac{L}{M\theta I}\left(\frac{V}{\sin i}\right)^2.$$

Even in complete ignorance of the shape and importance of the halo, the range of α is small and the variation of α with θ must also be small. Otherwise, the only quantity on the right-hand side that is not directly observable is the inverse of the mass-to-light ratio. In the present application of a test for q_0, the value for M/L is not needed, but we do need to know how the mean value of M/L for a statistical sample of galaxies changes with redshift, either because of selection effects or because of evolution of the population of galaxies. If some independent index of M/L were available, it would also help to reduce the scatter in the distance; such distance-independent quantities as rest-frame color, degree of central concentration of the light, and surface brightness are expected to be related to the stellar population and therefore to M/L.

As outlined above, the use of line-width measurements as an evolutionary or cosmological probe depends on an emission line profile that is bi-cusped, like the 21-cm line. However, it is not out of the question to consider that distant galaxies could be observed with at least a few elements of spatial resolution along the major axis. For example, at $z = 0.5$, 20 kpc corresponds to 2.8 arcsec; hence under conditions of very good seeing it should be possible to measure a rotation curve (Kron 1987). (Remember, all that is needed is a measure of the amplitude of the rotation curve, as opposed to the details of the shape in the central region.) The feasibility of such measurements has been demonstrated by Vogt *et al.* (1993) for two spiral galaxies at $z = 0.2$. Moreover, Franx (1993) has measured velocity dispersions from line widths in a cluster at $z = 0.18$, which can be exploited in much the same way.

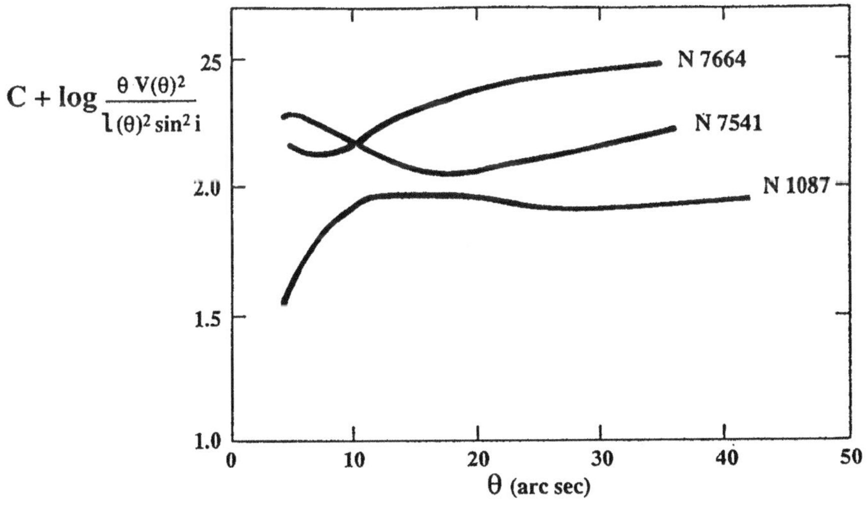

Fig. 9.1. The quantity $C + \log[(\theta/l(\theta))(V(\theta)/\sin i)^2]$, which is proportional to $\log(M/L)$. The angular information for NGC 7664 and NGC 1087 was scaled to the distance of NGC 7541.

We can test the effectiveness of the technique and evaluate the scatter from published data for nearby galaxies. Kent (1986) has obtained r-band radial profiles for many of the galaxies in the sample of Sc galaxies for which Rubin *et al.* (1980) had obtained rotation curves. This sample should be representative of luminous Sc galaxies. Figure 9.1 shows a quantity that is proportional to $\log(M/L)$ for three example galaxies, chosen to have a range of shapes for their rotation curves. The angular information for NGC 7664 and NGC 1087 was scaled to the distance of NGC 7541, but no other correction of any kind has been made to the data. While Figure 9.1 shows variations of M/L between the three galaxies, at least in each individual case the curves are flat over a substantial range, which means that for a fixed M/L, the same distance would be derived from data regardless of radial angle θ.

9.2 Adaptive Optics Imaging of Distant Galaxies

Diffraction-limited imaging of faint objects with ground-based telescopes can be achieved with a system that senses and corrects for atmospheric distortions. A good summary of the principal ideas behind the technique of *adaptive optics* is given by Beckers (1993); here the topic is outlined from the perspective of what is required to measure distant galaxies at high angular resolution.

Temperature fluctuations in Earth's atmosphere induce variations in the local index of refraction, with the result that what was initially a plane wavefront at the top of the atmosphere arrives as a corrugated wavefront at the telescope pupil. A characteristic scale length for the size of these distortions in the wavefront, called the *Fried parameter* r_0, can be derived assuming the turbulence is described by a Kolmogorov spectrum. Physically, r_0 is the characteristic distance between two points in the pupil where the wavefront has a phase difference of order one radian; at visible wavelengths, $r_0 = 10$ cm produces a long-exposure seeing image size of 1 arcsec. Since the index of refraction varies less at longer wavelengths, the seeing is better there and the parameter r_0 is larger. The time scale for variations in the wavefront depends on the distance of the distorting atmospheric element and its transverse velocity, being typically 0.01 sec at visible wavelengths and longer at longer wavelengths. Thus, to correct the wavefront in 1-arcsec seeing requires sampling at least every 10 cm in the pupil plane and every 0.01 second in time. To get enough photons in this collecting area and in this time interval to determine the shape of the wavefront requires a sixth-magnitude source! However, the object that is used to sample the wavefront need not be the object of interest, since there is a correlation of the atmospheric distortions between neighboring points in the sky. The solid angle over which there is significant coherence is called the *isoplanatic patch.* Unfortunately, the isoplanatic patch is quite small, being only a few arcseconds in diameter at visible wavelegths.

The quantities given above depend sensitively on the prevailing seeing. Adaptive optics should not be regarded as a technique that makes bad seeing good, but rather a technique which, if used in the best possible seeing, can result in significant gains in power at high spatial frequencies. The image of a star delivered by an adaptive optics system will have a core with the Airy width and a halo that is roughly the size of the original seeing disk. The figure of merit is the ratio of the energy in the central spike to the total energy, called the Strehl ratio. Adaptive optics systems are now being built that can achieve a Strehl ratio of about 0.15 at optical wavelengths. At a wavelength of $\sim$ 2 microns, the performance should be much enhanced, perhaps up to a Strehl ratio of 0.7. A star as faint as $K = 13^{m}$ can serve to determine the shape of the wavefront, and the isoplanatic patch is much larger in area. The combined result is that, compared to the optical, the K-band is much less constrained from the point of view of the availability of nearby bright references sources.

The hardware necessary to correct the wavefront has the following generic parts. The pupil of the telescope is re-imaged so that much smaller physical elements actually do the correction. A beam splitter sends part of the light to a wavefront sensor, which measures the tilt (or curvature) of each part of the wavefront on the pupil, with spatial resolution on the pupil at least equal to r_0. The rest of the light goes to an adaptive mirror placed in the re-imaged pupil. It has independently controllable elements of size r_0 or smaller, which enable the overall surface to be instantaneously shaped to take out the wavefront distortions of the atmosphere. Finally, hardware and software is needed to analyse the output of the wavefront sensor, construct a global picture of the shape of the wavefront, and send the appropriate servo signal to each element of the adaptive mirror. Systems now under development have of order 100 elements, which means that they could be used on telescopes with apertures up to 2 m in 0.5-arcsec seeing ($r_0 = 20$ cm at $\lambda 5000$) – conditions which might happen 10% of the time at a good observatory site. With these parameters, the Airy resolution $\lambda/D = 0.05$ arcsec, which is good enough to motivate the substantial efforts being invested in this technology.

Let's now explore what could be done with an 8 m telescope at near-infrared wavelengths. A value of $r_0 = 20$ cm at $\lambda 5000$ corresponds to $r_0 = 130$ cm at 2.4 μm (again, assuming a Kolmogorov model of atmospheric turbulence). Thus if each "region of influence" on the adaptive mirror corresponds to this dimension on the pupil, nominally only $\sim$40 adaptive mirror elements would be required. The wavefront requires correction on a time scale of about 0.13 sec, well within the frequency response of current wavefront sensing and correcting systems. The diameter of the isoplanatic patch at 2.4 μm is about 40 arcsec, which means that the reference star must be within this region. The fraction of the sky at high latitudes with sufficiently bright reference stars within this area is a few percent (Beckers 1993).

At 2.4 μm, $\lambda/D = 0.062$ arcsec for an 8-meter aperture, and pixels should be half this size. An array of 1024 $\times$ 1024 pixels would be 32 arcsec on a side and would thus sample well the isoplanatic patch. It will be assumed

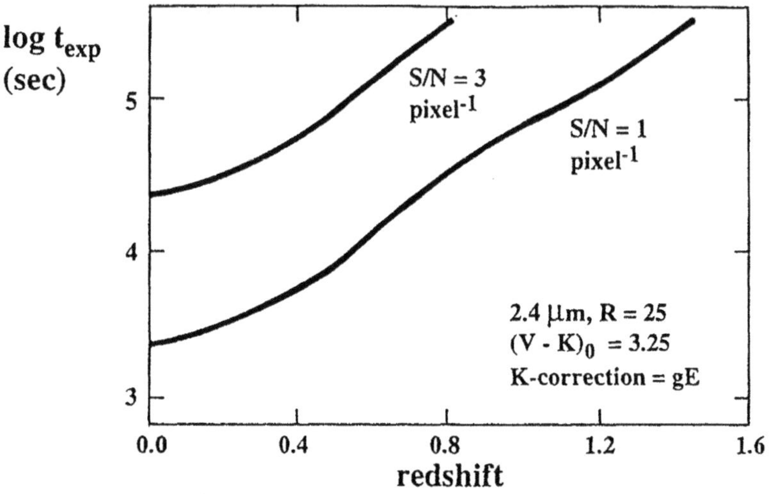

Fig. 9.2. Detection of faint surface brightness in the near-infrared with angular sampling at the diffraction limit. The example is an unevolving elliptical galaxy with surface brightness at $z = 0$ of $\mu_V = 22\overset{m}{.}5$ arcsec^{-2}. The curves give the integration time required to detect $S/N = 3$ and $S/N = 1$ per 0.031 arcsec pixel, as a function of redshift, for the instrument parameters described in the text.

that a total system efficiency $\eta = 0.2$ can be achieved and that the sky-plus-telescope background is thermal at $T = 263$ K with an emissivity of 0.15. This background is 1300 μJy arcsec^{-2} at 2.4 μm; assuming $\lambda/2D$ pixels and $R = 25^{\mathrm{m}}$, there would be 7 counts sec^{-1} pixel^{-1} per 8-meter aperture (Eq. 5.1), much larger than the leakage current for these detectors. The detector read noise is exceeded for exposures longer than a few minutes.

An example of the study of galaxy morphology at large redshift is the detection of our standard surface brightness, $\mu_V = 22\overset{m}{.}5$ arcsec^{-2} in the rest frame, which corresponds to 0.08 detected photons sec^{-1} pixel^{-1} per diffraction-limited aperture in the 2.4 μm band for a galaxy at rest with $V - K = 3\overset{m}{.}25$. At higher redshifts the intensity in the band depends on the source spectrum. Here I adopt the energy distribution of an elliptical galaxy, which will make little difference at lower redshifts and which will be conservative at higher redshifts. No evolutionary corrections are applied. Figure 9.2 shows the results of the calculation of the total exposure time required at this surface brightness to reach $S/N = 3$ and $S/N = 1$ per pixel, as a function of redshift. The dominant effect is the relativistic dimming of surface brightness. Galaxy morphology can usually be recognized even at $S/N = 1$ per pixel if there are enough pixels. The conclusion is that it is indeed possible to measure galaxy morphology at high redshift, but substantial observing time is required. As pointed out in Lecture 5, the photon count rate per pixel when sampling at the diffraction limit is independent of the telescope aperture – hence these very long exposure times are a fundamental fact of life.

9.3 Detection of Distant Supernovae

Since Type II supernovae have massive progenitors, their observed rate in some volume gives a direct measure of the current rate of star formation, and this will vary according to the characteristics of that time and place. For reference, the Milky Way has about one million supernovae per rotation period, and a protogalaxy could have 100 times this number, or about one per year (Matteucci, Tornambè, and Vettolani 1988).

The detection of supernovae in protogalaxies at high redshift also requires adaptive optics, observations at near-infrared wavelengths, and an 8-m class telescope. To estimate the observability of distant supernovae, a simple thermal model for the emergent flux is considered. The supernova envelope is assumed to have a radius $R = 2 \times 10^{15}$ cm, which would be appropriate a couple of weeks after the beginning of the expansion. Two temperatures are assumed, $T_{\rm eff} = 8000$ K and $T_{\rm eff} = 6000$ K. The hotter model has a luminosity $L = 3 \times 10^9 L_\odot$. As before, the cosmological parameters adopted are $q_0 = 0.5$, $H_0 = 50$ km sec^{-1} Mpc^{-1}.

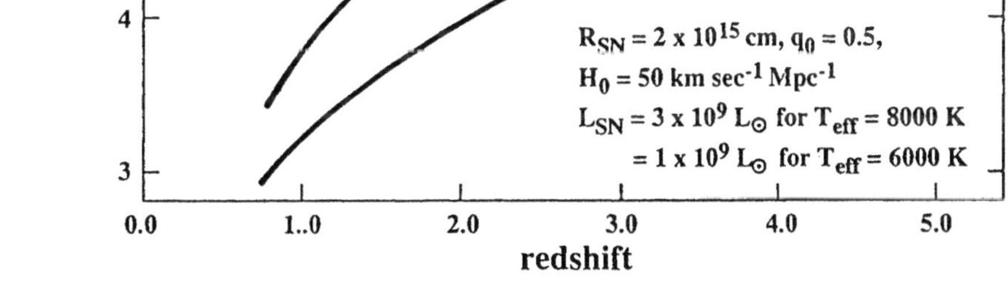

Fig. 9.3. Detection of a faint point source in the near-infrared with angular sampling at the diffraction limit. The example is a distant supernova, modeled separately as a blackbody with $T = 8000$ K and $T = 6000$ K. The curves give the integration time required to detect the supernova at $S/N = 10$, as a function of redshift. Noise-equivalent area = 15 pixels, other parameters same as for Figure 9.2.

The assumed noise-equivalent area is 15 pixels (King 1983), which is the number of pixels contributing to the background noise when the diffraction-limited image is sampled at $\lambda/2D$. At the peak of the light curve one would like to have $S/N = 10$. Figure 9.3 gives the exposure time required to achieve this signal-to-noise ratio as a function of redshift, using the same parameters as before (8-m telescope, 2.4 μm band). The conclusion is that supernovae can be detected to redshifts $z > 4$, perhaps higher than the galaxy formation epoch. Primeval galaxies may be revealed by their population of supernovae, rather than by the underlying star light which is harder to detect because it is diffuse. This is a good example of how sensitivity to high spatial frequencies can enhance the detection threshold for faint objects. (It is also an example of the detection regime where the time required to achieve a particular signal-to-noise ratio is inversely proportional to the *fourth power* of the telescope aperture.)

The volume enclosed at large redshift is sensitive to the adopted cosmological model, and the predicted rate of detection of supernovae varies accordingly. Assuming no change in the actual supernova rate per unit co-moving volume and the Einstein-de Sitter cosmological model, very roughly one new supernova would be detected per year per arcmin^2 assuming the same patch of sky were repeatedly observed every few weeks. (Recall that the field size per exposure is about 0.5 arcmin.) As mentioned earlier, the supernova rate at large redshift could be dramatically higher than what it is now.

A modest observing program would be to establish an upper limit to the number of very faint supernovae by observing a relatively small field over an interval of one or two years. Such limits to the rate of new supernovae as a function of the limiting flux would rule out classes of models that predict very large numbers of supernovae. A more extended observing program would be aimed at detecting at least a dozen supernovae with fluxes indicating distances greater than $z = 2$. This detection would provide strong constraints on stellar populations at large redshifts. A still more ambitious program would be to monitor individual supernovae in a number of wavebands to establish light curves. These data could be used in a cosmological test in the spirit proposed by Wagoner (1977).

9.4 *K*-Band Imaging in Antarctica

At visible wavelengths, the median redshift for galaxies at $B = 22^{\mathrm{m}}$ is only 0.25 or so (Lecture 7) because the cosmological expansion has the effect of diluting the radiation stream from the more distant objects. However, corresponding samples observed in the near-infrared are expected to include more galaxies at higher redshifts because 1) a larger fraction of the bolometric flux appears in the K-band, whereas the opposite is true at visible wavelengths (Lecture 4); 2) for galaxies with large spheroidal components, a robust result of evolutionary models is that the bolometric luminosity was of the order of 1^{m} brighter at $z = 1$ than at $z = 0$ (Lecture 2); 3) in general, redder galaxies are more luminous than bluer galaxies (Lecture 3); 4) longer wavelengths suffer less extinction from interstellar dust inside the galaxies.

The potential of deep K-band imaging can only be fully realized if the sky background is substantially reduced. In the near-infrared band, the background is high and is due partly to thermal emission from the atmosphere and from the telescope optics, and partly due to airglow lines. The former source of noise can be reduced by two orders of magnitude by operating the telescope from the South Pole, where the ambient temperature is about 210 K. (Due precaution must still be taken that the emissivity of the telescope is low.) The latter source of noise can also be reduced by noting that longward of 2.27 μm the "forest" of intense OH emission lines apparently ends. Thus from the South Pole there is a window from 2.27 μm to about 2.45 μm that is expected to be exceptionally dark. If the atmospheric models are correct, then it may be possible to reduce the background to about the level of the zodiacal light – this is the best that can be done (short of leaving the Solar System, of course).

This opportunity is the basis of a project within the Center for Astrophysical Research in Antarctica to place a 60-cm telescope at the South Pole to test the potential of the site (Hereld 1994). The telescope will also be large enough, and have adequate instrumentation, to conduct a survey of a few square degrees of sky that will be deep enough to be extraordinarily interesting. If all of these plans are successful, the next step would be to build a 2.5 m telescope at the same site.

To compute the expected flux limits, I assumed a net background (atmosphere and telescope) of 6.6 μ Jy arcsec^{-2} and a pixel scale of 0.8 arcsec (which is sufficient to ensure that the sky flux is greater than the detector noise for integration times of a few minutes). Given these assumed parameters, it is straightforward using Eq. 5.1 to compute the S/N for a point source. For a single exposure of 1250 sec, a 60-cm telescope, and Poisson noise in 9 pixels, the derived signal-to-noise ratio is 6.9 at $K_{AB} = 22^m$. This is amazingly good, considering the short exposure time and the small size of the telescope.

Accurate photometry requires accurate knowledge of the level of the background, which is determined from "blank sky" between the sources. At the great depth of this survey, image crowding is a problem (Lecture 5), especially for sampling at 0.8 arcsec per pixel. The advantage of the 2.5 m telescope in Antartica is that it will provide much better sampling, as well as allowing spectroscopic measurements.

References

Beckers, J.M. 1993, ARA&A **31**, 13

Franx, M. 1993, PASP **105**, 1058

Kent, S.M. 1986, AJ **91**, 1301

King, I.R. 1983, PASP **95**, 163

Kron, R.G. 1987, in *Nearly Normal Galaxies*, ed. S.M. Faber (New York: Springer-Verlag), p. 300

Hereld, M. 1994, in *Infrared Astronomy with Arrays: The Next Generation*, ed. I.S. McLean (Dordrecht: Kluwer)

Matteucci, F., Tornambè, A., and Vettolani, P. 1988, in *Towards Understanding Galaxies at Large Redshift*, ed. R.G. Kron and A. Renzini (Dordrecht: Kluwer), p. 85

Öpik, E. 1922, ApJ **55**, 406

Rubin, V.C., Ford, W.K., and Thonnard, N. 1980, ApJ **238**, 471

van der Kruit, P.C. and Pickles, A.J. 1988, in *Towards Understanding Galaxies at Large Redshift*, ed. R. G. Kron and A. Renzini (Kluwer: Dordrecht), p. 339

Vogt, N.P., Herter, T., Haynes, M.P., and Courteau, S. 1993, ApJ **415**, L95

Wagoner, R.V. 1977, ApJ **214**, L5

The Physics of Background Radiation

Malcolm S. Longair

University of Cambridge, Dept. of Physics, Cavendish Laboratory, Madingley Road, Cambridge CB3 0HE, England

1 Observations of the Extragalactic Background Radiation

1.1 A Prospectus

In introductory lecture courses on cosmology, it is traditional to refer to the study of background radiation as the 'oldest problem in cosmology'. What has come to be known as *Olbers' paradox* revolves around the question

Why is the sky dark at night?

Olbers was, in fact, only one of many distinguished scientists of the past who realised that the darkness of the night sky provides us with some general information about the large scale distribution of matter and radiation in the Universe. Edward Harrison in his book *Darkness at Night: the Riddle of the Universe* (1987) provides a delightful history of Olbers' paradox which is full of wonderful material — this book is essential reading for anyone who wishes to obtain insight into the process of discovery in astronomy and cosmology.

Recently, on a car journey from Oxford to Cambridge, the topic came up in conversation with Hermann Bondi, the author of the classic text *Cosmology* (1952) and to whom Harrison (1990) attributes the term *Olbers' Paradox*. Bondi gave what is probably the most general interpretation of the darkness of the night sky — namely that, in some very general sense, the Universe must be very far from equilibrium. The paradox on its own cannot, however, specify in what way the Universe is far from equilibrium. All that we can state with certainty is that the Universe is not in a state of thermodynamic equilibrium at the same temperature as the surface of the stars.

The attitude of astronomers to studies of the background radiation is strongly wavelength dependent. In those wavebands in which the extragalactic background radiation is difficult to detect or for which only upper limits are available — the radio, infrared, optical and ultraviolet wavebands — cosmologists normally turn their attention to other problems and use the observations as constraints upon physical processes in the Universe. In those wavebands in which the background radiation is very intense — the centimetre, millimetre, X-ray and γ-ray wavebands — the origin of the background assumes a central role in cosmological studies. Of these, pride of place must go to the Cosmic Microwave Background Radiation which is the cornerstone of much of modern cosmology.

The objectives of this School are to study *The Deep Universe* and in some sense, which I will define, my responsibility is to study the deepest Universe of all. As we will find out, this is indeed the case but not quite in the way in which the organisers of the School may have anticipated. The organisation of my lectures is therefore as follows:

1. *Observations of the background radiation in all accessible wavebands* (Chapter 1) We begin by discussing the problems of disentangling the extragalactic background radiation in each waveband from confusing sources and give some general indications of its origin.

2. *The theoretical infrastructure for undertaking cosmological studies of the background radiation* (Chapters 2 and 3) These chapters describe the theoretical framework within which cosmological studies are carried out. They are complementary to Dr. Sandage's description of the fundamentals of observational cosmology. My lectures are strongly pedagogical in nature and are concerned with establishing the physical principles underlying cosmological models. These are the essential tools needed for undertaking cosmological investigations. I would emphasise that my approach is only one of many which can be taken to derive the tools of classical cosmology. Through teaching this material over a number of years, I have found this approach to be the simplest way of understanding the formulae and techniques which are the bread and butter of the cosmologist.

3. *The contribution of discrete sources to the background radiation* (Chapter 4) In addition to identifying the principal contributors to the background radiation, we study how much can be learned from fluctuations in the spatial distribution of the background radiation in different wavebands. These ideas are applied to the particular cases of the X and γ-ray backgrounds in Chapter 5.

4. *The Standard Big Bang* In Chapter 6, a brief survey of some of the key features of the standard Big Bang model of the Universe is given. Chapters 7 and 8 are concerned with the problems of understanding the origin of the large scale structure of the distribution of galaxies. The theory of galaxy formation in its simplest form is the subject of Chapter 7. The reasons for taking dark matter models of galaxy formation very seriously are outlined in Chapter 8 and then in Chapter 9 the theory is confronted with observations of fluctuations in the Cosmic Microwave Background Radiation. These observations provide unique information about the very early evolution of the structures out of which galaxies formed.

5. *The Intergalactic Gas* In Chapter 10, we discuss what is known about the present and past state of the intergalactic gas out of which the galaxies must have condensed in the first place.

6. *Galaxy formation and evolution* Finally, in Chapter 11, we study some aspects of the origin and evolution of galaxies from the point of view of the background radiation and the origin of the heavy elements. We will find that the background radiation provides useful constraints about the origin of the heavy elements — I will describe some recent work which Andrew Blain and I have completed on the millimetre background radiation which is of direct relevance to these studies.

The organisers have requested that the pedagogical aspects of these studies be emphasised and I therefore ask expert readers to bear with me if I set out the arguments in somewhat more detail than is usual. My excuse is that these are the points which can sometimes cause students trouble.

1.2 Recommended Reading

There is a vast literature on cosmology and the background radiation and I will aim to include most of the important references as we go along. It may be helpful, however, if I list some of the volumes which I have found most useful in teaching and in preparing these lectures.

Two symposia have covered the complete range of studies of the background radiation:

The Galactic and Extragalactic Background Radiation (1990). IAU Symposium No. 139, (eds. S. Bowyer and C. Leinert). Dordrecht: Kluwer Academic Publishers.

The Extragalactic Background Radiation (1994). STScI Symposium in honour of Riccardo Giaconni held in May 1993, (eds. M. Livio, M. Fall, D. Calzetti and P. Madau). Cambridge: Cambridge University Press (in press).

The basic texts on cosmology and the physics of the expanding Universe which I use most often are as follows. The list is roughly in order of increasing level of sophistication.

Principles of Cosmology and Gravitation, by M.V. Berry (1989). Bristol: Adam Hilger.

Essential Relativity: Special, General, and Cosmological by W. Rindler (1977). New York: Springer-Verlag.

Gravitation and Cosmology by S. Weinberg (1972). New York: John Wiley and Co.

Relativistic Astrophysics by Ya.B. Zeldovich and I.D. Novikov. Volume 1 (1971) *Stars and Relativity*; Volume 2 (1983) *The Structure and Evolution of the Universe*. Chicago: Chicago University Press.

Principles of Physical Cosmology by P.J.E. Peebles (1993). Princeton: Princeton University Press.

Physics of the Early Universe (1990), (eds. J.A. Peacock, A.F. Heavens and A.T. Davies). Edinburgh: SUSSP Publications. The articles by Simon White (*Physical Cosmology*) and George Efstathiou (*Cosmological Perturbations*) can be particularly strongly recommended.

The Early Universe by E.W. Kolb and M.S. Turner (1990). Redwood City, California: Addison-Wesley Publishing Co.

I have written up my own versions of some of the basic material needed to understand current research in two publications:

Theoretical Concepts in Physics (1992). Cambridge: Cambridge University Press. Chapter 14 provides a gentle introduction to general relativity and Chapter 15 is an introduction to its application to cosmology.

Galaxy Formation in *Evolution of Galaxies: Astronomical Observations* (1989), (eds. I. Appenzeller, H.J. Habing and P. Lena). Heidelberg: Springer-Verlag. My chapter in this volume (pages 1 – 93) is a simple introduction to the basic physics of galaxy formation. I will recycle some of that material in this presentation but with significant enhancements to bring the story up-to-date.

I have also written a general introduction to recent developments in astrophysics as a whole in an Chapter entitled *The New Astrophysics* in the volume *The New Physics*, (ed. P.C.W. Davies) (1989). Cambridge: Cambridge University Press. This may be a useful source of background reading for the astrophysical context of the present study.

1.3 Observations of the Extragalactic Background Radiation – An Overview

Let us begin with an overview of the background radiation. During the academic year 1968-69, I worked in Moscow with Rashid Sunyaev and during that time we began a major review of the extragalactic background radiation which was published in Uspekhi Fizicheskikh Nauk in 1971 (Longair and Sunyaev 1971). I reproduce our spectrum in Fig. 1.1 and it is gratifying that it is still a remarkably good representation of the major features of the background spectrum.

Let us make a number of notes about Fig. 1.1. First of all, the spectrum shown in Fig. 1.1 is plotted in terms of the *intensity* of the radiation, that is, the power per unit frequency interval arriving per unit area at the observer from 1 steradian of sky, W m^{-2} Hz^{-1} sr^{-1}. This is a standard SI unit and I will rigorously use SI units throughout the text although I may on occasion

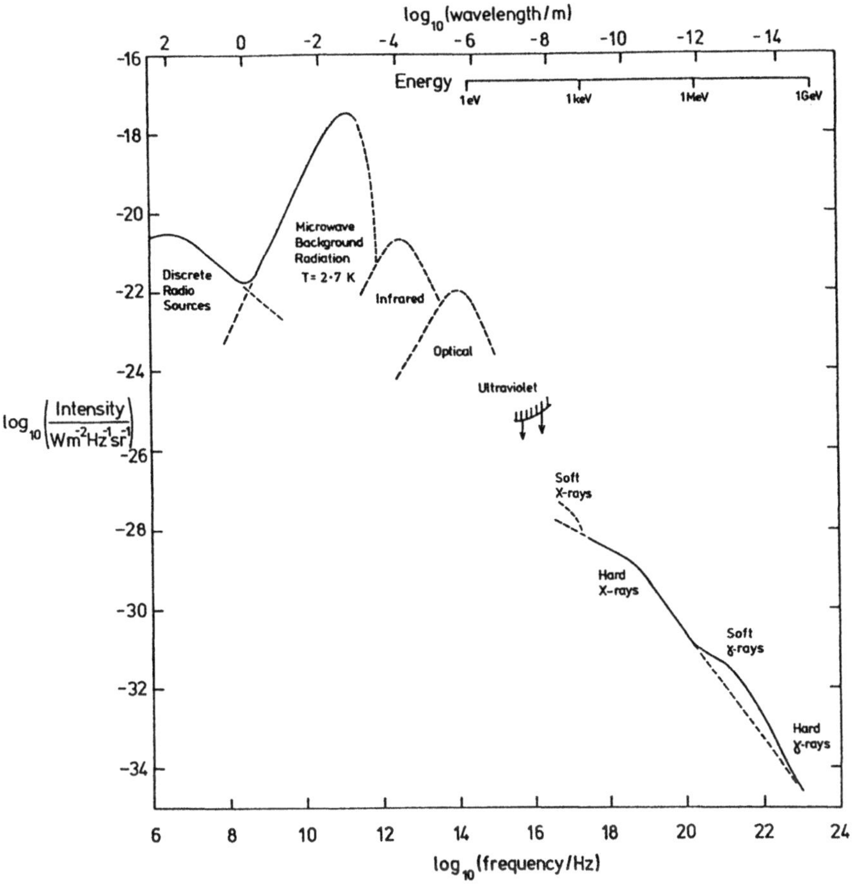

Fig. 1.1 The spectrum of the extragalactic background radiation as it was known in 1969 (Longair and Sunyaev 1971). The solid lines indicate regions of the electromagnetic spectrum in which extragalactic background radiation had been measured. The dashed lines were theoretical estimates of the background radiation due to discrete sources and should not be taken very seriously.

have to use non-SI units when displaying the results of others. Let me make the standard plea that everyone adopt SI units — there is no excuse for using other units which simply act as an unnecessary block to comprehension for those not directly involved in the field. Note that in some wavebands the intensities are expressed in terms of the power per unit wavelength interval I_λ. Since the same power is obtained in the interval $d\lambda$ corresponding to $d\nu$, $I_\nu d\nu = I_\lambda d\lambda$, it follows that $I_\lambda = (\nu^2/c)I_\nu$.

Expressing the background radiation in terms of an intensity is the best way of describing the spectrum of the radiation but it does not give a proper impression of the relative amounts of energy associated with different regions of the spectrum. This is given by

$$I = \int_{\nu(\min)}^{\nu(\max)} I_\nu \mathrm{d}\nu$$

where ν(max) and ν(min) are the maximum and minimum frequencies of the waveband concerned. It is often convenient to give an impression of the *power* or *energy* associated with each wavelength by approximating $I \sim \nu I_\nu = \lambda I_\lambda$ where I_λ is the intensity of the background per unit wavelength interval. Fig. 1.1 is redrawn in terms of νI_ν in Fig. 1.2.

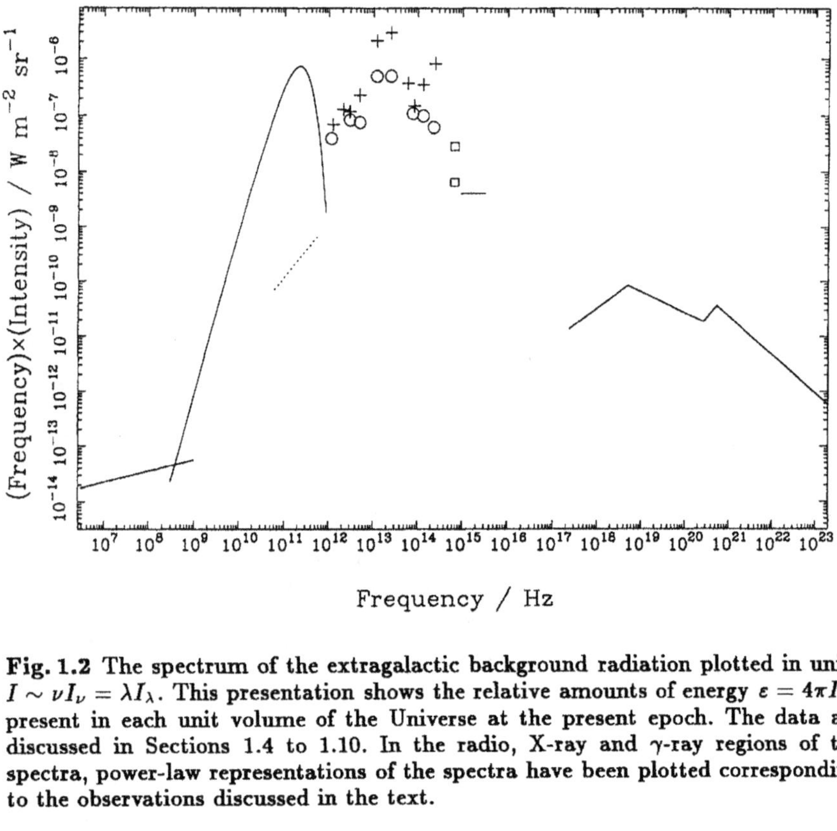

Fig. 1.2 The spectrum of the extragalactic background radiation plotted in units $I \sim \nu I_\nu = \lambda I_\lambda$. This presentation shows the relative amounts of energy $\varepsilon = 4\pi I/c$ present in each unit volume of the Universe at the present epoch. The data are discussed in Sections 1.4 to 1.10. In the radio, X-ray and γ-ray regions of the spectra, power-law representations of the spectra have been plotted corresponding to the observations discussed in the text.

The extragalactic background radiation is isotropic in all wavebands and so it is straightforward to convert the background intensity into an *energy density*. Per unit frequency interval, this is

$$\varepsilon_\nu = \frac{4\pi I_\nu}{c} \qquad \mathrm{J\ m^{-3}\ Hz^{-1}} \tag{1.1}$$

and this energy density is present in each cubic metre of the Universe at the present epoch. In terms of the total intensity, we find likewise

$$\varepsilon = \frac{4\pi I}{c} \quad \mathrm{J\ m^{-3}} \tag{1.2}$$

When Rashid and I produced Fig. 1.1 in 1969, background radiation had been detected in the radio waveband, the centimetre and millimetre waveband and in the X and γ-ray wavebands. Only upper limits were available in the ultraviolet, optical and infrared wavebands. We made theoretical estimates of the intensities due to galaxies in the latter wavebands but these were simply indicative of what might be expected. We will revise these estimates in what follows.

Overall, the situation is not so different today from our best efforts in 1969 but the limits are now very much better than before and of course there have been the spectacular measurements of the Cosmic Microwave Background Radiation made by the COBE satellite. Let us review the current state of the observations in order of increasing frequency.

1.4 The Radio Background Radiation

The story begins in 1933 with Jansky's discovery of the radio emission from the Galaxy. It was immediately apparent that, on large angular scales, the radio sky is dominated by diffuse Galactic emission. This great discovery caused little stir in the astronomical community and it was only after the Second World War that the nature and origin of the radio background emission became the subject of astronomical interest. By the late 1940s, the emission mechanism had been identified as synchroron radiation and, at about the same time, the first discrete radio sources were identified.

At that period, one of the principal motivations for attempting to extract the diffuse extragalactic component of the radio background radiation concerned the question of the distances and luminosities of the unidentified radio sources which continued to be discovered as the sky surveys extended to fainter and fainter flux densities. The argument goes as follows. Suppose the sources have typical luminosities L and space densities N_0. Then, from the results of Section 4.2 (expression 4.17), the diffuse background emission due to a uniform cosmological distribution of these sources is

$$I_\nu \approx \frac{c}{H_0} N_0 L \tag{1.3}$$

On the other hand, if we also measure the number of sources per steradian brighter than a given flux density S, $N(\geq S)$, that number is given by the expression (4.2)

$$N(\geq S) = \frac{1}{3} N_0 \left(\frac{L}{4\pi}\right)^{3/2} S^{-3/2} \tag{1.4}$$

Since the observed background intensity I_ν is an upper limit to the integrated intensity due to discrete sources and $N(\geq S)$ is observed, we can find a lower limit to L. This was the argument used by Martin Ryle to demonstrate reasonably convincingly that the bulk of the discrete radio sources had to be distant extragalactic objects. It was also the motivation for attempting to disentangle the intensity of the isotropic radio background from the anisotropic Galactic radio emission which is much more intense. This was a very difficult observational programme and several generations of Cambridge research students were almost broken in attempting to find a credible result.

The problem is that the radio sky is dominated by the synchrotron emission of our own Galaxy as is beautifully demonstrated by the all-sky radio map due to Glyn Haslam and his colleagues at the Max Planck Institute for Radio Astronomy at Bonn (see, for example, Longair 1989). As a result, wherever one looks in the sky, there is always intense radiation in the far-out sidelobes of the radio telescope. The best one can do is to map the sky at different wavelengths using *geometrically scaled antennae* so that, although the sidelobe problem is not eliminated, at least it is the same at different frequencies. What is observed on the sky is

$$I_\nu(\alpha,\delta) = I_{\rm gal}(\alpha,\delta) + I_0(\nu)$$

where the first term on the right-hand side represents the anisotropic component associated with the Galaxy and the term $I_0(\nu)$ represents the isotropic extragalactic component. The procedure is to map the sky at different frequencies, assume that the anisotropic component has the same radio spectrum in all directions and then find $I_0(\nu)$. The procedure works because the Galactic continuum spectrum is different from that of the diffuse extragalactic component. Specifically, the spectrum of our Galaxy has the form $I_\nu \propto \nu^{-0.4}$ at frequencies less than about 200 MHz whereas extragalactic radio sources have much steeper spectra.

The best results are still those presented by Bridle in 1967. It is convenient to express them in terms of the brightness temperature of the radiation $T_{\rm b} = (\lambda^2/2k)I_\nu$. At the traditional wavelength of 178 MHz, the frequency of the revised 3C Catalogue, the results are a follows. The minimum sky temperature at 178 MHz is about 80 K and includes both the minimum Galactic component as well as the isotropic component. As the errors build up, it is not possible to determine both the intensity and spectrum of the extragalactic component and so the isotropic component is extracted assuming different values for the radio spectral index, defined by $I_\nu \propto \nu^{-\alpha}$. If $\alpha = 0.75$, the isotropic background temperature is 30 ± 7 K; if $\alpha = 0.9$, the intensity corresponds to 15 ± 3 K. The typical spectral index of radio sources at 178 MHz is about 0.8 and so an intensity of ~ 23 K is a reasonable estimate of the background.

These figures can be compared with the brightness temperature found when the counts of radio sources are integrated to the lowest flux densities.

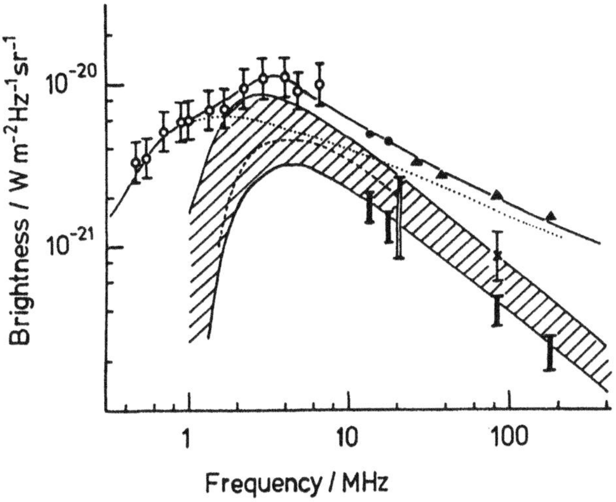

Fig. 1.3 The spectrum of the radio sky in the direction of the 'north halo minimum'. The solid line shows the best fit to the total background. The dotted line shows the Galactic contribution and the dashed line the estimated extragalactic contribution, the shaded region indicating the uncertainties in the latter estimate. Independent estimates of the extragalactic background at higher frequencies are also shown. (From Simon 1978)

The integrated background emission to sub-millijansky levels corresponds to about 20 K. We will discuss in much more detail the origin of this background intensity in Section 4.3. Thus, it seems that virually the whole of the radio background intensity can be attributed to discrete sources and there is not much room left for any other contribution to the background radio emission at low frequencies. One contribution of possible cosmological interest is the upper limit to the intensity of intergalactic bremsstrahlung which would have a flat radio spectrum, $I_\nu \propto \nu^0$. As a result, the best limit comes from observations of the minimum intensity of the radio background which occurs at about 400 MHz — at higher frequencies, the Cosmic Microwave Background Radiation becomes the dominant component. Once the discrete source component of the background and the Cosmic Microwave Background Radiation are removed, the upper limit to any residual diffuse component amounts to $T_{400\ \mathrm{MHz}} \leq 0.1$ K.

There are two footnotes to this story. The first is the touching story reported by Jasper Wall at the 1989 Heidelburg meeting on the Background Radiation (Wall 1990). In 1964, Jasper and Donald Chu were attempting to measure the background radiation at frequencies of 320 and 707 MHz. They found to their distress that they could not obtain the 'right' answer — their background spectrum was too flat (Wall, Chu and Yen 1970). As research students, the tacit assumption was made that they had simply made some error in the calibribation of their experiment. In the following year, 1965, the

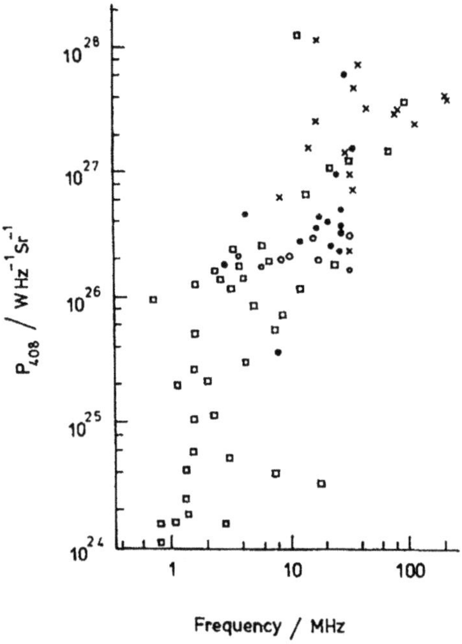

Fig. 1.4 The relation between radio luminosity at 408 MHz and the frequency at which the radio source is expected to exhibit synchrotron self-absorption. The radio sources form a representative sample of the radio sources in the 3CR catalogue. (From Simon 1978).

discovery of the Cosmic Microwave Background Radiation was announced and this was the origin of their excess antenna temperature.

The second footnote concerns the extragalactic background emission at very long wavelengths, $\lambda \sim 1 - 10$ MHz. This is an even more unfashionable waveband than the traditional radio waveband because the observations are very difficult to make as a result of ionospheric absorption and refraction. However, at certain locations in the auroral zone, it is possible to observe the sky at 10 MHz. In the 1960s Chris Purton and Alan Bridle did as good a job as could be done at that time at these very low frequencies from the Penticton Radio Observatory (Bridle and Purton 1968). The sky is still dominated by the synchrotron emission of the Galaxy at these very low frequencies but, because of the differences in spectral indices, the extragalactic component is relatively more important. The process which becomes important at these low frequencies is bremsstrahlung absorption so that, at 10 MHz, the Galactic plane is observed in absorption (Purton 1966). As observations are made at frequencies less than 10 MHz, the distance at which the bremsstrahlung optical depth becomes unity decreases and so the plane of the Galaxy becomes progressively more and more opaque. The spectrum of the background radiation in the region of the North Galactic Pole was, however, determined by the Canadian RAE1 satellite and the shape of the extragalactic component of the background could be estimated (Clark et al 1970). Evidence was found that the extragalactic spectrum displays a cut-off at low frequencies, $\nu \leq 3$ MHz (Fig. 1.3). The origin of this behaviour was discussed by Simon (1978). An obvious interpretation of the cut-off is that it is associated with synchrotron self-absorption in the discrete sources which make up the background. She studied the predicted spectra to very low frequencies of a complete sample of 3CR radio sources for which detailed radio structural information was available. Compact radio components and hot spots become synchrotron self-absorbed at frequencies $\nu \geq 100$ MHz and the only components which contribute to the $1-10$ MHz background radiation are the most diffuse components. Because of the strong inverse correlation between diffuse structure and radio luminosity, the greatest contributions to the background in the $1-10$ MHz waveband are made by relatively low luminosity sources (Fig. 1.4). Simon evaluated the predicted background spectrum when account was taken of the cosmological evolution of these sources (see Section 4.3) and found that she could account quite naturally for the inferred turn-over in the isotropic radio background spectrum.

1.5 The Cosmic Microwave Background Radiation

The Cosmic Microwave Background Radiation was discovered in 1965 by Penzias and Wilson whilst commissioning a very sensitive receiver system for centimetre wavelengths at the Bell Telephone Laboratories (Penzias and Wilson 1965). It was quickly established that this background radiation is remarkably uniform over the sky and that, in the wavelength range 1 m $> \lambda >$ 1 cm, the intensity spectrum had the form $I_\nu \propto \nu^2$, corresponding to the Rayleigh-Jeans region of a black-body spectrum at a radiation temperature of about 2.7 K. The maximum intensity of such a spectrum occurs at a wavelength of about 1 mm at which atmospheric emission makes a precise absolute measurement of the intensity impossible from the surface of the Earth. Several high-altitude balloon carrying millimetre and sub-millimetre spectrometers were flown during the 1970s and 1980s and evidence was found for the expected turn-over in the Wien region of the spectrum but there were discrepancies between the experiments and the absolute accuracy was not great (see Weiss (1980) for a discussion of the early spectral measurements). It was realised in the 1970s that the only way of studying the detailed spectrum and isotropy of the radiation over the whole sky was from above the Earth's atmosphere. After a long period of gestation, the Cosmic Background Explorer (COBE) was launched in November 1989 and is dedicated to studies of the background radiation, not only in the millimetre and submillimetre wavebands but throughout the infrared waveband as well.

The Far Infrared Absolute Spectrophotometer (FIRAS) has measured the spectrum of the background in the wavelength range 0.1 to 10 mm with very high precision during the first year of the mission. The FIRAS instrument involves cooling the detectors and the reference black-body source to liquid

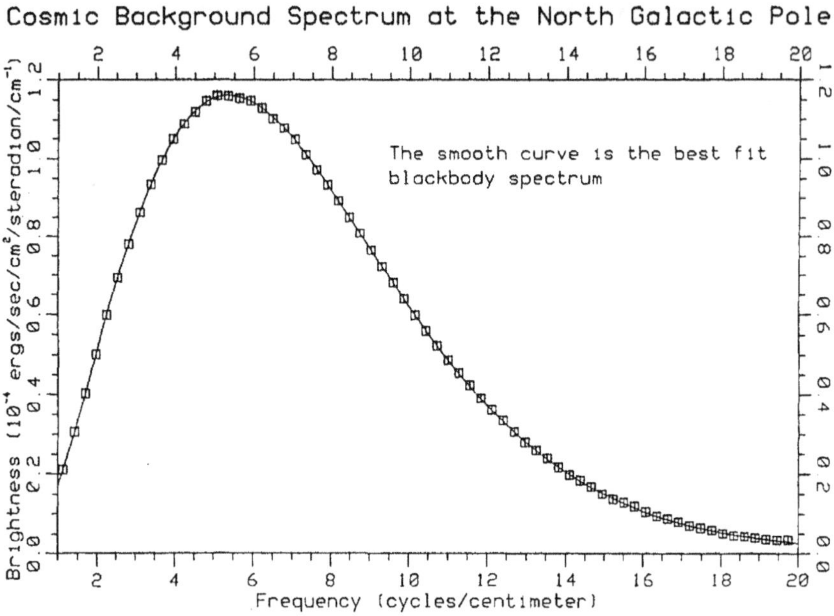

Fig. 1.5 The spectrum of the Cosmic Microwave Background Radiation as measured by the COBE satellite in the direction of the North Galactic Pole. Within the quoted errors, the spectrum is that of a perfect black body. (Mather et al 1990)

helium temperatures and there was only sufficient liquid helium for the first year of observation. It is a very beautiful experiment indeed and we have been teaching it in the Physics Department in Cambridge, not only as a key measurement for cosmology but also as a superb example of high precision experimental technique. The very first observations made by FIRAS revealed that the spectrum is very precisely of black-body form (Fig. 1.5). In the early spectrum shown in Fig. 1.5, it can be seen that the error boxes, which are shown as 1% of the peak intensity, are overestimates of the uncertainty because the black-body curve passes precisely through the centre of each error box. Much more data has become available since Fig. 1.5 was published in 1990 and, in particular, the thermometry involved in making the absolute measurments has been very carefully studied. More recent analyses reported by Hauser (1993) have shown that the spectrum is a perfect black body with a radiation temperature $T = 2.726 \pm 0.01$ K. Furthermore, the deviations from a perfect black-body spectrum in the wavelength interval $2.5 > \lambda > 0.5$ mm amount to less than 0.03% of the maximum intensity. This is the most beautiful example of which I am aware of a naturally occurring black-body radiation spectrum. We will use this result in Chapter 11 to constrain the amount of star formation and metal production in young galaxies at large redshifts.

The deviations from a black-body spectrum must be very small indeed. There are two useful ways of describing how large the deviations can be, both of which were pioneered by Zeldovich and Sunyaev (see Sunyaev and Zeldovich 1980). They showed that the injection of large amounts of thermal energy in the form of hot gas into the intergalactic medium can produce different types of distortion of the primordial black-body radiation spectrum. We will not go into the physics of these processes at this point, except to note the forms of distortion and the limits which can be set to the characteristic parameters. If there is early injection of energy prior to the epoch of recombination without any increase in the number of photons, the spectrum relaxes to an equilibrium distribution under Compton scattering which conserves photons but redistributes their energies. The resulting spectrum is of Bose-Einstein form with a dimensionless chemical potential μ.

$$I_\nu = \frac{2h\nu^3}{c^2}\left[\exp\left(\frac{h\nu}{kT_{\rm r}} + \mu\right) - 1\right]^{-1} \tag{1.5}$$

In the case of Compton scattering by hot electrons at late epochs, the mean energies of the photons are increased. Zeldovich and Sunyaev (1969) first showed that the distortion of the black-body spectrum takes the form

$$\frac{\Delta I_\nu}{I_\nu} = y\frac{xe^x}{(e^x - 1)}\left[x\left(\frac{e^x + 1}{e^x - 1}\right) - 4\right] \tag{1.6}$$

where y is the Compton scattering optical depth $y = \int (kT_{\rm e}/m_{\rm e}c^2)\sigma_{\rm T} N_{\rm e} dl$, $x = h\nu/kT_{\rm r}$ and $\sigma_{\rm T}$ is the Thomson scattering cross-section. In the limit of small distortions, $y \ll 1$, the intensity in the Rayleigh-Jeans region decreases as $\Delta I_\nu/I_\nu = -2y$. The total energy under the spectrum increases as $\varepsilon = \varepsilon_0 e^{4y}$. The most recent limits quoted by Mather (1993) are as follows:

$$y \leq 2.5 \times 10^{-5} \qquad \mu \leq 3.3 \times 10^{-4}$$

These are very powerful limits indeed and will prove to be of great astrophysical importance in studying the physics of the intergalactic gas (see Section 5.2 and Chapter 10).

Equally remarkable have been the observations of the isotropy of the Cosmic Microwave Background Radiation over the sky. The prime instruments for these studies are the Differential Microwave Radiometers which operate at 31.5, 53 and 90 GHz, thus sampling the Rayleigh-Jeans region of the spectrum. The angular resolution of the radiometers is 7°. In increasing levels of sensitivity, the results are as follows. At a sensitivity level about one part in 1000 of the total intensity, there is a large scale anisotropy over the whole sky associated with the motion of the Earth through the frame of reference in which the radiation would be the same in all directions. This is no more than the result of the Doppler effect due to the Earth's motion. As a result, the radiation is about one part in a thousand more intense in one direction and exactly the same amount less intense in the opposite direction. The intensity distribution has precisely the expected dipolar distribution around the sky and it is inferred that the Earth is moving at about 350 km s^{-1} with respect to the frame of reference in which the radiation would be 100% isotropic. At about the same level of intensity, the plane of our Galaxy can be observed as a faint band of emission over the sky. It is intriguing that, although not designed to undertake this task, exactly the same form of large scale anisotropy is observed by the FIRAS instrument.

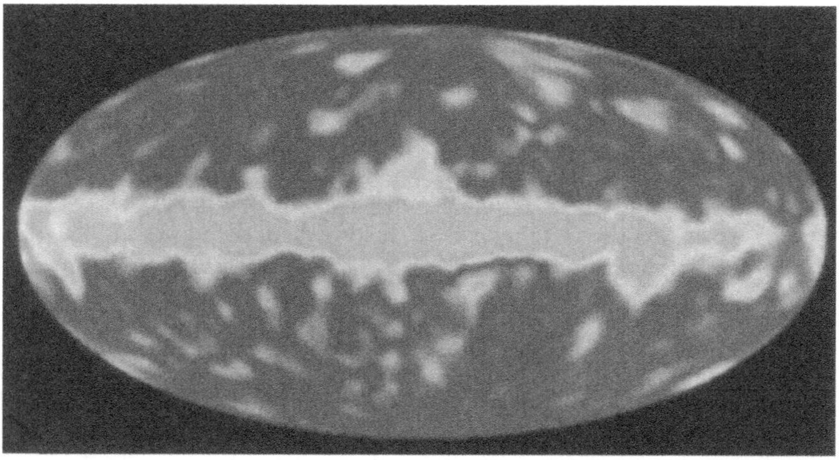

Fig. 1.6 The map of the whole sky in galactic coordinates as observed in the millimetre waveband at a wavelength of 5.7 mm (53 GHz) by the COBE satellite once the dipole component associated with the motion of the Earth through the background radiation has been removed. The residual radiation from the plane of the Galaxy can be seen as a bright band across the centre of the picture. The fluctuations seen at high galactic latitudes are noise from the telescope and the instruments, the rms value at each point being 36 μK but, when statistically averaged over the whole sky at high latitudes, an excess sky noise signal of 30 ± 5 μK is observed (Smoot et al 1992).

On angular scales 10° and greater, Smoot et al (1992) have achieved sensitivity levels of one part in 100,000 of the total intensity (Fig. 1.6). At this level, the radiation from the plane of the Galaxy is intense but is confined to a broad strip lying along the galactic equator. Away from this region, the sky appears quite smooth on a large scale but careful analysis of the variation of intensity from beam-area to beam-area on the sky has shown convincing evidence for tiny fluctuations in intensity over and above the instrumental noise. The signal amounts to only about 1 part in 100,000 of the total intensity and, when averaged over the clear region of sky at $|b| > 20°$ amounts to 30 ± 5 μK, a positive signal at the 6σ level. This is a very important result for cosmology as we will discuss in Chapter 9. These results were obtained from the first year of data and the senstivity of the experiment increases as more observations are taken. Unlike the FIRAS experiment, there are no cryogenics involved in the DRM experiment and so it can continue observing throughout the life of the mission which is expected to be three years.

The COBE experiment has dominated studies of the Cosmic Microwave Background Radiation but there are other key experiments underway from the ground concerning the fluctuations in background radiation. These are very important and will be discussed in much more detail in Chapter 9.

1.6 The Infrared Background Radiation

It has been known since the very successful flight of the IRAS satellite that the sky in the far infrared region of the spectrum $12 - 100$ μm is dominated by dust emission. The images of the sky in these wavebands bear more than a passing resemblance to the radio image of the sky with an intense ridge of emission along the Galactic plane, in the far infrared case associated with the dust emission of regions of star formation, giant molecular clouds and diffuse interstellar dust (see, for example, Fig. 6.2 of Longair (1989)). At the long wavelength end of this spectral region, $\lambda \sim 50 - 100$ μm, the background intensity in directions away from the Galactic plane is dominated by what is known as *cirrus*. This consists of diffuse, cool dust clouds at high latitudes which are similar in appearance to feathery cirrus clouds. It has been shown that they emit significant far infrared radiation even at high galactic latitudes.

At shorter wavelengths, the high latitude background is dominated by the emission of dust grains lying within the ecliptic plane of our own Solar System. These dust particles give rise to what is known as the *zodiacal light* which is sunlight scattered by the interplanetary dust grains. At wavelengths $\lambda \sim 12 - 25$ μm, the thermal emission of the zodiacal dust particles is a very striking feature of the IRAS maps of the sky and produces a significant background at high galactic latitudes. The result is that the limits to any isotropic background radiation are relatively poorly known compared with, say, the optical or millimetre wavebands.

The COBE satellite includes an instrument known as the Diffuse Infrared Background Experiment (DIRBE). Whereas the IRAS survey was designed to survey the far infrared sky for discrete sources, the objective of the DIRBE experiment is to make absolute brightness maps of the whole sky in ten photometric channels, 1.2, 2.2, 3.5, 4.9, 12, 25, 60, 100, 140 and 240 μm, all with an angular field of view of $0.7° \times 0.7°$. The preliminary results described by Hauser (1993) confirm the general picture outlined above. The maps of the sky in the near infrared wavebands, $1.2 - 3.5$ μm provide dust-free images of the large scale structure of our own Galaxy, showing clearly the disc and the central bulge.

Eventually, complete sky maps will be available at all these wavelengths and will contain a wealth of data for many branches of astronomy. From the perspective of the extragalactic background radiation, the hope is that, eventually, it will be possible to model the distributions of zodiacal light and high latitude cirrus sufficiently accurately that these components can be subtracted from the sky maps leaving the residual isotropic components, or at least upper limits to them. This is a very major undertaking but the quality of the data is so high that one must be optimistic about obtaining significantly improved limits as compared with the existing values.

In the meantime, Hauser (1993) has presented conservative DIRBE upper limits to the extragalactic component of the infrared background radiation from observations taken in the direction of the south ecliptic pole. According to Hauser, this is in general the most favorable direction for searching for extragalactic background components because of the foreground radiation due to interplanetary dust. The data are presented in energy units λI_λ in Table 1.1.

Table 1.1 The infrared sky brightness in the direction of the south ecliptic pole (Hauser 1993)

Wavelength λ (μm)	λI_λ ($\times 10^{-7}$ W m^{-2} sr^{-1})
1.2	8.3 ± 3.3
2.3	3.5 ± 1.4
3.4	1.5 ± 0.6
4.9	3.7 ± 1.5
12.0	29.0 ± 12
22.0	21.0 ± 8
55.0	2.3 ± 1
96.0	1.2 ± 0.5
151.0	1.3 ± 0.7
241.0	0.7 ± 0.4

There exist limits to the extragalactic infrared background intensities which I summarised in my review of all the extragalactic background data in 1989 (Longair 1990). These were obtained from rocket flights and from the IRAS survey and references to these data are given in the review. I have summarised the data in Fig. 1.7, the COBE data being indicated by crosses and the other limits by circles. It can be seen that some of them correspond to lower limits to the background intensity than those listed in Table 1.1.

One final point of interest concerns the wavebands which are most promising for the detection of a genuine extragalactic component of the background radiation. There are two spectral regions in which the interfering effect of

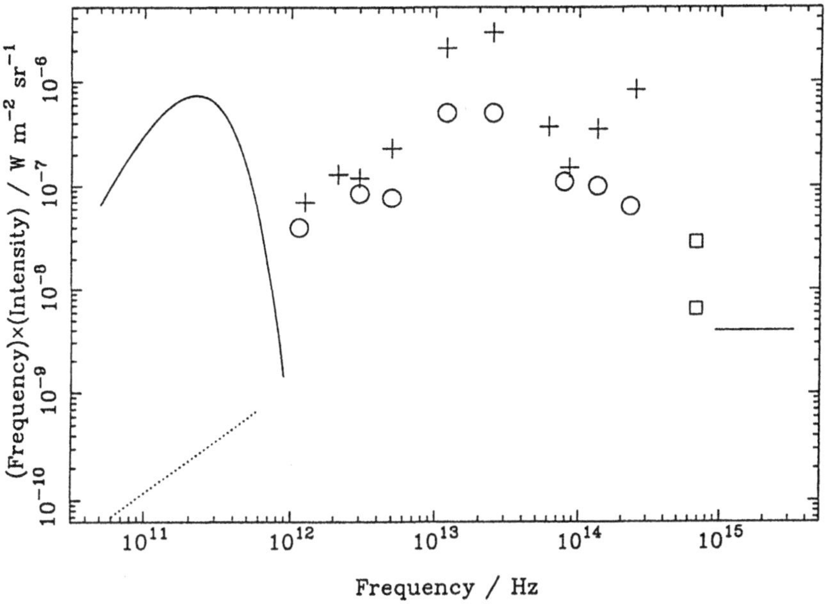

Fig. 1.7 Upper limits to the infrared, optical and ultraviolet background intensities from the data discussed in the text presented in energy units νI_ν. The COBE limits are shown as crosses and other infrared data by open circles. In the optical waveband, the measurements by Toller (1990) (lower box) and Mattila and Schnur (1990) (upper box) are shown. In the ultraviolet waveband an intensity corresponding to 200 photon units is shown. At millimetre wavelengths the spectrum of the Cosmic Microwave Background Radiation is shown, the dotted line showing a limit of 0.03% of the peak intensity in the wavelength range $0.5 < \lambda < 2.5$ mm.

interplanetary and interstellar dust are at a miminum. One is in the 'cosmological window' at about 3.4 μm. At this wavelength there is the minimum of scattered sunlight and the re-emission of the absorbed radiation by the heated dust. The other is at wavelengths longer than about 100 μm at which the spectrum of interstellar dust decreases very rapidly with increasing wavelength. Between 100 μm and the steep rise of the Wien region of the Cosmic Microwave Background Radiation, there is a minimum at about 200 – 300 μm in which sensitive searches can be made for diffuse extragalactic infrared radiation. The former window will be accessible by the Infrared Space Observatory mission of the European Space Agency and the latter by the Far Infrared Space Telescope (FIRST), also of ESA.

1.7 The Optical Background Radiation

The measurement of the extragalactic optical background emission is one of the most difficult observations in cosmology because the background is very faint and is swamped by the emission from stars in the Galaxy, scattered light and so on. Very deep exposures in the direction of the Galactic poles find very large surface densities of galaxies so that there is not much unobscured background radiation left. Fig. 1.8 shows one of Tyson's deepest images and it can be seen that roughly 25% of the sky is covered by the images of faint galaxies. However, the process of flat-fielding, which is essential in analysing the CCD images at such very faint magnitudes, means that any diffuse uniform component is eliminated from these images. In addition, the whole point of background observations is that they include the sum of all galaxies which are even fainter than those seen in images such as that of Fig. 1.8. The most recent analyses of observations of the background radiation are included in the 1989 Heidelberg symposium (Bowyer and Leinert 1990), in particular, in the reviews by Toller, Mattila and Schnur, and Tyson.

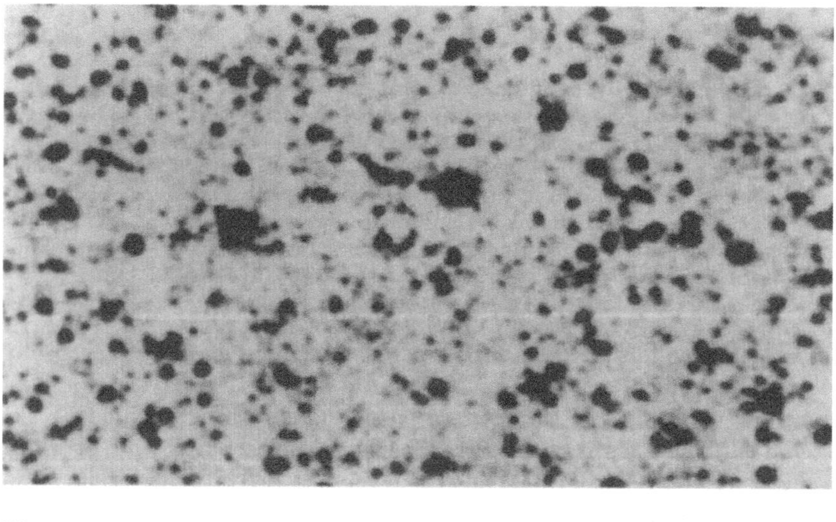

Fig. 1.8 A deep image of a small region of sky in the direction of the North Galactic Pole (Tyson 1990).

The direct approach is described by Toller. The problem is that, even at high galactic latitudes, there are many components which have to be eliminated before a reliable estimate of the isotropic extragalactic background can be found. In Table 1.2, I reproduce the intensities of the various interfering components in the direction of the galactic poles in units of 10^{-12} W m^{-2} nm^{-1} sr^{-1}. The origin of the various components are as follows:

Table 1.2 Typical intensities of selected astronomical sources as viewed from the Earth (after Toller 1990).

Component	Typical intensity ($\times 10^{-12}$ W m^{-2} nm^{-1} sr^{-1})
Full Moon	10^{12}
Airglow continuum	600
Zodiacal light (on ecliptic)	2500
Zodaical light (off ecliptic)	1000
Bright stars ($m_V < 6$)	250
Integrated starlight (Galactic plane)	2500
Integrated starlight (off Galactic plane)	600
Diffuse Galactic light	250
'Cosmic light'	10

1. The observations are made with photometers which subtend quite large angles on the sky so that a significant flux of the faint extragalactic light can be detected. Even at high galactic latitudes, there are some galactic stars in the beam and these have to be eliminated.
2. The IRAS observations of Galactic 'cirrus' show that there is a significant amount of dust at high galactic latitudes and this results in a diffuse component of scattered starlight.
3. There is dust within the plane of the Solar System and this scatters sunlight into the detector. This band of scattered light is seen clearly in the IRAS observations as well as in the optical waveband. It is known as the *zodiacal light.*
4. Even the dark night sky is not completely dark but contributes background light known as *airglow.*
5. Finally, there is light which is scattered by the dark atmosphere.

If observations are made from the surface of the Earth, the extragalactic component of the background has to be estimated in the presence of these interfering backgrounds. The best limits have been obtained from observations made from the Pioneer 10 spacecraft at a distance of 3 AU from the Sun as the vehicle was on its way to Jupiter. Carrying out these observations from space has the great advantage that components 3, 4 and 5 above can be eliminated at a stroke and it is only necessary to worry about the contributions from our own Galaxy. Toller (1990) quotes an estimate of the extragalactic background at a wavelength of 440 nm of:

$$I_\lambda = (15 \pm 15) \times 10^{-12} \quad \mathrm{W\ m^{-2}\ nm^{-1}\ sr^{-1}}$$

The corresponding value of νI_ν is

$$\nu I_\nu = \lambda I_\lambda = (6.6 \pm 6.6) \times 10^{-9} \quad \mathrm{W\ m^{-2}\ sr^{-1}}$$

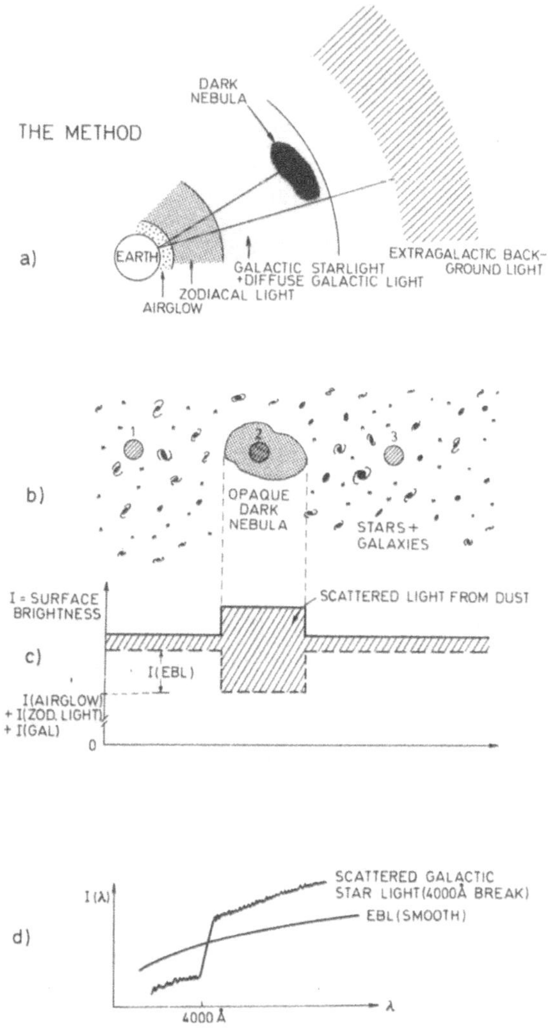

Fig. 1.9 **Illustrating the dark cloud method of estimating the extragalactic background optical emission (from Mattila and Schnur 1990).**

Incidentally, for those who enjoy non-SI units, this background intensity is often quoted as (1.6 ± 1.6) 10th magnitude solar type stars per square degree. This is not a particularly helpful unit but it does indicate just how faint the extragalactic background is. It is equivalent to about 28 magnitudes arcsec^{-2}.

One concern about the direct methods of determining the optical background radiation is the problem of determining the absolute intensity of the background. It is therefore of interest that there is available a differential

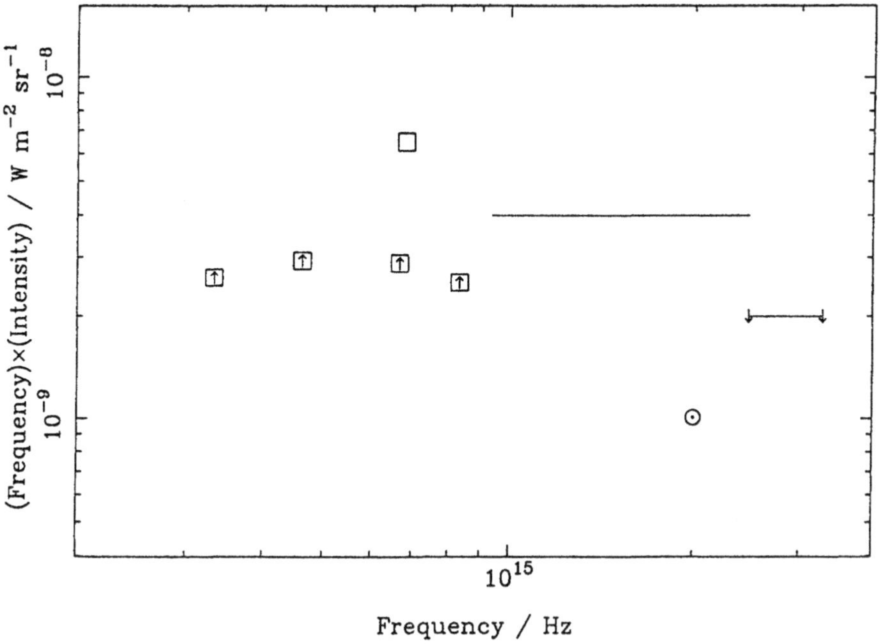

Fig. 1.10 Comparison of the integrated light of faint galaxies with upper limits to the extragalactic background light in the optical and ultraviolet regions of the spectrum. The open box is Toller's estimate of the optical background radiation. The boxes with upward pointing arrows represent the integrated background intensity due to galaxies in Tyson's deepest galaxy counts. The horizontal line corresponds to an intensity of 200 photon units in the ultraviolet waveband. The line with downward pointing arrows is the upper limit of 100 photon units between Lyman-α and the Lyman limit. The circled point corresponds to 50 photon units, the background due to galaxies found by Martin and Bowyer.

method for determining the extragalactic component of the background. The *dark cloud method* has been advocated by Mattila and his colleagues for a number of years. The method is illustrated in Fig. 1.9. If a very opaque dust cloud can be found, the component of the background originating beyond the dark cloud can be found by comparing the background intensities in the unobscured direction with the intensity observed in the direction of the cloud. In this way, all 5 interfering components of the background can be subtracted out leaving only the diffuse background originating from beyond the cloud. The only problem with this approach is that the dark cloud itself is a source of reflected stellar and Galactic light and this component has to be estimated as well. This contribution can be estimated by noting that the reflected light should display the Balmer discontinuity at 400 nm whereas the diffuse extragalactic light should not contain such a feature since the bulk of the background originates at redshifts of the order of 0.5.

Mattila and Schnur (1990) quote a measurement of the background intensity at 440 nm of

$$\nu I_\nu = \lambda I_\lambda = (2.9 \pm 1.1) \times 10^{-8} \quad \mathrm{W\ m^{-2}\ sr^{-1}}$$

This figure is within 2 standard deviations of the result found by Toller. At the very least this is probably a safe upper limit to the total extragalactic background light. The estimates of Toller (1990) and Mattila and Schnur (1990) are included in Fig. 1.7.

It is interesting to compare these limits with the integrated background emission due to galaxies in the very deep counts carried out by Tyson (1990). This comparison is shown in Fig. 1.10 in which it can be seen that the total light from galaxies is within about a factor of 2 of the value quoted by Toller (1990). We will return to these issues when we consider the background emission due to young galaxies and the formation of the heavy elements.

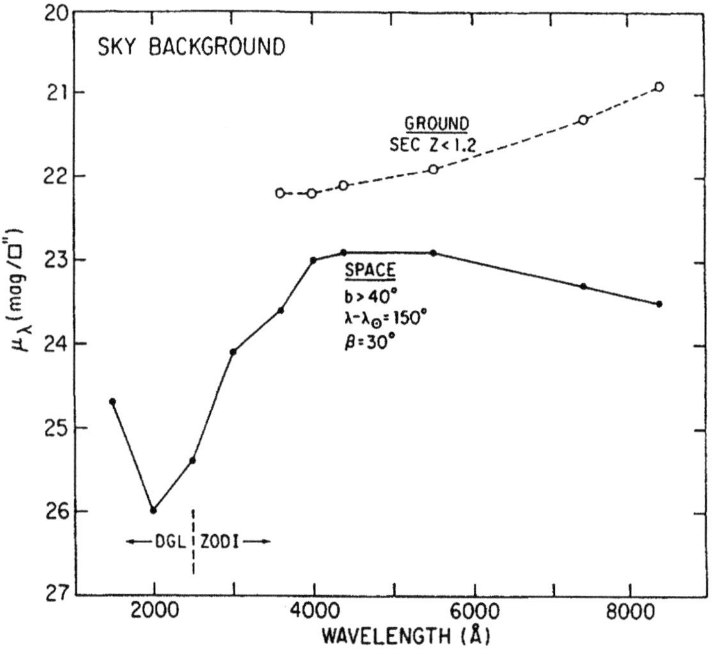

Fig. 1.11 Estimates of the energy distribution of the night-sky background from a good ground based site looking towards the zenith and from space. In the ultraviolet region of the spectrum, the main contributions are from zodiacal light at wavelengths longer than about 250 nm and from diffuse Galactic light at shorter wavelengths (from Henry 1991).

1.8 The Ultraviolet Background Radiation

The Earth's atmosphere is opaque at all wavelengths shorter than about 300 nm and so the ultraviolet Universe has to be studied from above the Earth's atmosphere. There are enormous benefits which result from having to make observations from space, in particular, the atmospheric effects which render these studies difficult in the optical waveband are eliminated. In addition, because the spectral energy distribution of the Sun decreases rapidly into the ultraviolet region of the spectrum, the zodiacal light becomes very weak in the ultraviolet waveband. Fig. 1.11 shows a comparison between the brightness of the sky as observed from the ground and in space (Henry 1991). It can be seen that there are clear advantages in observing in the ultraviolet waveband.

Observations of the background radiation in the wavelength region $121.6 < \lambda < 290$ nm have been surveyed by Bowyer (1991), Henry (1991) and Jakobsen (1993). It has been well known for some time that there is a strong correlation between the intensity of the ultraviolet background radiation and the column depth of neutral hydrogen $\int N_{\rm H} dl$ in the same direction. The standard interpretation of this result is that the ultraviolet radiation is the ultraviolet radiation of O and B stars scattered by diffuse interstellar dust. The observed correlation follows from the fact that the column densities of dust and neutral hydrogen are strongly correlated. When this correlation is extrapolated to zero column depth of neutral hydrogen, a finite ultraviolet background flux is still observed. Ultraviolet astronomers present their estimates of the background intensities in 'photon units' which are defined to be the photon intensity per unit wavelength interval in cgs units, that is, in photons s^{-1} cm^{-2} sr^{-1} $Å^{-1}$. In these units, the residual background intensity amounts to about $200 - 300$ photon units. Martin, Hurwitz and Bowyer (1990) have made observations of the ultraviolet background spectrum in a direction of very low neutral hydrogen column density, $\int N_{\rm H} dl = 1.0 \times 10^{20}$ cm^{-2}, using the Berkeley 'nebular' spectrometer on board the Space Shuttle. The background spectrum is shown in Fig. 1.12 and shows the presence of the intense line of CIV at 155 nm. It can be seen that the intensity decreases from about 400 photon units at 140 nm to about 250 photon units at about 190 nm. This spectrum represents an upper limit to the extragalactic ultraviolet background intensity. The question of interest is whether this spectrum truly represents the extragalactic background intensity or whether there are additional Galactic components still present along the line of sight in this direction. Examples of such processes include coronal line emission or two-photon emission from the ionised component of the interstellar medium. Martin, Hurwitz and Bowyer (1990) also discuss the possibility that a significant fraction of the background intensity may be associated with an 'anomalous' scattered component by dust which is not correlated with the neutral hydrogen column density.

The situation is far from clear at the moment but Jakobsen (1993) suggests that the extragalactic background intensity is probably within a factor

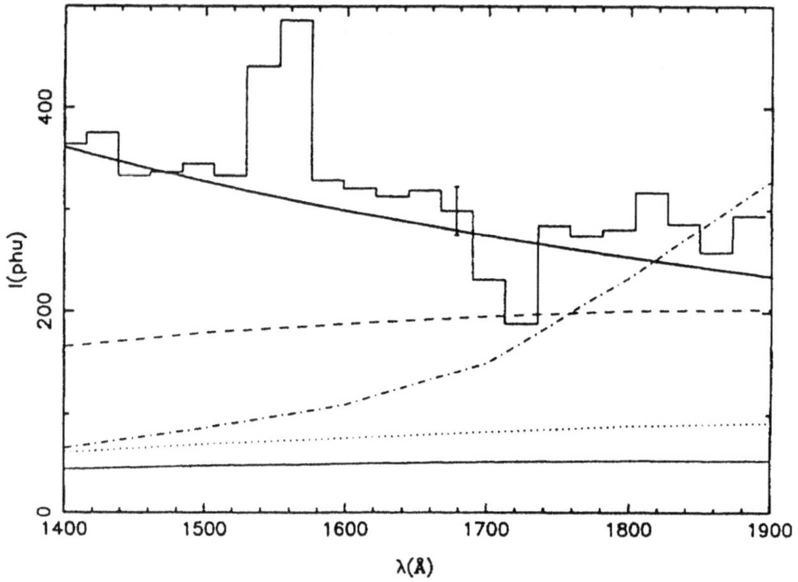

Fig. 1.12 The spectrum of the ultraviolet background radiation in a direction of very low neutral hydrogen column density ($\int N_{\rm H} dl = 10^{20}$ cm^{-2}) observed by Martin, Hurwitz and Bowyer from a Space Shuttle experiment. The upper solid line is a best power-law fit to the observed spectrum, $I_\lambda \propto \lambda^{-1.4}$. The prominent emission line at 155 nm is CIV. The lower solid line is the expected background due to a uniform population of non-evolving galaxies in the ultraviolet region of the spectrum and the other lines are various evolution models (see Martin, Hurwitz and Bowyer (1990) for details).

of 2 or 3 of 100 photon units in the wavelength region $130 < \lambda < 250$ nm. On Fig. 1.7, we have translated an intensity of $S_{\rm pu} = 200$ photon units into energy units νI_ν in SI units through the simple transformation $\nu I_\nu = 10^{14} ch\ S_{\rm pu} = 2 \times 10^{-11}\ S_{\rm pu}$ where νI_ν is expressed in W m^{-2} sr^{-1}, c is the speed of light and h is Planck's constant.

Two estimates are available of the background intensity associated with galaxies. One of these is the heroic study of Martin and Bowyer (1989) who used a version of the correlation function method described in Section 4.5 to estimate the number counts of discrete ultraviolet sources. The approach is similar to that of Shectman (1974) in the optical waveband and they established that a correlated signal is present among the photon counts recorded in their experiment which corresponds to the two-point correlation function for galaxies. Martin and Bowyer estimated the extragalactic intensity due to these galaxies to be about 50 photon units.

A second approach was taken by Armand, Milliard and Deharveng (1993) who made faint galaxy counts in the ultraviolet waveband at 200 nm from a

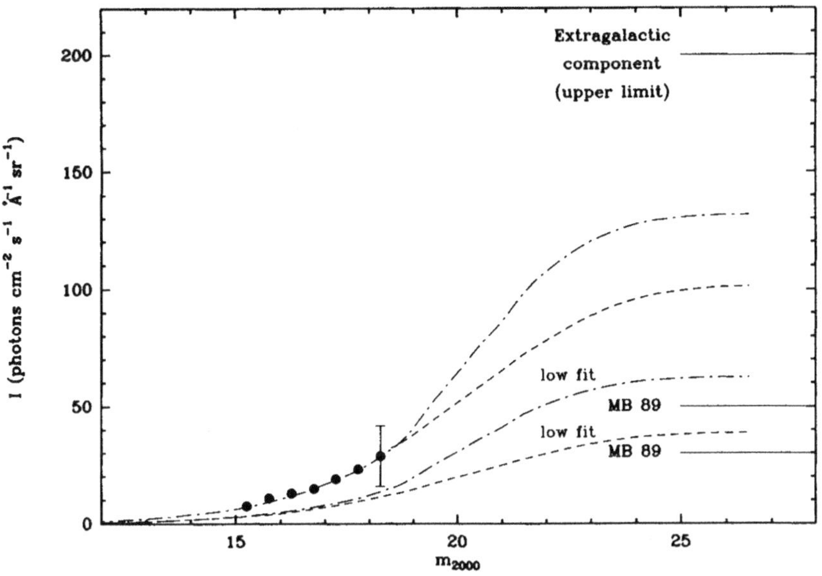

Fig. 1.13 The counts of galaxies at a wavelength of 200 nm observed by Armand, Milliard and Deharveng (1993). Galaxies brighter than $m_{uv} = 18.5$ account for about 30 photon units. Various extrapolations to fainter magnitudes and the estimates of Martin and Bowyer (1989) are also shown.

high flying balloon. The observed counts of galaxies to magnitude $m_{uv} = 18.5$ amounted to 30 photon units (Fig. 1.13). Thus, it seems inevitable that the ultraviolet background in this wavelength region must amount to about 100 photon units but, as noted by Jakobsen (1993), the exact value is uncertain. Martin, Hurwitz and Bowyer (1990) evaluated the expected background for a variety of models of the evolution of galaxies in the ultraviolet region of the spectrum and showed that they could account for much of the background although it seems difficult to account for the background intensity at the shortest wavelengths.

To the short wavelength side of Lyman-α, $91.2 < \lambda < 121.6$ nm, the background radiation has been studied from the Voyager 2 spacecraft. The observations reported by Holberg (1990) indicate that no background was detected at high galactic latitudes and an upper limit of 100 photon units has been derived. At even shorter wavelengths beyond the Lyman limit at 91.2 nm, the interstellar gas becomes opaque and direct measurements of the background radiation are no longer possible until the optical depth to photoelectric absorption becomes less than unity in the soft X-ray waveband.

1.9 The X-ray Background Radiation

The X-ray background emission was discovered in a pioneering X-ray rocket experiment by Giacconi and his colleagues in 1962. The objective of the experiment was to observe fluorescent X-rays from the Moon but instead the first discrete sources and the intense X-ray background emission were discovered. Remarkably, one of the first X-ray images taken by the ROSAT X-ray observatory shows the fluorescent emission from the Moon and also the occultation of the X-ray background by the Moon (Fig. 1.14).

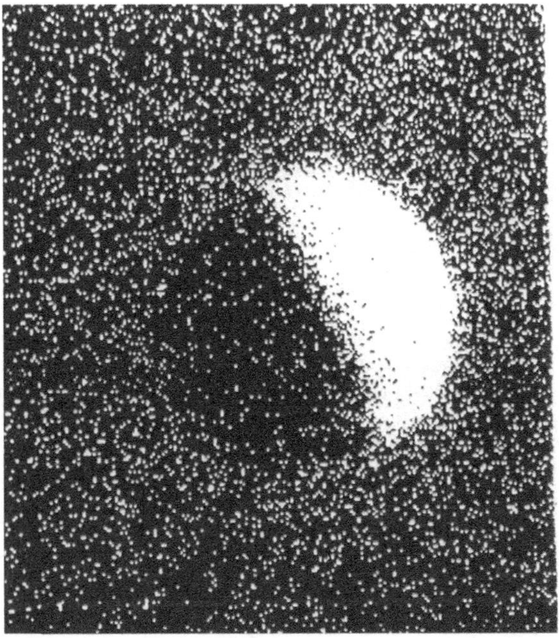

Fig. 1.14 An image of the Moon taken by the ROSAT Observatory showing the fluorescent X-ray emission from sunlight side of the Moon. It can also be seen that the Moon is occulting the diffuse X-ray background. (Courtesy of Prof. J. Trümper of the Max Planck Institute for Extraterrestrial Physics, Garching.)

There was therefore relatively little problem in determining the properties of the X-ray background. When observations are made with X-ray telescopes with angular resolution of about 5° or greater, the X-ray background emission is the prime source of background noise in the detector in the 1–10 keV energy band. Most of the observations of the diffuse X-ray background still date from before 1980 when the X-ray missions were still exploring the nature and content of the X-ray sky. Since that time, observational X-ray astronomy has moved on to higher sensitivity and greater angular resolution. Many results concerning the X-ray background radiation have been reviewed by Fabian and Barcons (1992) and in the conference proceedings which they edited (Barcons and Fabian 1992).

The X-ray background spectrum from 1 keV to 1 MeV is shown in Fig. 1.15(a). In the representation shown in Fig. 1.15, the spectrum can be described by two power-laws. At energies less than about 25 keV, the spectrum can be described by an intensity spectrum of the form

$$I(E) = 8.5\ E^{-0.40}\ \mathrm{keV\ cm^{-2}\ s^{-1}\ keV^{-1}\ sr^{-1}} \qquad 2.5 < E < 25\ \mathrm{keV}$$

At higher energies the data can be described by a power-law of the form

$$I(E) = 167 E^{-1.38}\ \mathrm{keV\ cm^{-2}\ s^{-1}\ keV^{-1}\ sr^{-1}} \qquad E > 25\ \mathrm{keV}$$

It is apparent from the spectrum that the 'break' at 25 keV is not sharp but nonetheless it is a prominent feature and the change in slope certainly takes place within less than a decade in photon energy. The origin of this feature of the X-ray background spectrum will prove to be one of the most stubborn problems in studies of the background radiation. The energy density of the X-ray background is quite large. Integrating the energy spectrum shown in Fig. 1.15 from 1 keV to 1 MeV, we find that the energy density is about 75 eV m^{-3}.

One form of spectrum which appears to provide a very good fit to the spectrum of the X-ray background is a thermal bremsstrahlung energy distribution at a temperature $kT = 40$ keV. The fit to the spectrum is illustrated in Fig. 1.15(b) from observations by Marshall et al (1980). In the discussion of Section 5.2, we will find that, although this is a remarkably good explanation for the form of the spectrum, the origin of the X-ray background in the 3 to 300 keV waveband remains a mystery. On the other hand, it is certain that the low energy X-ray background at 1 keV is the integrated emission of discrete X-ray sources as a result of the beautiful new deep X-ray surveys undertaken by the ROSAT observatory. We will recount this story in some detail in Section 5.1.

There is some evidence for an excess extragalactic background component at very low (soft) X-ray energies ($\epsilon < 1$ keV) but it is difficult to extract because of photoelectric absorption by the interstellar gas at energies less than about 1 keV.

1.10 The γ-ray Background Radiation

γ-ray emission from the plane of the Galaxy was discovered by the OSO-III satellite in 1967. This mission was followed by the SAS-2 satellite which discovered the diffuse γ-ray background radiation and by the COS-B satellite which made a detailed map of the Galactic γ-ray emission and discovered about 25 discrete γ-ray sources. Both the SAS-2 and COS-B observations were made at γ-ray energies greater than about 35 MeV and used arrays of

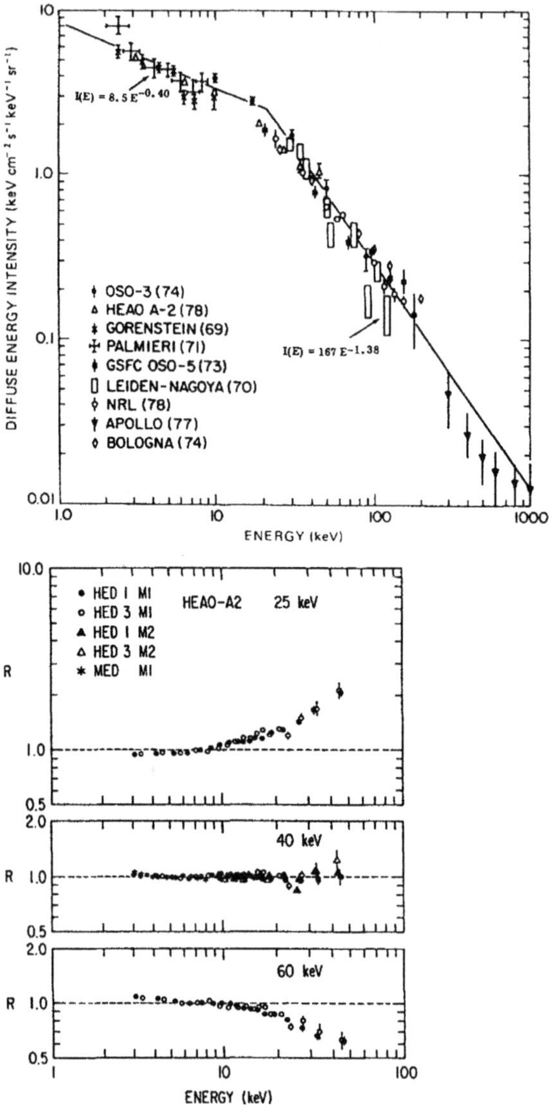

Fig. 1.15 (a) The energy spectrum of the diffuse X-ray background (From D. Schwartz in *Handbook of Space Astronomy and Astrophysics* (1990) by M. Zombeck, page 197. Cambridge: Cambridge University Press). (b) Illustrating the fits of thermal bremsstrahlung spectra at different temperatures to the X-ray background energy spectrum (Marshall et al 1980).

spark chambers to detect high energy γ-rays by the electron-positron pairs which they produce. The overall spectrum of the γ-ray background radiation from 1 to 300 MeV is shown in Fig. 1.16. Fichtel, Simpson and Thompson (1978) derived the spectra of the Galactic and isotropic γ-ray backgrounds and found that the high energy portion of the isotropic background spectrum can be represented by a power-law spectrum of the form

$$N(E) = 4 \times 10^{-2} E^{-2.7} \quad \text{photons cm}^{-2}\ \text{s}^{-1}\ \text{sr}^{-1}\ \text{MeV}^{-1}$$

in the energy range $35 < E < 150$ MeV where E is measured in MeV. The errors on the exponent are quite large, $2.7\ ^{+0.4}_{-0.3}$.

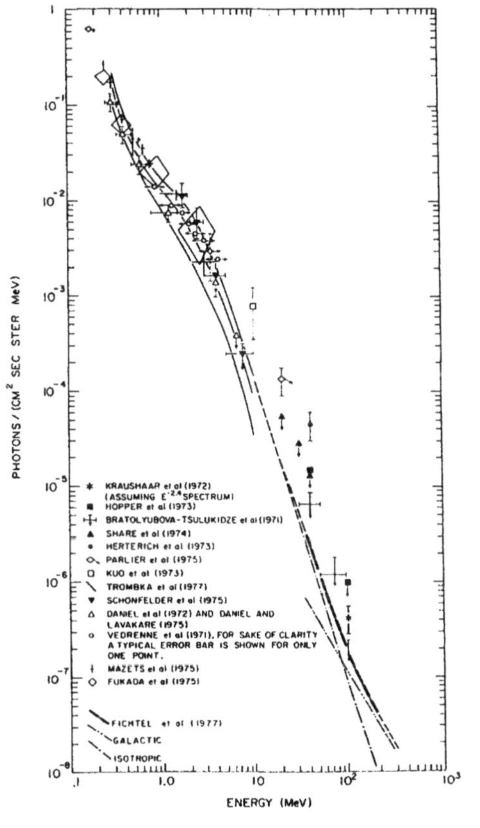

Fig. 1.16 The isotropic γ-ray background spectrum presented by Fichtel, Simpson and Thomson (1978). Their estimates of the spectra of the Galactic as well as the extragalactic γ-ray background are shown.

In the γ-ray energy range $1-10$ MeV, there are three competing processes of energy loss for the γ-ray photons — pair production at energies $E \geq 1$

MeV, Compton scattering and photoelectric absorption. As a result, it is a complex energy range experimentally. The spectrum of the γ-ray background shown in Fig. 1.16 suggests that there is an excess γ-ray background intensity in the $1-10$ MeV range. For the purposes of calculation, the γ-ray spectrum in the $1-3$ MeV range can be approximately represented by a power law of the form

$$N(E) = 10^{-2}E^{-1} \quad \text{photons cm}^{-2}\ \text{s}^{-1}\ \text{sr}^{-1}\ \text{MeV}^{-1}$$

where E is measured in MeV.

Thus, just as in the case of the X-ray waveband, there appears to be a feature in the low energy γ-ray background spectrum at about 1 MeV. It should be emphasised that there is considerable scatter in the data points in Fig. 1.16 and the new results from the Gamma Ray Observatory will be crucial in establishing more precisely the spectrum of the extragalactic background spectrum. Until these data are reduced, Fig. 1.16 is the best we have available. The energy density of radiation in the energy range $E \geq 1$ MeV is about 25 eV m^{-3}.

1.11 Summary

This survey of observations of the extragalactic background radiation gives some impression of the very wide range of observations, experimental techniques and astrophysics which come into play in disentangling the physics of the background radiation. From the point of view of astrophysics, the wavebands divide up naturally into those which are probably dominated by high energy astrophysical process, often referred to as 'non-thermal' processes, and those in which the physics involves 'thermal' processes.

The wavebands involving high energy astrophysical processes are the radio, X and γ-ray wavebands. The background spectra in these wavebands are probably all 'non-thermal' and the brightest extragalactic objects in these wavebands all involve 'non-thermal' high energy astrophysical activity.

In contrast, the millimetre, infrared, optical and ultraviolet wavebands are probably dominated by thermal processes. The Cosmic Microwave Background Radiation is the most perfect example of a thermal emission spectrum but it is in many ways quite unique in being the cooled thermal spectrum of the hot early stages of the Universe as a whole. In the infrared, optical and ultraviolet regions of the spectra, the basic processes involve the thermal emission of dust and stars. When we observe extragalactic systems, the spectral energy distributions are normally the integrated emission of all the thermal emitters in the system, that is, the dust and the stars. The most common extragalactic objects in these wavebands are galaxies and, by and large, the background radiation is the integrated emission from galaxies. In the ultraviolet region of the spectrum young massive stars dominate the spectrum of the galaxy whereas in the infrared spectral region it is the reradiated

Table 1.3 The energy densities and photon densities in different regions of the electromagnetic spectrum. Note that these are usually rough estimates which are useful for order-of-magnitude calculations only.

Waveband	Energy density of radiation eV m^{-3}	Number density of photons m^{-3}
Radio	$\sim 5 \times 10^{-2}$	$\sim 10^6$
Microwave	3×10^5	5×10^8
Infrared	?	?
Optical	$\sim 2 \times 10^3$	$\sim 10^3$
Ultraviolet	?	?
X-ray	75	3×10^{-3}
γ-ray	25	3×10^{-6}

emission of star-forming regions which contribute most of the background intensity. The intriguing questions concern how much we can learn about the evolution of these objects from studies of the background radiation.

To conclude this introduction, I list typical energy densities and number densities of photons in each of the wavebands in which a positive detection of the background radiation has been made. It must be emphasised that these are usually very rough estimates and, for any precise calcualation, integrations should be carried out over the relevant regions of the spectrum. There are gaps in this table in the infrared and far-ultraviolet regions of the spectrum. Fig. 1.2 can be used to interpolate appropriate values if they are needed. These rough figures are often useful for making simple estimates of the importance of physical processes in a cosmological setting.

References – Chapter 1

Armand, C., Milliard, B. and Deharveng, J.M. (1993). A&A, (in press).

Barcons, X. and Fabian, A.C. (eds) (1992). *The X-ray Background.* Cambridge: Cambridge University Press.

Berry, M.V. (1989). *Principles of Cosmology and Gravitation.* Bristol: Adam Hilger.

Bondi, H. (1952). *Cosmology.* Cambridge: Cambridge University Press.

Bowyer, S. (1991). ARA&A, **29**, 59.

Bowyer, S and Leinert, C. (eds) (1990). *The Galactic and Extragalactic Background Radiation.* Dordrecht: Kluwer Academic Publishers.

Bridle, A.H. (1967). MNRAS, **136**, 219.

Bridle, A.H. and Purton, C.R. (1968). AJ, **73**, 717.

Clark, T.A., Brown, L.W. and Alexander, J.K. (1970). Nat, **228**, 847.

Fabian, A.C. and Barcons, X. (1992). ARA&A, **30**, 429.

Fichtel, C.E., Simpson, G.A. and Thompson, D.J. (1978). ApJ, **222**, 833.

Giacconi, R., Gursky, H., Paolini, F. and Rossi, B. (1962). Phys. Rev. Lett., **9**, 439.

Harrison, E.R. (1987). *Darkness at Night: A Riddle of the Universe.* Cambridge, Massachusetts: Harvard University Press.

Harrison, E.R. (1990). In *The Galactic and Extragalactic Background Radiation*, (eds. S. Bowyer and C. Leinert), 3. Dordrecht: Kluwer Academic Publishers.

Hauser, J. (1993). In *Sky Surveys: Protostars to Protogalaxies*, (ed. T. Soifer), Vol. 43, 205. San Francisco: Astron. Soc. Pac. Conf. Ser.

Henry, R.C. (1991). ARA&A, **29**, 89.

Holberg, J.B. (1990). In *The Galactic and Extragalactic Background Radiation*, (eds. S. Bowyer and C. Leinert), 220. Dordrecht: Kluwer Academic Publishers.

Jakobsen, P. (1994). In *The Extragalactic Background Radiation*, (eds. M. Livio, M. Fall, D. Calzetti and P. Madau), (in press). Cambridge: Cambridge University Press.

Kolb, E.W. and Turner, M.S. (1990). *The Early Universe*. Redwood City, California: Addison-Wesley Publishing Co.

Livio, M., Fall, M., Calzetti, D. and Madau, P. (1994). *The Extragalactic Background Radiation*. Cambridge: Cambridge University Press.

Longair, M.S. and Sunyaev, R.A. (1971). Uspekhi Fiz. Nauk., **105**, 41. [English translation: Soviet Physics Uspekhi, **14**, 569.]

Longair, M.S. (1989). In *The New Physics*, (ed. P.C.W. Davies), 94. Cambridge: Cambridge University Press.

Longair, M.S. (1990). In *The Galactic and Extragalactic Background Radiation*, (eds. S. Bowyer and C. Leinert), 469. Dordrecht: Kluwer Academic Publishers.

Marshall, F.E., Boldt, E.A., Holt, S.S., Miller, R.B., Mushotzky, R.F., Rose, L.A., Rothschild, R.E. and Serlemitsos, P.J. (1980). ApJ, **235**, 4.

Martin, C. and Bowyer, S. (1989). ApJ, **338**, 667.

Martin, C., Hurwitz, M. and Bowyer, S. (1990). ApJ, **345**, 220.

Mather, J.C., Cheng, E.S., Eplee, R.E. Jr., Isaacman, R.B., Meyer, S.S., Shafer, R.A., Weiss, R., Wright, E.L., Bennett, C.L., Boggess, N.W., Dwek, E., Gulkis, S., Hauser, M.G., Janssen, M., Kelsall, T., Lubin, P.M., Moseley, S.H. Jr., Murdock, T.L., Silverberg, R.F., Smoot, G.F. and Wilkinson, D.T. (1990). ApJ, **354**, L37. (1990).

Mather, J. (1993). In *The Extragalactic Background Radiation*, (eds. M. Livio, M. Fall, D. Calzetti and P. Madau), (in press). Cambridge: Cambridge University Press.

Mattila, K. and Schnur (1990). In *The Galactic and Extragalactic Background Radiation*, (eds. S. Bowyer and C. Leinert), 257. Dordrecht: Kluwer Academic Publishers.

Peacock, J.A, Heavens, A.F. and Davies, A.T. (1990). *The Physics of the Early Universe*. Edinburgh: SUSSP Publications.

Peebles, P.J.E. (1993). *Principles of Physical Cosmology*. Princeton: Princeton University Press.

Penzias, A.A. and Wilson, R.W. (1965). ApJ, **142**, 419.

Purton, C.R. (1966). Ph.D. Dissertation, University of Cambridge.

Rindler, W. (1977). *Essential Relativity: Special, General and Cosmological*. New York: Springer-Verlag.

Shectman, S.A. (1974). ApJ, **188**, 233.

Simon, A.C.B. (1978). MNRAS, **180**, 429.

Smoot, G.F., Bennett, C.L., Kogut, A., Wright, E.L., Aymon, J., Boggess, N.W., Cheng, E.S., DeAmici, G., Gulkis, S., Hauser, M.G., Hinshaw, G., Lineweaver, C., Loewenstein, K., Jackson, P.D., Janssen, M., Kaita, E., Kelsall, T., Keegstra, P., Lubin, P., Mather, J.C., Meyer, S.S., Moseley, S.H., Murdock, T.L., Rokke, L., Silverberg, R.F., Tenorio, L., Weiss, R. and Wilkinson, D.T. (1992). ApJ, **396**, L1.

Sunyaev, R.A. and Zeldovich, Ya.B. (1980). ARA&A, **18**, 537.

Toller, G. In *The Galactic and Extragalactic Background Radiation*, (eds. S. Bowyer and C. Leinert), 21. Dordrecht: Kluwer Academic Publishers.
Tyson, A. (1990). In *The Galactic and Extragalactic Background Radiation*, (eds. S. Bowyer and C. Leinert), 245. Dordrecht: Kluwer Academc Publishers.
Wall, J.V. (1990). In *The Galactic and Extragalactic Background Radiation*, (eds. S. Bowyer and C. Leinert), 327. Dordrecht: Kluwer Academic Publishers.
Wall, J.V., Chu, T.Y. and Yen, J.L. (1970). Aust. J. Phys., **23**, 45.
Weiss, R. (1980). ARA&A, **18**, 489.
Weinberg, S. (1972). *Gravitation and Cosmology*. New York: John Wiley and Co.
Zeldovich, Ya.B. and Novikov, I.D. *Relativistic Astrophysics*, Vol. 1 (1971) *Stars and Relativity*; Vol. 2 (1983) *The Structure and Evolution of the Universe*. Chicago: Chicago University Press.
Zombeck, M. (1990). *Handbook of Space Astronomy and Astrophysics*. Cambridge: Cambridge University Press.

2 The Robertson-Walker Metric

This chapter is of a pedagogical nature. I will show how the key results for the standard cosmological models can be derived from simple physical arguments. This analysis is an enhancement of the approach described in *Theoretical Concepts in Physics* (Longair 1992). The homogeneous isotropic models can be derived without appeal to sophisticated mathematical procedures and the essence of the more complete derivations will become apparent as we proceed. Uniformly-expanding, isotropic, homogeneous world models are the natural point of departure for the construction of the standard world models. The observational bases for this assertion are the isotropy and homogeneity of the Universe on a very large scale and the velocity-distance relation for galaxies discovered by Hubble. My presentation will complement the approach taken by Dr. Sandage.

2.1 The Isotropy of the Universe

The Universe is obviously highly inhomogeneous on a small scale with matter condensed into stars which are congregated into galaxies which are themselves clustered, the associations ranging from small groups to giant regular clusters of galaxies. If we take our averages over larger and larger scales, however, the inhomogeneity becomes less and less. This statement can be formalised using 2-point correlation functions $\xi(r)$ which describe the average number density of galaxies at a distance r from any given galaxy. The function $\xi(r)$ is defined by the expression

$$N(r)\mathrm{d}V = N_0[1 + \xi(r)]\mathrm{d}V \tag{2.1}$$

where $N(r)\mathrm{d}V$ is the number of galaxies in the volume element $\mathrm{d}V$ at distance r from the galaxy and N_0 is a suitable average space density. It is found that the function $\xi(r)$ can be very well represented by a power-law of the form

$$\xi(r) = \left(\frac{r}{r_0}\right)^{-\gamma} \tag{2.2}$$

on physical scales from about 200 kpc to 20 Mpc in which the scale $r_0 = 5h^{-1}$ Mpc and the exponent $\gamma = 1.8$.[24] On scales greater than about $10h^{-1}$ Mpc the two-point correlation function decreases more rapidly that the power-law expression (2.2). Notice that the two-point correlation function is a circularly symmetric average about each galaxy and so is only a crude measure of the overall clustering of galaxies. It does, however, give the correct result

[24]The use of $h = H_0/(100\ \mathrm{km\ s^{-1}\ Mpc^{-1}})$ is a convenient device for adjusting the dimensions and luminosities of extragalactic objects for the reader's preferred value of Hubble's constant. If a value of $H_0 = 100\ \mathrm{km\ s^{-1}\ Mpc^{-1}}$ is preferred, $h = 1$; if the value $H_0 = 50\ \mathrm{km\ s^{-1}\ Mpc^{-1}}$ is preferred, $h = 0.5$ and so on.

that on physical scales $r > 5h^{-1}$ Mpc, the mean amplitude of the density perturbations is less than one and that the distribution of galaxies becomes smoother on the very largest scales. Notice that this means that the density perturbations on the largest scales are still in the linear regime, $\delta\rho/\rho \ll 1$.

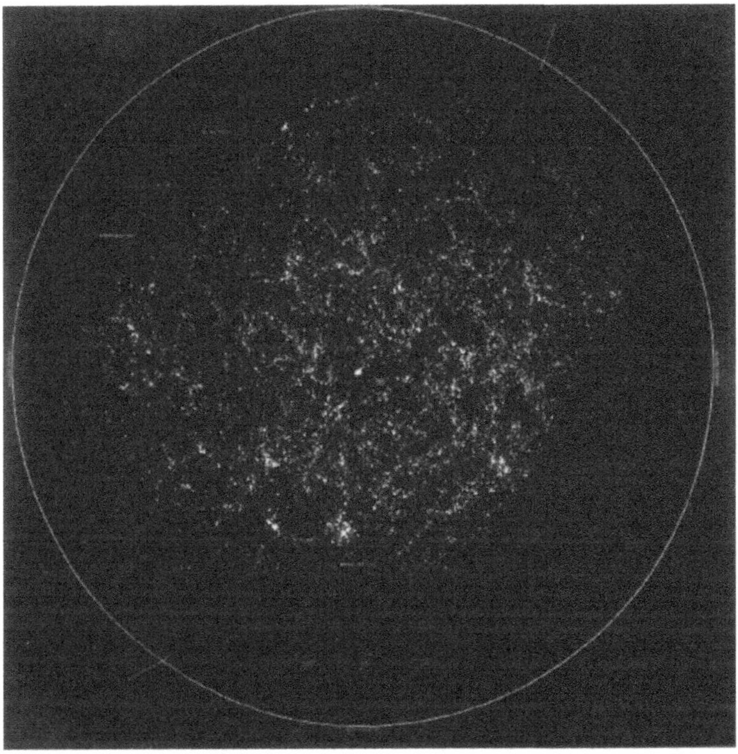

Fig. 2.1 The distribution of galaxies in the northern Galactic hemisphere derived from counts of galaxies undertaken by Shane, Wirtanen and their colleagues at the Lick Observatory in the 1960s. Over one million galaxies were counted in their survey. The northern Galactic pole is at the centre of the picture and the galactic equator is represented by the solid circle bounding the diagram. The projection of the sky onto the plane of the picture is an equal area projection. This photographic representation of the galaxy counts was made by Peebles and his colleagues. The large sector missing from the lower right-hand corner of the picture corresponds to an area in the southern celestial hemisphere which was not surveyed by the Lick workers. The decreasing surface density of galaxies towards the circumference of the picture, that is towards the galactic equator, is due to the obscuring effect of interstellar dust in the interstellar medium of our own Galaxy. The prominent cluster of galaxies close to the centre of the picture is the Coma cluster. (Seldner, Siebars, Groth and Peebles 1977).

A better representation of the overall distribution of galaxies is given in Fig. 2.1 which shows the distribution of galaxies in the Northern Galactic hemisphere once all the stars of our own Galaxy have been removed. The large 'bite' out of the picture in the bottom right corresponds to an area of the sky which was not observed in the Lick survey and the decrease in the numbers of galaxies towards the edges of the picture is due to extinction by interstellar dust in our own Galaxy. Therefore, only in the central region of Fig. 2.1 do we obtain a reasonably clean picture of the large-scale distribution of galaxies in the Universe. This is a picture of the distribution of galaxies on the grandest scale, a giant cluster such as the Coma cluster corresponding to the bright dot in the centre of the picture.

We now know that much of the obvious clumping, the holes and the stringy structures are real features of the distribution of galaxies. Fig. 2.2 shows the local distribution of galaxies derived from the Harvard-Smithsonian Astrophysical Observatory survey of bright galaxies. Over 14,000 galaxies are plotted in Fig. 2.2 and, if the galaxies were uniformly distributed in the local Universe, the points would be uniformly distributed over the diagram. It can be seen that there are gross inhomogeneities and irregularities in the local Universe. There are large 'holes' in which the local number density of galaxies is significantly lower than the mean and there are long 'filaments' of galaxies, including the feature known as the 'great wall' which extends from right ascensions 9^h to 17^h about half-way to the limit of the survey. Notice that there are a number of 'streaks' pointing toward our own Galaxy which lies at the centre of the diagram. These are bound clusters of galaxies, the lengths of the 'streaks' corresponding to the components of the velocity dispersion of the galaxies in the clusters along the line of sight. The scale of the large holes seen in Fig. 2.2 is about 30-50 times the scale of a cluster of galaxies. These are the largest known structures in the Universe and we should state immediately that one of the great cosmological problems is to reconcile the gross irregularity in the large-scale distribution of galaxies with the remarkable smoothness of the Cosmic Microwave Background Radiation described in Section 1.5. Nonetheless, the amplitude of these irregularities decreases with increasing scale so that on the very largest scales, one bit of Universe looks very much like another.

A big advance in these studies has been in describing the large-scale topology of the galaxy distribution. The approach taken by Gott and his colleagues has been to evaluate the topology of the distribution of holes and structures in the local distributions of galaxies (Gott et al 1986, Melott et al 1988). What they find is that the distribution of the galaxies on the large scale is 'sponge-like'. The material of the sponge represents the location of the galaxies and the holes represent the large voids seen in Fig. 2.2. Thus, both the holes and the distribution of galaxies can be thought of as being continuously connected throughout the local Universe. This topology is possible in three dimensions but not in two.

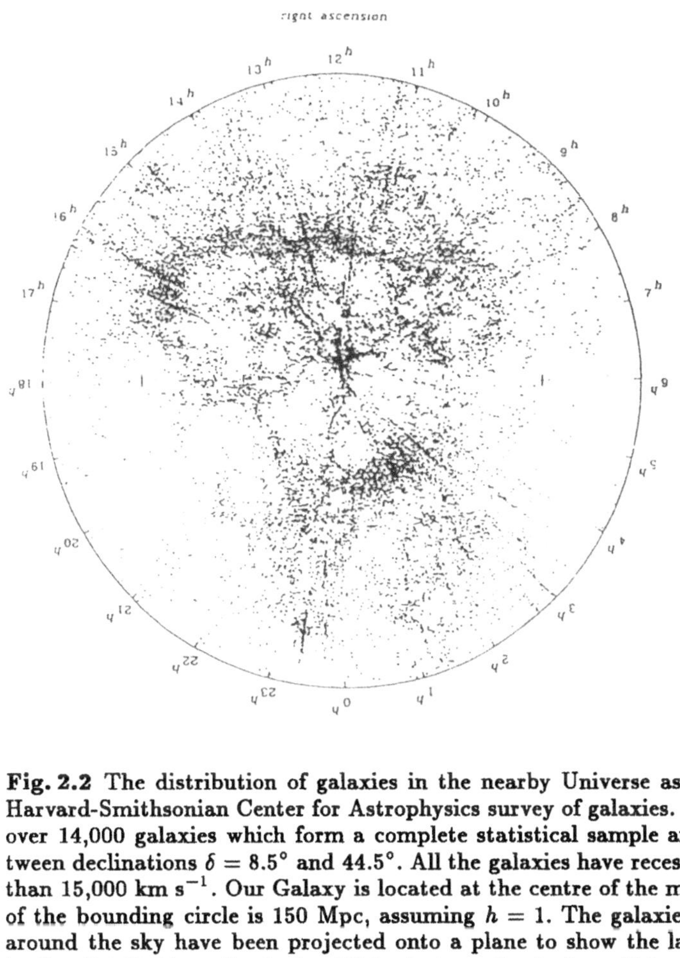

Fig. 2.2 The distribution of galaxies in the nearby Universe as derived from the Harvard-Smithsonian Center for Astrophysics survey of galaxies. The map contains over 14,000 galaxies which form a complete statistical sample around the sky between declinations $\delta = 8.5°$ and $44.5°$. All the galaxies have recession velocities less than 15,000 km s^{-1}. Our Galaxy is located at the centre of the map and the radius of the bounding circle is 150 Mpc, assuming $h = 1$. The galaxies within this slice around the sky have been projected onto a plane to show the large scale features in the distribution of galaxies. Rich clusters of galaxies which are gravitationally bound systems with internal velocity dispersions of about 10^3 km s^{-1} appear as 'fingers' pointing radially towards our Galaxy at the centre of the diagram. It can be seen that the distribution of galaxies is highly irregular with huge holes, filaments and clusters of galaxies throughout the local Universe. (Courtesy of Margaret Geller, John Huchra and the Harvard-Smithsonian Center for Astrophysics (1992).)

Even more impressive evidence comes from the distribution of extragalactic radio sources over the sky. It turns out that, when a survey of the radio sky is made, the objects which are easiest to observe are extragalactic radio sources associated with certain rare classes of galaxy at very great distances. Because they are rare objects, they sample the isotropy of the Universe on a large scale. Fig. 2.3 shows the distribution of the brightest 31,000 extragalactic radio sources in the northern hemisphere as observed at a wavelength of 6 cm which are included in the Greenbank Catalogue of radio sources (Gregory and Condon 1991). There is a hole in the centre of the distribution corresponding to a region which was not observed as part of the survey and two other holes in the vicinity of the intense sources Cygnus A and Cassiopeia A. There is a small excess of sources lying along the Galactic plane but otherwise the distribution is entirely consistent with the sources being distributed uniformly at random over the sky. The radio sources are ideal for probing the large scale distribution of discrete objects since they are so readily observed at large distances.

Fig. 2.3 The distribution of radio sources in the Greenbank Catalogue of radio sources at 6 cm (Gregory and Condon 1991). The picture includes 31,000 radio sources. In this equal area projection, the north celestial pole is in the centre of the diagram and the celestial equator around the solid circle. The area about the north celestial pole was not surveyed. There are 'holes' in the distribution about the bright sources Cygnus A and Cassiopeia A and a small excess of sources associated with the Galactic plane. Otherwise, the distribution does not display any significant departure from a random distribution (from Peebles 1993).

This is impressive enough, but it pales into insignificance compared with the recent results on the isotropy of the Cosmic Microwave Background Radiation. These remarkable results were described in detail in Section 1.5 and we will have much more to say about the interpretation of the fluctuation data in later lectures.

For the moment, however, our interest is in the isotropy of the Universe as a whole and we can state that there is certainly no evidence for any anisotropy in the distribution of the Cosmic Microwave Background Radiation at the level of one part in 100,000 when we look on the large scale. This is quite incredible precision for any cosmological experiment since one is normally lucky in cosmology if one knows anything within a factor of about 10. The obvious question is how the distribution of this radiation is related to the distribution of ordinary matter. The answer is not as straightforward as one would like and we need to understand the temperature history of the Universe to give the standard answer.

In the standard picture of the evolution of the Hot Big Bang, when the Universe was squashed to only about one thousandth of its present size, the temperature of the Cosmic Background Radiation must have been about one thousand times greater than it is now. The temperature of the background radiation was then $T = 2.726(1+z)$ K and so, at a redshift of 1500, the temperature of the radiation field was about 4000 K (see Sections 3.5 and 6.2). Then there were sufficient Lyman continuum photons in the Wien region of the black-body spectrum to photoionise all the neutral hydrogen in the Universe. When this occurred, there was very strong coupling between the Cosmic Background Radiation and the ionised matter by Thomson scattering. In fact, when we look back to these epochs, it is as if we were looking at the surface of a star surrounding us in all directions but the temperature of the radiation we observe has been cooled by the redshift factor of 1500 so that what we observe is redshifted into the millimetre waveband. This analogy also makes it clear that, because of the strong scattering of the radiation, we can only observe the very surface layers of our 'star'. We can therefore obtain no direct information about what was happening at earlier epochs as soon as we encounter the epoch at which the material of the Universe was ionised. This 'surface' at which the Universe becomes opaque to radiation is known as the *last scattering surface* and the fluctuations observed by COBE are interpreted as the very low intensity ripples present on that surface on angular scales of 10°. Thus, strictly speaking, in the standard interpretation, the COBE results provide information about the diffuse ionised intergalactic gas when the Universe was only about one thousandth of its present size. At that stage, the galaxies could not have formed and so all the ordinary matter, which was eventually to become galaxies as we know them, was still in the form of a remarkably smooth intergalactic ionised gas. Notice that the extragalactic radio sources provide complementary information to that provided by the Cosmic Microwave Background Radiation in that they refer to the large scale distribution of discrete objects such as galaxies once they had formed.

It should be emphasised that the standard interpretation assumes that the intergalactic gas was not reionised and heated at some later epoch. If that were to occur, the last scattering surface could occur at a significantly

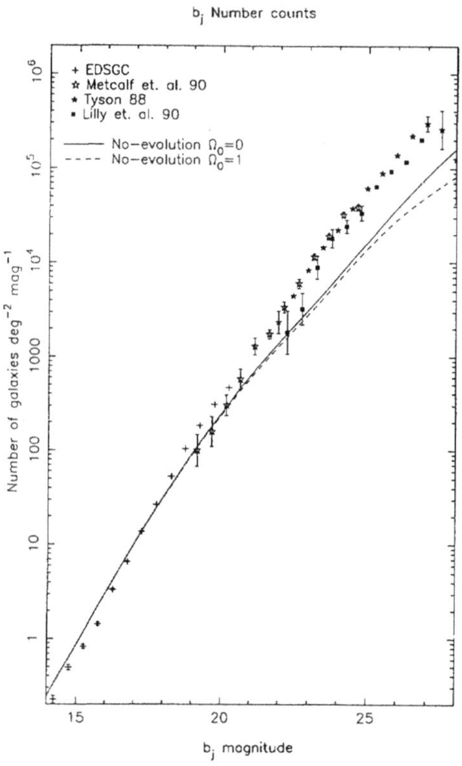

Fig. 2.4 The counts of faint galaxies observed in the blue (B) waveband compared with the expectations of uniform world models with $\Omega = 0$ and $\Omega = 1$ (after Jones et al 1991, Metcalfe et al 1991). It can be seen that the counts follow closely the expectations of uniform world models at magnitudes less than about 21 but that there is a excess of faint galaxies at greater magnitudes.

smaller redshift. This picture is not particularly plausible because the energy demands are very severe and the heating itself would probably result in much greater fluctuations than those observed.

2.2 The Homogeneity of the Universe and Hubble's Law

Both of these topics have been splendidly treated by Dr. Sandage and I need add little to what he has already stated.

2.2.1 Homogeneity

Hubble realised that a key test of the homogeneity of the Universe is provided by the counts of galaxies. As we will show in Chapter 4, in a homogeneous

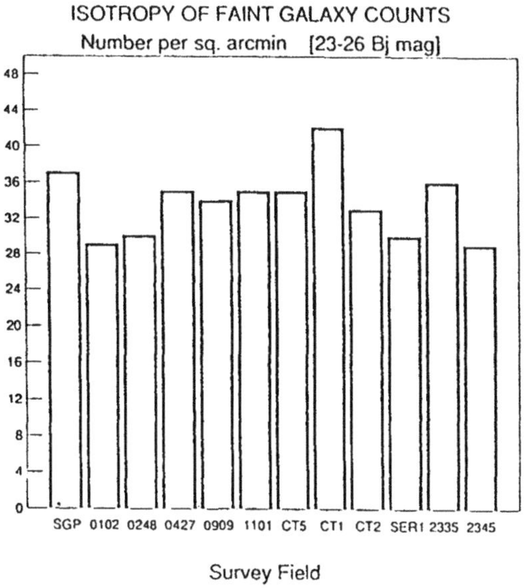

Fig. 2.5 **The isotropy of the faint number counts according to Tyson (1990). The galaxies counted have magnitudes in the range $23 < B_j < 26$ in 12 widely separated high latitude fields. The variation in number density of galaxies is about twice that expected of a random distribution.**

Universe it is expected that the number counts of galaxies follow the law $N(\geq S) \propto S^{-\frac{3}{2}}$ and this result is independent of the luminosity function of the sources so long as the counts do not extend to such large distances that the effects of the cosmological redshift have to be taken into account (Section 4.1). In terms of magnitudes, this relation is $N(\leq m) \propto 10^{0.6m}$. Hubble found that the counts of galaxies to about 19th magnitude followed this relation, showing directly that on average they are homogeneously distributed in space. The more recent counts of galaxies show exactly this same distribution (Fig. 2.4). Divergences from this relation only occur at much fainter magnitudes and an excess of faint blue galaxies is found. This story will be taken up by Dr. Kron.

One might worry about the effects of the large holes and inhomogeneities upon the distribution of galaxies on the largest scales. Although these must be present, on a large enough scale, the numbers seem to even out. Fig. 2.5 shows the counts of galaxies made by Tyson (1990) in the magnitude range $23 < B_j < 26$ in 12 widely separated high latitude fields. According to Tyson, the fluctuations in the surface densities found in these fields amount to about twice the dispersion expected if they were selected at random. Another way of expressing this result is to say that the faint galaxies are isotropic within about 2σ. An obvious interpretation of this result is that the counts are influenced by the large scale irregularities in the distribution of galaxies observed on the largest scales. To put it simply, the number counts depend upon exactly how many sheets and voids seen in Fig. 2.2 are present along the line of sight.

Another way of studying the homogeneity of the distribution of galaxies is to measure the two-point correlation function as a function of apparent magnitude. In this way, the statistical properties of the clustering of galaxies at different distances can be compared. Such an analysis has been carried out by Groth and Peebles (1977, 1986) and they showed that the two-point correlation functions determined from a bright sample of Zwicky galaxies, from the Lick counts of galaxies and from a deep sky survey plate in an area known as the Jagellonian field scale exactly as expected if the distribution of the galaxies displayed the same degree of correlation throughout the local Universe out to $z \sim 0.1$. A similar result has been found comparing the two-point correlation functions found at increasing apparent magnitude limits in the machine-scanned surveys carried out by the APM group at Cambridge (Maddox et al 1990). Fig. 2.6(a) shows the angular two-point correlation functions $w(\theta)$ measured at increasing apparent limits in the magnitude range $17.5 < m < 20.5$. In Fig. 2.6(b), all these functions are scaled to the angular correlation function found from the Lick survey. As Peebles (1993) expresses it, '...the correlation function analyes have yielded a new and positive test of the assumption that the galaxy space distribution is a stationary (statistically homogeneous) random process'. In terms of our 'sponge' picture of the distribution of galaxies in the local Universe, this result means that the sponge seems to have the same structure throughout the local Universe out to redshifts $z \sim 0.1$.

2.2.2 The Redshift–Apparent Magnitude Diagram

There is even less need to remark upon the *redshift-distance relation*, since this is comprehensively dealt with by Dr. Sandage. Simply for completeness, I note that the velocity-distance relation for galaxies can be written $v = H_0 r$ where v is the recession velocity of the galaxy and r is its distance from our Galaxy, H_0 being Hubble's constant. The one footnote I would add to Dr. Sandage's presentation is that the velocity-distance relation appears to apply for all classes of extragalactic system and my own personal favorite is the relation for the galaxies associated with strong radio sources, a recent example being shown in Fig. 2.7. It can be seen that the narrow dispersion in absolute magnitude for the radio galaxies extends to redshifts of 2 and greater. I still find this a quite remarkable result.

2.2.3 The Local Expansion of the Distribution of Galaxies

It is the combination of the observed isotropy and homogeneity of the Universe with Hubble's law which shows that the Universe as a whole is expanding uniformly at the present time. Let me show this formally by the following

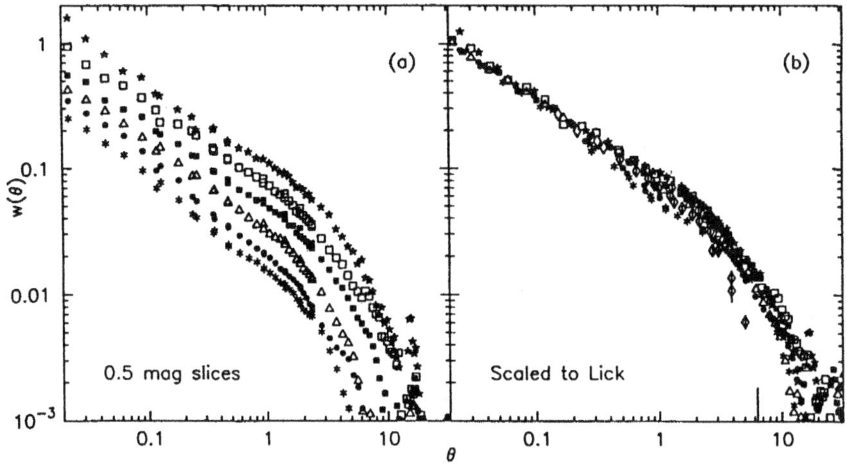

Fig. 2.6 The scaling test for the homogeneity of the distribution of galaxies using the correlation functions for galaxies derived from the APM surveys. (a) The correlation functions at increasing limiting apparent magnitudes in the range $17.5 < m < 20.5$ displayed at intervals of 0.5 magnitudes. (b) The two-point correlation functions scaled to the correlation function derived from the Lick counts of galaxies (Maddox et al 1990).

simple sum. Consider a uniformly expanding system of points (Fig. 2.8). Then, the definition of a uniform expansion is that the distances to any two points should increase by the same factor in a given time interval, that is, we require

$$\frac{r_1(t_2)}{r_1(t_1)} = \frac{r_2(t_2)}{r_2(t_1)} = \ldots = \frac{r_n(t_2)}{r_n(t_1)} = \ldots = \alpha = \text{constant} \tag{2.3}$$

for any set of points. The recession velocity of galaxy 1 from the origin is therefore

$$v_1 = \frac{r_1(t_2) - r_1(t_1)}{t_2 - t_1} = \frac{r_1(t_1)}{t_2 - t_1}\left[\frac{r_1(t_2)}{r_1(t_1)} - 1\right] = \frac{r_1(t_1)}{t_2 - t_1}(\alpha - 1) = H_0 r_1(t_1)$$

Similarly,

$$v_n = \frac{r_n(t_1)}{t_2 - t_1}(\alpha - 1) = H_0 r_n(t_1) \tag{2.4}$$

Thus, a uniformly expanding distribution of galaxies automatically results in a velocity-distance relation of the form $v \propto r$.

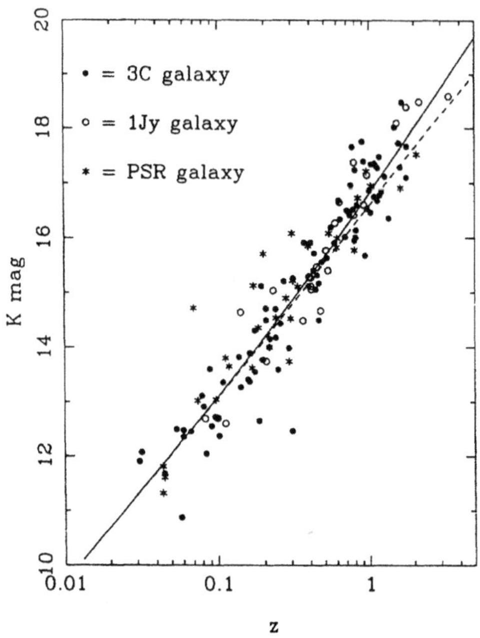

Fig. 2.7 The redshift-K-magnitude relation for radio galaxies associated with strong radio sources. The radio galaxies are selected from the 3CR, 1 Jy and PKS samples of extragalactic radio sources (Dunlop and Peacock 1990).

2.3 Isotropic Curved Spaces

Continuing with the pedagogial exposition, let us discuss briefly the concept of isotropic curved spaces. In flat space, we can write the distance between two points separated by $\mathrm{d}x$, $\mathrm{d}y$, $\mathrm{d}z$ as

$$\mathrm{d}l^2 = \mathrm{d}x^2 + \mathrm{d}y^2 + \mathrm{d}z^2 \tag{2.5}$$

Now, let us consider the simplest example of an isotropic *two-dimensional* curved space, namely the surface of a sphere. This *two-space* is isotropic because the radius of curvature of the sphere R_c is the same at all points in the two-space. We can set up an orthogonal frame of reference at each point locally on the surface of the sphere. It is convenient to work in spherical polar coordinates to describe positions on the surface of the sphere as indicated in Fig. 2.9. In this case, the orthogonal coordinates are the angular coordinates θ and ϕ, and we can write the expression for the increment of distance $\mathrm{d}l$ between two neighbouring points on the surface as

$$\mathrm{d}l^2 = R_\mathrm{c}^2 \mathrm{d}\theta^2 + R_\mathrm{c}^2 \sin^2\theta \mathrm{d}\phi^2 \tag{2.6}$$

The expression (2.6) is known as the *metric* of the two-dimensional surface and can be written more generally in tensor form

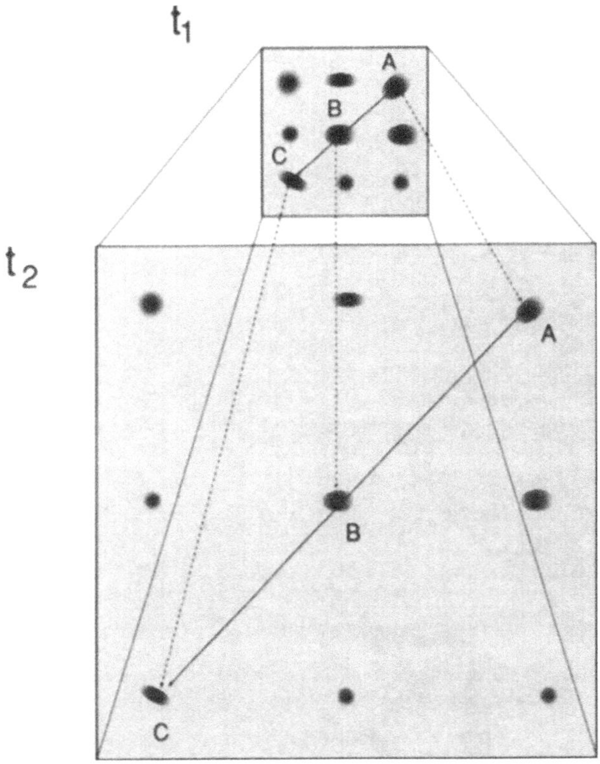

Fig. 2.8 Illustrating the origin of the velocity-distance relation for an isotropically expanding distribution of galaxies. The distribution of galaxies expands uniformly between the epochs t_1 and t_2. If, for example, we consider the motions of the galaxies relative to the galaxy A, it can be seen that galaxy C travels twice as far as galaxy B between the epochs t_1 and t_2 and so has twice the recession velocity of galaxy B relative to A. Since C is always twice the distance of B from A, it can be seen that the velocity-distance relation is a general property of isotropically expanding Universes.

$$dl^2 = g_{\mu\nu} dx^\mu dx^\nu \tag{2.7}$$

It is a fundamental result of differential geometry that the *metric tensor* $g_{\mu\nu}$ contains all the information about the intrinsic geometry of the space. Gauss first showed how it is possible to determine the local curvature of space from the metric tensor (see Weinberg 1972, Berry 1989). I will not repeat this calculation for the metric (2.6) but it is straightforward to show that in this case the local curvature of space $\kappa = R_c^{-2}$. Notice the pleasant result that the surface of a sphere which has the same radius of curvature at all points on the surface is automatically an isotropic curved 2-space. Notice that, in the general case, the curvature κ varies from point to point in the space.

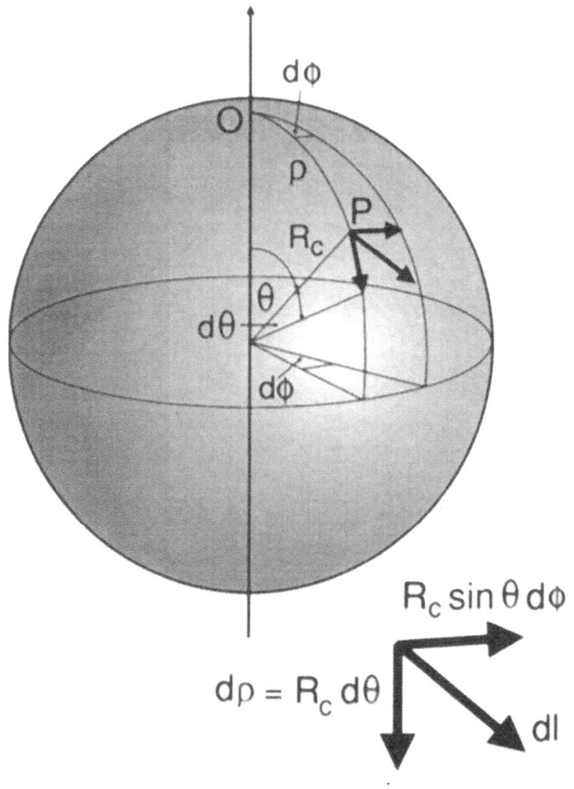

Fig. 2.9 The surface of a sphere as the simplest example of a two-dimensional curved space.

The extension to isotropic three-spaces is straightforward if we remember that any two-dimensional section through an isotropic three-space must be an isotropic two-space and we already know the metric tensor for this case. We have already worked out the length of the distance increment dl (expression 2.6). The natural system of coordinates for an isotropic two-space is a spherical polar system in which a radial distance ρ round the sphere is measured from the pole and the angle ϕ measures angular displacements at the pole. From Fig. 2.9, we can define the distance ρ round the arc of a great circle from the point O to P by $\theta = \rho/R_c$ and so the metric can be written

$$dl^2 = d\rho^2 + R_c^2 \sin^2\left(\frac{\rho}{R_c}\right) d\phi^2 \tag{2.8}$$

Notice that this distance ρ is the shortest distance between O and P on the surface of the sphere since it is part of a great circle. It is therefore known as the *geodesic distance* between O and P in the isotropic curved space. Geodesics play the role of straight lines in curved space.

We can write the metric in an alternative form if we introduce a distance measure

$$x = R_c \sin\left(\frac{\rho}{R_c}\right) \tag{2.9}$$

Differentiating and squaring, we find

$$dx^2 = \left[1 - \sin^2\left(\frac{\rho}{R_c}\right)\right] d\rho^2 \qquad d\rho^2 = \frac{dx^2}{1 - \kappa x^2} \tag{2.10}$$

where $\kappa = 1/R_c^2$ is the curvature of the two space. Therefore, we can rewrite the metric in the form

$$dl^2 = \frac{dx^2}{1 - \kappa x^2} + x^2 d\phi^2 \tag{2.11}$$

Notice the interpretation of the distance measure x. It can be seen from the metric that $dl = x d\phi$ is a metric distance perpendicular to the radial coordinate ρ and that it is the correct expression for the length of a line segment which subtends the angle $d\phi$ at geodesic distance ρ from O. It is therefore what is known as an *angular diameter distance* since it is guaranteed to give the correct answer for the length of a line segment perpendicular to the line of sight. We can use either ρ or x in our metric but notice that, if we use x, the increment of geodesic distance is $d\rho = dx/(1 - \kappa x^2)^{1/2}$.

We have dealt with the simplest example of an isotropic curved space and it can be shown that we have derived the general form of the metric for any isotropic curved space. I have given a simple derivation of this result (Longair 1992). All we have to add to what we have derived so far is that the curvature $\kappa = 1/R_c^2$ can be *positive* as in the spherical 2-space discussed above, *zero* in which case we recover flat Euclidean space ($R_c \to \infty$) and *negative* in which case the geometry becomes *hyperbolic* rather than spherical. If κ is negative, R_c is imaginary. We can recover the standard results for isotropic curved spaces if we substitute $R_c = iR'_c$ where R'_c is real. We find

$$x = R'_c \sinh\left(\frac{\rho}{R'_c}\right) \tag{2.12}$$

We can now write down the expression for the spatial increment in any isotropic three-dimensional curved space. By a straightforward extension of the above formalism, we can write the spatial increment

$$dl^2 = d\rho^2 + R_c^2 \sin^2\left(\frac{\rho}{R_c}\right) [d\theta^2 + \sin^2\theta \, d\phi^2] \tag{2.13}$$

in terms of the spherical polar coordinates (ρ, θ, ϕ). Notice that, so long as we allow R_c to be real or imaginary, we can encompass all three cases in the one formalism. An exactly equivalent form is obtained if we write the spatial increment in terms of x, θ, ϕ in which case we find

$$dl^2 = \frac{dx^2}{1-\kappa x^2} + x^2[d\theta^2 + \sin^2\theta\, d\phi^2] \tag{2.14}$$

We are now in a position to write down the *Minkowski metric* in any isotropic three-space. It is given by

$$ds^2 = dt^2 - \frac{1}{c^2}dl^2 \tag{2.15}$$

where dl is given by either of the above forms of the spatial increment, (2.13) or (2.14). We can now proceed to derive from this metric the *Robertson-Walker metric.*

2.4 The Robertson-Walker Metric

Formally, we begin by making an assumption which is called the *cosmological principle* — this is the statement that we are not located at any special place in the Universe. The corollary is that we are at a typical location in the Universe and that a suitably chosen observer located at any other point at the same cosmological epoch would observe the same large-scale features that we observe. Isotropic, homogeneous, expanding Universes satisfy this requirement since every observer who partakes in the uniform expansion observes the Universe to expand uniformly.

We now introduce a set of *fundamental observers*, defined as observers who move in such a way that the Universe appears to be isotropic to them.

Each of them has a clock and proper time measured by that clock is called *cosmic time.* There are no problems of synchronisation of the clocks carried by fundamental observers because, for example, they can be instructed to set their clocks to the same time when the Universe has a certain density.

We can now write down the metric for such Universes from the considerations of Section 2.3. For reasons which will become apparent in a moment, I will write the metric in the form

$$ds^2 = dt^2 - \frac{1}{c^2}[d\rho^2 + R_c^2 \sin^2(\rho/R_c)(d\theta^2 + \sin^2\theta d\phi^2)] \tag{2.16}$$

Notice that, in this form, t is cosmic time and $d\rho$ is an increment of proper distance in the radial direction.

Now there is a problem in putting this simple metric into a form which is useful for comparing observables at different distances, and consequently different epochs, in the Universe. This is illustrated by the simple space-time diagram shown in Fig 2.10. Since light travels at a finite velocity, we observe all astronomical objects along our *past light cone* which is centred on the Earth at the present epoch t_0. Therefore, when we observe distant objects, we do not observe them at the present epoch but rather at an earlier epoch when the Universe was still homogeneous and isotropic but the distances between fundamental observers were smaller and the spatial curvature different. The

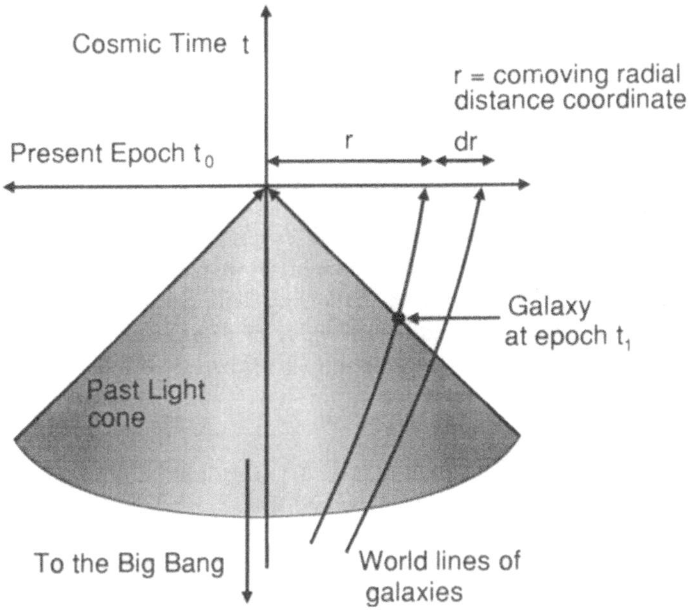

Fig. 2.10 A simple space-time diagram illustrating the definition of the radial co-moving coordinate distance

problem is that we can only apply the metric (2.16) to an isotropic curved space defined at a single epoch.

To resolve this problem, we perform the following thought experiment. To measure a proper distance which can be included in the metric (2.16), we line up a set of fundamental observers between the Earth and the galaxy whose distance we wish to measure. The observers are all instructed to measure the distance $d\rho$ to the next fundamental observer at a particular cosmic time t which they read on their own clock. By adding together all the $d\rho$s, we can find a proper distance ρ which is measured at a single epoch and which can be used in the metric (2.16). Notice that ρ is a *fictitious distance* in that we cannot actually measure distances in this way. We can only observe distant galaxies as they were at some epoch earlier than the present and we do *not* know how to project their positions relative to us forward to the present epoch without a knowledge of the kinematics of the expanding Universe. In other words, in practice, the distance measure ρ depends upon the choice of cosmological model. We will work out in a moment how to relate ρ to measurable quantities.

Let us now look at what happens to the ρ coordinates of galaxies in a uniformly expanding Universe. The definition of a uniform expansion is that between two cosmic epochs, t_1 and t_2, the distances of any two fundamental observers, i and j, change such that

$$\frac{\rho_i(t_1)}{\rho_j(t_1)} = \frac{\rho_i(t_2)}{\rho_j(t_2)} = \text{constant} \tag{2.17}$$

that is,

$$\frac{\rho_i(t_1)}{\rho_i(t_2)} = \frac{\rho_j(t_1)}{\rho_j(t_2)} = \ldots = \text{constant} = \frac{R(t_1)}{R(t_2)} \tag{2.18}$$

$R(t)$ is a universal function known as the *scale factor* which describes how the relative distance between any two fundamental observers changes with cosmic time t. Let us therefore adopt the following definitions. We set $R(t)$ equal to 1 at the present epoch t_0 and let the value of ρ at the present epoch be called r, that is, we can rewrite the relation (2.18) as

$$\rho(t) = R(t)r \tag{2.19}$$

r thus becomes a *distance label* which is attached to a galaxy or fundamental observer for all time and the variation in proper distance in the expanding Universe is taken care of by the scale factor $R(t)$; r is called the *comoving radial distance coordinate.*

Now let us apply the same condition (2.18) to proper distances which expand perpendicular to the line of sight between the epochs t and t_0.

$$\frac{\Delta l(t)}{\Delta l(t_0)} = R(t)$$

and hence, from the metric (2.16),

$$R(t) = \frac{R_c(t) \sin\left[\rho/R_c(t)\right] \, d\theta}{R_c(t_0) \sin[r/R_c(t_0)] d\theta}.$$

Reorganising this equation and using (2.19), we see that

$$\frac{R_c(t)}{R(t)} \sin[R(t)r/R_c(t)] = R_c(t_0) \sin[r/R_c(t_0)]$$

This is only true if

$$R_c(t) = R_c(t_0)R(t) \tag{2.20}$$

that is, the radius of curvature of the spatial sections is just proportional to $R(t)$. Thus, in order to preserve isotropy and homogeneity, the curvature of space changes as the Universe expands as $\kappa = R_c^{-2} \propto R^{-2}$.

Let us call the value of $R_c(t_0)$, that is, the radius of curvature at the present epoch, $\Re$. Then

$$R_c(t) = \Re R(t) \tag{2.21}$$

Substituting relations (2.19) and (2.21) into the metric (2.16), we obtain

$$ds^2 = dt^2 - \frac{R^2(t)}{c^2}[dr^2 + \Re^2 \sin^2(r/\Re)(d\theta^2 + \sin^2\theta d\phi^2)] \tag{2.22}$$

This is the *Robertson-Walker metric* in the form we will use in all our future analyses. Notice that it contains one unknown function $R(t)$, the scale factor, which describes the dynamics of the Universe and an unknown constant $\Re$ which describes the spatial curvature of the Universe at the present epoch.

It is possible to rewrite the metric in different ways. For example, if we use a *comoving angular diameter distance* $r_1 = \Re \sin(r/\Re)$, the metric becomes

$$ds^2 = dt^2 - \frac{R^2(t)}{c^2}\left[\frac{dr_1^2}{1-\kappa r_1^2} + r_1^2(d\theta^2 + \sin^2\theta\, d\phi^2)\right] \tag{2.23}$$

where $\kappa = 1/\Re^2$. Evidently, by a suitable rescaling of the r_1 coordinate $\kappa r_1^2 = r_2^2$, the metric could equally well be written

$$ds^2 = dt^2 - \frac{R_1^2(t)}{c^2}\left[\frac{dr_2^2}{1-kr_2^2} + r_2^2(d\theta^2 + \sin^2\theta\, d\phi^2)\right] \tag{2.24}$$

with $k = +1, 0$ and -1 for universes with spherical, flat and hyperbolic geometries respectively. Notice that, in this rescaling, the value of $R_1(t_0)$ at the present epoch is $\Re$ and not unity. This is a rather popular form for the metric and is the version used by Dr. Sandage in his lectures but I will use (2.22) because the r coordinate has an important (and clear) physical meaning.

The importance of the metrics (2.22), (2.23) and (2.24) is that they enable us to define the invariant interval ds^2 between events at any epoch or location in the expanding Universe. Let us recall the meanings of the various elements in the metric (2.22).

- t is cosmic time;
- $R(t)dr$ is the element of proper (or geodesic) distance in the radial direction;
- $R(t)[\Re \sin(r/\Re)]d\theta = R(t)r_1 d\theta$ is the element of proper distance perpendicular to the radial direction subtended by the angle $d\theta$ at the origin;
- Similarly, $R(t)[\Re \sin(r/\Re)] \sin\theta d\phi = R(t)r_1 \sin\theta d\phi$ is the element of proper distance in the ϕ-direction.

Notice that we have so far specified nothing about the physics of the expanding Universe. All of this has been absorbed into the function $R(t)$.

2.5 Observations in Cosmology

It is useful to produce a catalogue of results which are independent of the particular form of $R(t)$. First of all, let us elucidate the real meaning of redshift in cosmology.

2.5.1 Redshift

By redshift, we mean the shift of spectral lines to longer wavelength. If λ_e is the wavelength of the line as emitted and λ_0 the observed wavelength, the redshift z is defined to be

$$z = \frac{\lambda_0 - \lambda_e}{\lambda_e} \tag{2.25}$$

According to special relativity, the radial velocity inferred from the redshift is given by the standard relation

$$1 + z = \left(\frac{1 + v/c}{1 - v/c}\right)^{1/2} \tag{2.26}$$

In the limit of small redshifts, $v/c \ll 1$, the relation (2.26) reduces to

$$v = cz \tag{2.27}$$

This is the type of velocity which Hubble used in deriving the velocity-distance relation $v = H_0 r$. I will argue that this was an unfortunate step.

Let us now consider a wave packet of frequency ν_1 emitted between cosmic times t_1 and $t_1 + \Delta t_1$ from a distant galaxy. This wave packet is received by the observer at the present epoch in the cosmic time interval t_0 to $t_0 + \Delta t_0$. The signal propagates along null-cones, that is, $\mathrm{d}s^2 = 0$ and so, considering radial propagation from source to observer, $\mathrm{d}\theta = 0$ and $\mathrm{d}\phi = 0$, the metric (2.22) gives us

$$\mathrm{d}t = -\frac{R(t)}{c}\mathrm{d}r \tag{2.28}$$

that is,

$$\frac{c\,\mathrm{d}t}{R(t)} = -\mathrm{d}r \tag{2.29}$$

The minus sign appears because the origin of the r coordinate is the observer. Therefore, considering first the leading edge of the wave packet, the integral of (2.29) is

$$\int_{t_1}^{t_0} \frac{c\,\mathrm{d}t}{R(t)} = -\int_r^0 \mathrm{d}r \tag{2.30}$$

The end of the wave packet must travel the same distance in units of comoving distance coordinate since the r coordinate is fixed to the galaxy for all time. Therefore,

$$\int_{t_1+\Delta t_1}^{t_0+\Delta t_0} \frac{c\mathrm{d}t}{R(t)} = -\int_r^0 \mathrm{d}r$$

that is,

$$\int_{t_1}^{t_0} \frac{cdt}{R(t)} + \frac{c\Delta t_0}{R(t_0)} - \frac{c\Delta t_1}{R(t_1)} = \int_{t_1}^{t_0} \frac{cdt}{R(t)}$$

Since $R(t_0) = 1$, we find that

$$\Delta t_0 = \frac{\Delta t_1}{R(t_1)} \tag{2.31}$$

This is the cosmological expression for the phenomenon of *time dilation*. When we observe distant galaxies, they have $R(t_1) < 1$ and so phenomena are observed to take longer in our frame of reference than in that of the source. This is exactly the phenomenon which is observed for relativistic muons propagating through the atmosphere.

The result (2.31) provides us with an expression for *redshift*. If we take $\Delta t_1 = \nu_1^{-1}$ to be the period of the emitted waves and $\Delta t_0 = \nu_0^{-1}$ to be the observed period, we find that

$$\nu_0 = \nu_1 R(t_1) \tag{2.32}$$

Rewriting this result in terms of redshift z we find that

$$z = \frac{\lambda_0 - \lambda_e}{\lambda_e} = \frac{\lambda_0}{\lambda_e} - 1 = \frac{\nu_1}{\nu_0} - 1$$

that is,

$$1 + z = \frac{1}{R(t_1)} \tag{2.33}$$

This is one of the most important relations in cosmology and displays the real meaning of redshift. *Redshift is simply a measure of the scale factor of the Universe when the source emitted its radiation.* Thus, when we observe a galaxy with redshift $z = 1$, the scale factor of the Universe when the light was emitted was $R(t) = 0.5$, that is, the Universe was half its present size. Note, however, that we obtain no information about *when* the light was emitted. If we did, we could determine directly from observation the function $R(t)$. We do not understand the physics of galaxies and quasars well enough to be able to estimate times or ages from observation. In this circumstance, we have to have some theory of the dynamics of the Universe in order to determine $R(t)$.

Thus, redshift does not really have anything to do with velocities at all in cosmology. The redshift is a beautiful dimensionless number which, as $(1 + z)^{-1}$, tells us the relative distance between galaxies when the light was emitted compared with that distance now. It is a great pity that Hubble multiplied z by c. I hope we will eventually get rid of the c.

One very useful result of this calculation is that it enables us to derive an expression for the comoving radial distance coordinate r. Equation (2.30) can be rewritten

$$r = \int_{t_1}^{t_0} \frac{c\mathrm{d}t}{R(t)} \tag{2.34}$$

Thus, once we know $R(t)$ we can immediately find r by integration. This integral emphasises the point that r is an artificial distance which depends upon how the Universe expanded between the emission and reception of the radiation.

2.5.2 Hubble's Law

Using our original prescription of proper distances, we can write Hubble's Law, $v = H_0\rho$ as

$$\frac{\mathrm{d}\rho}{\mathrm{d}t} = H_0\rho \tag{2.35}$$

Now, substituting $\rho = R(t)r$, we find that

$$r\frac{\mathrm{d}R(t)}{\mathrm{d}t} = H_0 R(t) r$$

that is,

$$H_0 = \dot{R}/R \tag{2.36}$$

Since we measure Hubble's constant at the present epoch, $t = t_0$, $R = 1$, we find that

$$H_0 = (\dot{R})_{t_0} \tag{2.37}$$

Thus, Hubble's constant H_0 defines the present expansion rate of the Universe. Notice, however, that we can define a value of Hubble's constant at any epoch through the more general relation

$$H(t) = \dot{R}/R \tag{2.38}$$

2.5.3 Angular Diameters

The great simplification which results from the use of a metric of the form (2.22) is apparent in working out the angular size of an object of proper length d perpendicular to the radial coordinate at redshift z. The relevent spatial component of the metric (2.22) is the term in $\mathrm{d}\theta$ and hence the proper length must be

$$d = R(t)D\,\Delta\theta = \frac{D\Delta\theta}{(1+z)} \tag{2.39}$$

$$\Delta\theta = \frac{d(1+z)}{D} \tag{2.40}$$

where we have introduced an *effective distance* $D = \Re \sin(r/\Re)$. Clearly, for small redshifts $z \ll 1$ and $r \ll \Re$, expression (2.40) reduces to the Euclidean relation $d = r\Delta\theta$.

The expression (2.40) can also be written in the form

$$\Delta\theta = \frac{d}{D_{\mathrm{A}}} \tag{2.41}$$

so that the relation between d and $\Delta\theta$ looks like the standard Euclidean relation. To achieve this, we have to introduce the distance measure $D_{\mathrm{A}} = D/(1+z)$ which is known as an *angular diameter distance* and which is often used in the literature.

Another useful calculation is the angular diameter of an object which continues to partake in the expansion of the Universe. This is the case for infinitesimal perturbations in the expanding Universe. A good example is the angular diameter which large scale structures present in the Universe today would have subtended at the epoch of recombination if they had simply expanded with the Universe. This calculation is used to work out physical sizes corresponding to the angular scale of the fluctuations observed in the Cosmic Microwave Background Radiation on an angular scale of 10°. If the physical size of the object is $d(t_0)$ now, its physical size at redshift z was $d(t_0)/(1+z)$. Therefore, the object subtended an angle

$$\Delta\theta = \frac{d(t_0)}{D}$$

Notice that in this case the $(1+z)$ factor has disappeared from the expression (2.40).

2.5.4 Apparent Intensities

Suppose a source at redshift z has luminosity $L(\nu_1)$ (measured in W Hz^{-1}), that is, the total energy emitted over 4π steradians per unit time per unit frequency interval. Let us suppose that $N(\nu_1)$ photons of energy $h\nu_1$ are emitted by the source in the bandwidth $\Delta\nu_1$ in the proper time interval Δt_1 and that it has redshift z. Then the luminosity of the source is

$$L(\nu_1) = \frac{N(\nu_1)h\nu_1}{\Delta\nu_1\Delta t_1} \tag{2.42}$$

These photons are distributed over a 'sphere' centred on the source at epoch t_1 and when the 'shell' of photons arrives at the observer at the epoch t_0, a certain fraction of them is intercepted by the telescope. We need to know how the photons spread out over a sphere between the epochs t_1 and t_0. The photons are observed at t_0 with frequency $\nu_0 = R(t_1)\nu_1$, in a proper time interval $\Delta t_0 = \Delta t_1/R(t_1)$ and in the waveband $\Delta\nu_0 = R(t_1)\Delta\nu_1$. The only complication is that we must relate the diameter of our telescope Δl to the angular diameter $\Delta\theta$ which it subtends at the source at epoch t_1. Again, the metric (2.22) provides an elegant answer. The proper distance Δl refers to the present epoch at which $R(t) = 1$ and hence

$$\Delta l = D\Delta\theta \tag{2.43}$$

where $\Delta\theta$ is the angle measured by the appropriate fundamental observer located at the source. Notice the difference between relations (2.40) and (2.43). They correspond to angular diameters measured in opposite directions along the light cone. In fact, the factor of $(1+z)$ difference between them is part of a more general relation concerning angular diameter measures along light cones which is known as the *reciprocity theorem.*

Therefore, the surface area of the telescope is $\pi\Delta l^2/4$ and the solid angle subtended by this area at the source is $\Delta\Omega = \pi\Delta\theta^2/4$. The number of photons incident upon the telescope in time Δt_0 is

$$N(\nu_1)\Delta\Omega/4\pi \tag{2.44}$$

but they are now observed with frequency ν_0. Therefore, the flux density of the source, that is, the energy received per unit time, per unit area and per unit bandwidth (W m^{-2} Hz^{-1}) is

$$S(\nu_0) = \frac{N(\nu_1)h\nu_0\Delta\Omega}{4\pi\Delta t_0\Delta\nu_0(\pi/4)\Delta l^2} \tag{2.45}$$

We can now relate the quantities in the expression (2.45) to properties of the source, using the above relations (2.32) and (2.33).

$$S(\nu_0) = \frac{L(\nu_1)R(t_1)}{4\pi D^2} = \frac{L(\nu_1)}{4\pi D^2(1+z)} \tag{2.46}$$

If the spectra of sources are of power law form $L(\nu) \propto \nu^{-\alpha}$, this relation becomes

$$S(\nu_0) = \frac{L(\nu_0)}{4\pi D^2(1+z)^{1+\alpha}} \tag{2.47}$$

We can repeat the analysis for *bolometric* luminosities and flux densities. In this case, we consider the total energy emitted in a finite bandwidth $\Delta\nu_1$ which is received in the bandwidth $\Delta\nu_0$, that is

$$L_{\rm bol} = L(\nu_1)\Delta\nu_1 = 4\pi D^2 S(\nu_0)(1+z) \times \Delta\nu_0(1+z) = 4\pi D^2(1+z)^2 S_{\rm bol}$$

where the bolometric flux density is $S_{\rm bol} = S(\nu_0)\Delta\nu_0$. Therefore,

$$S_{\rm bol} = \frac{L_{\rm bol}}{4\pi D^2(1+z)^2} = \frac{L_{\rm bol}}{4\pi D_{\rm L}^2} \tag{2.48}$$

The bolometric luminosity can be integrated over any suitable bandwidth so long as the corresponding redshifted bandwidth is used to measure the bolometric flux density at the present epoch.

$$\sum_{\nu_0} S(\nu_0)\Delta\nu_0 = \frac{\sum_{\nu_1} L(\nu_1)\Delta\nu_1}{4\pi D^2(1+z)^2} = \frac{\sum_{\nu_1} L(\nu_1)\Delta\nu_1}{4\pi D_{\mathrm{L}}^2} \tag{2.49}$$

The quantity $D_{\mathrm{L}} = D(1+z)$ is often called a *luminosity distance* since this definition makes the relation between S_{bol} and L_{bol} look like an inverse square law. A key point about relations such as (2.46) and (2.49) is the presence of time dilation factors $(1+z)$ which reduce the flux density by more than the 'inverse square law'. However, this is a somewhat misleading way of expressing the result because D itself does not behave like a Euclidean distance with increasing redshift.

The formula (2.46) is the best expression for relating the observed intensity $S(\nu_0)$ to the intrinsic luminosity of the source $L(\nu_1)$. We can also write the expression (2.46) in terms of the luminosity of the source at the observing frequency ν_0 as

$$S(\nu_0) = \frac{L(\nu_0)}{4\pi D_{\mathrm{L}}}\left[\frac{L(\nu_1)}{L(\nu_0)}(1+z)\right] \tag{2.50}$$

The last term in square brackets is known as the *K-correction*. The history of the origin of this term has been dealt with by Dr. Sandage. The K-correction was introduced by the pioneer optical cosmologists in the 1930s in order to 'correct' the apparent magnitude of distant galaxies for the effects of redshifting the spectrum when observations are made at a fixed observing frequency ν_0. Taking logarithms and multiplying by -2.5, we can convert the terms in square brackets into a correction to the apparent magnitude of the galaxy and then we find

$$K(z) = -2.5\log_{10}\left[\frac{L(\nu_1)}{L(\nu_0)}(1+z)\right] \tag{2.51}$$

Notice that, in the form (2.51), the K-correction is correct for *monochromatic* flux densities and luminosities. In the case of observations in the optical waveband, in which magnitudes are measured through standard filters, averages have to be taken over the spectral energy distributions of the objects within the spectral windows in the emitted and observed wavebands. This is a straightforward calculation once the spectrum of the object is known.

Personally, I prefer not to work with K-corrections but rather to work directly with the expression (2.46) and take appropriate averages. K-corrections are, however, rather firmly embedded in the literature and it is often convenient to use the term to describe the effects of shifting the emitted spectrum through the observing wavelength window.

2.5.5 Number Densities

We often need to know the number of sources in a particular redshift range, z to $z + \mathrm{d}z$. Since there is a one-to-one relation between r and z, which is still undefined at this stage in our development, the problem is very simple because, by definition, r is a radial proper distance defined at the present epoch and hence the number of objects in the interval of radial comoving coordinate distance r to $r + \mathrm{d}r$ is given by results already obtained in Section 2.3. The simple space-time diagram shown in Fig. 2.10 illustrates how we can evaluate the numbers of objects in the comoving distance interval $\mathrm{d}r$ entirely by working out volumes at the present epoch. The volume of a spherical shell of thickness $\mathrm{d}r$ at comoving distance coordinate r is

$$\mathrm{d}V = 4\pi\Re^2 \sin^2(r/\Re)dr = 4\pi D^2 \mathrm{d}r \tag{2.52}$$

Therefore, if N_0 is the present space density of objects and their number is conserved in the expanding Universe,

$$\mathrm{d}N = 4\pi N_0 D^2 \mathrm{d}r \tag{2.53}$$

The definition of comoving coordinates automatically takes care of the expansion of the Universe.

2.5.6 The Age of the Universe

Finally, let us work out an expression for the age of the Universe, T_0. We can do this from a rearranged version of equation (2.28). The basic differential relation is

$$-\frac{c\mathrm{d}t}{R(t)} = \mathrm{d}r$$

and hence

$$T_0 = \int_0^{t_0} \mathrm{d}t = \int_0^{r_{\max}} \frac{R(t)\mathrm{d}r}{c} \tag{2.54}$$

where $r_{\max}$ is the comoving distance coordinate corresponding to $R = 0, z = \infty$.

2.6 Conclusion

We have made real progress and the results we have derived can be used to work out the relations between intrinsic properties and observables for any isotropic, homogeneous world model. Let me summarise the procedure by the following cook-book recipe.

1. First work out from theory the function $R(t)$ and the radius of curvature of the space at the present epoch $\Re$.

2. Now work out the expression for the *comoving coordinate distance* using the integral

$$r = \int_{t_1}^{t_0} \frac{c\mathrm{d}t}{R(t)}$$

This integration yields an expression for r as a function of redshift z.

3. Next, work out the *effective distance* D which is given by

$$D = \Re \sin \frac{r}{\Re}$$

This relation determines D as a function of redshift z.

4. If so desired, the *luminosity distance* $D_{\mathrm{L}} = D(1+z)$ and *angular diameter distance* $D_{\mathrm{A}} = D/(1+z)$ can be introduced.

5. The number of objects $\mathrm{d}N$ in the redshift interval $\mathrm{d}z$ and solid angle Ω can be found from the expression

$$\mathrm{d}N = \Omega N_0 D^2 \, \mathrm{d}r$$

where N_0 is the number density of objects which is assumed to be conserved as the Universe expands.

We will develop some explicit solutions for these functions in the next chapter.

References – Chapter 2

Berry, M.V. (1989). *Principles of Cosmology and Gravitation.* Bristol: Adam Hilger.

Dunlop, J.S. and Peacock, J.A. (1990). MNRAS, **247**, 19.

Gott, J.R. III, Melott, A.L. and Dickenson, M. (1986). ApJ, 306, 341.

Gregory, P.C. and Condon, J.J. (1991). ApJS, **75**, 1011.

Groth, E.J. and Peebles, P.J.E. (1977). ApJ, **217**, 385.

Groth, E.J. and Peebles, P.J.E. (1986). ApJ, **310**, 507.

Jones, L.R., Fong, R., Shanks, T., Ellis, R.S. and Peterson, B.A. (1991). MNRAS, **249**, 481.

Longair, M.S. (1992). *Theoretical Concepts in Physics.* Cambridge: Cambridge University Press.

Maddox, S.J., Efstathiou, G., Sutherland, W.J. and Loveday, J. (1990). MNRAS, **242**, 43P.

Melott, A.L., Weinberg, D.H. and Gott, J.R. III (1988). ApJ, **306**, 341.

Metcalfe, N., Shanks, T., Fong, D. and Jones, L.R. (1991). MNRAS, **249**, 498.

Peebles, P.J.E. (1993). *Principles of Physical Cosmology.* Princeton: Princeton University Press.

Seldner, M., Siebars, B., Groth, E.J. and Peebles, P.J.E. (1977). AJ., **82**, 249.

Tyson, A. (1990). In *The Galactic and Extragalactic Background Radiation,* (eds. S. Bowyer and C. Leinert), 245. Dordrecht: Kluwer Aademic Publishers.

Weinberg, S. (1972). *Gravitation and Cosmology.* New York: John Wiley and Co.

3 World Models

We will consider the dynamics of world models in three parts. First of all we consider the standard dust models, then radiation-dominated models and finally a brief look at inflationary models. The standard dust models begin with dynamical equations derived from Einstein's General Theory of Relativity. This is by far the best relativistic theory of gravity we possess. Let us look briefly at the observational evidence which supports this assertion.

3.1 Experimental and Observational Tests of General Relativity

Gravity is the only large scale force we know of which acts upon all forms of matter and radiation. When Friedman first solved the field equations of General Relativity for isotropically expanding Universes, the evidence for the theory was good but not perhaps overwhelming. The most remarkable result was the prediction of the exact perihelion shift of Mercury which had remained an unsolved problem in the celestial mechanics of the the Solar System since the time of its discovery by Le Verrier in 1859.

Most of the tests of General Relativity involve the observation of astronomical objects and there has been excellent progress in testing the predictions of the theory, for example, by measuring the deflection of electromagnetic signals from distant astronomical objects as they are occulted by the Sun and the 'fourth' test of General Relativity discovered in 1964 by Shapiro of the time delay as the electromagnetic radiation from a distant object passes through the gravitational potential of a massive body such as the Sun.

In my view, the most spectacular results have come from radio observations of pulsars. These radio sources are rotating, magnetised neutron stars and they emit beams of radio emission from their magnetic poles as shown schematically in Fig. 3.1. The typical parameters of a neutron star are given in the figure. Observations by Taylor and his colleagues using the Arecibo radio telescope have demonstrated that these are the most stable clocks we know of in the Universe. They have shown up variations in the time-keeping of the most accurate laboratory clocks, as has been demonstrated by comparing the times measured by two pulsars against a standard clock.

The pulsars have enabled a wide variety of very sensitive tests to be carried out of General Relativity and the possible existence of a background flux of gravitational radiation but the most intriguing systems are those pulsars which are members of binary systems. More than 20 of these are now known, the most important being those in which the other member of the binary system is also a neutron star and in which the neutron stars form a close binary system. The first of these to be discovered was the binary pulsar PSR 1913 + 16 which is illustrated schematically in Fig. 3.2. The system has a binary period of only 7.75 hours and the orbital eccentricity is large, $e = 0.617$. This system is a pure gift for the relativist. To test General Relativity,

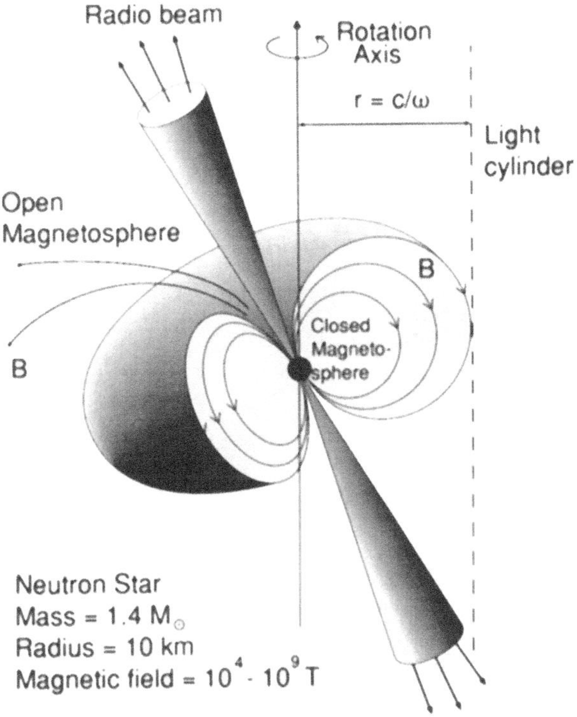

Fig. 3.1 A schematic diagram of a pulsar showing the displacement between the axis of the magnetic dipole and the rotation axis of the neutron star. The radio pulses are assumed to be due to beams of radio emission from the poles of the magnetic field distribution and are associated with the passage of the beam across the line of sight to the observer. Typical parameters of the neutron stars are indicated on the diagram.

we require a perfect clock in a rotating frame of reference and systems such as PSR 1913+16 are ideal for this purpose. The neutron stars are so inert and compact that the binary system is very 'clean' and so can be used for some of the most sensitive tests of General Relativity yet devised. To give just a few examples of the precision which can be obtained, I reproduce with the kind permission of Professor J. Taylor some of the recent tests which have been made of General Relativity.

In Fig. 3.3, the determination of the masses of the two neutron stars in the binary system PSR 1913+16 is shown assuming that General Relativity is the correct theory of gravity. Various parameters of the binary orbit can be measured very precisely and these provide different estimates of functions involving the masses of the two neutron stars. In Fig. 3.3, the various parameters of the binary orbit are shown, those which have been measured

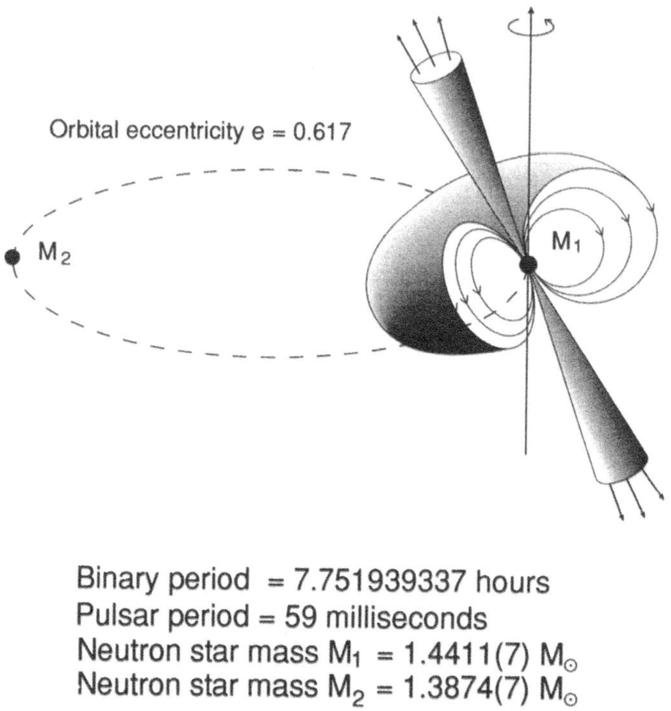

Fig. 3.2 A schematic diagram showing the binary pulsar PSR 1913+16. As a result of the ability to measure precisely many parameters of the binary orbit from ultra-precise pulsar timing, the masses of the two neutron stars have been measured with very high precision.(Data courtesy of Professor J. Taylor.)

with very good accuracy being indicated by an asterisk. It can be observed that the different loci intersect very precisely at a single point in the m_1/m_2 plane. Some measure of the precision with which the theory is known to be correct can be obtained from the accuracy with which the masses of the neutron stars are known as indicated in Fig. 3.2. These are the most accurately known masses for any extra solar-system object.

A second remarkable measurement has been the rate of loss of orbital rotational energy by the emission of gravitational waves. The binary system loses energy by the emission of gravitational radiation and the rate at which energy is lost can be precisely predicted once the masses of the neutron stars and the parameters of the binary orbit are known. The rate of change of the angular frequency Ω of the orbit due to gravitational radiation energy loss is precisely known, $-\mathrm{d}\Omega/\mathrm{d}t \propto \Omega^5$. The change in orbital phase due to the emission of gravitational waves has been observed over a period of 17 years and the observed changes over that period agree precisely with the

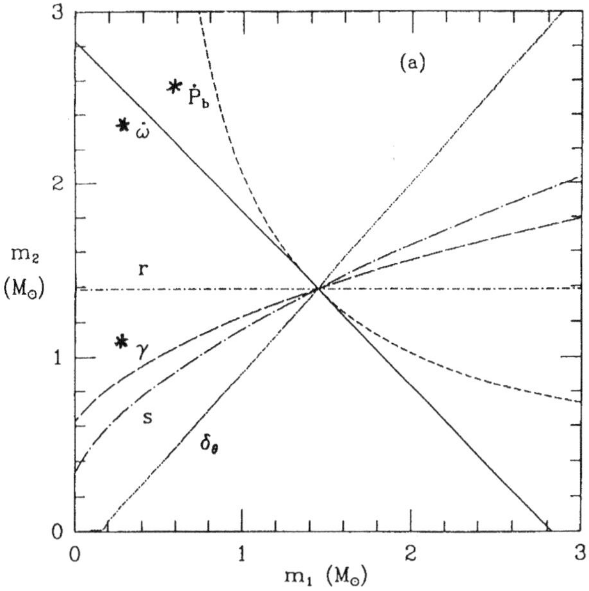

* = accurately measured

Fig. 3.3 The measurement of the masses of the neutron stars in the binary system PSR 1913+16 resulting from very precise timing of the arrival times of the pulses at the Earth. The different parameters of the neutron star's orbit depend upon different combinations of the masses m_1 and m_2 of the neutron stars. It can be seen that the lines intersect very precisely at a single point in the $m_1 - m_2$ plane (Courtesy of Professor J. Taylor).

predictions of General Relativity (Fig. 3.4). Thus, although the gravitational waves themselves have not been detected, exactly the correct energy loss rate from the system has been measured — it is generally assumed that this is convincing evidence for the existence of gravitational waves and this observation acts as a spur to their direct detection by future generations of gravitational wave detectors. This is a very important result for the theory of gravitation since this result alone enables a wide range of alternative theories of gravity to be eliminated. For example, since General Relativity predicts only quadrupolar emission of gravitational radiation, any theory which, say, involve the dipole emission of gravitational waves can be eliminated.

Thus, General Relativity has passed every test which has been made of the theory and we can have much greater confidence than in the past that it is an excellent description of relativistic gravity. The same techniques of accurate pulsar timing can also be used to determine whether or not there is any evidence for the gravitational constant G changing with time. These tests are slightly dependent upon the equation of state used to describe the interior of the neutron stars but for the complete range of possible equations

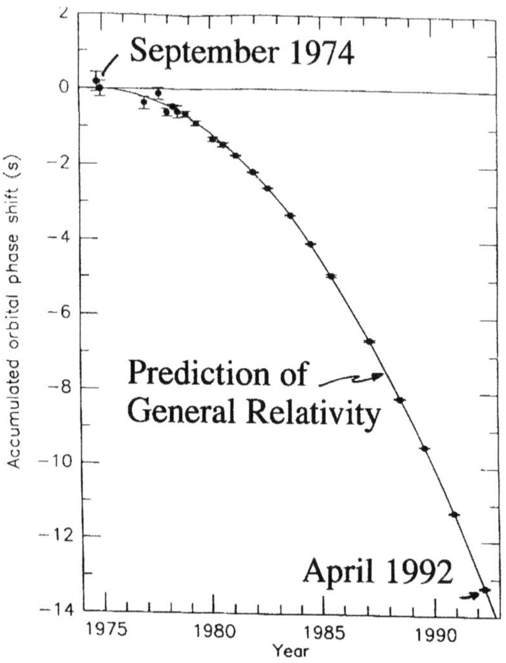

Fig. 3.4 The change of orbital phase as a function of time for the binary neutron star system PSR 1913+16 compared with the expected changes due to gravitational radiation energy loss by the binary system (Courtesy of Professor J. Taylor).

of state, the limits of $\dot{G}/G$ are less than about 10^{-11} year^{-1}. Thus, there can have been little change in the value of the gravitational constant over typical cosmological timescales which are about $(1-2)\times 10^{10}$ years. Continued observations of certain of the binary pulsars should enable this limit to be improved by an order of magnitude. For cosmological studies, there is therefore no reason to use any theory other than General Relativity for describing the large-scale dynamics of the Universe.

3.2 The Einstein Field Equations

According to the general theory of relativity, the Einstein field equations for an isotropically expanding Universe under the influence of gravity can be written as follows:

$$\ddot{R} = -\frac{4\pi GR}{3}\left(\rho + \frac{3p}{c^2}\right) + \left[\frac{1}{3}\Lambda R\right] \tag{3.1}$$

$$\dot{R}^2 = \frac{8\pi G\rho}{3}R^2 - \frac{c^2}{\Re^2} + \left[\frac{1}{3}\Lambda R^2\right] \tag{3.2}$$

In these equations, R is the scale factor, ρ is the inertial mass density of the matter and radiation content of the Universe and p is its pressure. Notice that the pressure term in equation (3.1) is a relativistic correction to the inertial mass density so that the quantity in large round brackets represents the total inertial mass density. Unlike normal pressure forces which depend upon the gradient of the pressure and, for example, hold up stars, this pressure term depends linearly on the pressure and, since it contributes to the inertial mass, increases the gravitational force. $\Re$ is the radius of curvature of the geometry of the world model at the present epoch and so the term $-c^2/\Re^2$ in equation (3.2) is simply a constant of integration.

I have included the famous (or infamous) *cosmological constant* Λ in equations (3.1) and (3.2) in large square brackets. This term has had a chequered history in that it was originally introduced by Einstein in order to produce static solutions of the field equations more than 10 years before it was discovered that the Universe is in fact non-static in the sense that it is expanding uniformly. As we will discuss in Section 3.6, it has had a new lease of life with the development of inflationary models for the early history of the Universe.

3.3 The Standard Dust Models – The Friedman World Models

This analysis is performed in all the standard text-books. By *dust*, we mean a pressureless fluid, $p = 0$. In addition, we set the cosmological constant $\Lambda = 0$. It is convenient to refer the density of matter to its value at the present epoch ρ_0. Because of conservation of mass, $\rho = \rho_0 R^{-3}$ and so the pair of equations reduces to the following simple form

$$\dot{R}^2 = \frac{8\pi G\rho_0}{3}R^{-1} - \frac{c^2}{\Re^2} \tag{3.3}$$

Milne and McCrea (1934) first showed that a relation of this form can be derived using non-relativistic Newtonian dynamics. We will perform this calculation because the ideas implicit in this argument can be used to understand some of the problems which arise in the theory of galaxy formation.

We consider a galaxy at distance x from the Earth and work out its deceleration due to the attraction of matter inside the sphere of radius x centred on the Earth (Fig. 3.5). By Gauss's theorem, because of the spherical symmetry of the distribution of matter within x, we can replace that mass $M = (4\pi/3)\rho x^3$ by a point mass at the centre of the sphere and so the deceleration of the galaxy is

$$m\ddot{x} = -\frac{GMm}{x^2} = -\frac{4\pi x\rho m}{3}$$

Notice that the mass of the galaxy m cancels out on either side of the equation, showing that the deceleration refers to the dynamics of the Universe as a whole rather than to any particular galaxy. Now, we make the same substitutions as before — replace x by the comoving value r using the scale

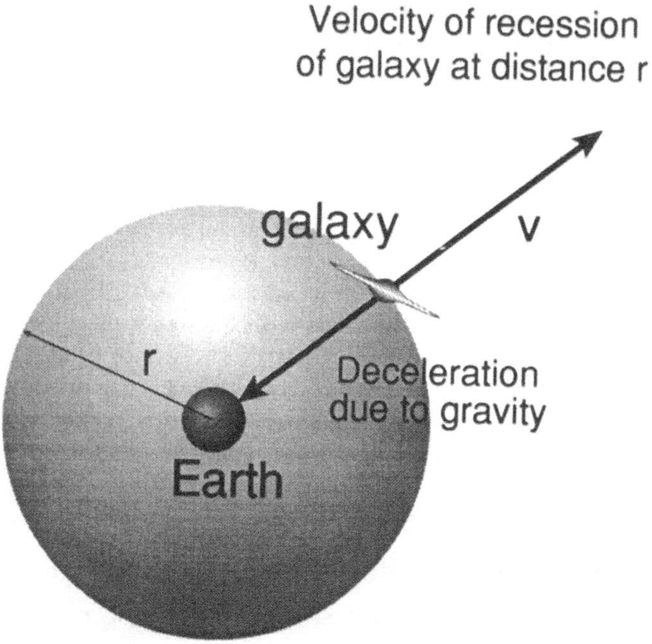

Fig. 3.5 Illustrating the dynamics of Newtonian world models.

factor R, $x = Rr$, and express the density in terms of its value at the present epoch, $\rho = \rho_0 R^{-3}$. Therefore,

$$\ddot{R} = -\frac{4\pi G\rho_0}{3}\frac{1}{R^2} \tag{3.4}$$

which is identical to equation (3.1) for dust models with $\Lambda = 0$. Integrating this equation, we find

$$\dot{R}^2 = \frac{8\pi G\rho_0}{3}R^{-1} + \text{constant} \tag{3.5}$$

This result is identical to equation (3.2) if we identify the constant with $-c^2/\Re^2$.

The above analysis brings out a number of important points about the world models of General Relativity. First of all, note that, because of the assumption of isotropy, local physics is also global physics. This is why the Newtonian argument works. The same physics which defines the local behaviour of matter also defines its behaviour on the largest scales. For example, the curvature of space within one cubic metre is exactly the same as that on the scale of the Universe itself.

A second point is to note that, although we might appear to have placed the Earth in a rather special position in Fig. 3.5, an observer located on any galaxy would perform exactly the same calculation to work out the deceleration of our Galaxy relative to that galaxy because of the cosmological principle which asserts that all fundamental observers should observe the same large scale features of the Universe at the same epoch. In other words, the Newtonian calculation applies for all observers who move in such a way that the Universe appears isotropic to them which is, by definition, for all fundamental observers.

Third, notice that at no point in the argument did we ask over what physical scale the calculation was to be valid. It is a remarkable fact that this calculation describes correctly the dynamics of the Universe on scales which are greater than the *horizon scale* which we take to be $r = ct$, that is, the maximum distance between points which can be causally connected at the epoch t. The reason for this is again the same as for the first two points – local physics is also global physics and so, if the Universe were set up in such a way that it had uniform density on scales far exceeding the horizon scale, the dynamics on these very large scales would be exactly the same as the local dynamics. We will find this idea very helpful in understanding the theory of the evolution of small perturbations in the expanding Universe.

The solutions of Einstein's field equations were discovered by A.A. Friedman in the period 1922-24 (see Tropp, Frenkel and Chernin 1993). He died of typhoid during the civil war in Leningrad in 1925 and did not live to see what have become the standard models of the Universe bear his name. Georges Lemaître brought Friedman's work to the wider notice of astronomers and cosmologists during the 1930s. The solutions of the field equations are appropriately referred to as the *Friedman models* of the Universe. To begin with, it is convenient to express the density of the world model in terms of a *critical density* ρ_c which is defined to be $\rho_c = (3H_0^2/8\pi G)$ and then to refer the actual density of the model ρ to this value through a *density parameter* $\Omega_0 = \rho/\rho_c$. Thus, the density parameter is given by

$$\Omega_0 = \frac{8\pi G\rho}{3H_0^2} \tag{3.6}$$

The subscript 0 has been attached to Ω because the critical density ρ_c changes with cosmic epoch as does Ω. The dynamical equation (3.3) therefore becomes

$$\dot{R}^2 = \frac{\Omega_0 H_0^2}{R} - \frac{c^2}{\Re^2} \tag{3.7}$$

Several important results can be deduced from this equation. If we set $t = t_0$, $R = 1$, that is, their values at the present epoch, we find that

$$\Re = \frac{c/H_0}{(\Omega_0 - 1)^{1/2}} \qquad \text{and} \qquad \kappa = \frac{(\Omega_0 - 1)}{(c/H_0)^2} \tag{3.8}$$

This last result shows that there is a one-to-one relation between the density of the Universe and its spatial curvature, one of the most beautiful results of the standard Friedman world models. The solutions of equation (3.7) are displayed in Fig. 3.6 which shows the well-known relation between the dynamics and geometry of the Friedman world models.

1. The models with $\Omega_0 > 1$ have closed, spherical geometry and they collapse to an infinite density in a finite time, an event sometimes referred to as the 'big crunch';
2. The models having $\Omega_0 < 1$ have open, hyperbolic geometries and expand forever. They would reach infinity with a finite velocity.
3. The model with $\Omega_0 = 1$ separates the open from the closed models and the collapsing models from those which expand forever. This model is often referred to as the *Einstein-de Sitter model* or the *critical model.* The velocity of expansion tends to zero as R tends to infinity. It has a particularly simple variation of $R(t)$ with cosmic epoch,

$$R = \left(\frac{3}{2}H_0 t\right)^{2/3} \qquad \kappa = 0 \tag{3.9}$$

Another important result is the function $R(t)$ for the empty world model, $\Omega_0 = 0$, $R(t) = H_0 t$, $\kappa = -(H_0/c)^2$. This model is sometimes referred to as the *Milne model.* It is an interesting exercise to show why it is that, in the completely empty world model, the global geometry of the Universe is hyperbolic. The reason is that in the empty model, the galaxies partaking in the universal expansion are undecelerated and any individual galaxy always has the same velocity relative to the same fundamental observer. Therefore, the cosmic times measured in different frames of reference are related by the standard Lorentz transform $t' = \gamma(t - vr/c^2)$ where $\gamma = (1 - v^2/c^2)^{-1/2}$. The key point is that the conditions of isotropy and homogeneity have to be applied at constant cosmic time t' in the frames of all the fundamental observers. The Lorentz transform shows that this cannot be achieved in flat space but is is uniquely satisfied in hyperbolic space with $\kappa = -(H_0/c)^2$. I have give a simple derivation of this result (Longair 1992).

The general solution of equation (3.7) is most conveniently written in parametric form

$$R = a(1 - \cos\theta) \qquad t = b(\theta - \sin\theta) \tag{3.10}$$

where

$$a = \frac{\Omega_0}{2(\Omega_0 - 1)} \quad \text{and} \quad b = \frac{\Omega_0}{2H_0(\Omega_0 - 1)^{3/2}}$$

We observe that, just like Hubble's constant H_0, which measures the local expansion rate of the distribution of galaxies, so we can define the local deceleration of the Universe at the present epoch, $\ddot{R}(t_0)$. It is conventional to

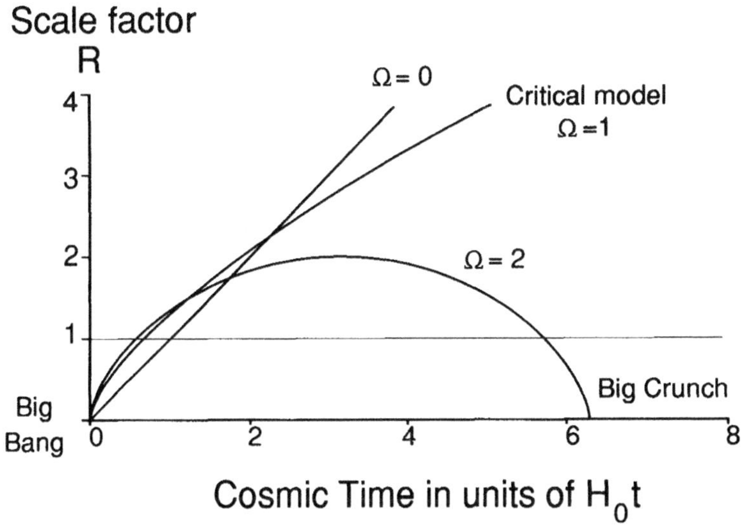

Fig. 3.6 The dynamics of the classical Friedman models parameterised by the density parameter $\Omega_0 = \rho/\rho_{crit}$. If $\Omega_0 > 1$, the Universe collapses to $R = 0$ as shown; if $\Omega_0 < 1$, the Universe expands to infinity and has a finite velocity of expansion as R tends to infinity. In the case $\Omega_0 = 1$, $R = (t/t_0)^{2/3}$ where $t_0 = (2/3)H_0^{-1}$. The time axis is given in terms of the dimensionless time $H_0 t$. At the present epoch $R = 1$ and in this presentation, the three curves have the same slope of 1 at $R = 1$, corresponding to a fixed value of Hubble's constant. If t_0 is the present age of the Universe corresponding to $R = 1$, then for $\Omega_0 = 0$ $H_0 t_0 = 1$, for $\Omega_0 = 1$ $H_0 t_0 = 2/3$ and for $\Omega_0 = 2$ $H_0 t_0 = 0.57$.

define the *deceleration parameter* q_0 to be the dimensionless deceleration at the present epoch through the expression

$$q_0 = -\left(\frac{\ddot{R}}{\dot{R}^2}\right)_{t_0} \tag{3.11}$$

Substituting into equation (3.1), we can show immediately that the deceleration parameter q_0 is directly proportional to the density parameter Ω_0,

$$q_0 = \Omega_0/2 \tag{3.12}$$

Note that this result is only true if the cosmological constant Λ is zero. In general, we find

$$q_0 = \frac{\Omega_0}{2} - \frac{1}{3}\frac{\Lambda}{H_0^2} \tag{3.13}$$

In the same way, if $\Lambda \neq 0$, the relation between the curvature and the density parameter becomes

$$\kappa = \frac{(\Omega_0 - 1)}{(c/H_0)^2} + \frac{1}{3}\frac{\Lambda}{c^2} \tag{3.14}$$

An important result for many aspects of cosmology is the relation between redshift z and cosmic time t. It is straightforward to show from equations (3.3) and (3.8) that, because $R = (1+z)^{-1}$

$$\frac{\mathrm{d}z}{\mathrm{d}t} = -H_0(1+z)^2(\Omega_0 z + 1)^{\frac{1}{2}} \tag{3.15}$$

Cosmic time t measured from the Big Bang follows immediately by integration

$$t = \int_0^t dt = -\frac{1}{H_0}\int_\infty^z \frac{dz}{(1+z)^2(\Omega_0 z + 1)^{1/2}} \tag{3.16}$$

It is a simple exercise to show that the present age of the Universe t_0 is H_0^{-1} if $\Omega_0 = 0$ and $(2/3)H_0^{-1}$ if $\Omega_0 = 1$.

Just as it is possible to define Hubble's constant at any epoch by $H = \dot{R}/R$, so we can define a density parameter Ω at any epoch through the definition $\Omega = 8\pi G\rho/3H^2$. Since $\rho = \rho_0(1+z)^3$, it follows that

$$\Omega H^2 = \frac{8\pi G}{3}\rho_0(1+z)^3$$

It is a useful exercise to show that this relation can be rewritten

$$\left(1 - \frac{1}{\Omega}\right) = (1+z)^{-1}\left(1 - \frac{1}{\Omega_0}\right) \tag{3.17}$$

This is an important result because it shows that, whatever the value of Ω_0 now, because $(1+z)^{-1}$ becomes very small at large redshifts, Ω tends very closely to the value 1 in the distant past. There are two ways of looking at this result. On the one hand, it is very convenient that the dynamics of all world models tend to those of the Einstein-de Sitter model in the early stages of the dust filled models. On the other hand, we observe that it is remarkable that the Universe is within roughly a factor of ten of the value $\Omega_0 = 1$ at the present day. If the value of Ω_0 were significantly different from 1 in the distant past, then it would be very widely different from 1 now as can be seen from equation (3.17). The fact that the curvature of space κ must be close to zero now results in what is often referred to as the *flatness problem*. The problem is that our Universe must have been very finely tuned indeed to the value $\Omega = 1$ in the distant past if it is to end up close to $\Omega_0 = 1$ now. Some argue that it is so remarkable that our Universe is within a factor of ten of $\Omega_0 = 1$ now, the only reasonable value the Universe can have is Ω_0 precisely equal to 1. Proponents of the inflationary picture of the early Universe have very good reasons why this should be the case.

We can now complete our programme of finding expressions for the comoving coordinate distance r and the 'effective distance' D. We recall that the increment of comoving coordinate distance is

$$dr = \frac{cdt}{R(t)} = -cdt(1+z) \tag{2.29}$$

Therefore,

$$dr = \frac{cdz}{H_0(1+z)(\Omega_0 z+1)^{1/2}} \tag{3.18}$$

By integrating from redshifts 0 to z, we can find the expression for r.

$$r = \frac{2c}{H_0(\Omega_0-1)^{1/2}}\left[\tan^{-1}\left(\frac{\Omega_0 z+1}{\Omega_0-1}\right)^{1/2} - \tan^{-1}(\Omega_0-1)^{-1/2}\right] \tag{3.19}$$

Finally, we need to find D by evaluating $D = \Re\sin(r/\Re)$ where $\Re$ is given by the expression (3.8). It is a straightforward calculation to show that

$$D = \frac{2c}{H_0\Omega_0^2(1+z)}\left\{\Omega_0 z + (\Omega_0-2)[(\Omega_0 z+1)^{1/2}-1]\right\} \tag{3.20}$$

This is the famous formula first derived by Mattig (1959), the history of which was described by Dr. Sandage. Although I have derived it using the formulae for spherical geometry, it has the great advantage of being correct for all values of Ω_0. It is an extremely useful formula and can be used directly in all the expressions derived in Chapter 2 for relating intrinsic properties to observables.

3.4 Testing the Friedman Models

Dr. Sandage has dealt in great detail with this topic and I only wish to add one footnote to that story. It will be noted that one of the key tests of the Friedman models is provided by the relation $\Omega_0 = 2q_0$. It is important to recognise that this equation states that the deceleration of the Universe at the present epoch can be entirely attributed to the decelerating influence of the inertial mass density of matter. Now, the deceleration of the Universe and its present mass density are separately measurable quantities and so a key test of the Freidmann models is to find out whether or not this equation is correct. The density parameter Ω_0 can be found from the virial theorem in its various disguises and so the question is how to measure the deceleration of the Universe from the classical cosmological tests. The answer is that, at small redshifts, the differences between the world models depend upon the deceleration parameter and not upon the density parameter. Let us demonstrate this by a simple argument which was given by Dr. James Gunn at the 8th Saas-Fee Course (Gunn 1978).

The key point is to work out the dependence of the 'effective distance' D upon q_0. Let us work this out entirely in terms of *kinematics* rather than through the dynamical equations proceeding from Einstein's field equations. We can write the variation of the scale factor R with cosmic epoch in terms of a Taylor series as follows

$$\begin{aligned} R &= R(t_0) + \dot{R}(t_0)\ dt + \frac{1}{2}\ddot{R}(t_0)(dt)^2 + \ldots \\ &= 1 - H_0\tau - \frac{1}{2}q_0 H_0^2\tau^2 + \ldots \end{aligned} \tag{3.21}$$

where we have introduced the look-back time τ which is given by $\tau = t_0 - t$ where t_0 is the present epoch and t is some earlier epoch. The above expansion can be written in terms of $x = H_0\tau$ and so, writing $R = (1+z)^{-1}$, we find

$$\frac{1}{1+z} = 1 - x - \frac{q_0}{2}x^2 + \ldots \tag{3.22}$$

Now, we want to express the redshift z to second order in the look-back time τ. This is achieved by making a further Taylor expansion of $[1 - x - (1/2)q_0x^2]^{-1}$ to second order in x. Carrying out this expansion, we find

$$z = x + (1 + \frac{q_0}{2})x^2 + \ldots \tag{3.23}$$

We can now find the expression for the comoving coordinate distance r by taking the integral

$$\begin{aligned} r &= \int_0^\tau \frac{c\mathrm{d}\tau}{R} = \int_0^\tau c(1+z)\mathrm{d}\tau \\ &= \frac{c}{H_0}\left[x + \frac{x^2}{2} + \left(1 + \frac{q_0}{2}\right)\frac{x^3}{3}\ldots\right] \end{aligned} \tag{3.24}$$

Finally, we can express r to second order in z by dividing the expression (3.24) successively through by expression (3.23). We find

$$r = \left(\frac{c}{H_0}\right)\left[z - \frac{z^2}{2}(1+q_0) + \ldots\right] \tag{3.25}$$

The last step is to evaluate $D = \Re\sin(r/\Re)$ but since we are only interested in small values of $r/\Re$, we see that $D = r$.

The next step is to rewrite the expression (3.20) in terms of q_0 rather than Ω_0 and then expand that expression to second order in z, that is,

$$\begin{aligned} D &= \frac{c}{H_0 q_0^2(1+z)}\left\{q_0 z + (q_0 - 1)[(2q_0 z + 1)^{1/2} - 1]\right\} \\ &= \left(\frac{c}{H_0}\right)\left[z - \frac{z^2}{2}(1+q_0) + \ldots\right] \qquad \text{for } z \ll 1 \end{aligned} \tag{3.26}$$

Thus, we obtain exactly the same result as that obtained from the kinematic argument. What this means is that to second order in redshift, the 'effective distance' D does not depend upon the density parameter Ω_0 at all — it only depends upon the deceleration. To express this result another way, at the same look-back time, a galaxy in an $\Omega_0 = 0$ Universe has a smaller recessional velocity than in a Universe with $\Omega_0 = 1$ because, in the latter case, the expansion of the Universe was greater in the past. Alternatively, at the same redshift z or scale factor R, the look-back time is smaller in an $\Omega_0 = 1$ Universe than in an $\Omega_0 = 0$ model as can be observed by inspection of Fig. 3.6. This explains why the same object is expected to be brighter in an $\Omega_0 = 1$ model as compared with an $\Omega_0 = 0$ model. The significance of these remarks is that the prime reason for the difference between the world models is the deceleration of the system of galaxies, rather than the curvature of space or the effects of gravity.

The procedure is therefore clear but observationally very demanding — we seek to determine the deceleration of the Universe by determining the 'effective distance' D at small redshifts, say $z \leq 0.5$, at which there are small but appreciable differences between the world models but which determine directly the local deceleration of the Universe rather than its mean density. My hope would be that this programme might be possible through one of the 'physical methods' of determining D.

The importance of this analysis is that it provides a test of the validity of General Relativity on the largest scales we have accessible to us in the Universe at the present epoch. It would represent an enormous step forward if we were able to show that indeed $\Omega_0 = 2q_0$ with an order of magnitude improved accuracy.

3.5 Radiation Dominated Universes

At the opposite extreme from dust-filled Universes are those in which radiation

contributes all the inertial mass. In this case, we cannot neglect the pressure term in equation (3.1). For a gas of photons, massless particles or a relativistic gas in the ultrarelativistic limit, $E \gg mc^2$, pressure p is related to energy density ε by $p = \frac{1}{3}\varepsilon$ and the inertial mass density of the radiation ρ_{rad} is related to its energy density ε by $\varepsilon = \rho_{\text{rad}} c^2$. We can now work out simply how the energy density of radiation varies with redshift. If $N(h\nu)$ is the number density of photons of frequency ν, then the energy density of radiation is found by summing over all frequencies

$$\varepsilon = \sum_{\nu} h\nu N(h\nu)$$

Now the number density of photons varies as $N = N_0 R^{-3} = N_0(1+z)^3$ and the energy of each photon changes with redshift by the usual redshift factor

$\nu = \nu_0(1+z)$. Therefore, the variation of the energy density of radiation with epoch is

$$\varepsilon = \sum_{\nu_0} h\nu_0 N_0(h\nu_0)(1+z)^4$$

$$\varepsilon = \varepsilon_0(1+z)^4 = \varepsilon_0 R^{-4} \tag{3.27}$$

A case of particular interest is that of black-body radiation. The energy density is given by the Stefan-Boltzmann law, $\varepsilon = aT^4$ and its spectral energy density by the Planck distribution

$$\varepsilon(\nu)\mathrm{d}\nu = \frac{8\pi h\nu^3}{c^3}\frac{1}{e^{h\nu/kT}-1}\mathrm{d}\nu$$

It immediately follows that for black body radiation the radiation temperature T_r varies with redshift as

$$T_\mathrm{r} = T_0(1+z)$$

Correspondingly, the spectrum of the radiation changes as

$$\begin{aligned}\varepsilon(\nu_1)\mathrm{d}\nu_1 &= \frac{8\pi h\nu_1^3}{c^3}[(e^{h\nu_1/kT_1}-1)]^{-1}\mathrm{d}\nu_1\\ &= \frac{8\pi h\nu_0^3}{c^3}[e^{h\nu_0/kT_0}-1)^{-1}](1+z)^4\mathrm{d}\nu_0\\ &= (1+z)^4\ \varepsilon(\nu_0)\ \mathrm{d}\nu_0\end{aligned}$$

Thus, it can be seen that upon redshifting, a black body spectrum preserves its form but the radiation temperature changes as $T_\mathrm{r} = T_0(1+z)$ and the frequency of each photon as $\nu = \nu_0(1+z)$. Another way of looking at these results is in terms of the adiabatic expansion of a gas of photons. The adiabatic index, that is, the ratio of specific heats γ, for radiation and a relativistic gas in the ultrarelativistic limit is $\gamma = 4/3$. It is a simple exercise to show that, in an adiabatic expansion, $T_\mathrm{r} \propto V^{-(\gamma-1)} = V^{-\frac{1}{3}}$ which is exactly the same as the above result.

The variations of p and ρ with R are now substituted into equations (3.1) and (3.2). We find

$$\ddot{R} = \frac{8\pi G\varepsilon_0}{3c^2}\frac{1}{R^3} \tag{3.28}$$

$$\dot{R}^2 = \frac{8\pi G\varepsilon_0}{3c^2}\frac{1}{R^2} - \frac{c^2}{\Re^2} \tag{3.29}$$

We will show in a moment that the Universe becomes radiation-dominated at early epochs corresponding to values to $R \leq 10^{-3} - 10^{-4}$. At these early epochs we can neglect the constant term $c^2/\Re^2$ and then the integration of equation (3.29) is straightforward.

$$R = \left(\frac{32\pi G\varepsilon_0}{3c^2}\right)^{\frac{1}{4}} t^{\frac{1}{2}} \tag{3.30}$$

Thus, the dynamics of the radiation-dominated models are very simple, $R \propto t^{\frac{1}{2}}$, and depend only upon the total inertial mass density in relativistic or massless forms. Notice that we have to add all the contributions to ε at the relevant epochs.

3.6 Inflationary Models

These models have come into prominence as a result of the deepening understanding of elementary particle physics and its application to the early stages of the Hot Big Bang. Kolb and Turner (1990) provide an excellent survey of these ideas. These considerations lead to physical processes which are non-intuitive from the stand point of classical physics.

First of all, we consider the dynamical equations according to Einstein's original prescription but retain the cosmological constant Λ. Suppose the Universe is empty, $\rho = 0$. Then, equation (3.1) becomes

$$\ddot{R} = \frac{1}{3}\Lambda R \tag{3.31}$$

As Zeldovich (1968) has remarked, this equation shows that the cosmological constant describes the *repulsive effect of a vacuum* — any test particle introduced into the vacuum acquires an acceleration simply by virtue of being located there. According to classical physics, there is no simple physical picture for this process but there is a natural interpretation in the context of quantum field theory. The stress-energy tensor of a vacuum leads to a negative energy equation of state, $p = -\rho c^2$. This pressure may be thought of as a 'tension' rather than a pressure. When such a vacuum expands from V to $V + \mathrm{d}V$, the work done $p\mathrm{d}V$ is just $-\rho c^2 \mathrm{d}V$ so that, during the expansion, the mass-density of the negative energy field ρ_{vac} remains constant. Carroll, Press and Turner (1992) show how the theoretical value of Λ can be evaluated using simple concepts from quantum field theory and they find the mass density of the repulsive field to be $\rho_{\mathrm{vac}} = 10^{95}$ kg m^{-3}. This is a problem. This density corresponds to a value of the cosmological constant which is about 10^{122} times greater than acceptable values which correspond to $\rho_{\mathrm{vac}} \leq 10^{-27}$ kg m^{-3} at the present epoch.

This represents a rather large discrepancy but it is not one we should pass over lightly. In the inflationary model of the very early Universe, it is exactly this force which causes the exponential expansion. If the inflationary picture is adopted, we have to explain why ρ_{vac} decreased by a factor of at least 10^{120} at the end of the inflationary era. Within this context, 10^{-120} looks remarkably close to zero which would correspond to the standard Friedman picture with $\Lambda = 0$.

What is the current status of the cosmological constant? To quote Carroll, Press and Turner (1992)

> *The cosmological constant, Λ, is an idea whose time has come ... and gone ... and come ... and so on*

Einstein's introduction of the Λ term predated Hubble's discovery of the expansion of the system of galaxies. Einstein realised that he had a relativistic theory of gravity which enabled him to construct self-consistent models for the Universe as a whole. In 1917, he had no reason to seek non-stationary solutions of the field equations and the Λ term was introduced to produce static solutions which he believed, incorrectly, enabled Mach's principle to be incorporated into his model of the Universe. Once the expansion of the system of galaxies was discovered, Einstein is reported to have stated that the introduction of the cosmological constant was 'the greatest blunder of my life' (Gamow 1970). Since then it has been in and out of fashion in response to different cosmological problems. It was very popular during the 1930s when it seemed that the age of the Universe as estimated by $T_0 \sim H_0^{-1}$ was significantly less than the age of the Earth. The Eddington-Lemâitre model was of special interest because the cosmological constant was chosen so that the Universe had a long coasting phase during which the attractive force of gravity was almost perfectly balanced against the repulsive effect of the Λ-term. The conflicting age estimates could be reconciled in this model. This particular problem disappeared when it was found that the value of Hubble's constant had been greatly overestimated. There would be the same type of problem today if high values of Hubble's constant were adopted and large ages for the oldest globular clusters are accepted. I leave this story to Dr. Sandage. At the present time, there is no positive observational evidence that $\Lambda \neq 0$ but it cannot be excluded that it is non-zero.

Let us show very briefly how some of the concepts of particle physics can provide a physical realisation of the cosmological constant and a few of its consequences for cosmology. A key development has been the introduction of Higgs fields into the theory of weak interactions. These and other ideas in quantum field theory are very clearly described by Zeldovich (1986) in an article aimed at astronomers and astrophysicists. The Higgs field was introduced into electro-weak theory in order to eliminate high order singularities in the theory and to endow the $W^{\pm}$ and Z^0 bosons with masses. The precise measurement of the masses of these particles at CERN has confirmed the theory very precisely. The Higgs fields have the property of being *scalar* fields, unlike the vector fields of electromagnetism or the tensor fields of General Relativity. Zeldovich shows that scalar fields result in negative energy equations of state $p = -\rho c^2$. Fields of this nature, but associated with the phase transition when the strong force decoupled from the electro-weak force in the very early Universe, are the prime candidates for the source of the cosmological negative energy equation of state.

Let us substitute this negative energy equation of state into equation (3.1). Then,

$$\ddot{R} = \frac{8\pi G R}{3}\rho_{\text{vac}} \tag{3.32}$$

It can be seen that this result is formally identical to equation (3.31) and provides a physical basis for the inclusion of the cosmological constant in the field equations. Formally, we can obtain the exact equivalence by setting $\Lambda = 8\pi G\rho_{\text{vac}}$.

Let us show how at least some of the basic problems of the standard Friedman models can be overcome in this picture. We can begin with the first integral of the dynamical equations (equation 3.2)

$$\dot{R}^2 = \frac{1}{3}\Lambda R^2 - \frac{c^2}{\Re^2} \tag{3.33}$$

We recall that $\Re$ is the radius of curvature of the geometry of the space at a fixed reference epoch which was chosen to be the present epoch in the analysis of Section 2.4. As was shown in Section 2.4, this radius of curvature changes with scale factor as $\Re(t) = \Re R$ and so we can find how $\Re$ changes with epoch by solving equation (3.33). The solution is

$$R(t) = \frac{c}{\Re}\frac{1}{(\Lambda/3)^{1/2}}\left\{\cosh\left[(\Lambda/3)^{1/2}t\right] - 1\right\} \tag{3.34}$$

This corresponds to an exponential increase in $R(t)$ and correspondingly $\Re(t)$ with cosmic epoch. Since $\Re(t) = \Re R(t)$, it follows that, for $(\Lambda/3)^{1/2}t \gg 1$

$$\Re(t) = \frac{c}{2(\Lambda/3)^{1/2}}\exp\left[(\Lambda/3)^{1/2}t\right] \tag{3.35}$$

Thus, the radius of curvature of the geometry of the Universe grows exponentially with time under the influence of the negative energy equation of state.

Kolb and Turner (1990) describe how these ideas might be realised in the early Universe. If this process is associated with spontaneous symmetry breaking during the Grand Unified Theory phase transition, the characteristic energies are $E \sim 10^{14}$ GeV at a time 10^{-34} seconds after the Big Bang. In a typical realisation of the inflation picture, the Universe continues to be driven by the negative energy equation of state until about 100 times this time. By that point, there has been an enormous energy release associated with the phase transition which reheats the Universe to a very high temperature and from that point onwards the dynamics become those of the standard radiation dominated Universe.

The result is that the radius of curvature of the geometry grows by a huge factor of $\exp(100) = 10^{43}$ over this short time period. If, for example, we consider the horizon scale after 10^{-34} seconds, $r \sim ct = 3 \times 10^{-26}$ m, this dimension would have expanded to 3×10^{17} m by the end of the inflation epoch. If the Universe had spatial curvature on the scale of the horizon at the earlier epoch, $\Re(t) \sim ct$, it can be seen that geometry would have been 'straightened out' by the later epoch. The radius of curvature $\Re$ continues to scale as R, which is proportional to $t^{1/2}$, throughout the subsequent radiation dominated epochs. The result is that the radius of curvature of the spatial geometry of the Universe would now be 3×10^{42} m which far exceeds the present dimension of the Universe, $\sim 10^{26}$ m. In other words, for all practical purposes, the geometry would be flat and correspond to a Universe with $\Omega_0 = 1$. Fig. 3.7 from Kolb and Turner (1990) shows clearly the distinction between the standard Friedman picture and the inflationary scenario — note, in particular, the factor of 10^{43} which is somewhat abbreviated in the lower diagram.

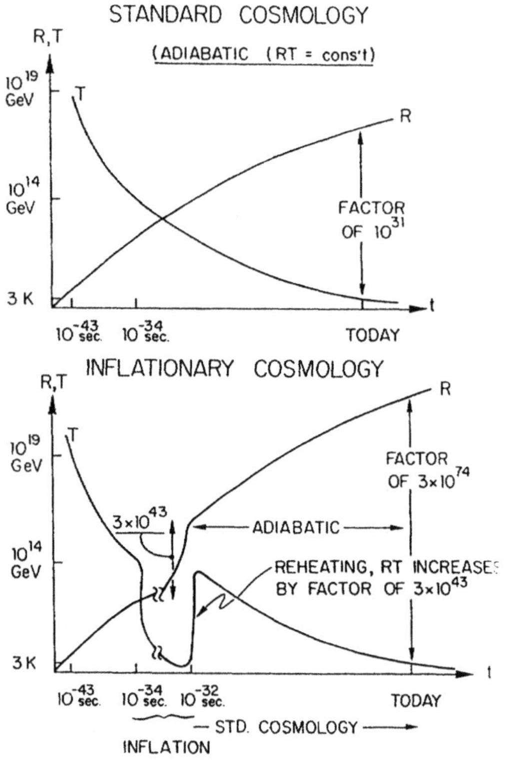

Fig. 3.7 Comparison of the evolution of the scale factor and temperature in the standard Big Bang and inflationary cosmologies (From Kolb and Turner 1990).

In addition to giving a reason why Ω_0 should be unity at the present day, the above analysis also shows why the inflationary model can account for the large scale isotropy of the Universe now. A patch on the scale of the horizon at the GUT phase transition is 'stretched' to scales far beyond the present horizon of the Universe. In consequence, there was time in the very early Universe for isotropy and homogeneity to be set up on scales less than the horizon which are then inflated by an enormous factor to account for the isotropy of the Universe today.

These are intriguing ideas and appear to offer a plausible explanation for two of the fundamental problems of modern cosmology, the flatness problem and that of accounting for the overall isotropy and homogeneity of the Universe. The proponents of the inflationary scenario also believe that it can explain the origin of the fluctuations from which galaxies form and the absence of magnetic monopoles in the Universe now. My exposition does scant justice to this whole field and the reader is recommended to consult Kolb and Turner (1990) for an authoritative and enjoyable exposition of these ideas.

References – Chapter 3

Carroll, S.M., Press, W.H. and Turner, E.L. (1992). ARA&A, **30**, 499.
Gamow, G. (1970). *My World Line*. New York: Viking Press.
Gunn, J.E. (1978). In *Observational Cosmology* by J.E. Gunn, M.S. Longair and M.J. Rees. Geneva: Geneva Observatory Publications.
Kolb, E.W. and Turner, M.S. (1990). *The Early Universe*, Redwood City, California: Addison-Wesley Publishing Co.
Longair, M.S. (1992). *Theoretical Concepts in Physics*. Cambridge: Cambridge University Press.
Mattig, W. (1959). Astr. Nach., **285**, 1.
Milne, E.A. and McCrea, W.H. (1934). Q.J. Math., **5**, 73.
Tropp, E.A., Frenkel, V.Ya. and Chernin, A.D. (1993). *Alexander A. Friedman; the Man who Made the Universe Expand*. Cambridge: Cambridge University Press.
Zeldovich, Ya.B. (1968). Usp. Fiz. Nauk, **95**, 209. [Translation - Sov. Phys. - Usp., **11**, 381.]
Zeldovich, Ya.B. (1986). Astrophys. Sp. Phys. Rev. **5**, 1.

4 Number Counts and the Background Radiation

At last, we can get down to the business of working out the source counts and background radiation from a cosmological distribution of discrete sources or from the diffuse intergalactic medium.

4.1 Number Counts of Discrete Sources

Consider first of all a uniform distribution of a single class of source with space density N_0 and luminosity $L(\nu)$. We can consider $L(\nu)$ to be the spectral energy distribution of this class of source. From the considerations of sections 2.5.4 and 2.5.5, the flux density of such a source at frequency ν_0 is

$$S(\nu_0) = \frac{L(\nu_1)}{4\pi D^2(1+z)} \tag{2.47}$$

where $\nu_1 = \nu_0(1+z)$ and the number of sources per steradian in the interval of comoving coordinate distance $\mathrm{d}r$ is

$$\mathrm{d}N = N_0 D^2 \mathrm{d}r \tag{2.53}$$

The *integral source counts*, $N(\geq S)$, are defined to be the numbers of sources per steradian with flux densities greater than or equal to some limiting value S at frequency ν_0. For a complete sample of sources with lower limiting flux density S, a class of sources with identical spectral energy distributions $L(\nu)$ can be observed to a limiting comoving distance coordinate r_{max} which is the solution of the expression (2.47). Then, if the sources are uniformly distributed in space, that is, the sources have a constant *comoving space density* N_0, the integral number counts are

$$N(\geq S) = \int_0^{r_{\mathrm{max}}} N_0 D^2 \mathrm{d}r \tag{4.1}$$

For any given cosmological model, $R(t)$ is defined and so there is a known relation between r and redshift z. The only point to beware of is that the above formalism assumes that S and $\mathrm{d}N(r)$ are monotonic functions of redshift and a little care has to be taken if they are not (see, for example, Chapter 11).

In the limit of small redshifts, $z \to 0$, $D \to r$ and so

$$N(\geq S) = N_0 \int_0^{r_{\mathrm{max}}} r^2 \mathrm{d}r = \frac{1}{3} N_0 r_{\mathrm{max}}^3$$

Since, in this limit, $S = L/4\pi r_{\mathrm{max}}^2$, it follows that

$$N(\geq S) = \frac{1}{3} N_0 \left(\frac{L}{4\pi}\right)^{3/2} S^{-3/2} \tag{4.2}$$

Now, we need to integrate over all luminosity classes of source. This is found by integrating over the *luminosity function* of the sources which is defined such that $N_0(L)\mathrm{d}L$ is the comoving space density of sources with luminosities in the luminosity range L to $L+\mathrm{d}L$. We therefore have to integrate (4.1) over all luminosities, that is, we need to take the double integral

$$N(\geq S) = \int_L \int_0^{r_{\max}(L)} N_0(L) D^2 \mathrm{d}r \mathrm{d}L \tag{4.3}$$

In the small redshift limit, $z \rightarrow 0$, we find

$$N(\geq S) = \frac{1}{3} S^{-3/2} \int_L N_0(L) \left(\frac{L}{4\pi}\right)^{3/2} \mathrm{d}L \tag{4.4}$$

This is the famous 'three-halves' power law, $N(\geq S) \propto S^{-3/2}$ and is often known as the *Euclidean source counts.* Notice that the Euclidean source counts are independent of the form of the luminosity function of the sources since it only appears inside the integral in (4.4). This result can be written in alternative ways. Optical counts of galaxies are normally expressed in terms of the number of galaxies brighter than a given limiting magnitude m. Since $m = \text{constant} - 2.5 \log_{10} S$, it follows that

$$N(\leq m) \propto 10^{0.6m} \tag{4.5}$$

As described by Dr. Sandage, Hubble realised that this relation provided a test of the homogeneity of the distribution of galaxies in the Universe (see Section 2.2.1 and Fig. 2.4).

One of the problems with the expression (4.3) is the fact that, in counting *all* the sources brighter than each limiting flux density S, the numbers counted are not independent at different flux densities and, to avoid this, it is preferable to work in terms of *differential source counts* which are defined to be the number of sources in the flux density interval S to $S + \mathrm{d}S$. For the Euclidean source counts, this can be found by differentiating the expression (4.4). Then,

$$\mathrm{d}N_0(S) \propto S^{-5/2}\mathrm{d}S \tag{4.6}$$

The use of integral versus differential source counts became an issue when they extended to low flux densities at which the counts began to converge (see Jauncey 1975).

The various forms of the Euclidean source count can be regarded as *null hypotheses* since such counts are expected at small redshifts in all cosmological models. In consequence, it is often helpful to normalise the observed or theoretical counts to the Euclidean prediction, that is, if $\Delta N(S)$ is the observed or predicted numbers of sources in $\mathrm{d}S$, it is often convenient to work in terms of *normalised, differential source counts* which are defined as $\Delta N(S)/\Delta N_0(S)$ where $\Delta N_0(S)$ is the Euclidean prediction.

In real world models, the source counts deviate from the Euclidean prediction because D and r are not linear functions of redshift z. As an illustration, it is useful to work out the normalised differential source count for a single luminosity class of source with a power-law spectral index $S(\nu) \propto \nu^{-\alpha}$. For the standard Friedman world models, it is can be shown that

$$\frac{\Delta N(S)}{\Delta N_0(S)} = \frac{2c(1+z)^{-\frac{3}{2}(1+\alpha)}}{H_0(\Omega_0 z+1)^{1/2}\left[D(1+\alpha)+2(1+z)\frac{\mathrm{d}D}{\mathrm{d}z}\right]} \tag{4.7}$$

In the case of the critical model, $\Omega_0 = 1$, this expression reduces to

$$\frac{\Delta N(S)}{\Delta N_0(S)} = \frac{(1+z)^{-\frac{3}{2}(1+\alpha)}}{[(1+\alpha)(1+z)^{1/2}-\alpha]} \tag{4.8}$$

The expression (4.7) is plotted in Fig. 4.1 for world models having $\Omega_0 = 0$, 1 and 2 assuming a spectral index $\alpha = 0.75$. The redshifts at which the number counts have different values of $\Delta N/\Delta N_0$ are indicated on each curve. These curves make the important point that, for a uniform distribution of sources, the slope of the differential counts departs from the Euclidean value at remarkably small redshifts. For example, for the critical model $\Omega_0 = 1$, the slope of the differential source counts at a redshift of 0.5 is -2.08, corresponding to an integral source count with slope -1.08. Even at quite small redshifts, say $0.2 - 0.3$, the departures from the -1.5 law are significant. Notice that this calculation has been undertaken for a simple power-law spectrum but the result is generally true unless the source spectra are highly inverted, an example of which is discussed in Chapter 11.

The above calculation has been carried out for a single luminosity class of source and, to find more realistic predictions, these counts have to be convolved with the luminosity function of the sources. It is apparent, however, that, in general, any population of sources which extends to redshifts $z \sim 1$ is expected to have a source count which has slope significantly less steep than -1.5.

Finally, we have to take account of the effects of cosmological evolution of the population of sources. It is simplest to regard the evolutionary changes as changes to the luminosity function of the sources with cosmic epoch. These can be written as a modification of the *comoving space density* of sources as a function of cosmic epoch. Formally, we can write

$$N(L,z) = N_0(L) f(L,z,\alpha,\mathrm{type},\ldots) \tag{4.9}$$

where the *evolution function* f takes the constant value 1 if the source distribution does not change with cosmic epoch. The evolution function can be made as complicated as is necessary. For example, to work out the optical counts of galaxies, separate luminosity functions have to be evaluated for each class of galaxy and then, in place of simple power law spectra, the typical spectrum of each class of galaxy has to be used. This is often achieved

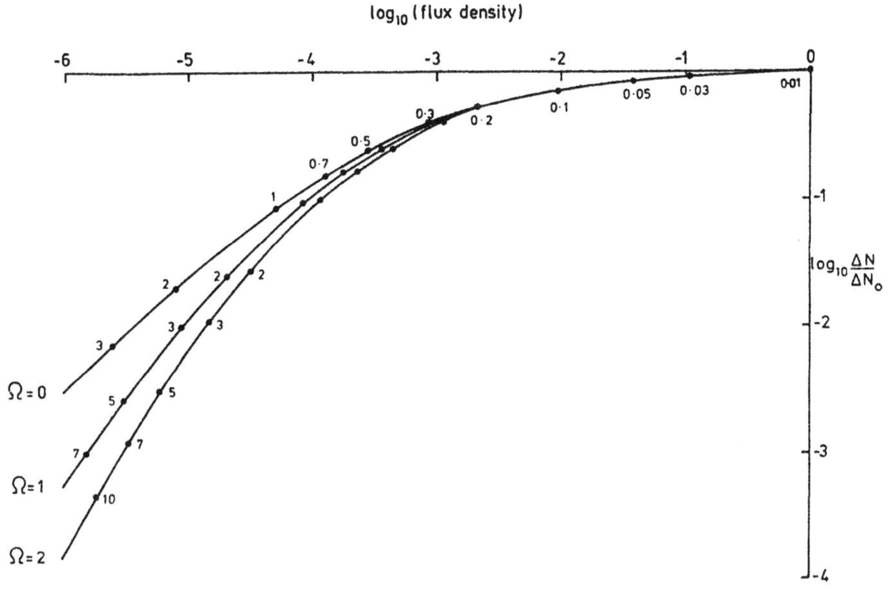

Fig. 4.1 The predicted normalised differential counts of sources for a single luminosity class of source with a power-law spectrum $S \propto \nu^{-0.75}$. The redshifts at which the sources are observed are indicated on each of the curves.

by working out the *K-corrections* for the different classes of galaxy which contribute to the counts (see the expression 2.50). In addition, evolutionary changes of the spectral energy distributions of galaxies with cosmic epoch can be built into the above evolution function. Formally, the integral source counts for any form of evolution function can be written

$$N(\geq S_0) = \int_L \int_{S \geq S_0} N_0(L) f(L, z, \alpha, \text{type}, \ldots) D^2 \mathrm{d}r \mathrm{d}L \qquad (4.10)$$

This formalism is exactly the same as that recommended by Drs. Kron and Sandage. An obvious concern is the great variety in the properties of galaxies. In the extreme case in which one wished to take account of the individual properties of *every* galaxy, it is possible to work out the space density corresponding to each of them by selecting a complete flux density limited sample of objects and then evaluating the volume of space $V_{\max}$ within which each object could have been observed and still remain within the sample. The local space density of such a object is $V_{\max}^{-1}$ and the spectral energy distribution of the galaxy can be used to work out K-corrections.

Notice also that, if it is known that cosmological evolutionary effects are influencing the counts, these effects have to be taken into account when evaluating the local space density of sources from magnitude or flux density limited samples.

4.2 The Background Radiation

At last, we are able to work out the background intensity due to a cosmological distribution of discrete sources. For the sake of illustration, we will assume that the sources have power-law spectra of the form $S \propto \nu^{-\alpha}$ and then the flux density-luminosity relation becomes

$$S(\nu_0) = \frac{L(\nu_0)}{4\pi D^2(1+z)^{1+\alpha}} \tag{4.11}$$

Now, the numbers of sources per steradian in an increment of comoving coordinate distance $\mathrm{d}r$ in the case of a uniform distribution of sources is

$$\mathrm{d}N = N_0 D^2 \mathrm{d}r \tag{4.12}$$

Therefore, background intensity $I(\nu_0)$ due to this uniform distribution of sources is

$$\begin{aligned} I(\nu_0) &= \int S(\nu_0)\mathrm{d}N \\ &= \int_0^\infty \frac{L(\nu_0)}{4\pi D^2(1+z)^{1+\alpha}} N_0 D^2 \mathrm{d}r \\ &= \frac{L(\nu_0)N_0}{4\pi} \int_0^\infty (1+z)^{-(1+\alpha)}\mathrm{d}r \end{aligned} \tag{4.13}$$

For the Friedman world models, we find from the expression (3.18)

$$\mathrm{d}r = \frac{c\mathrm{d}z}{H_0(\Omega_0 z+1)^{1/2}(1+z)} \tag{3.18}$$

and so we obtain the important result

$$I(\nu_0) = \frac{c}{H_0}\frac{L(\nu_0)N_0}{4\pi} \int_0^\infty \frac{\mathrm{d}z}{(\Omega_0 z+1)^{1/2}(1+z)^{2+\alpha}} \tag{4.14}$$

This result can be compared with the Newtonian version of the same calculation which, from the small redshift limit, $z \to 0$, $r = cz/H_0$, becomes

$$I(\nu_0) = \frac{L(\nu_0)N_0}{4\pi} \int_0^\infty \mathrm{d}r \tag{4.15}$$

This is the naive version of what is referred to as *Olbers' paradox*, namely, that, in a isotropic, infinite, stationary Euclidean Universe, the background radiation diverges. This naive sum has not taken account of the finite sizes of the sources and eventually we have to take into account of the overlapping of their images. Nor does the argument take account of thermodynamics since in such an infinite static Universe eventually all the matter comes into thermodynamic equilibrium at the same temperature.

Unlike the Olbers' sum, it is apparent that the integral over redshift in (4.14) converges provided $\alpha > -1.5$. Any realistic spectrum must eventually turn over at a high enough frequency and so a finite integral is always obtained. It is useful to work out the background intensity for two typical world models. For the cases $\Omega_0 = 0$ and $\Omega_0 = 1$, we find

$$\begin{aligned} \Omega_0 = 0 \qquad & I(\nu_0) = \frac{c}{(1+\alpha)H_0}\frac{L(\nu_0)N_0}{4\pi} \\ \Omega_0 = 1 \qquad & I(\nu_0) = \frac{c}{(1.5+\alpha)H_0}\frac{L(\nu_0)N_0}{4\pi} \end{aligned} \tag{4.16}$$

Thus, it can be seen that, for typical values of α, to order of magnitude, the background intensity is just that originating within a typical cosmological distance (c/H_0), that is,

$$I(\nu_0) \sim \frac{c}{H_0}\frac{L(\nu_0)N_0}{4\pi} \tag{4.17}$$

A combination of factors leads to the convergence of the integral for the background intensity. Inspection of the integral (4.13) shows that part of the convergence is due the redshift factor $(1+z)^{-(1+\alpha)}$ which is associated with the redshifting of the emitted spectrum of the sources. The second is the dependence of r upon redshift z. This relation is only linear at small redshifts, $z \ll 1$. In the case of the critical model, $r = (2c/H_0)[1-(1+z)^{-1/2}]$ which converges to the value $2c/H_0$ as $z \to \infty$. This convergence is associated with the fact that the Friedman models of the Universe have a finite age and consequently there is a finite maximum distance from which electromagnetic waves can reach the Earth.

Let us look in a little more detail at the origin of the background radiation in the uniform models. We take as an example the critical model $\Omega_0 = 1$ with $\alpha - 1$. Then, the background intensity out to redshift z is

$$I(\nu_0) = \frac{2c}{5H_0}L(\nu_0)N_0\left[1-(1+z)^{-5/2}\right] \tag{4.18}$$

From this it is easy to show that half of the background intensity originates at redshifts $z \leq 0.31$. A similar calculation for the case of the empty world model, $\Omega_0 = 0$, shows that half the intensity comes from redshifts less than 0.42. So much for the cosmological significance of the background radiation! I am sure it must come as a disappointment to the organisers of a school entitled 'The Deep Universe' that the background radiation mostly originates at small redshifts. What is more to the point is the fact that, because half of the background is expected to originate at redshifts less than about 0.5, the principal contributors to the background radiation are not difficult to identify nowadays, provided their positions are accurately known. If the main sources of the background are associated with galaxies, there should be no difficulty in discovering the principal contributors to the background radiation, provided the sources are uniformly distributed in space. This statement is not correct if the properties of the sources have evolved strongly with cosmic epoch and we take up that topic now.

4.3 The Effects of Evolution – The Case of the Radio Background Emission

Just as in the case of the source counts, we can write the expression for intensity of the background radiation if the properties of the sources evolve with cosmic epoch in terms of the evolution function. By the same type of analysis as in the case of the source counts (expression 4.10), we find

$$I(\nu_0) = \frac{c}{H_0}\frac{L(\nu_0)N_0}{4\pi}\int_0^\infty \frac{f(L, z, \mathrm{type}, \ldots)\mathrm{d}z}{(\Omega_0 z + 1)^{1/2}(1+z)^{2+\alpha}} \tag{4.19}$$

The simplest example of the effects of evolution upon the background radiation and the source counts is the radio background emission. This is a well known story (see, for example, Wall 1990, Peacock 1993). The counts of radio sources and optically selected quasars show an excess of faint sources as compared with the expectations of uniform world models (Fig. 4.2). At high flux densities, the source count is roughly $N(\geq S) \propto S^{-1.8}$ which represents an excess of faint sources even compared with the Euclidean prediction. It is now known that the radio galaxies and quasars which display the excess of faint sources have redshifts about 1 and so the difference between the uniform models and the observations is very significant indeed. This is illustrated by the comparison of the expected counts of sources with the observations in Fig. 4.2. Notice that, at the faintest flux densities, there is a flattening of the source counts and these are probably low luminosity radio sources associated with starburst galaxies (Rowan-Robinson et al 1993).

The interpretation of radio source counts has been the subject of many studies, the most complete analysis of a very large body of high quality data being due to Dunlop and Peacock (1990). They employed the free-form modelling techniques developed by Peacock and Gull (1981) and Peacock (1985) to determine best-fits models for the evolution of the radio source populations. An example of the results of the modelling procedures is shown in Fig. 4.3 which shows how the *comoving* luminosity functions of the radio galaxies and quasars change with redshift. It appears as though the luminosity function is shifted to higher luminosities out to redshifts of about 2 and then begins to shift back again. According to these analyses, it looks as though there is a 'cut-off' to the evolving source distribution beyond redshifts of 2 but this is not well established because of the statistical difficulty of finding sufficient sources at large redshifts (Peacock 1993). The volume elements decrease rapidly with increasing redshift and so it becomes progressively more and more difficult to find large redshift sources even if there is no cut-off.

In his most recent analyses of the observations, Peacock has concluded that the simplest description of the forms of evolution necessary to account

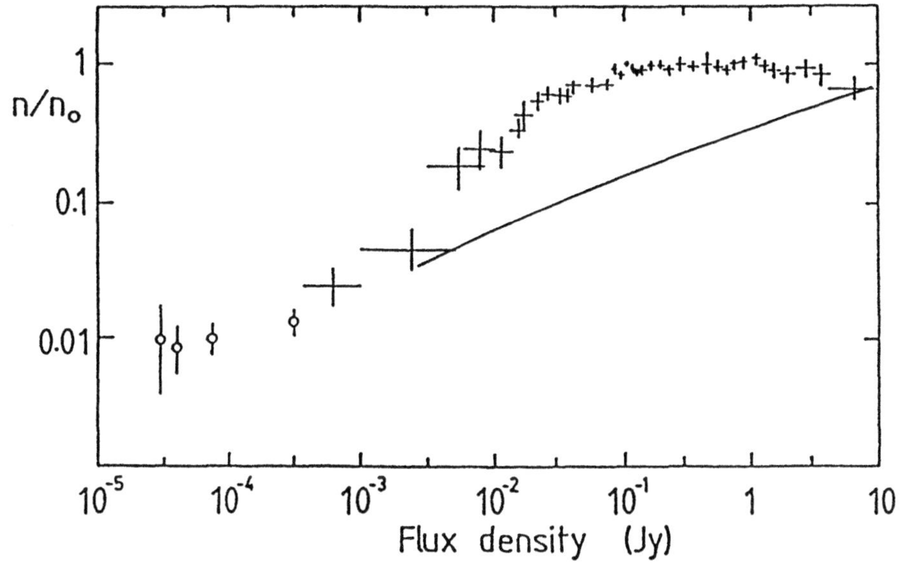

Fig. 4.2 Comparison of the counts of radio sources at 5 GHz with the expectations of uniform world models based upon the radio luminosity function of radio sources determined from complete high flux density samples (After Wall 1990).

for both the radio source counts and the optical counts of quasars (Boyle 1993) have the same form. Both sets of data can be satisfactorily described by 'luminosity evolution' models in which it is assumed that the luminosities of the sources change in the following manner with redshift:

$$\begin{aligned} L(z) &= L_0(1+z)^3 \qquad 0 < z < 2 \\ L(z) &= 27 L_0 \qquad z > 2 \end{aligned} \tag{4.20}$$

Notice that this represents very strong evolution of the source population between redshifts 0 and 2.

Let us illustrate by a simple calculation how such evolution can strongly influence the intensity of the background radiation. It is a simple calculation to work out the integrated background emission from a population of sources which locally has luminosity L_0 and space density N_0 with and without this form of evolution. From the expressions (4.13) and (3.18), the integrated background intensity is

$$I(\nu_0) = \frac{c}{H_0}\frac{N_0}{4\pi}\int_0^\infty \frac{L(z)\mathrm{d}z}{(1+z)^{7/2}} \tag{4.21}$$

where we have assumed that the spectral index of the sources α is 1 and that $\Omega_0 = 1$.

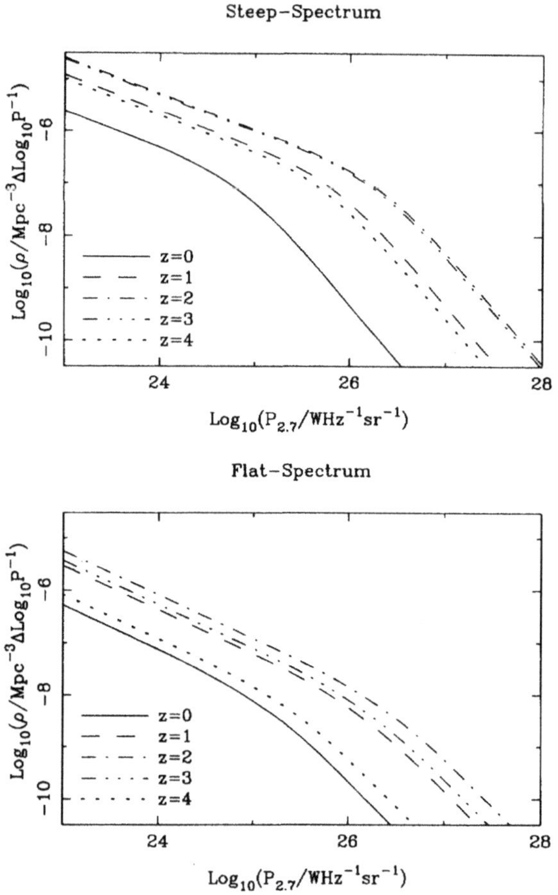

Fig. 4.3 Illustrating the evolution of the luminosity function of extragalactic radio sources with steep and flat radio spectra with redshift (or with cosmic epoch). Note that these luminosity functions are presented by per unit *comoving volume* so that the changes in the functions are over and above the changes in number density associated with the expansion of the Universe (Dunlop and Peacock 1990)

In the *no evolution* case, $L(z) = L_0$ and the background intensity is

$$I(\nu_0) = \frac{2}{5}\frac{c}{H_0}\left(\frac{N_0 L_0}{4\pi}\right) \tag{4.22}$$

In contrast, in the *evolution* case, adopting the variation of luminosity with redshift given by the expressions (4.20), the background intensity is

$$I(\nu_0) = \frac{12\sqrt{3}}{5}\frac{c}{H_0}\left(\frac{N_0 L_0}{4\pi}\right) \tag{4.23}$$

Inclusion of the effects of cosmological evolution into the calculation results in a background $6\sqrt{3} \sim 10$ times greater than without evolution. This simple calculation is entirely consistent with the discussion presented in Section 1.4.1 on the extragalactic radio background emission in which it was stated that the background due to strong radio sources would amount to only $1-2$ K if there were no evolution but amounts to about $16-19$ K when the effects of evolution are taken into account.

Thus, in this case, the background emission *does* come from the 'deep Universe', the bulk of the background originating at redshifts of the order 2. Note, however, that this only occurs because of the *very* strong effects of cosmological evolution. Inspection of Fig. 4.3 shows that, although the luminosities of the sources only increase by about 27 between redshift 0 and 2, the comoving space density of high luminosity sources increases by about a factor of 1000. These calculations make the point that the evolution has to be very drastic, which it is for radio galaxies and quasars, to make a significant impact upon the intensity of the background emission due to discrete sources.

4.4 The Background Radiation and the Source Counts

Let us look at the relation between the observed source counts and the background radiation. The background radiation from a population of sources with differential source count $\mathrm{d}N \propto S^{-\beta}\mathrm{d}S$ is

$$I \propto \int_{S_{\min}}^{S_{\max}} S\mathrm{d}N \propto \int_{S_{\min}}^{S_{\max}} S^{-(\beta-1)}\mathrm{d}S = \frac{1}{2-\beta}\left[S^{(2-\beta)}\right]_{S_{\min}}^{S_{\max}} \tag{4.24}$$

Thus, there is a critical value $\beta = 2$ for the slope of the differential source counts. If the slope of the counts is steeper than $\beta = 2$, the background intensity $I_\nu \propto S_{\min}^{(2-\beta)}$. On the other hand, if the slope of the differential source counts is less than $\beta = 2$, the background intensity is proportional to $S_{\max}^{2-\beta}$. Thus, most of the background radiation originates from that region of the counts with slope $\beta = 2$.

Now, for a Euclidean population of sources $\beta = 2.5$. In real world models, the slope is 2.5 at small redshifts but decreases at larger redshifts as discussed in Section 4.1. From the considerations which led to the expression (4.8), we showed that the slope of the differential counts is about 2 by a redshift of 0.5, showing again that the bulk of the background emission originates from redshifts $z < 1$.

4.5 Fluctuations in the Background Radiation due to Discrete Sources

Another topic of interest is the amplitude of *fluctuations* in the background radiation due to discrete sources. This is a well-known problem, first solved for the more difficult case of observations made with a radio interferometer

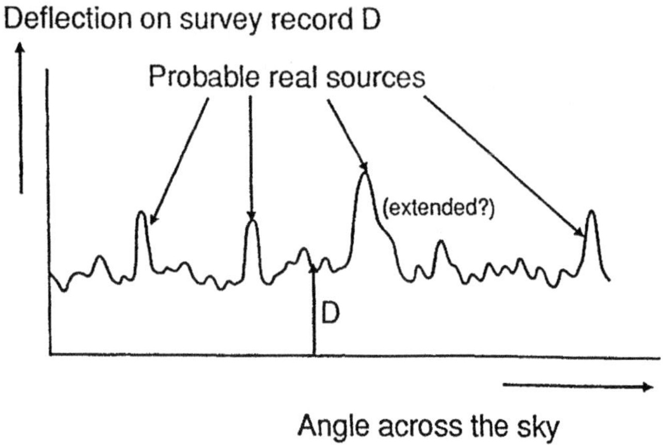

Fig. 4.4 Illustrating the fluctuations in the intensity of the background radiation due to the superposition of faint sources.

by Scheuer (1957). The problem may be stated as follows. Suppose the sky is observed with a telescope of beamwidth θ and the integral counts of sources are given by $N(\geq S) \propto S^{-\beta}$. Then, if the survey extends faint enough, eventually a flux density is reached at which there is one source per beam area and fainter, more numerous sources cannot be detected individually. In this circumstance, the noise level of the survey is due to the random superposition of faint sources within the beam of the telescope. The situation is illustrated schematically in Fig. 4.4. The problem of making observations of radio sources when the 'noise' is due to the presence of faint unresolved sources in the beam is often referred to as *confusion*. This problem afflicted the early radio surveys and is the source of fluctuations in the X-ray background emission when observed at low angular resolution (Fig. 4.5).

Scheuer (1957) provided the complete solution to the problem of determining the source counts in radio surveys which are confusion limited. In his analysis, he found the correct slope for the counts of radio sources from the early radio surveys, $N(\geq S) \propto S^{-1.8}$. This result was only in apparent contradiction with the counts of sources themselves which suggested a much steeper slope but which were very badly affected by confusion. Subsequent surveys carried out with radio telescopes with narrower beam patterns and consequently much less subject to the problems of confusion confirmed his result. I consider this to be a very important and original paper but one which is scarcely known. It was the first paper to find the correct result for the slope of the source counts at high flux densities. The steep slope showed that the counts were inconsistent with the Euclidean world model and with

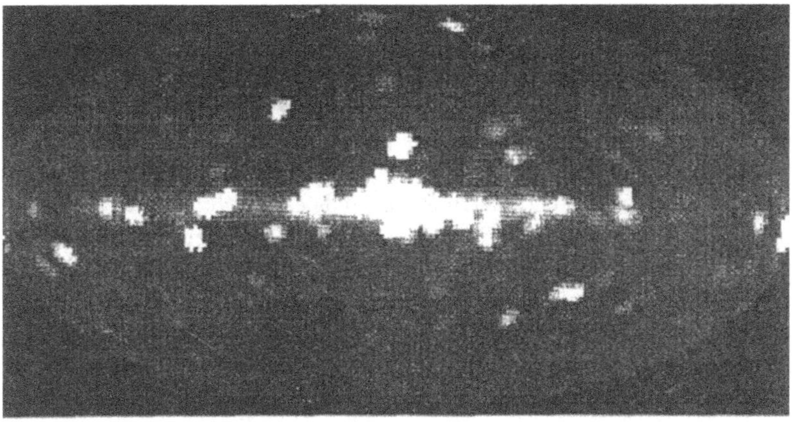

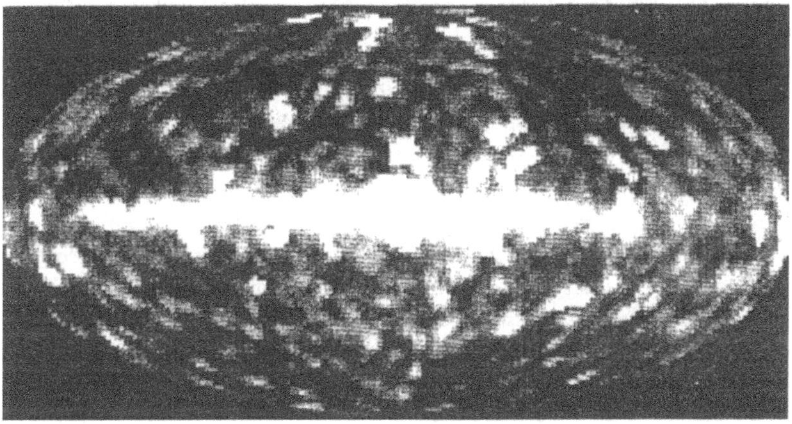

Fig. 4.5 (a) The X-ray map of the sky observed by the HEAO-1 A2 telescope at 3° resolution in Galactic coordinates. (b) A contrast enhanced image of the X-ray sky as observed by the HEAO-1 A2 experiment in the 2 – 10 keV energy band, showing flucutations in the X-ray background intensity at high Galactic latitudes (Fabian and Barcons 1992).

uniform Friedman models. I should declare an interest in that Peter Scheuer was my Ph.D. supervisor.

The simplest presentation I know of the problem for a single beam telescope is also by Scheuer (1974) who went on to solve the easier problem for surveys of the sky made by early X-ray telescopes. His approach was to forget all about the detection of individual sources but to deal directly with the amplitude of the intensity fluctuations on the map. If the map is sampled at the information rate for observations with a telescope of beam-width θ, that is, at twice per beam-width, a probability distribution is found of the amplitude of the deflection D of the record from some zero level. The term 'deflection'

D was used since the original radio astronomy surveys were recorded on strip charts and the deflections really were the deflections of the recording pen. Typical probability distributions, known as $P(D)$, are illustrated in Fig. 4.6. It can be seen that the distributions are non-Gaussian but, according to the central limit theorem, the noise level is given by the standard deviation of the probability distribution $P(D)$. The very large deflections are identified as discrete sources and the $P(D)$ distribution tends asymptotically to the differential source count $P(D) \propto D^{-\beta}$. A criterion for their identification as real sources has to be established. Normally some criterion such as 5 times the standard deviation of the confusion noise or one source per twenty or thirty beam-areas is selected. The problem is that, in a confusion limited survey, the flux densities of sources are systematically overestimated because of the random presence of faint sources in each beam. It was this effect which led to the overestimation of the flux densities of faint radio sources in the early radio source catalogues and hence to an excessively steep source count (see Scheuer 1990 for the history of these problems).

By carrying out a statistical analysis of the expected function $P(D)$ for sources selected randomly from a differential source count of the form $\mathrm{d}N(S) \propto S^{-\beta}$, Scheuer showed how the slope of the source counts could be found. Fig. 4.6 shows the normalised $P(D)$ distributions for different value of β. It can be seen that the shape of the $P(D)$ distribution provides a means of determining the form of the source counts. To order of magnitude, the most probable value of $P(D)$ corresponds to the flux density of those sources which have surface density of roughly one source per beam-area — Scheuer (1974) gives a simple statistical argument to show why this should be so. At higher flux densities, the sources are too rare to make a large contribution to the beam-to-beam variation in background signal. At lower flux densities, many faint sources add up statistically and so contribute to the background intensity but the fluctuations are dominated by the brightest sources present in each beam. At roughly one source per beam area, the fluctuations can be thought of as arising from whether or not the source is by chance within the beam. Thus, whereas the reliable detection of individual sources can only be made to about 5 or 6 times the confusion noise level, statistical information concerning the source counts can be obtained to about one source per beam area.

Scheuer's analysis was entirely analytic but it is nowadays much simpler to use Monte Carlo methods to work out the functions $P(D)$ for the assumed form of source count. The first of these statistical studies using Monte Carlo modelling procedures was carried out by Hewish (1961) in his analysis of the original records of the 4C survey. He found the first evidence for the convergence of the radio source counts at low flux densities at a frequency of 178 MHz. These procedures have been used to determine the source counts to the very faintest flux densities in the radio waveband (see, for example,

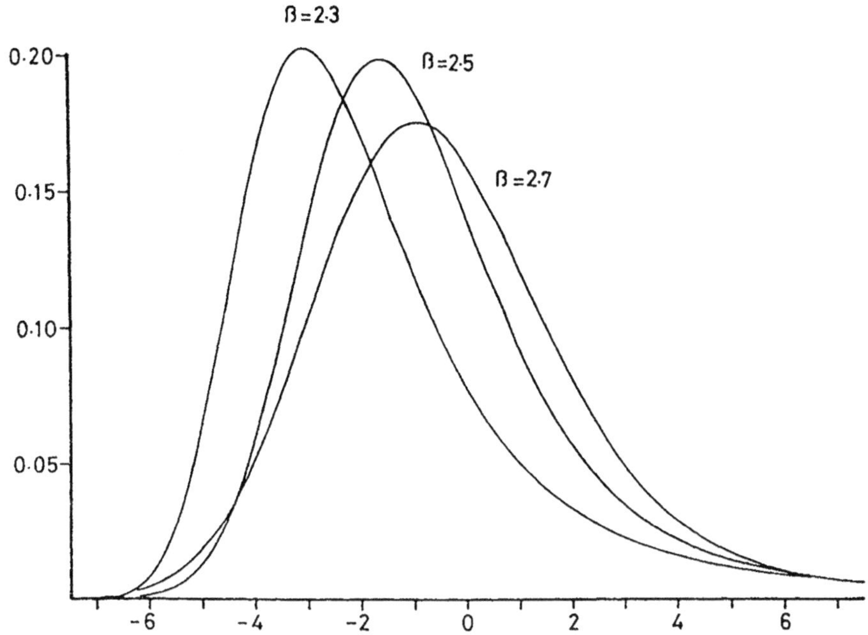

Fig. 4.6 Examples of the theoretical $P(D)$ distributions for observations made with a single-beam telescope for different assumed slopes of the differential source counts $dN(S) \propto S^{-\beta} dS$. The zero point of the abscissa is the mean amplitude $\bar{D}$ and the areas under the probability distributions have been normalised to unity. The distributions tend asymptotically to $dN \propto D^{-\beta}\, dD$ at large deflections D. (From Scheuer 1974).

Fomalont et al 1988) and in the deep ROSAT surveys by Hasinger et al (1993) (see Section 5.1).

Another approach to the interpretation of fluctuations in the background radiation is to look for a signal in the correlation function of the fluctuations. This has been carried out succesfully in the optical waveband by Shectman (1974) who found a clear signature corresponding to the two-point correlation function for galaxies. The observed fluctuation spectrum is in quite remarkable agreement with the standard correlation function found in studies of large samples of galaxies. The most recent application of this approach is the heroic work of Martin and Bowyer (1989). In a short rocket flight, they were able to make a survey of a small region of sky and found a significant correlated signal among the spatial distribution of the counts. With a number of reasonable assumptions, they were able to show that they had detected the ultraviolet emission from galaxies (see Section 1.8).

Similar analyses have been carried out for the fluctuations in the X-ray background as observed by the HEAO 1 A-2 experiment (Persic et al 1989). The binning of the background counts was in pixels 3° in size. No significant signal was found in the two-dimensional autocorrelation function on all angular scales greater than 3°. This constrains the clustering of the sources which might make up the background radiation to scales less than about 50 Mpc. They showed that the observed upper limits to the clustering would be consistent with the observed cross-correlation function for galaxies

and clusters of galaxies. A similar analysis has been carried out by Barcons and Fabian (1989) who have studied fluctations in five deep *Einstein* IPC fields. A signal is observed on the scale of 5 arcmin but it is not certain that this is of astrophysical origin. A maximum comoving clustering scale of $10h^{-1}$ Mpc is found. This rules out models in which the background is associated with clusters of galaxies at low redshifts, $z \leq 1$.

It is probably true to say that the epoch of studies of fluctuations in the background radiation due to discrete sources comes to an end as soon as source counts extend to such faint flux densities that the bulk of the background emission can be accounted for. This is now the case for the radio waveband and for the X-ray background at 1 keV.

References – Chapter 4

Barcons, X. and Fabian, A.C. (1989). MNRAS, **237**, 119.
Boyle, B. (1993). In *The Nature of Compact Objects in Active Galactic Nuclei*, (eds. A. Robinson and R.J. Terlevich). Cambridge: Cambridge University Press (in press).
Dunlop, J.S. and Peacock, J.A. (1990). MNRAS, **247**, 19.
Fabian, A.C. and Barcons, X. (1992). ARA&A, **30**, 429.
Fomalont, E.B., Kellermann, K.I., Anderson, M.C., Weistrop, D., Wall, J.V., Windhorst, R.A. and Kristian, J.A. (1988). AJ, **96**, 1187.
Hasinger, G., Burg, R., Giacconi, R., Hartner, G., Schmidt, M., Trümper, J. and Zamorani, G. (1993). A&A, **275**, 1.
Hewish, A. (1961). MNRAS, **123**, 167.
Jauncey, D.L. (1975). ARA&A, **13**, 23.
Martin, C. and Bowyer, S. (1989). ApJ, **338**, 677.
Peacock, J.A. (1985). MNRAS, **217**, 601.
Peacock, J.A. (1993). In *The Nature of Compact Objects in Active Galactic Nuclei*, (eds. A. Robinson and R.J. Terlevich). Cambridge: Cambridge University Press (in press).
Peacock, J.A. and Gull, S.F. (1981). MNRAS, **196**, 611.
Persic, M., de Zotti, G., Boldt, E.A., Marshall, F.E., Danese, L., Franceschini, A. and Palumbo, G.G.C. (1989). ApJ, **336**, L47.
Rowan-Robinson, M., Benn, C.R., Lawrence, A., McMahon, R.G. and Broadhurst, T.J. (1993). MNRAS, **263**, 123.
Scheuer, P.A.G. (1957). Proc. Camb. Phil. Soc., **53**, 764.
Scheuer, P.A.G. (1974). MNRAS, **167**, 329.

Scheuer, P.A.G. (1990). In *Modern Cosmology in Retrospect*, (eds. B. Bertotti, R. Balbinot, S. Bergia and A. Messina), 331. Cambridge: Cambridge University Press.
Shectman, S.A. (1974). ApJ, **188**, 233.
Wall, J.V. (1990). In *The Galactic and Extragalactic Background Radiation*, (eds. S. Bowyer and C. Leinert), 327. Dordrecht: Kluwer Academic Publishers.

5 The Origin of the X-ray and Gamma-ray Backgrounds

In this chapter, we tackle the origin of the extragalactic X and γ-ray background radiation which has recently been reviewed in detail by Fabian and Barcons (1992). As was emphasised in Sections 1.9 and 1.10, it is important to remember that the X- and γ-ray backgrounds span about 7 decades in energy compared with, say, the half decade of the optical waveband. It is now clear that each decade has its own individuality and the principal contributors to the background intensity are likely to be different in different energy ranges. Let us begin with the one region of the high energy background spectrum which we can account for by discrete sources with a high degree of confidence.

5.1 The Origin of the Soft X-ray Background: 0.5–2 keV

The German-US-UK ROSAT mission is one of the most important missions in X-ray astronomy. One of its principal objectives is to carry out a complete survey of the sky in the X-ray energy band 0.1 to 2.4 keV. The survey will contain about 60,000 sources and information on their X-ray spectra will be available in 4 X-ray 'colours'. The last complete survey of the X-ray sky was the HEAO-1 survey which was carried out in the late 1970s. The flux density limit of the ROSAT survey will be about 100 times fainter than that of HEAO-1. In addition to the sky survey, 3400 pointed observations to carry out specific astrophysicial programmes had been completed by the spring of 1993. Among these observations have been very deep observations of a small region of sky to define the X-ray source counts to the faintest achievable flux densities.

Earlier this year, the results of the first deep ROSAT surveys were published by Hasinger et al (1993) and two separate surveys are involved. The first is derived from 26 fields observed as part of the ROSAT medium deep survey and discrete sources in these fields have been catalogued. The second consists of one very long, deep exposure of duration 42 hours in a region known as the 'Lockman Hole' which is a region of sky in which the neutral hydrogen column density has a very low value, $N_{\rm H} = 5.7 \times 10^{19}$ cm^{-2}. As a result, in this direction, there is minimum photoelectric absorption by the neutral component of the interstellar gas which becomes important at X-ray energies $\epsilon < 1$ keV (see, for example, section 4.2 of Longair 1992).

The X-ray source counts were derived in two different ways from these observations. In the case of the medium deep survey, particular care was taken to understand the effects of source confusion at low X-ray flux densities. The very deep survey was analysed using a $P(D)$ analysis as described in section 4.5. The resulting source counts are shown in Fig. 5.1.

It can be seen that the differential X-ray source counts bear a strong resemblance to the differential counts of radio sources (Fig. 4.2). The slopes of

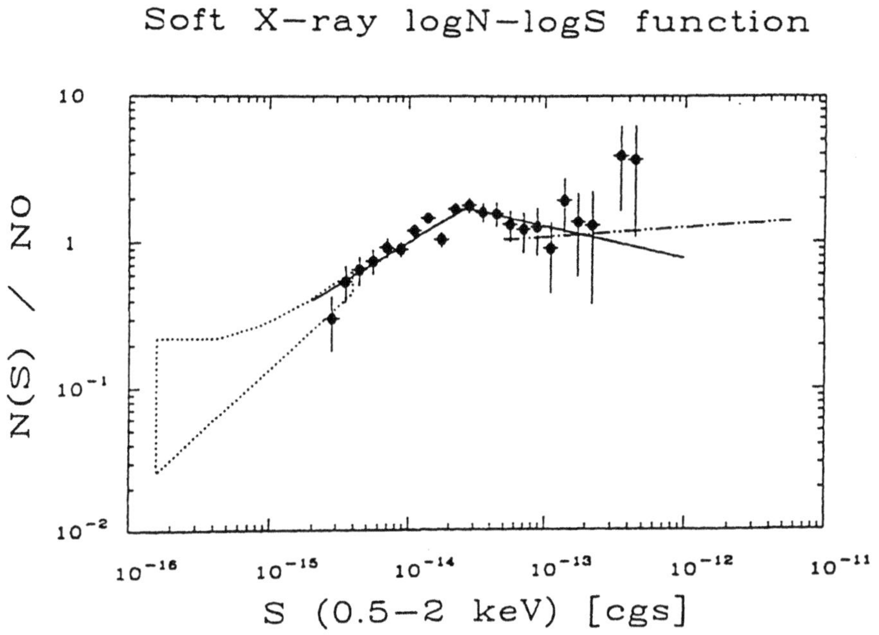

Fig. 5.1 The differential, normalised counts of faint X-ray sources observed by the ROSAT X-ray observatory. The filled circles are sources from the ROSAT Medium Deep Sky Survey, corrected for systematic effects. The dot-dash line is the best-fit source count from the Einstein Observatory surveys. The dotted area at faint flux densities shows the 90% confidence limits from the fluctuation analysis of the deepest ROSAT survey in the Lockman hole (Hasinger et al 1993).

the differential source counts quoted by Hasinger et al (1993) are as follows. At high X-ray flux densities, $S > 3 \times 10^{-14}$ erg s^{-1} cm^{-1}, the *differential* counts have slope $\beta = 2.72 \pm 0.27$ and below this flux density the counts from the medium deep survey have slope $\beta = 1.94 \pm 0.19$. The best-fit slope of the very deep survey, which extends statistically to flux densities just greater than 10^{-16} erg s^{-1} cm^{-2}, is 1.8. The integral form of the X-ray source counts is shown in Fig. 5.2. It is a useful exercise, which I leave to the reader, to use the order of magnitude rules described in Section 4.6 to work out the number density of sources at the confusion limit of the ultra-deep survey in the Lockman Hole. It will be found that the result is in satisfactory agreement with the number densities derived from the Monte Carlo simulations which are indicated by the dotted area in Figs. 5.1 and 5.2. The preliminary results concerning the identifications of the sources in the medium deep survey are entirely consistent with a picture in which the X-ray sources follow the same type of cosmological evolutionary behaviour as the radio galaxies, radio quasars and optically selected quasars. Thus, the form of evolution described by the expression (4.20) is a useful simple description of the type of evolution function needed to account for the X-ray number counts.

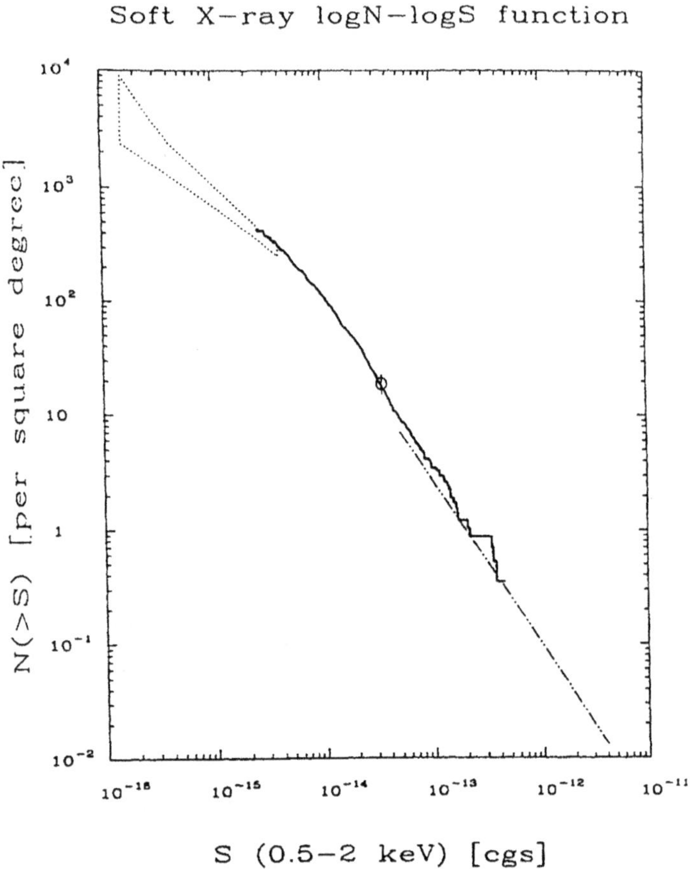

Fig. 5.2 The integral X-ray source counts in the $0.5-2$ keV energy band obtained from ROSAT medium deep survey (solid line) and the very deep survey (dotted area). The dot-dash line shows the counts derived from the Einstein Source Survey and the circle the counts from the Einstein deep survey (Hasinger et al 1993).

It is now straightforward to work out how much of the X-ray background in the $0.5-2$ keV energy band can be attributed to discrete X-ray sources. According to Hasinger et al (1993), the background intensity which can be attributed to discrete sources with flux densities greater than 2×10^{-15} erg $\mathrm{cm}^{-2}\ \mathrm{s}^{-1}$ amounts to 59% of the total background intensity in this energy range. Extending the counts to the bottom of the deep survey, about 75% of the background intensity can be accounted for. Thus, not more than 25%

of the background radiation in this waveband is not associated with discrete sources and it is entirely plausible that essentially all the background can be attributed to discrete sources. For example, if the slope of the source counts found in the very deep survey is extrapolated to zero flux density, the total background intensity can be explained.

5.2 The X-ray Background Radiation: 2–100 keV

The ROSAT deep surveys can account convincingly for the X-ray background intensity in the energy range 0.5 – 2 keV but this is very far from the end of the story. As discussed in section 1.4.3, the spectrum of the background in the 1 – 20 keV waveband has the form $I(\epsilon) \propto \epsilon^{-0.4}$. This is in contrast to the spectra of the quasars and X-ray sources observed in bright source samples which have energy spectra $S(\epsilon) \propto \epsilon^{-0.7}$ or steeper. According to Hasinger et al (1993), the fainter ROSAT sources have spectra of the form $S(\epsilon) \propto \epsilon^{-0.96}$. These data suggest that, while the background intensity at 1 keV can be attributed to the integrated emission of discrete sources such as active galaxies and quasars, there must be some other class of source which makes up the background at harder X-ray energies. This is a long-standing, unresolved problem. Let us consider some of the possibilities.

5.2.1 Thermal Bremsstrahlung from Hot Intergalactic Gas

It was pointed out by Marshall et al (1980) that the spectrum of the X-ray background in the energy range 1 – 50 keV can be very well described by a thermal bremsstrahlung spectrum at a temperature of 40 keV (Fig. 1.15(b)). If the emission resulted from diffuse hot gas at redshift z, the temperature of the gas would have to be $40(1+z)$ keV. There are several major problems with this seemingly straightforward solution. One of these is that, to obtain the observed intensity, the density of the diffuse intergalactic gas would have to be high, corresponding to a density parameter in baryons of $\Omega_b \geq 0.23$. This value is greater than the upper limit inferred from studies of the primordal nucleosynthesis of deuterium and the light elements, $\Omega_b \leq 0.03h^{-2}$ (see Section 6.5). Furthermore, it is natural in this picture to attribute the high degree of ionisation of the intergalactic gas to the same process reponsible for heating the gas to a very high temperature and so, because the heating would then have to take place at a redshift $z \sim 3$ (see Chapter 10), the energy requirements for heating the gas would be very great. For example, the thermal energy density of intergalactic gas at a temperature $kT = 160$ keV and density corresponding to $\Omega_0 = 0.23$ at a redshift of 3 is 7.6×10^7 eV m^{-3} which is the same as the energy density of the Cosmic Microwave Background Radiation at that redshift. This energy density corresponds to a density parameter at that redshift of $\Omega = 3.3 \times 10^{-3}$.

There is, however, a more serious problem and that is that such large quantities of hot gas result in a large Compton scattering optical depth

$$y = \int \frac{kT_e(z)}{m_e c^2} \frac{\sigma N_e(z)}{(1+z)} \mathrm{d}r \tag{5.1}$$

where $\mathrm{d}r$ is the element of comoving coordinate distance. Sunyaev and Zeldovich (1970) first showed that such large quantities of hot gas along the line of sight lead to Compton scattering of the photons of the Cosmic Microwave Background Radiation and consequently to distortions from a pure black-body spectrum. The Compton scattering process is statistical in that the photons encounter hot electrons with temperature $T_e \gg T_{rad}$. There is no net energy gain in first order Compton scatterings since there are as many head-on as following collisions with the hot electrons but there is a net energy gain to second order in v/c. I have given a discussion of the origin of this result in my text-book *High Energy Astrophysics, Volume 1*, section 4.3.4 (Longair 1992). The form of the distortion of the spectrum is given by the expression (1.6) and is illustrated in Fig. 5.3.

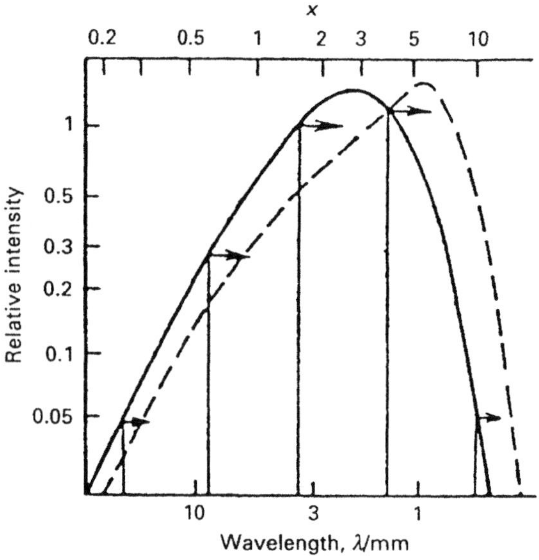

Fig. 5.3 Illustrating the Compton scattering of a Planck distribution by hot electrons in the case in which the Compton optical depth $y = 0.15$. The intensity decreases in the Rayleigh-Jeans region of the spectrum and increases in the Wien region (Sunyaev 1980).

The most recent analyses of the COBE spectrum of the Cosmic Microwave Background Radiation reported by Mather (1993) have given an upper limit to the Compton scattering optical depth $y \leq 2.5 \times 10^{-5}$. The most conservative estimate we can make for y is to assume that the hot gas is only present at the present epoch in which case we can approximate

$y = (c/H_0)\sigma_T N_0(kT_e/m_e c^2)$. Assuming $\Omega_b = 0.23$, and $kT = 40$ keV, we find $y = 5 \times 10^{-4}$, already in conflict with the observations. If we adopt any of the more physically realistic models in which the heating takes place at a larger redshift, the value of y is greater by factor of roughly $(1+z)^{2.5}$ resulting in an even greater conflict with the observed value of y. More detailed models of the heating of the intergalactic gas which can account for the observed X-ray background spectrum come to exactly the same conclusion (Taylor and Wright 1989).

Fabian and Barcons (1992) argue convincingly that, although the value of Ω_b can be reduced if the hot gas is clumped, the clumps also result in large fluctuations in the Cosmic Microwave Background Radiation due to the Sunyaev-Zeldovich effect when the radiation passes through the clumps. We can therefore conclude that the X-ray background is not due to hot diffuse intergalactic gas. To complicate matters further, it has to be recalled that most of the X-ray background at 1 keV can be associated with discrete sources with energy spectra $I(\varepsilon) \propto \varepsilon^{-0.7}$ and so this component has to be subtracted from the X-ray background spectrum in the $1-20$ keV energy range. This has the effect of flattening the spectrum as compared with the spectrum of thermal bremsstrahlung in this energy range. Fabian and Barcons were led to conclude that 'the perfect bremsstrahlung shape of the X-ray background is just a cosmic conspiracy'.

5.2.2 The X-ray Spectra of Seyfert Galaxies and the X-ray Reflection Spectra of Cool Clouds

One obvious approach is to study the X-ray spectra of active galaxies in the energy range $1-20$ keV in order to find out if any of them have spectra similar in form to that of the X-ray background. One possible clue comes from the spectra of a number of Seyfert-type galaxies observed by the Japanese GINGA satellite. Pounds et al (1990) found evidence for significant distortions from a simple power-law spectrum in this energy range when they added together the spectra of a number of Seyfert galaxies (Fig. 5.4). Fig. 5.4(a) shows the summed spectrum for the 12 Seyfert galaxies and the residuals observed when a best-fitting power-law is compared with the observations. Fig. 5.4(b) shows the improved fit obtained when a reflected X-ray component labelled (b) is added to the power-law spectrum. Fabian et al (1990) noted that the reflected component has a spectrum which could provide the type of spectral feature needed to account for the break in the spectrum of the background at about 30 keV.

The process of X-ray reflection involves illuminating a cool gas cloud with a power-law X-ray energy spectrum and then working out the reflected X-ray spectrum. There are two competing processes, Compton scattering which is important at high energies and photoelectric absorption which becomes dominant at low X-ray energies. In a single Compton scattering of high energy photons by stationary electrons, the average decrease in energy of each photon

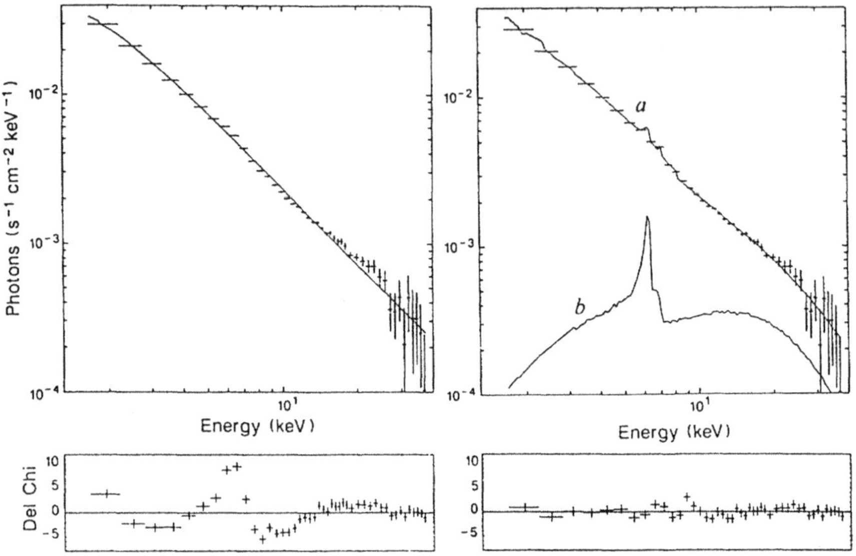

Fig. 5.4 (a) The summed X-ray spectra of 12 Seyfert galaxies compared with a best-fitting power-law spectrum. The goodness-of-fit to this energy distribution is displayed in the lower panel in which it can be seen that there are significant deviations from a power-law spectrum. (b) The addition of a reflection component to the power-law distribution results in a much improved fit to the summed spectra (From Pounds et al 1990).

per Compton collision is

$$\left\langle \frac{\Delta\epsilon}{\epsilon} \right\rangle = \frac{h\nu}{m_e c^2} \tag{5.2}$$

Thus, the high energy photons lose more energy per collision than the lower energy photons, resulting in a progressive steepening of the spectrum with increasing energy. At low energies, photoelectric absorption is the dominant process and has the characteristic power-law dependence of the absorption cross-section $\sigma_{\mathrm{ph}}(\epsilon) \propto \epsilon^{-3}$ at energies greater than the characteristic absorption edges. The most important absorption edge in the 2–10 keV range is the K-edge of iron which occurs about 8 keV. Thus, at low X-ray energies there is strong absorption due to photoelectric absorption and a prominent feature at energies just greater than the K absorption edge of iron. The net result is that there is a maximum in the reflected spectrum at about 20 keV at which the combined energy losses due to Compton scattering and photoelectric absorption are at a minimum. To solve the problem properly, it is necessary

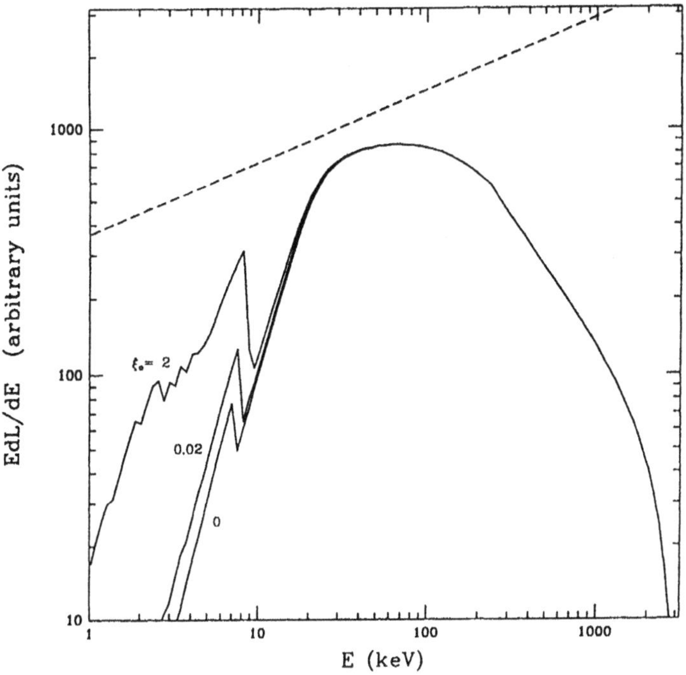

Fig. 5.5 The reflection spectra from a semi-infinite, plane-parallel cold medium assuming the cosmic abundance of the elements. The prominent feature at about 8 keV is associated with the K-absorption edge of iron. The input X-ray spectrum is of power-law form $I(\varepsilon) \propto \varepsilon^{-0.7}$. The units on the ordinate are $\varepsilon I(\varepsilon)$ in our notation. ξ is the differential ionisation parameter which describes how the ionisation state of the cold cloud is modified by the incident X-ray spectrum. In the case $\xi = 0$, the matter remains cold. According to Lightman and White, the value $\xi = 2$ is at the upper limit of acceptable values for active galactic nuclei. (Lightman and White 1988).

to carry out a radiative transfer analysis of the competing absorption and scattering processes and this has been carried out by Lightman and White (1988) whose results are shown in Fig. 5.5.

It can be seen that qualitatively the reflected spectrum has the correct signature to account for the features observed in Seyfert galaxy spectra (Fig. 5.4). Zdziarski et al (1993) have found it difficult, however, to find wholly acceptable spectra for the X-ray background from this reflection process. Another problem is why it should be that only the reflected component is observed. Expressed alternatively, there has to be some mechanism by which the reflected component is enhanced at the expense of the direct component.

Another possibility for flattening the X-ray spectra in the 1 — 10 keV waveband would be if the sources have strong photoelectric absorption within the nuclear regions. It is found that there there are about three times more sources in the 2 – 10 keV waveband than expected from observations at 1 – 2 keV and these could be sources with column depths of neutral hydrogen much greater than 10^{21} cm^{-2}. Some X-ray galaxies are known with very large column depths of neutral hydrogen, up to about $N_{\mathrm{H}} \sim 10^{24}$ cm^{-2}, which means that the optical depth for photoelectric absorption would be about 1 at an energy of 10 keV (see, for example, Pounds 1990). Such sources would not be detected in sky surveys carried out at soft X-ray energies such as the ROSAT survey. The superposition of sources with different large column depths at different redshifts could lead to a flattening of the background relative to the typical X-ray spectra of active galaxies.

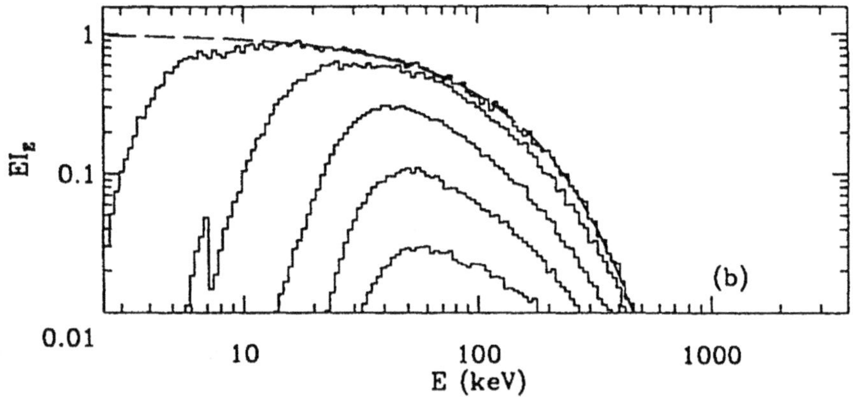

Fig. 5.6 Monte Carlo calculations of the intensity spectra transmitted through a spherical cloud of cosmic abundance of the elements. The electron temperature is assumed to be zero. The incident spectrum (dashed line) is of the form $I_\varepsilon = \varepsilon^{-\alpha}\exp(-\varepsilon/kT)$ with $\alpha = 1$ and $kT = 100$ keV. The histograms show the transmitted spectra for Thomson optical depths of $\tau_{\mathrm{T}} = 0.1, 1, 3, 5$ and 7 (From Madau, Ghisellini and Fabian 1993).

A variant of this type of model has been presented by Madau, Ghisellini and Fabian (1993). They start from the observation that, according to unified models of Seyfert galaxies, there is a thick torus surrounding the active galactic nucleus, the broad lines from the nucleus only being observed if the torus is observed more or less along its axis (Antonucci and Miller 1985). Objects observed in such directions would be identified as Seyfert I galaxies while those observed at an angle to the axis of the torus would not display the broad emission lines because of obscuring effect of the torus and so be classified as Seyfert II galaxies. Madau, Ghisellini and Fabian propose that the X-rays from the nuclei of such Seyfert II galaxies would be strongly modified because they have to pass through a column depth of roughly 10^{24} cm^{-2}.

Once again the competing processes of photoelectric absorption at the low energies and Compton energy losses at high energies result in a modified transmitted spectrum. An example of the resulting spectra expected from tori of different hydrogen column depths is shown in Fig. 5.6, assuming an input energy spectrum of the form $I_\varepsilon = \varepsilon^{-\alpha} \exp(-\varepsilon/kT)$.

Rees (1993) has pointed out that the process of Compton energy losses can result quite naturally in a feature in the spectra of X-ray sources in the $10-50$ keV energy range. If the optical depth to Thomson scattering is τ, the number of scatterings undergone by the photons in escaping form the region is $\sim \tau^2$. Since the energy loss per collision is $\Delta\varepsilon/\varepsilon \sim (\varepsilon/m_e c^2)$, there are expected to be significant Compton distortions of the energy spectra for those photons for which $\Delta\varepsilon \sim \varepsilon$ after τ^2 collisions, that is for energies greater than $\varepsilon \approx m_e c^2/\tau^2$. Thus, if $\tau \sim 3-5$, a feature should be observed in the hard X-ray spectra of these sources at $\sim 20-50$ keV.

5.2.3 X-ray Sources with the Correct Forms of X-ray Spectra

Are there any sources known with the correct form of spectrum to account for the X-ray background? Sunyaev (1993) has offered two possibilities. Observations with the SIGMA/GRANAT hard X-ray telescope have shown that X-ray accreting pulsars have energy spectra which are very hard (see also Ballet 1993). All of them have spectra similar to that of the X-ray background with a sharp cut-off at energies $\varepsilon \sim 50-100$ keV. The numbers of such sources present in our Galaxy today are not nearly sufficient to account for the intensity of the background emission if it is assumed that the background were due to the integrated emission of such sources in normal galaxies. It would be necessary to appeal to a very much greater number of X-ray accreting pulsars in galaxies at earlier cosmological epochs.

The spectrum of an extragalactic X-ray source which seems to have exactly the correct form is NGC 4151. Its spectral index is $\alpha = 0.4$ in the energy range $1-30$ keV and there is a spectral cut-off at energies greater than 60 keV. According to Sunyaev (1993), this form of spectrum can be attributed to Comptonisation of the radiation from the nucleus. Other Seyfert I galaxies observed by the Compton Gamma Ray Observatory are reported to have similar spectra (see references in Madau et al 1993).

It is evident that the origin of the spectrum of the X-ray background at energies $\epsilon \geq 2$ keV remains an unsolved problem. Further observations by the Compton Gamma Ray Observatory may shed much more light on this problem. There is a clear need for sky surveys at hard X-ray energies.

5.3 The Gamma-ray Background

Prior to the launch of the Compton Gamma Ray Observatory, knowledge of extragalactic γ-ray sources was fragmentary in the extreme. A few extragalactic sources were known such as NGC 4151 and 3C273 and there was some evidence that there is a feature in the γ-ray background spectrum at about 1 MeV.

Following the launch of the Compton Observatory, many more extragalactic sources are now known, although the picture is still far from complete. Since the launch of the Observatory in April 1991, the whole sky has been surveyed in high energy γ-rays although at the time of writing (July 1993) the map corresponding to the famous COS-B map of the sky has not been published. To summarise some of the highlights of the Compton Observatory programme to date related to extragalactic astrophysics (Kniffen 1993):

- Over 500 γ-ray bursts have been observed and their distribution seems to be isotropic over the sky. Furthermore, their source counts, as measured by the value of $V/V_{\rm max}$ shows a deficit of faint sources. The value of $V/V_{\rm max}$ is only 0.324 ± 0.016 compared with the value of 0.5 expected for a uniform distribution. This suggests that the burst sources either originate in an extended halo distribution about our Galaxy or might be a cosmological population. In the later case, the γ-ray bursts would be the most violent events we know of in the Universe.
- The extragalactic γ-ray sources are highly variable. The source 3C 279 was detected during a violent outburst soon after launch when it became the brightest source in the sky but has since disappeared.
- One of the most remarkable results has been the detection of strong γ-ray emission at energies $\varepsilon \sim 100$ MeV from 24 extremely active galactic nuclei. The remarkable feature of these sources is that all of them are radio-loud compact quasars of the most violently variable types known. The energetics of the γ-ray emission are so great that the radiation has to be beamed. It is pleasing that, in half of these compact radio sources, there is evidence for superluminal motion of the radio components, thus providing independent evidence for relativistic beaming.
- γ-rays have also been detected from the Magellanic Clouds. Of special interest are the observations of the Small Cloud because, knowing the amount of cold interstellar gas present, upper limits can be found for the flux of cosmic ray protons and nuclei in the Cloud. The observed γ-ray intensity from the Small Magellanic Cloud indicates that the flux of cosmic rays is significantly less than the local flux of cosmic rays in our Galaxy, thus showing that the local cosmic ray flux is not universal.

It is probably too early to make detailed computations of the cosmological significance of these results but the recent papers by Padovani et al (1993) and Stecker et al (1993) give a flavour for the type of science which can be expected. They have noted that all the known high energy γ-ray sources are core-dominated flat spectrum radio sources. For each of the sources they have evaluated the ratio of the γ-ray to radio luminosity. If these ratios apply to all compact radio sources, the background emission can be evaluated adopting models for the cosmological evolution of the radio source population. Carrying out this calculation, they found that they could already account for about 65% of the 100 MeV background intensity and possibly for all of it. There does not seem to be any room for the γ-ray emission of the radio-quiet counterparts of the radio quasars.

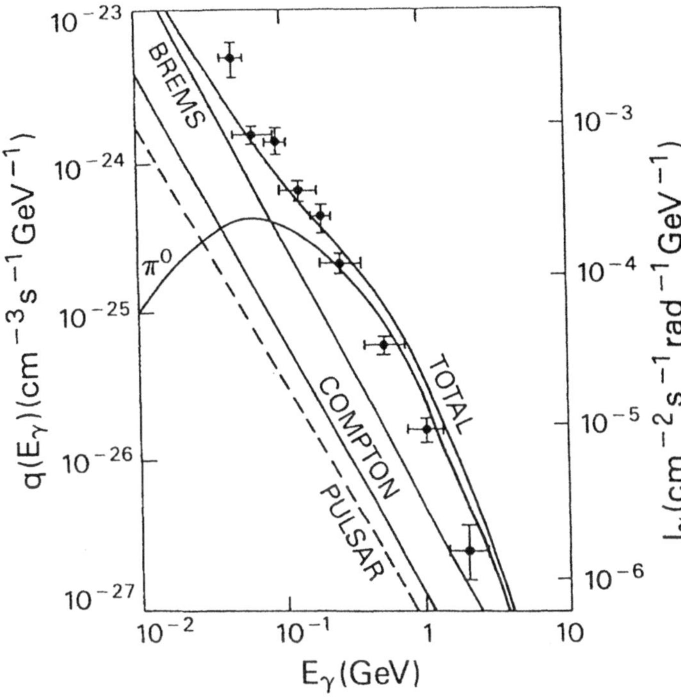

Fig. 5.7 The diffuse high energy γ-ray spectrum of our Galaxy from observations by the COS-B and SAS-2 satellites. The observed spectrum has been accounted for in this representation by π^0-decays as well as relativistic bremsstrahlung from Galactic cosmic ray electrons and inverse Compton scattering of starlight. (From Stecker 1989).

These conclusions are, in fact, updated versions of the arguments discussed by Setti and Woltjer (1979) who noted that not more than 10% of all quasars could emit as intense γ-rays as 3C273 or else the 100 MeV γ-ray background intensity would be exceeded. Most authors are of the consensus that the high energy γ-ray background is probably associated with the superposition of discrete sources such as quasars and active galaxies.

There remains the problem of the γ-ray feature at $1 - 3$ MeV. In this spectral region, the observations by the Compton Observatory can provide unique information. The spectrum of 3C273 displays a spectral break at about 2 MeV of the form which would be needed to account for the background as the integrated emission of discrete sources.

Just as a reminder of the potential richness of this area for high energy astrophysics, I reproduce in Fig. 5.7 the spectrum of the plane of our Galaxy in high energy γ-rays as observed by the COS-B and SAS-2 satellites (Stecker 1989). It can be observed that it is quite plausible that the high energy γ-ray emission of the Galaxy consists of contributions from π^0-decays as well as relativistic bremsstrahlung from Galactic cosmic ray electrons and inverse Compton scattering of starlight. Observations by the Compton Observatory should show how typical our Galaxy is of other extragalactic systems and which of these processes might make significant contributions to the γ-ray background radiation. More details of these processes and observations of the γ-ray background are given in the monograph *Gamma-Ray Astronomy* by Ramana-Murthy and Wolfendale (1993).

References – Chapter 5

Antonucci, R.R. and Miller, J.S. (1985). ApJ, **297**, 621.

Ballet, J. (1993). In *The Frontiers of Space and Ground-based Astronomy*, 27th ESLAB Symposium, (eds. W. Wamsteker, M.S. Longair and Y. Kondo), (in press). Dordrecht: Kluwer Scientific Publishers.

Fabian, A.C. and Barcons, X. (1992). ARA&A, **30**, 429.

Fabian, A.C., George, I.M., Miyoshi, S. and Rees, M.J. (1990). MNRAS, **242**, 14P.

Hasinger, G., Burg, R., Giacconi, R., Hartner, G., Schmidt, M., Trümper, J. and Zamorani, G. (1993). A&A, **275**, 1.

Kniffen, D. (1993). In *The Frontiers of Space and Ground-based Astronomy*, 27th ESLAB Symposium, (eds. W. Wamsteker, M.S. Longair and Y. Kondo), (in press). Dordrecht: Kluwer Scientific Publishers.

Lightman, A.P. and White, T.R. (1988). ApJ, **335**, 57.

Longair, M.S. (1992). *High Energy Astrophysics. Vol. 1.* Cambridge: Cambridge University Press.

Madau, P., Ghisellini, G. and Fabian, A.C. (1993). ApJ, **410**, 7.

Mather, J. (1993). In *The Extragalactic Background Radiation*, (eds. M. Livio, M. Fall, D. Calzetti and P. Madau), (in press). Cambridge: Cambridge University Press.

Marshall, F.E., Boldt, E.A., Holt, S.S., Miller, R.B., Mushotzky, R.F., Rose, L.A., Rothschild, R.E. and Serlemitsos, P.J. (1980). ApJ, **235**, 4.

Padovani, P., Ghisellini, G., Fabian, A.C. and Celotti, A. (1993). MNRAS, **260**, L21.

Pounds, K.A. (1990). MNRAS, **242**, 20P.

Pounds, K.A., Nandra, K., Stewart, G.C., George, I.M. and Fabian, A.C. (1990). Nat, **344**, 132.

Ramana-Murthy, P.V. and Wolfendale, A.W. (1993). *Gamma-Ray Astronomy.* Cambridge: Cambridge University Press.

Rees, M.J. (1993). In *The Extragalactic Background Radiation*, (eds. M. Livio, M. Fall, D. Calzetti and P. Madau), (in press). Cambridge: Cambridge University Press.
Setti, G. and Woltjer, L. (1979). A&A, **76**, L1.
Stecker, F.W. (1989). In *Cosmic Gamma Rays, Neutrinos and Related Astrophysics*, (eds M.M. Shapiro and J.P. Wefel), 85. Dordrecht: Kluwer Academic Publishers.
Stecker, F.W., Salamon, M.H. and Malkan, M.A. (1993). ApJ, **410**, 71.
Sunyaev, R.A. (1980). Sov. Astron. Lett., **6**, 213.
Sunyaev, R.A. (1993). In *The Extragalactic Background Radiation*, (eds. M. Livio, M. Fall, D. Calzetti and P. Madau), (in press). Cambridge: Cambridge University Press.
Sunyaev, R.A. and Zeldovich, Ya.B. (1970). Astrophys. Sp. Sci., **7**, 20.
Taylor, G.B. and Wright, E.L. (1989). ApJ, **339**, 619.
Zdziarski, A.A., Zycki, P.T., Svensson, R. and Boldt, E. (1993). ApJ, **405**, 125.

6 A Brief Thermal History of the Universe

We now turn to the role of the background radiation, in particular the Cosmic Microwave Background Radiation, in the study of the Big Bang model of the Universe. This is a very well known story and so I will only highlight the essential features of the story which are necessary to interpret the observations of the Cosmic Microwave Background Radiation. For more details reference should be made to the texts recommended in Section 1.2 (for example, Longair 1989).

6.1 The Matter and Radiation Content of the Universe

First, we have to estimate the major contributors to the inertial mass density of the Universe and this is found by comparing the inertial mass densities in matter and radiation. As demonstrated in Section 1.11, the Cosmic Microwave Background Radiation provides by far the greatest contribution to the energy density of radiation in intergalactic space and so we can write

$$\frac{\rho_{\rm rad}}{\rho_{\rm matter}} = \frac{aT^4(z)}{\Omega_0 \rho_{\rm c}(1+z)^3 c^2} = \frac{2.6 \times 10^{-5}(1+z)}{\Omega_0 h^2} \tag{6.1}$$

Thus, the Universe is expected to be *matter-dominated* at redshifts $z \leq 4 \times 10^4 \Omega_0 h^2$ and the dynamics are described by the standard Friedman models of Section 3.3, $R \propto t^{2/3}$ provided $\Omega_0 z \gg 1$. At redshifts $z \geq 4 \times 10^4 \Omega_0 h^2$, the Universe is *radiation-dominated* and then the dynamics are described by the solutions of Section 3.5, $R \propto t^{1/2}$.

Another important number is the present photon to baryon ratio. Assuming $T_{\rm rad} = 2.726$ K,

$$\frac{N_\gamma}{N_{\rm B}} = \frac{3.6 \times 10^7}{\Omega_0 h^2} \tag{6.2}$$

If photons are neither created nor destroyed during the expansion of the Universe, this number is an invariant. This ratio is a measure of the factor by which the photons outnumber the baryons in the Universe at the present epoch and is also proportional to the specific entropy per baryon during the radiation-dominated phases of the expansion (see, for example, Kolb and Turner 1990).

As shown in Section 3.5, the spectrum of the Cosmic Microwave Background Radiation preserves its black-body form during the expansion of the Universe but is shifted to higher photon energies and radiation temperatures by a factor $(1+z)$, $T(z) = 2.726(1+z)$ K, and the total energy density increases by a factor $(1+z)^4$. We can therefore identify certain epochs which are of special significance in the temperature history of the Universe.

6.2 The Epoch of Recombination

At a redshift $z \approx 1,500$, the radiation temperature of the Microwave Background Radiation $T_r \approx 4,000$ K and then there are sufficient photons with energies $h\nu \geq 13.6$ eV in the tail of the Planck distribution to ionise all the neutral hydrogen in the intergalactic medium. It may at first appear strange that the temperature is not closer to 150,000 K, at the temperature at which $\langle h\nu \rangle = kT = 13.6$ eV, the ionisation potential of neutral hydrogen. The important points to note are that the photons far outnumber the baryons in the intergalactic medium (expression 6.2) and there is a broad range of photon energies present in the Planck distribution. It needs only about one in 10^8 of the photons present to have energy greater than 13.6 eV to have as many ionising photons as hydrogen atoms. Thus, the ionisation of the intergalactic gas occurs at a lower temperatures than would be predicted by equating $h\nu$ to kT. This phenomenon occurs in a number of astrophysical problems and is due to the same feature of broad equilibrium distributions such as the Planck and Maxwell distributions — there is always a very small number of very energetic particles or photons present in any equilibrium distribution. Examples include the photoionisation of the regions of ionised hydrogen, the temperature at which nuclear burning is initiated in the cores of stars and the temperature at which dissociation of light nuclei by background thermal photons takes place in the early Universe. In all these cases, the physical processes take place at temperatures about a factor of 10 less than expected.

The result is that at redshifts $z \geq 1,500$, the intergalactic gas is an ionised plasma and for this reason the redshift $z_r = 1,500$ is referred to as the *epoch of recombination*. At earlier epochs, $z \approx 6,000$, helium is fully ionised as well. The most important consequence is that the Universe is opaque to Thomson scattering. This is the simplest of the scattering processes which impede the propagation of photons from their sources to the Earth through an ionised plasma. The photons are scattered without any loss of energy by free electrons. The optical depth of the intergalactic gas to Thomson scattering can be written in the usual way

$$d\tau_T = \sigma_T N_e(z) \frac{dr}{(1+z)} = \sigma_T N_e(z) c \frac{dt}{dz} dz \tag{6.3}$$

where σ_T is the Thomson scattering cross-section $\sigma_T = 6.665 \times 10^{-29}$ m^2. Detailed calculations of the ionisation state of the intergalactic gas with redshift are discussed in Section 9.1 in which it is shown that the optical depth of the intergalactic gas becomes unity at a redshift close to 1,000. Let us work out the optical depth in the limit of large redshifts, assuming that the Universe is matter-dominated so that the cosmic-time redshift relation (3.15) can be written $dz/dt = -H_0 \Omega_0^{1/2} z^{5/2}$. Then,

$$\tau_T = \frac{2}{3}\frac{c}{H_0}\frac{\sigma_T \rho_c \Omega_0^{1/2}}{m_p}[z^{3/2} - z_0^{3/2}] = 0.04\,(\Omega_0 h^2)^{1/2}[z^{3/2} - z_0^{3/2}] \tag{6.4}$$

It can be seen that the optical depth to Thomson scattering becomes very large as soon as the intergalactic hydrogen becomes fully ionised at redshift z_0. As a result, the Universe beyond a redshift of about 1,000 becomes unobservable because any photons originating from larger redshifts are scattered many times before they propagate to the Earth. Consequently all the information they carry about their origin is rapidly lost. There is therefore a *photon barrier* at a redshift of 1,000 beyond which we cannot obtain information directly using photons. We will return to the process of recombination and the variation of the optical depth of the intergalactic gas to Thomson scattering with redshift in Chapter 9 because it is crucial in evaluating the amplitude of fluctuations in the Cosmic Microwave Background Radiation. If there is no further scattering of the photons between the epoch of recombination and now, the redshift of about 1,000 becomes the *last scattering surface* and therefore the fluctuations we observe on the sky today were imprinted on the Cosmic Background Radiation at that epoch.

6.3 The Epoch of Equality of Matter and Radiation Inertial Mass Densities

At a redshift $z = 4 \times 10^4 \Omega_0 h^2$, the matter and radiation make equal contributions to the inertial mass density and at larger redshifts the Universe is radiation-dominated. The difference in the variation of the scale factor with cosmic epoch has already been discussed. There are two other important changes. First, after the intergalactic gas recombines, and specifically at redshifts $z \leq 100$, there is negligible coupling between the matter and the photons of the Microwave Background Radiation because the matter is neutral. This statement would of course be incorrect if the intergalactic medium were ionised at some later epoch. At redshifts greater than 1,000, however, the intergalactic hydrogen is ionised and the matter and radiation have a very large optical depth for Thomson scattering as shown by equation (6.4).

If the matter and radiation were not thermally coupled, they would cool independently, the hot gas having ratio of specific heats $\gamma = 5/3$ and the radiation $\gamma = 4/3$. These result in adiabatic cooling which depends upon the scale factor R as $T_{\rm m} \propto R^{-2}$ and $T_{\rm r} \propto R^{-1}$ for the matter and radiation respectively. We would therefore expect the matter to cool much more rapidly than the radiation. This is not the case, however, because the matter and radiation are coupled by Compton scattering. In particular, there are sufficient Thomson scatterings of the electrons by the photons of the background radiation to maintain the matter at the same temperature as the radiation.

The exchange of energy between photons and electrons is an enormous subject and has been treated by Weymann (1965), Sunyaev and Zeldovich (1980) and Pozdnyakov, Sobol and Sunyaev (1983). The equation for the rate of exchange of energy between a thermal radiation field at radiation temperature $T_{\rm r}$ and a plasma at temperature $T_{\rm e}$ interacting solely by Compton scattering has been derived by Weymann (1965).

$$\frac{\mathrm{d}\varepsilon_{\mathrm{r}}}{\mathrm{d}t} = 4N_{\mathrm{e}}\sigma_{\mathrm{T}}c\varepsilon_{\mathrm{r}}\left(\frac{kT_{\mathrm{e}} - kT_{\mathrm{rad}}}{m_{\mathrm{e}}c^2}\right) \tag{6.5}$$

where ε_{r} is the energy density of radiation. This equation expresses the fact that, if the electrons are hotter than the radiation, the radiation is heated up by the matter and, contrariwise, if the radiation is hotter than the matter, the matter is heated by the radiation. The astrophysical difference between the two cases arises because of the enormous difference in the number densities between the photons and the electrons $N_\gamma/N_{\mathrm{e}} = 3.6 \times 10^7(\Omega_0 h^2)^{-1}$.

Let us look at this difference from the point of view of the optical depths for the interaction of an electron with the radiation field and of a photon with the electrons of the intergalactic gas. In the first case, the optical depth for interaction of an electron with the radiation field is $\tau_{\mathrm{e}} = \sigma_{\mathrm{T}}cN_\gamma t$ whereas that of the photon with the electrons is $\tau_\gamma = \sigma_{\mathrm{T}}cN_{\mathrm{e}}t$ where σ_{T} is the Thomson cross-section and t is the age of the Universe. This means that it is much more difficult to modify the spectrum of the photons as opposed to the energy distribution of the electrons because in the time any one photon is scattered by an electron, the electron has been scattered many times by the photons. Another way of expressing this result is to say that the heat capacity of the radiation is very much greater than that of the matter.

We consider two applications of these formulae. In the first, we consider the heating of the electrons by the Compton scattering of the photons of the Microwave Background Radiation. The collision time between electrons, protons and atoms is always much shorter than the age of the Universe and hence, when energy is transferred from the radiation field to the electrons, it is rapidly communicated to the matter as a whole. This is the process by which the matter and radiation are maintained at the same temperature in the early Universe. Following Peebles (1968), let us work out the smallest redshift at which Compton scattering can maintain the matter and radiation at the same temperature. At epoch t, the total energy transfer to the matter is

$(\mathrm{d}\varepsilon_{\mathrm{r}}/\mathrm{d}t)t$ where $(\mathrm{d}\varepsilon_{\mathrm{r}}/\mathrm{d}t)$ is given by the equation (6.5). When the total energy transfer is of the same order as the energy density in the radiation field, no more energy can be transferred and the heating ceases. At this point $(T_{\mathrm{rad}} - T_{\mathrm{e}})/T_{\mathrm{rad}} \approx 1$ and hence the condition becomes $4T_{\mathrm{rad}}N_{\mathrm{e}}\sigma_{\mathrm{T}}ctk/m_{\mathrm{e}}c^2 \approx 1$. We can write $N_{\mathrm{e}} = 11x\Omega_0 h^2(1+z)^3\ \mathrm{m}^{-3}$ where x is the degree of ionisation of the intergalactic gas. The thermal contact between the photons and the plasma is very strong up to the epoch of recombination and it is the energy transfer after recombination which is of interest. We need the variation of x with redshift following recombination and this has been evaluated by Peebles (1968). Inspecting his tables for the degree of ionisation of the intergalactic gas, we find that the decoupling of matter and radiation occurred at a redshift $z \approx 100$ when $x \approx 10^{-5}$. Thus, Compton scattering maintains the matter and radiation at the same temperature even at epochs well after the epoch of recombination.

In the second case, we study the necessary conditions for significant distortion of the spectrum of the Cosmic Microwave Background Radiation. If, by some process,
the electrons are heated to a temperature greater than the radiation temperature and if no photons are created, the spectrum of the radiation is distorted from its black-body form by Compton scattering. The interaction of the hot electrons with the photons results in an average frequency change of $\Delta\nu/\nu \approx kT_e/m_e c^2$. Thus, to obtain a significant change in the energy of the photons, $\Delta\nu/\nu \approx 1$, we require the Compton optical depth

$$\tau_C = \int \left(\frac{kT_e}{m_e c^2}\right) \sigma_T c N_e dt \tag{6.6}$$

to be one or greater. If we take $T_e = T_r(1+z)$ K, we find that $\tau = 1$ at a redshift $z = 2\times 10^4(\Omega_0 h^2)^{-1/5}$. Zeldovich and Sunyaev (1980) have discussed in detail the forms of distortion of the spectrum of the Cosmic Microwave Background Radiation. In particular, if there had been a major release of energy at these large redshifts which heated the matter to temperatures greater than T_{rad}, Compton scatterings would result in a Bose-Einstein spectrum with a finite chemical potential μ (expression 1.5). The limits to the dimensionless chemical potential obtained from the FIRAS experiment on COBE indicate that $\mu \leq 3.3\times 10^{-4}$ so that there cannot have been a major release of energy at these epochs. Such an energy release might have been associated, for example, with the dissipation of low mass adiabatic fluctuations (see Section 7.6).

Once the thermal history of the intergalactic gas has been worked out, the variation of the speed of sound with cosmic epoch can be determined. All sound speeds are roughly the square root of the ratio of total energy density to total inertial mass density. More precisely, the speed of sound c_s is given by

$$c_s^2 = \left(\frac{\partial p}{\partial \rho}\right)_S$$

where the subscript S means 'at constant entropy', that is, we consider adiabatic sound waves. The complication is that, from the epoch when the energy densities of matter and radiation were equal to beyond the epoch of recombination and the subsequent neutral phase, the dominant contributors to p and ρ change dramatically as the Universe changes from being radiation-dominated to matter-dominated, the coupling between the matter and the radiation becomes weaker and finally the plasma recombines at redshifts of about 1,000. As long as the matter and radiation are closely coupled, the square of the sound speed can be written

$$c_s^2 = \frac{(\partial p/\partial T)_{rad}}{(\partial \rho/\partial T)_{rad} + (\partial \rho/\partial T)_{mat}} \tag{6.7}$$

where the partial derivatives are taken at constant entropy. It is straightforward to show that this reduces to the following result:

$$c_s^2 = \frac{c^2}{3} \frac{4\rho_{\text{rad}}}{4\rho_{\text{rad}} + 3\rho_{\text{mat}}} \tag{6.8}$$

Thus, in the radiation-dominated phases, $z \geq 4 \times 10^4 \Omega_0 h^2$, the speed of sound tends to the relativistic sound speed, $c_s = c/\sqrt{3}$. However, at smaller redshifts, the sound speed decreases as the contribution of the inertial mass density in the matter becomes more important. After recombination, the sound speed is just the thermal sound speed of the matter which, because of the close coupling between the matter and the radiation, has temperature $T \approx 4,000$ K at $z = 1,500$.

6.4 Early Epochs

We can now extrapolate the Hot Model back to much earlier epochs. First, we can extrapolate back to redshifts $z \approx 10^8$ when the radiation temperature was about $T_r = 3 \times 10^8$ K. These temperatures are sufficiently high for the background photons to have γ-ray energies, $\varepsilon = kT_r = 25$ keV. At this high temperature, the high energy photons in the tail of the Planck distribution are energetic enough to dissociate light nuclei such as helium and deuterium. At earlier epochs, all nuclei are dissociated. We will study briefly the process of primordial nucleosynthesis of the light elements in the next sub-section.

At a redshift, $z \approx 10^9$, electron-positron pair production from the thermal background radiation can take place and the Universe was then flooded with electron-positron pairs, one pair for every pair of photons present in the Universe now. At a slightly earlier epoch the opacity of the Universe for weak interactions becomes unity.

We can extrapolate even further back in time to $z \approx 10^{12}$ when the temperature of the background radiation was sufficiently high for baryon-antibaryon pair production to take place from the thermal background. Just as in the case of the epoch of electron-positron pair production, the Universe was flooded with baryons and antibaryons, one pair for every pair of photons present in the Universe now.

We can carry on this process of extrapolation further and further back into the mists of the early Universe as far as we believe we understand high energy particle physics. Probably most particle physicists would agree that the standard model of elementary particles has been tried and tested to energies of at least 100 GeV and so we can probably trust laboratory physics back to epochs as early as 10^{-6} s although the more conservative of us would probably be happier to accept 10^{-3} s. I show schematically in Fig. 6.1 the standard thermal history of the Hot Big Bang. How far back one is prepared to extrapolate is largely a matter of taste. The most ambitious theorists have no hesitation in extrapolating back to the very earliest Planck eras, $t_P \sim (Gh/c^5)^{1/2} = 10^{-43}$ s, when the relevant physics was certainly very different from the physics of the Universe from redshifts of about 10^{12} to the present day (see Section 3.6).

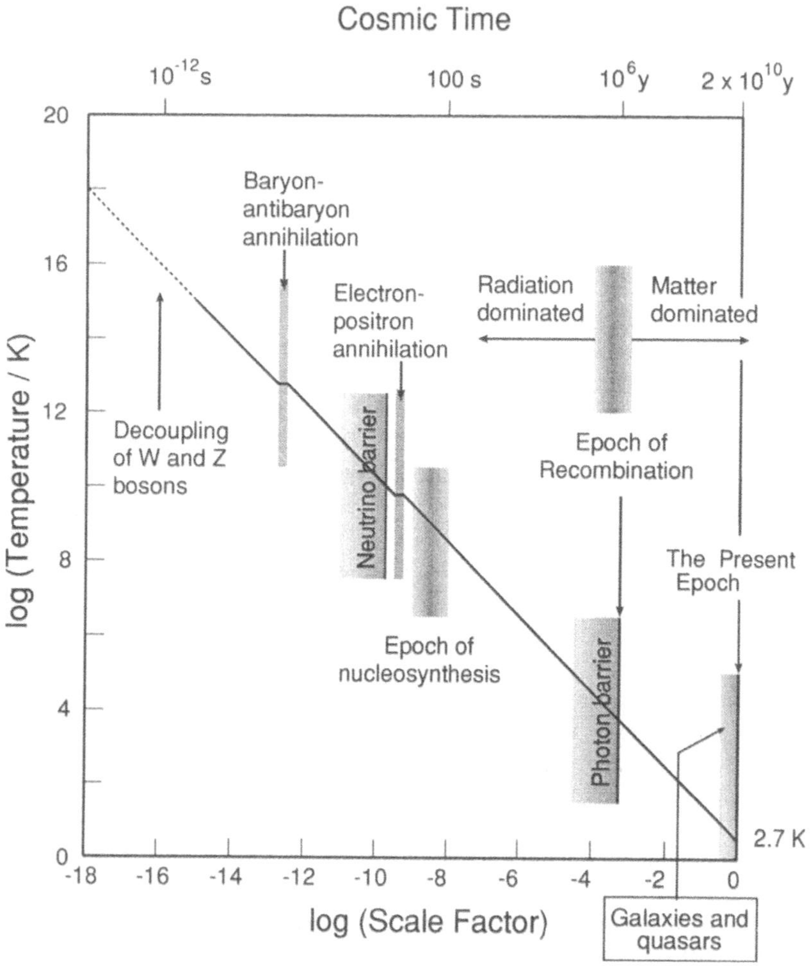

Fig. 6.1 The thermal history of the standard Hot Big Bang. The radiation temperature decreases as $T_r \propto R^{-1}$ except for abrupt jumps as different particle-antiparticle pairs annihilate at $kT \approx mc^2$. Various important epochs in the standard model are indicated. An approximate time scale is indicated along the top of the diagram. The neutrino and photon barriers are indicated. In the standard model, the Universe is optically thick to neutrinos and photons prior to these epochs.

6.5 Nucleosynthesis in the Early Universe

Primordial nucleosynthesis provides one of the most important constraints upon the density parameter of matter in the form of baryons Ω_b and this is a key part of our story. The basic physics of nucleosynthesis in the Big Bang is as follows. Consider a particle of mass m at very high temperatures such that its total energy is much greater than its rest mass energy, $kT \gg mc^2$.

If the timescales of the interactions which maintain this species in thermal equilibrium with all the other species present at temperature T are shorter than the age of the Universe at that epoch, statistical mechanics tells us that the equilibrium number densities of the particle and its antiparticle are

$$N = \bar{N} = \frac{4\pi g}{h^3}\int_0^{\infty} \frac{p^2 dp}{e^{E/kT} \pm 1} \tag{6.9}$$

where g is the statistical weight of the particle, p is its momentum and the $\pm$ sign depends upon whether the particles are fermions (+) or bosons (−). It will be recalled that photons are massless bosons for which $g = 2$, nucleons and antinucleons are fermions for which $g = 2$ and neutrinos are fermions which possess the helicity and hence $g = 1$. As a result, the following equilibrium number densities N and energy densities ε are found:

$$g = 2 \quad N = 0.244\left(\frac{2\pi kT}{hc}\right)^3 \mathrm{m}^{-3} \quad \varepsilon = aT^4 \quad \text{photons}$$

$$g = 2 \quad N = 0.183\left(\frac{2\pi kT}{hc}\right)^3 \mathrm{m}^{-3} \quad \varepsilon = \frac{7}{8}aT^4 \quad \text{nucleons}$$

$$g = 1 \quad N = 0.091\left(\frac{2\pi kT}{hc}\right)^3 \mathrm{m}^{-3} \quad \varepsilon = \frac{7}{16}aT^4 \quad \text{neutrinos}$$

To find the total energy density, we have to add all the equilibrium energy densities together, that is

$$\text{Total Energy Density} = \varepsilon = \chi(T)aT^4 \tag{6.10}$$

When the particles become non-relativistic, $kT \ll mc^2$, and the abundances of the different species are still maintained by interactions between the particles, the non-relativistic limit of the integral (6.9) gives an equilibrium number density

$$N = g\left(\frac{mkT}{h^2}\right)^{3/2} \exp - \left(\frac{mc^2}{kT}\right) \tag{6.11}$$

Let us consider the decoupling of protons and neutrons in the early Universe. At redshifts less than 10^{12}, the neutrons and protons are non-relativistic, $kT \ll mc^2$, and their equilibrium abundances are maintained by the electron-neutrino weak interactions

$$e^+ + n \rightarrow p + \bar{\nu}_e \qquad \nu_e + n \rightarrow p + e^- \tag{6.12}$$

For the neutrons and protons the values of g are the same and so the relative abundances of neutrons to protons is

$$\left[\frac{n}{p}\right] = \exp\left(-\frac{\Delta mc^2}{kT}\right) \tag{6.13}$$

where Δmc^2 is the mass difference between the neutron and the proton. This abundance ratio freezes out when the neutrino interactions can no longer maintain the equilibrium abundances of neutrons and protons. The condition for 'freezing out' is that the timescale of the weak interactions becomes greater than the age of the Universe. The timescale for the weak interactions is $t_{\rm weak} = (\sigma N c)^{-1}$ where σ is the weak interaction cross-section which is proportional to the square of the energy $\sigma \propto E^2$. N is the number density of nucleons which decreases as the Universe expands as R^{-3}. Since $R \propto T^{-1}$ and $E \propto T$, it follows that the weak interaction cross section decreases as $t_{\rm weak} \propto T^{-5}$.

This timescale can be compared with the timescale of the expansion of the Universe which is given by the expression (3.30). We have to include all the contributors to the energy density of the matter and radiation and so we have to use equation (6.9) to relate energy density and cosmic time. We therefore find

$$\varepsilon = \chi(T) a T^4 = \frac{3c^2}{32\pi G} t^{-2}$$

$$t \propto T^{-2}$$

Thus, the time scale of the weak interactions decreases much more rapidly with temperature than does the expansion time scale. Decoupling takes place when $t = t_{\rm weak}$. Substituting $\sigma = 3 \times 10^{-49} (E/m_e c^2)^2$ m^2 into the above formula, we find that decoupling takes place at an energy $kT \approx 1$ MeV. Since the difference in rest masses of the neutron and proton corresponds to $\Delta mc^2 = 1.28$ MeV, substituting into equation (6.13), we find that, at a temperature $kT = 1$ MeV when the Universe is only 1 s old, the neutron fraction is

$$\left[\frac{n}{n+p}\right] = 0.21$$

The neutron fraction does not decrease according to equation (6.13) after this epoch but only very slowly because the reactions (6.7) can no longer maintain the equilibrium abundances. Detailed calculations by Peebles (1966) quoted by Weinberg (1972) show that after 300 s the neutron fraction has fallen to 0.123. It is at this epoch that the bulk of the formation of the light elements takes place as shown in Fig. 6.2 (Wagoner 1973). In the nuclear reactions, almost all the neutrons are combined with protons to form ^{4}He nuclei so that for every pair of neutrons a helium nucleus is formed. The predicted helium to hydrogen mass ratio is therefore just twice the neutron fraction

$$\left[\frac{^4\mathrm{He}}{\mathrm{H}}\right] \approx 0.25$$

The detailed evolution of the light elements during the epoch of nucleosynthesis illustrated in Fig. 6.2 is the result of detailed calculations by Wagoner (1973). It turns out that in addition to ^{4}He, which is always produced with an abundance of about 23 to 25%, there are traces of the light elements deuterium (D), helium-3 (^{3}He) and lithium-7 (^{7}Li).

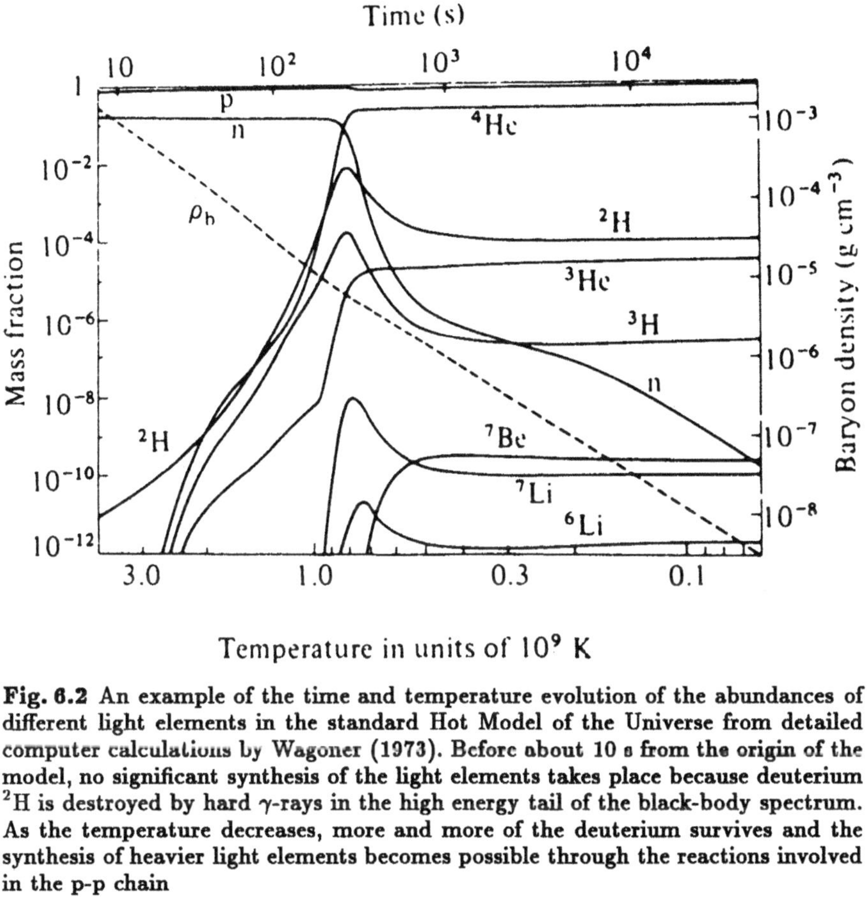

Fig. 6.2 An example of the time and temperature evolution of the abundances of different light elements in the standard Hot Model of the Universe from detailed computer calculations by Wagoner (1973). Before about 10 s from the origin of the model, no significant synthesis of the light elements takes place because deuterium ^{2}H is destroyed by hard γ-rays in the high energy tail of the black-body spectrum. As the temperature decreases, more and more of the deuterium survives and the synthesis of heavier light elements becomes possible through the reactions involved in the p-p chain

$$p + n \to D \quad D + D \to {}^3He + n \quad {}^3He + n \to T + p$$
$$D + D \to T + p \quad T + D \to {}^4He + n \quad {}^3He + {}^3He \to {}^4He + 2p$$

Notice that the synthesis of elements such as D, ^{3}He, ^{4}He, ^{7}Li and ^{7}Be is completed after about 15 m.

These are quite remarkable results. It has always been a great problem to understand why the abundance of helium is so high wherever it can be observed in the Universe. Its chemical abundance always appears to be greater than about 24%. The optical luminosity of our Galaxy is principally associated with hydrogen burning in stars and it can be easily shown that, if it has remained unchanged over cosmological time-scales, only about 1% of the mass of the Galaxy would have been converted from hydrogen into helium. Furthermore, once the central 10% of the mass of the star has been converted into helium, the star moves off the main sequence and becomes a red giant. This limit is known as the Schönberg-Chandrasekhar limit. During subsequent evolution on the giant branch, helium is burned into heavier elements and so it is difficult to understand why the helium abundance should be as high as 25% by mass and why the same value should be found wherever one looks. In addition, it has always been a mystery where the deuterium in the Universe could have been synthesised. It is a very fragile nucleus and is destroyed rather than created in stellar interiors. The same argument applies to the isotope of helium, ^{3}He, and to ^{7}Li. It is remarkable that it is precisely these elements which are synthesised in the early stages of the Hot Big Bang. The reason is simple. In stellar interiors, nucleosynthesis takes place in roughly thermodynamic equilibrium over very long timescales whereas in the early stages of the Hot Big Bang the 'explosive' nucleosynthesis is all over in a few minutes. The distinction is between stationary and non-stationary nucleosynthesis.

Notice that the physics which determines the abundance of ^{4}He is different from the processes responsible for the synthesis of the other light elements. The synthesis of ^{4}He is essentially thermodynamic, in that it is fixed by the ratio of neutrons to protons when the neutrinos decoupled from the nuclear reactions which maintained equilibrium between the protons and neutrons. In other words, the ^{4}He abundance is a measure of the *temperature* of the Universe at the epoch of decoupling of the neutrinos. In contrast, the abundances of the other light elements are determined by how far through the p-p chain the reactions can proceed before the temperature falls below that at which nucleosynthesis can take place. Thus, in a high density Universe, there is time for essentially all the neutrons to be combined into deuterium nuclei which then combine to form ^{4}He nuclei. On the other hand, if the matter density is low, there is not time for all the intermediate stages in the synthesis of helium to be completed and the result is a much higher abundance of deuterium and ^{3}He. Thus, the abundances of the deuterium and ^{3}He are measures of the *density* of the Universe. This has been quantified by Wagoner's calculations which are displayed in Fig 6.3. It can be seen that, for the standard Hot Big Bang, the ^{4}He abundance is remarkably insensitive to the present mass density in the Universe, in contrast to that of the other light elements.

The deuterium and ^{3}He abundances provide strong constraints upon the present baryon density of the Universe. It is found that the observed deuterium abundance relative to hydrogen is always about $[D/H] \approx 1.5 \times 10^{-5}$ (see, for example, Linsky 1993). Therefore, since we only know of ways of de-

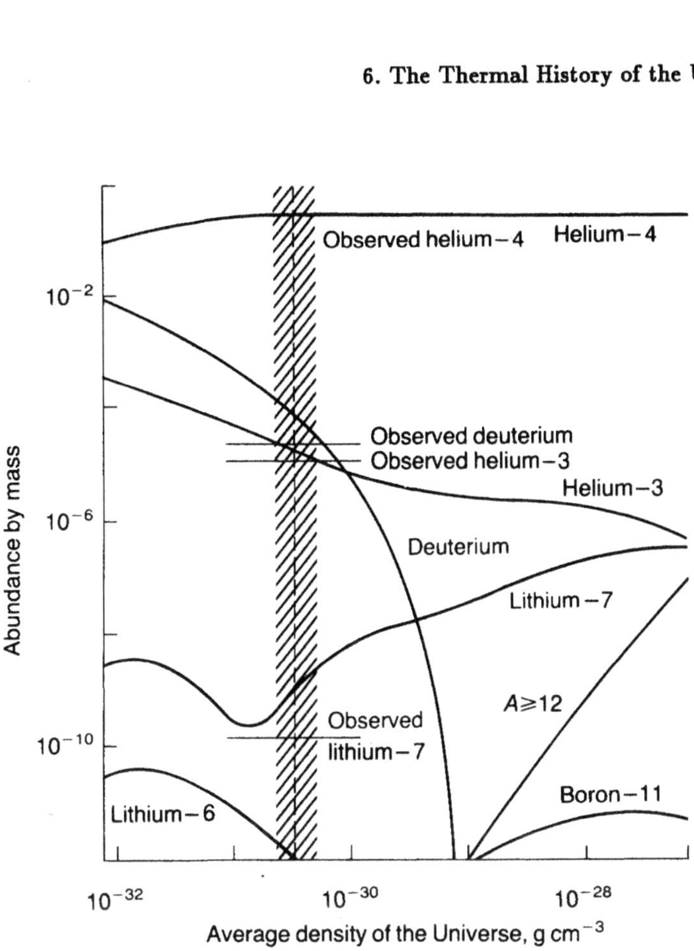

Fig. 6.3 The predicted primordial abundances of the light elements compared with their observed abundances. The present density of the Universe is shown along the abscissa. The observed abundances are in good agreement with models which have $\Omega_0 h^2 \leq 0.015$. This value is significantly smaller than the closure value $\Omega = 1$ (after Audouze 1986).

stroying deuterium rather than creating it, this figure provides a firm lower limit to the amount of deuterium which should be produced by primordial nucleosynthesis. In turn, this sets an upper limit to the present baryon density of the Universe. The figure which results from the most recent analyses is

$$\Omega_B h^2 \leq 0.015 \tag{6.14}$$

(see, for example, Audouze 1986). Thus, even adopting a small value of $h = 0.5$, this argument strongly suggests that baryonic matter cannot close the Universe. This will prove to be a key part of the story of the problems of galaxy formation.

As a by-product of the above analysis, we have derived the epoch at which the Universe becomes opaque to neutrinos, namely, the epoch when the neutrinos could no longer maintain the neutrons and protons in thermodynamic equilibrium. Just as there is a barrier for photons at a redshift of about 1,500, so there is a *neutrino barrier* at an energy $kT \approx 1$ MeV. This means that, if it were possible to undertake neutrino astronomy, we would expect the background neutrinos to be last scattered at the epoch corresponding to $kT = 1$ MeV.

There is an interesting piece of physics associated with the evaluation of the temperature of the neutrino background. We have shown above that the weak

interactions maintain the equilibrium abundances of the neutrons and protons until $kT \approx 1$ MeV. Prior to this epoch, the photons, neutrinos, electrons and their antiparticles were the only species which were relativistic and they all had the same thermal temperature. However, at an energy $kT \approx 0.5$ MeV, the electrons and positrons annihilated creating γ-ray photons. These high energy photons were rapidly thermalised by Compton scattering and so the temperature of the radiation became greater than that of the neutrinos. From that epoch onwards, the expansion was adiabatic and the temperatures of the neutrinos and photons can be worked out assuming that all the entropy of the electrons and positrons is transferred to the radiation field. The result is that, if the radiation temperature is 2.726 K at the present epoch, the temperature of the neutrinos is expected to be $(4/11)^{1/3}$ of this value, that is, $T_\nu = 1.94$ K. This is another genuinely cosmological background component which must be present according to the theory of the standard Big Bang. Unfortunately, the neutrino background is far below the detection threshold for neutrino detectors.

One of the more remarkable results of these studies is that they enabled limits to be set to the number of neutrino species which could have been present during the epochs when the light elements were being synthesised. If there had been more than three species of neutrino present, they would have contributed to the inertial mass density of massless particles and so would have speeded up the early expansion of the Universe. Consequently, the decoupling of the neutrinos would have taken place at a higher temperature resulting in the overproduction of helium (see the expression 6.13). From this type of cosmological argument, it was shown that there could not be more than three families of neutrinos, a result subsequently confirmed by the width of the decay spectrum of $W^\pm$ and Z^0 bosons measured by LEP at CERN.

There is a related point concerning the influence of the neutrinos upon the dynamics of the Universe. The dynamics during the radiation-dominated

phase is determined by the total inertial mass density of massless particles and this should include the neutrinos. In fact, throughout the radiation dominated phase from $kT \approx 1$ MeV to the epoch of equality of the energy densities of massless and cold matter, the appropriate value of χ is 1.7. This has the effect of changing slightly the epochs at which various important changes took place during the history of the Universe.

References – Chapter 6

Audouze, J. (1986). In *Nucleosynthesis and Chemical Evolution* by J. Audouze, C. Chiosi and S.E. Woosley, 429. Geneva: Geneva Observatory Publications.
Kolb, E.W. and Turner, M.S. (1990). *The Early Universe.* Redwood City, California: Addison-Wesley Publishing Co.
Linsky, J. (1993). In *The Frontiers of Space and Ground-based Astronomy,* 27th ESLAB Symposium, (eds. W. Wamsteker, M.S. Longair and Y. Kondo), (in press). Dordrecht: Kluwer Scientific Publishers.
Longair, M.S. (1989). *Galaxy Formation* in *Evolution of Galaxies - Astronomical Observations,* (eds. I. Appenzeller, H.J. Habing and P. Lena), 1. Heidelberg: Springer-Verlag.
Peebles, P.J.E. (1966). ApJ, **146**, 542.
Peebles, P.J.E. (1968). ApJ, **153**, 1.
Pozdnyakov, L.A., Sobol, I.M. and Sunyaev, R.A. (1983). Astrophys. Sp. Phys. Rev., **2**, 263.
Sunyaev, R.A. and Zeldovich, Ya.B. (1980). ARA&A, **18**, 537.
Wagoner, R.V., ApJ, **179**, 343.
Weinberg, S. *Gravitation and Cosmology.* New York: John Wiley and Co.
Weymann, R. (1965). Phys. Fluids, **8**, 2112.

7 The Origin of the Large-Scale Structure of the Universe

The considerations of Chapter 6 show that the standard Big Bang model is remarkably successful in accounting for four independent large-scale features of our Universe

- Hubble's law and the expansion of the distribution of galaxies
- The remarkable isotropy of the Universe on a large scale
- The black-body spectrum of the Cosmic Microwave Background Radiation
- The abundances of the light elements and their isotropes, deuterium, helium and probably lithium

This suggests that the framework of the Big Bang models is the best starting point for the construction of more complete models of the origin of structure in the Universe.

Despite the success of the Big Bang model, we have created a very dull Universe. There is no structure, there are no stars, galaxies or clusters and certainly no astronomers. The best way of approaching the problem is to regard the standard Big Bang as the first approximation to our Universe. The next step is to ask what happens when we perturb it — out of perturbations of a perfectly isotropic Universe we may hope to recreate the diversity of objects we see in the Universe today.

Galaxies are complex systems but the aim of the cosmologist is not to explain all their detailed features. This task is left to the astrophysicist who has to account for the detailed astrophysics of galaxies and how they evolve. The aims of the cosmologist are much more modest. These are to explain the origin of structure in the Universe in the sense that regions of *density contrast* $\delta\rho/\rho$ reach amplitude 1 from initial conditions which must have been remarkably isotropic and homogeneous; $\delta\rho$ is the density perturbation over the mean background density ρ. Once the initial perturbations have grown in amplitude to $\delta\rho/\rho \approx 1$, their development becomes non-linear and they rapidly evolve towards bound structures during which star formation and other astrophysical processes lead to galaxies as we know them. The cosmologist's objective is therefore to account for the initial conditions necessary for the formation of galaxies and other large scale structures. In its simplest form, the cosmologist seeks to explain how fluctuations can grow to amplitude $\delta\rho/\rho \approx 1$ in the expanding Universe. This may appear to be a rather modest goal but it turns out to give rise to some of the most difficult problems of modern cosmology. This problem forces us to investigate seriously processes in the very early Universe.

Structures such as galaxies, clusters and other large scale structures must have formed relatively late in the Universe. We can deduce this from the typical mean densities of these objects now. To order of magnitude, the density contrasts $\delta\rho/\rho$ for galaxies, clusters of galaxies and superclusters are

$\sim 10^6$, 1,000 and a few respectively at the present epoch. Since the density of matter in the Universe changes as $(1+z)^3$, it follows that galaxies could not have separated out as discrete objects at redshifts greater than about 100, and the corresponding redshifts for clusters and superclusters are $z \sim 10$ and 1 respectively — if they had, they would now have much greater mean densities than those observed. We conclude that the galaxies and larger scale structures must have separated out from the expanding gas at redshifts less than 100 which is well into the matter-dominated phase of the Big Bang and after the epoch of recombination. This is an important conclusion since it means that galaxies as we know them were not formed in the inaccessible remote past but in the redshift range which should, in principle, be accessible to observation.

I have already given an elementary derivation of many of the results of the following sections (Longair 1989) and so at a number of points I will simply summarise the essential points. A more detailed exposition of many of these ideas is presented in the book *Structure Formation in the Universe* by Padmanabhan (1993) and in Peebles' monumental study *Principles of Physical Cosmology* (1993).

7.1 The Non-relativistic Wave Equation for the Growth of Small Perturbations in the Expanding Universe

The analysis of the growth of small perturbations under gravity in the expanding Universe is one of the classics of theoretical astrophysics. The origin of the problem dates back to the work of Jeans in the first decade of this century and then to a classic paper by Lifshitz in 1946.

The problem gets off to a very bad start. We first write down the equations of gas dynamics for a compressible fluid under gravity. These consist of three partial differential equations which describe the conservation of mass, the equation of motion for an element of the fluid and the equation for the gravitational potential in the presence of a density distribution ρ. These are:

$$\textit{Equation of Continuity}: \frac{\partial \rho}{\partial t} + \nabla \cdot (\rho \mathbf{v}) = 0 \tag{7.1}$$

$$\textit{Equation of Motion}: \frac{\partial \mathbf{v}}{\partial t} + (\mathbf{v} \cdot \nabla)\mathbf{v} = -\frac{1}{\rho}\nabla p - \nabla \phi \tag{7.2}$$

$$\textit{Gravitational Potential}: \nabla^2 \phi = 4\pi G \rho \tag{7.3}$$

These equations describe the dynamics of a fluid of density ρ and pressure p in which the velocity distribution is $\mathbf{v}$. The gravitational potential ϕ at any point is given by Poisson's equation (7.3) in terms of the density distribution ρ. It is important to remember exactly what the partial derivatives mean. In equations (7.1), (7.2) and (7.3), the partial derivatives describe the variation of the quantities *at a fixed point in space*. These coordinates are often referred to as *Eulerian coordinates*. There is another way of writing the equations of fluid dynamics in which the motion of a particular fluid element is followed. These are known as *Lagrangian coordinates*. Derivatives which follow the fluid element are written as total derivatives $\mathrm{d}/\mathrm{d}t$ and it is straightforward to show that

$$\frac{\mathrm{d}}{\mathrm{d}t} = \frac{\partial}{\partial t} + (\mathbf{v} \cdot \nabla) \tag{7.4}$$

The equations of motion can therefore be written in Lagrangian form in which we follow the behaviour of one element of the fluid.

$$\frac{\mathrm{d}\rho}{\mathrm{d}t} = -\rho \nabla \cdot \mathbf{v} \tag{7.5}$$

$$\frac{\mathrm{d}\mathbf{v}}{\mathrm{d}t} = -\frac{1}{\rho}\nabla p - \nabla \phi \tag{7.6}$$

$$\nabla^2 \phi = 4\pi G \rho \tag{7.7}$$

Notice that, for the problem we are analysing, we can think of the equations (7.5), (7.6) and (7.7) as being written in comoving coordinates, that is, the behaviour of a particular element of the expanding Universe is followed rather than what would be observed if one sat at a fixed point in space and watched the galaxies expand past it.

It is standard practice now to establish the zero order solution for the unperturbed medium, that is, a uniform state in which ρ and p are the same everywhere and $\mathbf{v} = 0$. Unfortunately this solution does not exist. Equation (7.7) shows that if everything is uniform and the velocity is zero, we only obtain solutions if $\rho = 0$! This is a problem since it means that there is no static solution with finite density and pressure and this is a worry for those who insist upon mathematical rigour.

Fortunately, we want to treat the growth of fluctuations in an expanding medium and this eliminates this particular problem. We are interested in the zero order solutions for the velocity $\mathbf{v}$, the density ρ, the pressure p and the gravitational potential ϕ. The zero order solutions are $\mathbf{v}_0$, ρ_0, p_0 and ϕ_0 and these satisfy the above equations (7.5), (7.6) and (7.7).

$$\frac{\mathrm{d}\rho_0}{\mathrm{d}t} = -\rho_0 \nabla \cdot \mathbf{v}_0 \tag{7.8}$$

$$\frac{\mathrm{d}\mathbf{v}_0}{\mathrm{d}t} = -\frac{1}{\rho_0}\nabla p_0 - \nabla \phi_0 \tag{7.9}$$

$$\nabla^2 \phi_0 = 4\pi G \rho_0 \tag{7.10}$$

The next step is to write down the equations including first order perturbations to the uniformly expanding medium and so we write

$$\mathbf{v} = \mathbf{v}_0 + \delta\mathbf{v} \quad \rho = \rho_0 + \delta\rho \quad p = p_0 + \delta p \quad \phi = \phi_0 + \delta\phi \tag{7.11}$$

These are substituted into equations (7.5), (7.6) and (7.7). The equations are expanded to first order in small quantities and then equations (7.8), (7.9) and (7.10) are subtracted from each of them in turn.

I will not repeat the analysis of these equations which I have already given in detail (Longair 1989). We simply note that it is convenient to introduce the *density contrast* $\Delta = \delta\rho/\rho$ and the speed of sound in the medium $c_s^2 = \delta p/\delta\rho$. The last step in the analysis is to seek wave solutions for Δ of the form $\Delta \propto \exp i(\mathbf{k}_c \cdot \mathbf{r} - \omega t)$ and hence derive a wave equation for Δ.

$$\frac{d^2\Delta}{dt^2} + 2\left(\frac{\dot{R}}{R}\right)\frac{d\Delta}{dt} = \Delta(4\pi G\rho_0 - k^2 c_s^2) \tag{7.12}$$

where $\mathbf{k}_c$ is the wavevector in comoving coordinates. The proper wavevector $\mathbf{k}$ is related to $\mathbf{k}_c$ by $\mathbf{k}_c = R\mathbf{k}$. Equation (7.12) is the result we have been seeking and from it follow a number of important conclusions. It is as important as any equation in astrophysical cosmology.

7.2 The Jeans' Instability

Let us return first of all to the problem studied by Jeans. We obtain the differential equation for gravitational collapse in a static medium by setting $\dot{R} = 0$ in equation (7.12). Then, for waves of the form $\Delta = \Delta_0 \exp i(\mathbf{k} \cdot \mathbf{r} - \omega t)$, the dispersion relation is

$$\omega^2 = c_s^2 k^2 - 4\pi G\rho_0 \tag{7.13}$$

It is intriguing that this relation was first derived by Jeans in 1902. The corresponding equation in the electrostatic case was derived by Langmuir in the 1920s and describes the dispersion relation for longitudinal plasma oscillations

$$\omega^2 = c_s^2 k^2 + \frac{N_e e^2}{m_e \epsilon_0}$$

where N_e is the electron density and m_e is the mass of the electron. The formal similarity of the physics may be appreciated from comparison of the attractive gravitational acceleration in a region of mass density ρ_0 and the repulsive electrostatic acceleration in a region of electron charge density $N_e e$. The equivalence of $-G\rho_0$ and $N_e e^2/4\pi\epsilon_0 m_e$ is apparent.

The dispersion relation (7.13) describes oscillations or instability depending upon the sign of its right-hand side.

(a) If $c_s^2 k^2 > 4\pi G\rho_0$, the right-hand side is positive and the perturbations are oscillatory, that is, they are sound waves in which the pressure gradient is sufficient to provide support for the region. Writing the inequality in terms of wavelength, stable oscillations are found for wavelengths less than the critical *Jeans' wavelength* λ_J

$$\lambda_J = \frac{2\pi}{k_J} = c_s\left(\frac{\pi}{G\rho_0}\right)^{\frac{1}{2}} \tag{7.14}$$

(b) If $c_s^2 k^2 < 4\pi G\rho_0$, the right-hand side is negative, corresponding to unstable modes. The solutions can be written

$$\Delta = \Delta_0 \exp(\Gamma t + i\mathbf{k} \cdot \mathbf{r})$$

where

$$\Gamma = \pm\left[4\pi G\rho_0\left(1 - \frac{\lambda_J^2}{\lambda^2}\right)\right]^{\frac{1}{2}}$$

Notice that the positive solution corresponds to exponentially growing modes. For wavelengths much greater than the Jeans' wavelength, the growth rate Γ becomes $\pm(4\pi G\rho_0)^{1/2}$. Thus, the characteristic growth time for the instability is

$$\tau = \Gamma^{-1} = (4\pi G\rho_0)^{-1/2} \approx (G\rho_0)^{-1/2}$$

This is the famous *Jeans' Instability* and the time scale τ is the typical collapse time for a region of density ρ_0. Notice that the expression for the Jeans' length is just the distance a sound wave travels in a collapse time.

The physics of this result is very simple. The instability is driven by the self-gravity of the region and the tendency to collapse is resisted by the internal pressure gradient. We can easily derive the Jeans' instability criterion by considering the pressure support of a region of internal pressure p, internal densityρ and radius r. The equation for hydrostatic support for the region can be written

$$\frac{\mathrm{d}p}{\mathrm{d}r} = -\frac{G\rho M(< r)}{r^2}$$

To order of magnitude, $\mathrm{d}p/\mathrm{d}r \approx -p/r$ and $M \approx \rho r^3$. Therefore, since $c_s^2 \approx p/\rho$, the critical scale is $r \approx c_s(G\rho)^{-1/2}$. Thus, the Jeans' length is the scale which is just stable against gravitational collapse. If the region were any larger, the gravitational forces would overwhelm the internal pressure gradients and the region would collapse under gravity. This classical Jeans' instability is of importance for the processes of star formation in galaxies.

7.3 The Jeans' Instability in an Expanding Medium

The results of the analysis which follows are so important that I propose to present three different versions of the final result. My excuse is the same as that given by the Bellman in *The Hunting of the Snark* by Lewis Carroll (1876).

Just the place for a Snark! I have said it thrice:
What I tell you three times is true.

7.3.1 Small Perturbation Analysis

We now return to the full version of equation (7.12).

$$\frac{\mathrm{d}^2\Delta}{\mathrm{d}t^2} + 2\left(\frac{\dot{R}}{R}\right)\frac{\mathrm{d}\Delta}{\mathrm{d}t} = \Delta(4\pi G\rho_0 - k^2c_s^2) \tag{7.12}$$

The second term $2(\dot{R}/R)(\mathrm{d}\Delta/\mathrm{d}t)$ modifies the classical Jeans' analysis in crucial ways. It is apparent from the right-hand side of equation (7.12) that the Jeans' instability criterion applies in this case as well but the growth rate is significantly modified. Let us work out the growth rate of the instability in the long wavelength limit $\lambda \gg \lambda_J$ in which case we can neglect the pressure term $c_s^2k^2$. We therefore have to solve the equation

$$\frac{\mathrm{d}^2\Delta}{\mathrm{d}t^2} + 2\left(\frac{\dot{R}}{R}\right)\frac{\mathrm{d}\Delta}{\mathrm{d}t} = 4\pi G\rho_0\Delta \tag{7.15}$$

Rather than deriving the general solution, let us consider the special cases $\Omega_0 = 1$ and $\Omega_0 = 0$ for which the scale factor–cosmic time relations are $R = (3H_0t/2)^{2/3}$ and $R = H_0t$ respectively.

1. $\Omega_0 = 1$ In this case,

   $$4\pi G\rho_0 = \frac{2}{3t^2} \qquad \text{and} \qquad \frac{\dot{R}}{R} = \frac{2}{3t}$$

   Therefore

   $$\frac{\mathrm{d}^2\Delta}{\mathrm{d}t^2} + \frac{4}{3t}\frac{\mathrm{d}\Delta}{\mathrm{d}t} - \frac{2}{3t^2}\Delta = 0 \tag{7.16}$$

   We seek power-law solutions of the form $\Delta = at^n$. Substituting into equation (7.14), we find

   $$n(n-1) + \frac{4}{3}n - \frac{2}{3} = 0$$

   which has solutions $n = 2/3$ and $n = -1$. The latter solution corresponds to a decaying mode. The $n = 2/3$ solution corresponds to the growing mode we are seeking $\Delta \propto t^{2/3} \propto R = (1+z)^{-1}$. This is the key result

   $$\frac{\delta\rho}{\rho} \propto (1+z)^{-1} \tag{7.17}$$

   In contrast to the *exponential* growth found in the static case, the growth of the perturbation in the case of the expanding Universe is *algebraic*. This is the origin of the problems of forming galaxies by gravitational collapse.

2. $\Omega_0 = 0$ In this case,

$$\rho_0 = 0 \qquad \text{and} \qquad \frac{\dot{R}}{R} = \frac{1}{t}$$

and hence

$$\frac{\mathrm{d}^2\Delta}{\mathrm{d}t^2} + \frac{2}{t}\frac{\mathrm{d}\Delta}{\mathrm{d}t} = 0 \tag{7.18}$$

Again, seeking power-law solutions of the form $\Delta = at^n$, we find $n = 0$ and $n = -1$, that is, in this case there is a decaying mode and one of constant amplitude $\Delta =$ constant.

These simple results describe the evolution of small amplitude perturbations, $\delta\rho/\rho \ll 1$. In the early stages of the matter-dominated phase, the dynamics of the world models approximate to the Einstein-de Sitter model, $R \propto t^{2/3}$ and so the amplitude of the density contrast grows linearly with R. In the late stages, when the Universe may approximate to the $\Omega_0 = 0$ model, the amplitude of the perturbations grow very slowly and in the limit $\Omega_0 = 0$ do not grow at all. This last result is not particularly surprising since if $\Omega_0 = 0$ there is no gravitational driving force to make the perturbation grow!

7.3.2 Perturbing the Friedman Solutions

In our second approach, we investigate the physical reason for this behaviour from considerations of the dynamics of the Friedman world models. We demonstrated in Section 3.3 how the dynamics of these models could be understood in terms of a simple Newtonian model. We can model the development of a spherical perturbation in the expanding Universe by embedding a spherical region of density $\rho + \delta\rho$ in an otherwise uniform Universe of density ρ (Fig. 7.1). Using the same logic as in Section 3.3, the spherical region behaves like a Universe of slightly higher density. We can therefore use the parametric solutions (3.10) for the dynamics of the world models

$$R = a(1 - \cos\theta) \qquad t = b(\theta - \sin\theta) \tag{7.19}$$

where

$$a = \frac{\Omega_0}{2(\Omega_0 - 1)} \qquad \text{and} \qquad b = \frac{\Omega_0}{2H_0(\Omega - 1)^{3/2}}$$

The trick is now to look at solutions for small values of θ, corresponding to early epochs of the matter-dominated phase. Expanding to third order in θ, $\cos\theta = 1 - \frac{1}{2}\theta^2$, $\sin\theta = \theta - \frac{1}{6}\theta^3$, we find the solution

$$R = \Omega_0^{1/3}\left(\frac{3H_0t}{2}\right)^{2/3} \tag{7.20}$$

This solution corresponds to the conclusion derived from equation (3.17) that, in the early stages, the dynamics of the world models tend towards those of

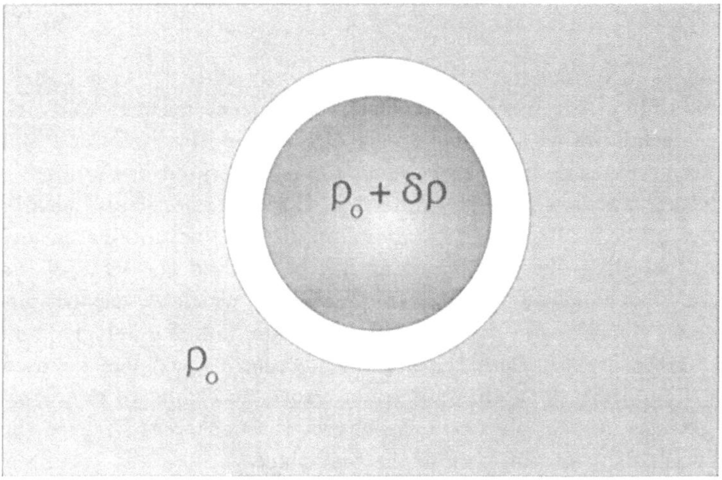

Fig. 7.1 Illustrating a spherical perturbation with slightly greater density than the average in a uniformly expanding Universe. The region with slightly greater density behaves dynamically exactly like a model Universe with density $\rho_0 + \delta\rho$.

the Einstein-de Sitter model, $\Omega_0 = 1$, that is, $R \propto t^{2/3}$, but with a different constant of proportionality. Now let us look at a region of slightly greater density embedded within the background model. To derive this behaviour, we expand the expressions for R and t to fifth order in θ, $\cos\theta = 1 - \frac{1}{2}\theta^2 + \frac{1}{24}\theta^4 \dots$, $\sin\theta = \theta - \frac{1}{6}\theta^3 + \frac{1}{120}\theta^5 \dots$ The solution follows in exactly the same manner as equation (7.18)

$$R = \Omega_0^{1/3}\left(\frac{3H_0 t}{2}\right)^{2/3}\left[1 - \frac{1}{20}\left(\frac{6t}{b}\right)^{2/3}\right] \tag{7.21}$$

We can immediately write down an expression for the evolution of the density of the model with cosmic epoch

$$\rho(R) = \rho_0 R^{-3}\left[1 + \frac{3}{5}\frac{(\Omega_0 - 1)}{\Omega_0}R\right] \tag{7.22}$$

Notice that if $\Omega_0 = 1$, there is no growth of the perturbation. The density perturbation may be considered to be a mini-Universe of slightly higher density than $\Omega_0 = 1$ embedded in an $\Omega_0 = 1$ model. Therefore, taking the density contrast to be the difference between the model with $\Omega_0 > 1$ and the critical model $\Omega_0 = 1$, we find

$$\frac{\delta\rho}{\rho} = \frac{\rho(R) - \rho_0(R)}{\rho_0(R)}$$

Therefore

$$\frac{\delta\rho}{\rho} = \frac{3}{5}\frac{(\Omega_0 - 1)}{\Omega_0}R \tag{7.23}$$

This result shows why the perturbation grows only linearly with cosmic epoch. The instability corresponds to the slow divergence between the variation of the scale factors with cosmic epoch of the model with $\Omega_0 = 1$ and one with slightly greater density. This behaviour is illustrated in Fig. 7.2.

This model has another very great merit in that it demonstrates clearly that this law of growth of the perturbations applies to fluctuations on any physical scale, including those of wavelength greater than the scale of the horizon, $r > ct$. This follows from the same reasoning which we used in discussing the global dynamics of the Universe in Section 3.3. If a perturbation is set up on a scale greater than the horizon, it behaves just like a closed Universe and the amplitude of the fluctuation grows according to $\delta\rho/\rho \propto R$ — again the physics is local physics and the growth is coherent because the perturbation was set up in that way in the first place.

7.3.3 Collapsing Poles

You may not like my third argument as much as the first two but it contains exactly the same physics. Consider what happens when we attempt to balance a long pole of length l and mass m on one end. We all know that the situation is unstable and that the pole falls over. This is no more than a gravitational instability in which there is no pressure or other force to prevent collapse. We can easily work out the growth rate of the instability by conservation of energy in a gravitational field. In Fig. 7.3, the pole is shown at an angle θ to the vertical and then the law of conservation of energy states that the loss of gravitational potential energy $(gml/2)(1 - \cos\theta)$ must equal the increase in rotational energy $(1/2)I\omega^2$ about the bottom end of the pole O where I is moment of inertia of the pole about O.

$$\frac{gml}{2}(1 - \cos\theta) = \frac{1}{2}I\omega^2 \tag{7.24}$$

Since $I = (1/3)ml^2$ and $\omega = \dot{\theta}$, it follows that

$$\dot{\theta}^2 = 3\frac{g}{l}(1 - \cos\theta) \tag{7.25}$$

There is an exact solution for this expression into the non-linear regime but let us only deal with the small angle approximation in which we write $\cos\theta = (1 - \theta^2/2 + \ldots)$. Then we obtain the simple exponential equation for the collapse of the pole

$$\dot{\theta} = \left(\frac{3g}{2l}\right)^{1/2}\theta \tag{7.26}$$

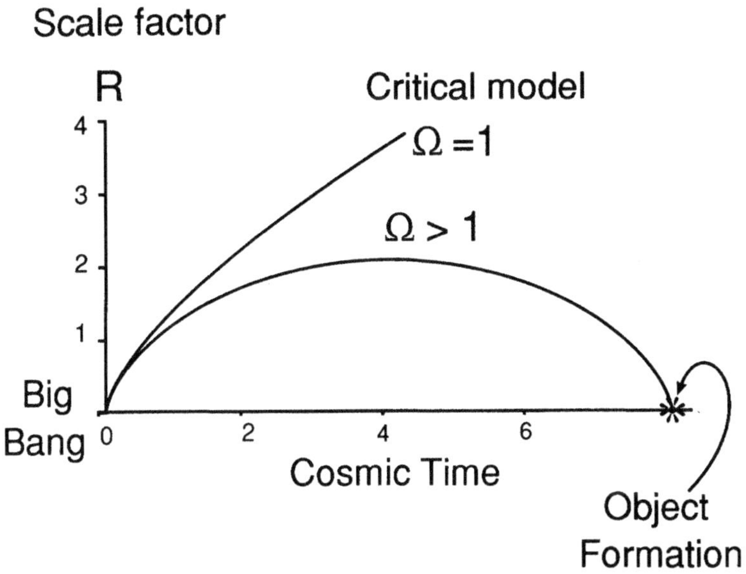

Fig. 7.2 Illustrating the growth of a spherical perturbation in the expanding Universe as the divergence between two Friedman models with slightly different densities.

The solution $\theta = \theta_0 \exp(\Gamma t)$ with $\Gamma = (3g/2l)^{1/2}$ is the exact analogue of the equation for the growth of the Jeans' instability in the absence of pressure gradients.

To modify this result for the case of the expanding Universe, we note that, in the absence of pressure gradients, the differential equation (7.12) for the growth rate of the instability is

$$\frac{d^2\Delta}{dt^2} + 2\left(\frac{\dot{R}}{R}\right)\frac{d\Delta}{dt} = \Delta(4\pi G\rho_0) \tag{7.27}$$

Notice that the force driving the instability on the right-hand side of this expression depends only upon the product of the gravitational constant G and the density ρ_0. Now, in the expanding Universe, $\rho_0 \propto R^{-3}$ and, in the case of the critical model $\Omega_0 = 1$, $R \propto t^{2/3}$ Therefore, the driving force is proportional $G\rho_0 \propto t^{-2}$. To simulate this case for our collapsing pole, we can assume that the gravitational acceleration is proportional to t^{-2}, in which case the equation of motion of the pole (7.26) becomes

$$\dot{\theta} = \left(\frac{A}{t^2}\right)^{1/2}\theta \tag{7.28}$$

Inspection of the equation (7.28) shows that the solutions are of power-law form $\theta \propto t$ rather than exponentially growing solutions. This simple calcula-

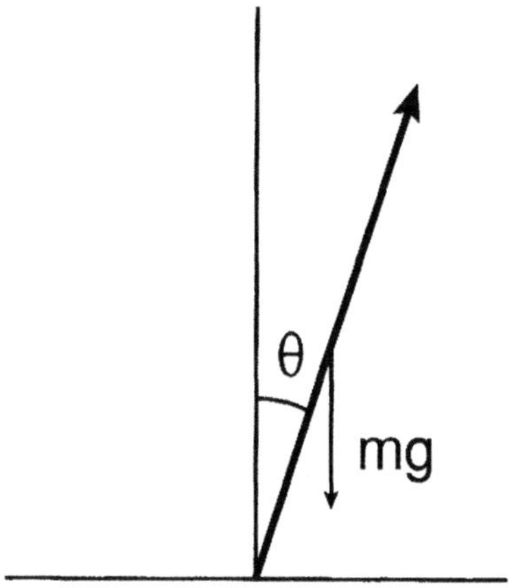

Fig. 7.3 Illustrating a falling pole.

tion illustrates the origin of the linear algebraic growth of the Jeans' instability in the expanding Universe. The effect of the gravitational driving force diminishes with time because the mean density of the Universe decreases as it expands.

7.3.4 The Relativistic Case

We investigate next the case of a relativistic gas because, in the early stages of the Big Bang, the primordial perturbations are in a radiation-dominated plasma for which the relativistic equation of state $p = \frac{1}{3}\varepsilon$ is applicable. We therefore require the relativistic generalisations of equations (7.5), (7.6) and (7.7). Equation (7.5), the equation of continuity, becomes an equation describing the conservation of energy. There is no simple way of demonstrating this except by using the general energy-momentum tensor for a fully relativistic gas.

The equation of energy conservation becomes

$$\frac{\mathrm{d}}{\mathrm{d}t}\left(\rho + \frac{p}{c^2}\right) = \frac{\dot{p}}{c^2} - \left(\rho + \frac{p}{c^2}\right)(\nabla \cdot \mathbf{v}) \tag{7.29}$$

Substituting $p = \frac{1}{3}\rho c^2$ into equation (7.29), we derive the relativistic continuity equation

$$\frac{\mathrm{d}\rho}{\mathrm{d}t} = -\frac{4}{3}\rho(\nabla \cdot \mathbf{v}) \tag{7.30}$$

The differential equation for the gravitational potential ϕ becomes

$$\nabla^2\phi = 4\pi G(\rho + \frac{3p}{c^2}) \tag{7.31}$$

For a fully relativistic gas, this becomes

$$\nabla^2\phi = 8\pi G\rho$$

Finally, the expression for the acceleration of an element of the fluid in the gravitational potential ϕ remains the same as before

$$\frac{\mathrm{d}\mathbf{v}}{\mathrm{d}t} = -\nabla\phi \tag{7.32}$$

where the pressure gradient term has been neglected. The net result is that the equation for the evolution of the perturbations of a relativistic gas are formally of exactly the same form as for the non-relativistic case but with slightly different constants. Therefore, an analysis essentially identical to that carried out in Sections 7.3.1 applies in the relativistic case as well. Going through the same analysis as before, neglecting the pressure gradient terms, we find the following differential equation for the growth of the instability

$$\frac{\mathrm{d}^2\Delta}{\mathrm{d}t^2} + 2\left(\frac{\dot{R}}{R}\right)\frac{\mathrm{d}\Delta}{\mathrm{d}t} - \frac{32\pi G\rho_0}{3}\Delta = 0 \tag{7.33}$$

This equation is formally identical to equation (7.12). Using exactly the same approach as in Section 7.3.1, we seek solutions of the form $\Delta = at^n$, recalling that in the radiation-dominated phases, the scale factor–cosmic time relation is given by equation (3.30) in which $R \propto t^{1/2}$. Going through precisely the same procedure, we find solutions $n = \pm 1$. The growing solution corresponds to

$$\Delta \propto t \propto R^2 \propto (1+z)^{-2} \tag{7.34}$$

Thus, once again, the unstable modes grow algebraically with cosmic time. It will be noted again that nowhere does the analysis describe the scale of the perturbation relative to the horizon scale.

7.4 The Basic Problem

The key results we have derived can be expressed as follows. At those redshifts at which $\Omega_0 z > 1$, the dynamics of the Universe are approximately those of the critical model $R \propto t^{2/3}$ and then the growth rate of the instability is

$$\Delta = \frac{\delta\rho}{\rho} \propto R = \frac{1}{1+z} \tag{7.35}$$

At redshift less than $1/\Omega_0$, the instability grows much more slowly and in the limit $\Omega_0 = 0$ does not grow at all. In the radiation-dominated phase, perturbations on scales greater than the Jeans' length grow as $(1+z)^{-2}$.

Since galaxies and astronomers certainly exist at a redshift $z = 0$, it follows that $\Delta \geq 1$ at $z = 0$ and so, at the last scattering surface, $z \sim 1,000$, the amplitude of the fluctuations must have been $\Delta = \delta\rho/\rho \geq 10^{-3}$. These density perturbations must have been present on the last scattering surface and the key question is whether or not they are observable. This is not a trivial sum and we need to look at the process of decoupling of the matter and the radiation in a little more detail.

I like to look at the results of this section in two ways.

- On the one hand, the slow growth of the fluctuations is the source of all our headaches in trying to understand the origin of galaxies — the large scale structures do not condense out of the primordial plasma by exponential collapse, as probably occurs in the formation of stars in the interstellar medium. It is a very much slower process.
- The other side of the coin is that, because of the slow growth of the fluctuations, we have the real possibility of learning about many aspects of the early Universe which would otherwise have been excluded. We have the opportunity of studying the formation of structure on the last scattering surface at a redshift $z \sim 1,000$ and, even more important, we can learn crucial information about the spectrum of fluctuations which must have been created in the very early Universe. Thus, thanks to the slow growth of the fluctuations, we have a direct probe of the very early Universe.

7.5 The Evolution of Adiabatic Baryonic Fluctuations in the Standard Hot Big Bang

We now have all the information we need to discuss the simplest case, that of the evolution of adiabatic baryonic perturbations in the standard Hot Big Bang model. I issue the warning immediately that this model will fail but there are many important features of the model which will recur in those which are in better accord with the observations.

We need the following information:

(1) *The Jeans' length* is the maximum scale for stable fluctuations at any epoch.
(2) *The horizon scale* is the maximum distance over which information can be communicated at a particular cosmic epoch t and hence is just $r_{\rm H} = ct$.
(3) *The growth rates* of the unstable models are algebraic with epoch. In the matter-dominated phase, the perturbation grows as R so long as $\Omega_0 z \gg 1$. The growth is much slower at smaller redshifts and becomes zero in the limit $\Omega_0 = 0$. In the early radiation-dominated phases, the growth rate is algebraic with $\Delta \propto R^2$.

Although there is some ambiguity about how to relate the wavelength $\lambda_{\rm J}$ to the mass of the object which ultimately forms from it, we will use for illustrative purposes the concept of the *Jeans' mass* which is the mass of baryons contained within a region of radius λ_J, $M_J = (4\pi\lambda_J^3/3)\rho_B$. The expectation is that this is roughly the ultimate mass of the object which forms from a perturbation of this scale. Let us consider first of all the radiation-dominated phases. The mass density in baryons is $\rho_B = 1.88\times10^{-26}\Omega_B h^2(1+z)^3$ kg m^{-3} where Ω_B is the density parameter in baryons at the present epoch. The Jeans' length in the radiation-dominated phase is

$$\lambda_J = \frac{c}{\sqrt{3}}\left(\frac{\pi}{G\rho}\right)^{1/2}$$

where ρ is the total mass density including both photons and neutrinos, that is, $\rho = 4.81\times10^{-31}\chi(1+z)^4$ kg m^{-3}, recalling that $\chi = 1.7$ when the neutrinos are taken into account. Therefore, the Jeans' mass in the early stages of the radiation-dominated phase, $z \gg 4\times10^4\Omega_0 h^2$, is

$$M_J = 2.8\times10^{30} z^{-3}\Omega_B h^2\ \mathrm{M_\odot} \tag{7.36}$$

There are several important conclusions which follow from this result. The first is that the Jeans' mass grows as $M_J \propto R^3$ as the Universe expands. Thus, the Jeans' mass is one solar mass $\mathrm{M_\odot}$ at a redshift $z = 10^{10}$ and increases to the mass of a large galaxy $M = 10^{11}\mathrm{M_\odot}$ at redshift $z = 3\times10^6$. The second conclusion follows from a comparison of the Jeans' length with the horizon scale $r_H = ct$. Using equation (3.30), the horizon scale can be written

$$r_H = ct = c\left(\frac{3}{32\pi G\rho}\right)^{1/2} \quad \text{and} \quad \lambda_J = c\left(\frac{\pi}{3G\rho}\right)^{1/2}$$

It is apparent that, in the radiation-dominated phase, the Jeans' length is of the same order as the horizon scale.

The physical meaning of these results is clear. If we consider a perturbation of galactic mass, say $M = 10^{11}\mathrm{M_\odot}$, in the early stages of the radiation-dominated phase, its scale far exceeds the horizon scale and hence the amplitude of the perturbation grows as R^{-2}. At a redshift $z \approx 3\times10^6$, the perturbation enters the horizon and, at the same time, the Jeans' length becomes larger than the scale of the perturbation. The perturbation is therefore stable against gravitational collapse and becomes a sound wave which oscillates at constant amplitude. As long as the Jeans' length remains greater than the scale of the perturbation, the perturbation does not grow in amplitude.

The variation of the Jeans' mass with redshift is shown in Fig. 7.4. The variation of the sound speed with redshift has been included in these calculations. At the epoch of equality of the rest mass energies in matter and radiation, the sound speed becomes less than the relativistic sound speed $c/\sqrt{3}$ according to equation (6.8). It is an interesting question whether or not this regime exists prior to the epoch of recombination. According to the analyses of Sections 6.2 and 6.3, if Ω_0 were as low as 0.1 and $h = 0.5$, the epoch of equality of the matter and radiation energy densities would occur about a redshift of 1,000, the epoch of recombination. In this case, there

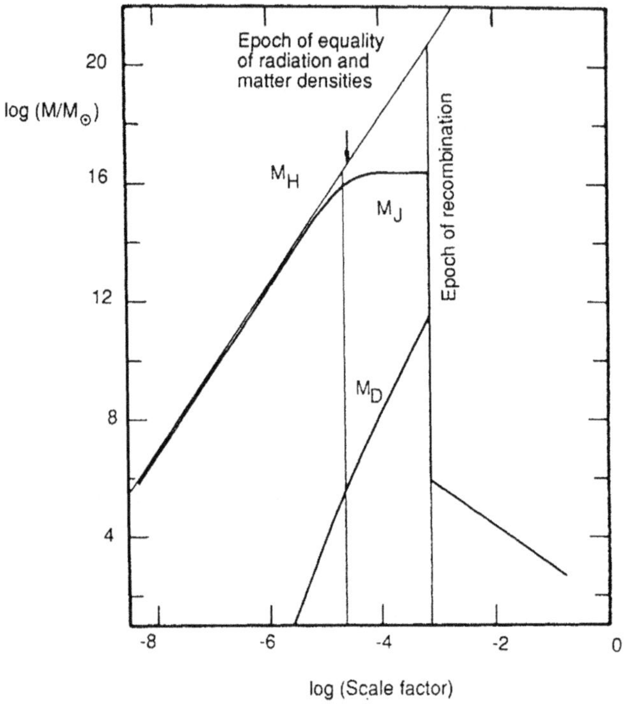

Fig. 7.4 The evolution of the Jeans' mass M_J with scale factor. Also shown is the evolution of the mass scales which are damped by photon diffusion M_D and the mass within the horizon M_H. The dependences of various scale lengths upon scale factor are shown.

would be a precipitous drop in the appropriate sound speed from $c/\sqrt{3}$ to the thermal sound speed of a gas at 4,000 K. If the values of Ω_0 and h are greater than these values, this intermediate regime would exist and the appropriate sound speed is $c_s = c(4\rho_{rad}/9\rho_B)^{1/2}$ and the mass density variation with scale factor is $\rho \propto R^{-3}$. A simple calculation shows that the Jeans' mass is independent of scale factor (or redshift) until the epoch of recombination. The next crucial epoch is the epoch of recombination when the primordial plasma recombines and there is an abrupt drop in the sound speed. The pressure within the perturbation is no longer provided by the radiation but by the internal thermal pressure. Because of the close coupling between the matter and radiation even at the redshift of 1,500, the matter and radiation temperatures are more or less the same. This means that the appropriate sound speed is the adiabatic sound speed for a gas at temperature 4000 K. The sound speed is therefore $c_s = (5kT/3m_H)^{1/2}$ and the Jeans' mass

$$M_{\mathrm{J}} = \frac{4\pi}{3}\left(\frac{\pi c_{\mathrm{s}}}{G}\right)^{3/2} \rho_{\mathrm{B}}^{-1/2} = 10^{6}(\Omega_0 h^2)^{-1/2}\ \mathrm{M}_{\odot} \tag{7.37}$$

Thus, the Jeans' mass drops abruptly to masses much less than typical galactic masses. This means that all perturbations with mass greater than about $10^6\ \mathrm{M}_\odot$ grow according to $\Delta \propto R$ until $\Omega_0 z \sim 1$. It is intriguing to note that the Jeans' mass immediately following recombination corresponds roughly to the mass of globular clusters which are known to be the oldest stellar systems in our Galaxy. The evolution of the Jeans' mass following recombination depends strongly upon the subsequent thermal history of the gas. If the gas continued to cool adiabatically as the Universe expands, $T \propto R^{-2}$, the Jeans' mass would decrease as $M_{\mathrm{J}} \propto R^{-1.5}$. However, it is unlikely that the gas cools in this simple fashion. We know that the intergalactic gas is very highly ionised at epochs corresponding to $z \sim 3$ from the absence of Lyman-α absorption in the spectra of distant quasars and so the intergalactic gas must have been strongly heated at some epoch between $z = 1,500$ and $z = 4$ (see Chapter 10).

7.6 Dissipation Processes in the Pre-Recombination Phases of the Hot Big Bang

To complete our discussion of the physics of adiabatic baryonic fluctuations in the standard model, dissipative processes in the radiation-dominated phase have to be considered. As we have discussed above, the matter and radiation are closely coupled in the radiation-dominated phases but the coupling is not perfect and the radiation can diffuse out of the the fluctuations. Since the radiation provides all the pressure support for the perturbation, the net result is that the perturbation is damped out if the radiation can diffuse out of the fluctuation. This process was first described by Silk (1968) and is often referred to as *Silk damping*. The photons are trapped within the region by Thomson scattering but eventually they can diffuse out of the region, in the process damping out the perturbation. A more complete discussion of this process is given by Weinberg (1972). The following simple arguments present the essence of the full calculation. At any epoch there is a photon mean free path for scattering by electrons $\lambda = (N_{\mathrm{e}}\sigma_{\mathrm{T}})^{-1}$ where $\sigma_{\mathrm{T}} = 6.665 \times 10^{-29}$ m^2 is the Thomson cross-section. We recall that the protons and electrons are closely coupled electrostatically as in a fully ionised plasma and so the photons are also closely coupled to the protons as well. We therefore ask how far the photons can diffuse in the cosmic time scale at epoch t. There are several ways of looking at this process. One can either think in terms of the diffusion coefficient for photons D which, according to kinetic theory, is related to the mean free path λ by $D = \frac{1}{3}\lambda c$. Then, the scale over which the photons can diffuse is $r \sim (Dt)^{1/2}$, that is, $r \approx (\frac{1}{3}\lambda ct)^{1/2}$ where t is cosmic time. Another way of looking at this same result is in terms of the total distance travelled in the diffusion process. The distance travelled from the point of origin by the photon is $r \approx N^{1/2}\lambda$ where N is the number of scatterings. We can now work out the mass of baryons contained within this scale r as a function of cosmic epoch. In the early pre-recombination phase, $z > 4 \times 10^4 \Omega_0 h^2$, the Universe is radiation-dominated and, from equation (3.30), $t = (3/32\pi G\rho)^{-1/2} = 2.3 \times 10^{19}(1+z)^{-2}$ s. The number density of electrons varies as $N_e = \rho(1+z)^3/m_p = 11\Omega_B h^2(1+z)^3$ m^{-3} where m_p is the mass of the proton. Thus the damping mass, sometimes referred to as the *Silk mass*, is

$$M_D = \frac{4\pi}{3} r^3 \rho_B = 2 \times 10^{26}(\Omega_B h^2)^{-1/2}(1+z)^{-9/2}\ M_\odot \tag{7.38}$$

In the redshift range just prior to recombination, the Universe may become matter-dominated if there exists a period during which $4 \times 10^4 \Omega_0 h^2 > z > 1,500$. In this case, the cosmic time–redshift relation is given by the expression (7.18) which can be rewritten $t = \frac{2}{3H_0}(1+z)^{-3/2}\Omega_0^{-1/2}$. Repeating the above analysis for this case, we find

$$M_D = 1.8 \times 10^{23}(\Omega_B h^2)^{-5/4}(1+z)^{-15/4}\ M_\odot \tag{7.39}$$

The key question is what range of masses survives to the epoch of recombination $z = 1,500$. If $\Omega_0 h^2 = 1$, the Universe is matter-dominated throughout the redshift range 4×10^4 to 1,500 and then using equation (7.38), we find $M_D = 2 \times 10^{11}\ M_\odot$; if $h = 0.5$, $M_D = 10^{12} M_\odot$. In the opposite extreme, if the Universe is of low density, $\Omega_B = 0.1$ and $h = 0.5$, the Universe remains radiation-dominated to $z = 1,500$ and then according to equation (7.39), the damping mass is $M_D = 6 \times 10^{12} M_\odot$. A more detailed treatment of the damping process is given by Weinberg (1972) with essentially the same results. The upshot of these calculations is that all perturbations with masses $M \leq 10^{12} M_\odot$ are damped out because of the diffusion of photons relative to the matter perturbations. According to the strict theory of adiabatic baryonic perturbations, only massive perturbations on the scales of massive galaxies and greater survive into the post-recombination eras. The perturbations which would have resulted in stars, star clusters and even normal galaxies such as our own are damped to exponentially small values of Δ and it is assumed that these structures must form by the process of fragmentation of the larger scale structures which survive to $z < 1,500$. The evolution of the damping mass M_D with cosmic epoch is included schematically in Fig. 7.4.

7.7 Baryonic Pancake Theory

We can now put these ideas together to produce the standard baryonic pancake theory of the origin of galaxies. The considerations of the last section show that, in the standard baryonic fluctuation picture, only large-scale perturbations with $M \geq 10^{12}\ M_\odot$ survive to the epoch of recombination. All the fluctuations on smaller mass scales are damped very effectively by photon diffusion. This was the theory which was developed in detail in the 1970s, principally by Zeldovich and his colleagues (see Doroshkevich et al 1974). I can recommend very strongly the recently published *Selected Works of Yakob Borisovich Zeldovich. Volume 2* (1993) which includes all his great astrophysical and cosmological papers.

We use the results developed in Section 7.3 for the subsequent evolution of the baryonic perturbations. All perturbations with masses greater than about $10^{12}M_\odot$ grow as $\Delta \propto (1+z)^{-1}$ until the epoch at which $\Omega_0 z \sim 1$. In the pure baryonic picture, the density parameter in baryons Ω_B must be less than $0.03h^{-2}$ because of the constraints provided by primordial nucleosynthesis (Section 6.5) and so, even in the case $h = 0.5$, the perturbations only grow very slowly at redshifts $z \leq 10$. We know that galaxies certainly exist at a redshift of 1 and it is likely that they had already formed by a redshift of about $2-3$. Therefore the fluctuations must have developed to large amplitude by this epoch, that is, $\Delta \sim 1$. Since $\Delta \propto (1+z)^{-1}$, the amplitude of the perturbations at the epoch of recombination must have been $\Delta \geq 3 \times 10^{-3}$. In the pre-recombination epochs, this fluctuation is a sound wave which oscillates with constant amplitude and, as we demonstrated in Section 7.5, this is also the amplitude with which it enters the horizon at earlier epochs. It is assumed that these finite-sized fluctuations on scales which are greater than the horizon scale in the early Universe are produced by elementary processes in the early Universe or that they are part of the initial conditions from which the Universe evolved.

The baryonic pancake scenario results from consideration of the ultimate development of the large scale perturbations when they become non-linear. The structures which survive on the scale of clusters and superclusters of galaxies are unlikely to be perfectly spherical and, in a simple approximation, they can be approximated by ellipsoids with three unequal axes. It can be shown that such ellipsoids collapse most rapidly along the shortest axis with the result that flattened structures like pancakes form (see Zeldovich, *Selected Papers, Volume 2*, 1993). The density becomes large in the plane of the pancake and the infalling matter is heated to a high temperature as the matter collapses into the pancake. It is assumed that galaxies form by fragmentation or thermal instabilities within the pancakes.

This theory has a number of successes but also a number of very serious problems. Among the successes we can include the fact that it accounts for the large-scale structure of the Universe rather naturally. The pancakes create flattened, stringy structures which form late in the Universe. The fact that quasar activity is at a maximum at epochs $z \sim 2-3$ might be explained by the fact that most galaxy formation only occurred at these epochs.

The problems are, however, rather great. In particular, the fluctuations in the matter density at the epoch of recombination must have been rather large. We discuss the theory of the origin of temperature fluctuations in the Microwave Background Radiation later but we can give a simple indication of the nature of the problem already. The fluctuations which survive to recombination are adiabatic and so there are temperature fluctuations corresponding to the density and pressure perturbations on the last scattering surface. The adiabatic perturbations result in temperature fluctuations

$$\left(\frac{\delta T}{T}\right) \approx \frac{1}{3}\left(\frac{\delta \rho}{\rho}\right) \tag{7.40}$$

Thus, even in the most favourable scenario, it can be seen that temperature fluctuations $\delta T/T \geq 10^{-3}$ are expected and this exceeds by two orders of magnitude the upper limits to the temperature fluctuations on the scale of clusters of galaxies at the epoch of recombination. We will give a much improved treatment of the origin of fluctuations in the Cosmic Microwave Background Radiation in Chapter 9. There is a way out of this problem and that is to assume that the intergalactic gas is fully ionised at redshifts less than 1,500 so that there is a large optical depth to Thomson scattering which would damp the amplitude of the temperature fluctuations. However, in this theory in which everything forms late in the Universe, there are no discrete sources which could release energy at these early epochs.

7.8 Concluding Remarks

It may seem curious that I have spent so much time developing a theory which ends up in serious conflict with the observations. The important point about the above analysis is that we have had to develop many of the tools which are necessary for the formulation of any theory of galaxy formation and we will return to them again and again in the subsequent Sections. It is also important to appreciate that this theory is the best one can do within the conventional scenario in which one assumes that all the matter in the Universe is in the form of baryonic matter. The fact that there are serious conflicts with the observations is one strong motivation for taking very seriously those models in which the Universe is dominated by dark matter and this is the topic which we have to address now. It is a pity that the simplest 'dull person's' view of the origin of structure in the Universe runs into these difficulties since the model would have a much more solid physical foundation if we did not have to introduce new physics. The other side of the coin is that we are forced to take seriously new pieces of physics which may well lead to new understanding of fundamental physical processes. This would be yet one more example of astronomical problems leading to fundamental physical processes which cannot yet be studied in the laboratory. This is the story we take up now.

References – Chapter 7

Doroshkevich, A.G., Sunyaev, R.A. and Zeldovich, Ya.B. (1974). *Confrontation of Cosmological Theories with Observational Data*, (ed. M.S. Longair), 213. Dordrecht: D. Reidel Publishing Co.

Jeans, J.H. (1902). *Phil. Trans. Roy. Soc.*, **199**, 1.

Lifshitz, E.M. (1946). J. Phys., USSR Acad. Sci., **10**, 116.

Longair, M.S. (1989). *Galaxy Formation* in *Evolution of Galaxies - Astronomical Observations*, (eds. I. Appenzeller, H.J. Habing and P. Lena), 1. Heidelberg: Springer-Verlag.

Padmanabhan, T. (1993). *Structure Formation in the Universe*. Cambridge: Cambridge University Press.

Peebles, P.J.E. (1993). *Principles of Physical Cosmology*. Princeton: Princeton University Press.

Silk, J. (1968). ApJ, **151**, 459.

Weinberg, S. (1972). *Gravitation and Cosmology*. New York: John Wiley and Co.

Zeldovich, Ya.B. (1993). *Selected Works of Yakob Borisovich Zeldovich. Vol. II*, (eds. J.P. Ostriker, G.I. Barenblatt and R.A. Sunyaev). Princeton: Princeton University Press.

8 Dark Matter and Galaxy Formation

8.1 Introduction

The considerations of section 7.7 demonstrate that the fluctuations in the Microwave Background Radiation should be of much greater amplitude than those observed if the simple baryonic adiabatic theory were correct. We will show how models in which the dominant mass contribution is in the form of non-baryonic dark matter can overcome this problem.

One of the most tantalising parts of this story concerns the value which should be adopted for Ω_0. On the theoretical side, there are several powerful arguments which suggest that Ω_0 should have the value 1. It was shown in section 3.3 that the Friedman world model with $\Omega_0 = 1$ is the only 'stable' world model in the sense that, if the Universe were set up initially with any other value, it would now be very far from the critical density. Furthermore, the value $\Omega_0 = 1$ comes naturally out of the inflationary model for the early Universe which can account for other independent features of our Universe, for example, its homogeneity on a large scale. On the observational side, the picture is not so simple. The density parameter corresponding to the visible light in galaxies, which is all associated with baryonic matter, is about 2 orders of magnitude less than the critical value. We know, however, that there is also dark matter in these systems and, from application of the virial theorem to galaxies and clusters, there must be about 10 times more dark matter than visible matter. The constraints on baryonic dark matter which come from primordial nucleosynthesis (section 6.5) correspond to $\Omega_B \leq 0.015h^{-2}$ and so it is possible, within the uncertainties in the estimates, that all the dark matter needed to bind galaxies and clusters could be baryonic.

There is, however, evidence that the value of Ω_0 found on scales larger than clusters of galaxies is greater than 0.1. This evidence comes from variants of the virial theorem applied to very large scale systems. For example, perturbations to the Hubble flow due to the presence of large scales systems such as the local supercluster or the 'Great Attractor' enable estimates of the average density of matter in the Universe within these regions to be found. Another variant is the 'cosmic virial theorem' in which the random velocities of field galaxies are related to large-scale density perturbations in the distribution of matter in the Universe. A similar approach has been used to estimate the average density of matter in the Universe from the surveys of IRAS galaxies. In all these cases, what is found from the observations is the quantity $\Omega_0^{0.6}/b$ where b is the bias parameter which describes how much more clumpy the distributon of galaxies is compared with the underlying distribution of matter in general. These techniques have all found values of $\Omega_0 > 0.1$ and, in a number of cases, it is found that $\Omega_0^{0.6}/b \sim 1$. The situation is by no means settled but it is evident that, if $\Omega_0 = 1$, then most of the matter in the Universe must be non-baryonic and the bulk of it must be located between the clusters of galaxies.

8.2 Forms of Dark Matter

There are many possibilities for the dark matter which must be present in the outer regions of large galaxies, in clusters of galaxies and other large scale systems. The fundamental problem is that we are limited by observation in the types of dark matter which can be easily detected. There are, for example, many forms of ordinary baryonic matter which would be very difficult to detect, let alone non-baryonic dark matter or black holes.

Forms of *baryonic dark matter* which are difficult to detect are those which are very weak emitters of electromagnetic radiation. For example, stars with $M \leq 0.05\ \mathrm{M}_\odot$, which are of too low mass to burn hydrogen into helium, are very faint objects and have proved very difficult to detect — these objects are collectively referred to as *brown dwarfs*. By the same token, low mass solid bodies such as planets, asteroids and other small lumps of rock are extremely difficult to detect. One amusing example which is often quoted is that the dark matter could all be in the form of standard bricks (or copies of the Astrophysical Journal!) and, even if they were sufficiently common to make $\Omega_0 = 1$, they would be extremely difficult to detect by either their emission or absorption properties. It is quite possible that the dark matter in the outer regions of galaxies and in clusters of galaxies is in the form of low mass stars since the inferred mass-to-light ratios in these systems are about a factor of 5 to 10 less than that necessary to close the Universe. We will, however, exclude baryonic dark matter from the rest of the discussion because of the constraint $\Omega_\mathrm{B} \leq 0.015\ h^{-2}$ which comes from primordial nucleosynthesis.

Another possible candidate for the dark matter is *black holes*. Useful limits to the number density of black holes in certain mass ranges come from studies of the number of gravitational lenses observed among large samples of extragalactic radio sources and from the absence of gravitational lensing by stellar mass black holes in the haloes of galaxies (Canizares 1987, Hewitt et al 1987). The limits for black holes with masses $M \sim 10^{10} - 10^{12}\mathrm{M}_\odot$ correspond roughly to $\Omega_0 \leq 1$ and similar limits are found for solar mass black holes. At the moment it cannot be excluded that the dark matter might consist of a very large population of low mass black holes but these would have to be produced by a rather special initial perturbation spectrum in the very early Universe before the epoch of primordial nucleosynthesis. To produce primordial black holes, the fluctuations would have to exceed $\delta\rho/\rho = 1$ on scales greater than the horizon. The fact that black holes of mass less than about 10^{12} kg evaporate by Hawking radiation on a cosmological timescale sets a firm lower limit to the possible masses of mini-black holes which could contribute to the dark matter at the present epoch (Hawking 1975).

An important programme to detect dark objects in the halo of our Galaxy has been described by Alcock et al (1993a). The idea is to search for the characteristic signature of gravitational microlensing events when a 'massive compact halo object' (or MACHO) passes in front of a background star. They describe a large programme to search for these rare events by making regular photometric observations of several million stars in the Magellanic Clouds and in the Galactic halo. The technique is sensitive to discrete objects with masses in the range $10^{-7} < M < 100$ $M_\odot$, be they brown dwarfs, planets, isolated neutron stars or black holes. According to their estimates, if such objects make up the dark matter in the halo of our Galaxy, about 10 or more gravitational microlensing events should be observed if their large sample of stars is precisely observed photometrically each night for a four-year period. The first experiments were begun in 1993 and the first candidate event was reported in October of the same year by Alcock et al (1993b). The characteristic signature of a gravitational lensing event was observed and the mass of the lensing object was estimated to lie in the range $0.03 < M < 0.5$ $M_\odot$. This may be an example of a 'brown dwarf' which could make up a significant fraction of the halo dark matter. One event proves little and it will be systematic observation of such events over several years which will provide definitive evidence for such objects in the Galactic halo.

The most fashionable form of dark matter is *non-baryonic dark matter*. These are often referred to collectively as 'weakly interacting massive particles' (or WIMPs). One of the attractions of these ideas is that they can be related to the types of particles which may exist according to current theories of elementary particles. There are many possibilities and I will only mention the most popular suggestions.

1. The smallest mass candidates are the *axions* which have rest mass energies about $10^{-2} - 10^{-5}$ eV. If they exist, they must have formed when the temperature of the Universe was about 10^{12} K but they never acquired thermal velocities as they were never in equilibrium. They remain 'cold' and behave like the very massive particles discussed in (3).
2. Another possibility is that the three types of neutrino have finite rest masses. The most interesting possibility is that their rest mass energies lie in the range $10 - 30$ eV. Laboratory experiments have provided upper limits to the rest mass of the electron antineutrino of the order of 20 to 30 eV (see, for example, Perkins 1987). There was great excitement when the Soviet group of Lyubimov et al (1980) reported the measurement of a finite rest mass of about 30 eV for the electron neutrino because this is almost exactly the value needed to close the Universe. The number density of neutrinos of a single type in thermal equilibrium at temperature T is $N = 0.091 \times 10^4 (2\pi kT/hc)^3$ m^{-3}. If there are N_ν neutrino types present all with rest mass m_ν, the present mass density of neutrinos in the Universe would be $\rho_\nu = N N_\nu m_\nu$. If this were to be equal to the critical density $\rho_c = 1.88 \times 10^{-26} h^2$ kg m^{-3}, the rest mass energy of the neutrino would be $m_\nu = 184 h^2/N_\nu$ eV. There are six neutrino species present with this number density, the electron, muon and tau neutrinos and their anti-neutrinos, $N_\nu = 6$ and hence the necessary rest mass of the neutrino is $30h^2$ eV. Since h lies in the range 0.5 to 1 with some preference for the lower end of this range, it follows that if the neutrino rest mass were about

10 to 20 eV, the neutrinos could close the Universe. This range of neutrino masses would just be consistent with the remarkable observations of the distribution of arrival times of the neutrinos associated with the supernova explosion in the Large Magellanic Cloud which occurred in February 1987. Further limits to the possible range of neutrino masses can be derived from the phase-space constraints for the numbers of neutrinos which would be necessary to bind large galaxies and clusters of galaxies (Tremaine and Gunn 1979). This analysis showed that neutrinos with rest masses in this range could bind giant galaxies and clusters of galaxies but they could not be responsible for binding the haloes of dwarf spiral galaxies. Other groups have not been able to confirm the result found by Lyubimov and his colleagues and the present laboratory limits to the rest mass of the electron neutrino correspond to $m_\nu c^2 \leq 9$ eV. This limit does not exclude the possibility that the μ and τ neutrinos could have greater rest masses.

3. A third possibility is that the dark matter is in some form of massive ultra-weakly interacting particle. Almost certainly these particles would have to have masses greater than about 40 GeV since they would have been expected to appear as products of the decay of the $W^\pm$ and Z^0 bosons and so broaden the width the decay spectrum of these particles. These particles might be gravitinos, the supersymmetric partner of the graviton, or photinos, the supersymmetric partner of the photon, or possibly some form of massive neutrino-like particle as yet unknown. The possible existence of these types of particles represents theoretical extrapolation far beyond the range of energies which have been explored experimentally but the ideas are sufficiently compelling that many particle theorists take seriously the possibility that cosmological studies will prove to be important in constraining theories of elementary processes at ultra-high energies.

How seriously should we take some of the more speculative ideas? There are two reasons why I believe we have to take them seriously. The simple baryonic picture of galaxy formation fails and this strongly suggests that some essential ingredient is missing. The above possibilities are the best that is on offer from the high energy particle physicists. A second reason is that it is now possible to detect certain types of non-baryonic dark matter in laboratory experiments. Particles with masses $mc^2 \geq 1$ GeV and velocities equal to the typical velocities of stars bound in the Galactic halo can now be detected in laboratory recoil experiments involving detectors cooled to about 10 mK. In these types of detector, the kinetic energy of the particle is absorbed by the lattice of a very pure semiconductor material and the very small temperature rise in the sample is measured. It has been shown that, if the halo of our galaxy were bound by 1 GeV particles, there would be a significant detection rates of events in such a cryogenic detector. If this class of experiment were to produce a positive result, it would have a very profound impact upon the theory of elementary particles.

8.3 Instabilities in the Presence of Dark Matter

It is conventional to consider three types of non-baryonic dark matter according to the rest masses of the species. The terms *hot* and *cold dark matter* are used to describe particles with rest masses about 10 eV and ≥ 1 GeV respectively. The terms refer to the velocity dispersions which the material would have now if it had simply cooled as the Universe expanded. Each species remains relativistic so long as $mc^2 \leq kT$ and therefore the least massive particles become non-relativistic latest in the Universe. They would therefore have the greatest thermal velocity dispersion now. From the comparison $kT = mc^2$, it can be seen that the hot and cold dark matter species were relativistic at redshifts $z \sim 4 \times 10^4$ and $\geq 4 \times 10^{12}$ respectively. Thus, the cold dark matter was very cold indeed by the time galaxy formation began.

A key result concerns the coupling of fluctuations in the dark matter and in the baryons. The ordinary matter and radiation are completely decoupled from the dark matter except through their mutual gravitational influence. Let us write down again the expression for the development of the gravitational instability, that is, equations (7.12) and (7.33), which we write

$$\ddot{\Delta} + 2\left(\frac{\dot{R}}{R}\right)\dot{\Delta} = A\rho_0\Delta \tag{8.1}$$

where $A = 4\pi G$ in the matter-dominated case and $A = 32\pi G/3$ in the radiation-dominated case. The following points should be noted. First, in the radiation dominated case, this equation applies to fluctuations on scales greater than the horizon scale. Second, even if cold dark matter is the dominant form of matter, in the radiation-dominated epochs during which $\varepsilon_{\rm rad}/c^2 \gg \rho_{\rm dark}$, its dynamical role is insignificant compared with that of the radiation and thus the gravitational perturbations are standard adiabatic fluctuations in the closely coupled radiation-dominated plasma. Third, throughout the matter-dominated era, most of the inertial mass is in the dark matter and therefore the evolution of these perturbations dominates the development of the baryonic perturbations. The fourth point is that, for all the dark matter perturbations, the non-baryonic particles are collisionless and hence there is no internal pressure to support the fluctuations.

Let us write the density contrast in the baryons and in the dark matter as $\Delta_{\rm B}$ and $\Delta_{\rm D}$ respectively. We consider first the epochs immediately after recombination. We have to solve the coupled equations

$$\ddot{\Delta}_{\rm B} + 2\left(\frac{\dot{R}}{R}\right)\dot{\Delta}_{\rm B} = A\rho_{\rm B}\Delta_{\rm B} + A\rho_{\rm D}\Delta_{\rm D} \tag{8.2}$$

$$\ddot{\Delta}_{\rm D} + 2\left(\frac{\dot{R}}{R}\right)\dot{\Delta}_{\rm D} = A\rho_{\rm B}\Delta_{\rm B} + A\rho_{\rm D}\Delta_{\rm D} \tag{8.3}$$

Rather than find the general solution, let us find the solution for the case in which the dark matter has $\Omega_0 = 1$ and the baryon density is negligible compared with that of the dark matter. Then equation (8.3) reduces to equation (8.1) for which we have already found the solution $\Delta_{\mathrm{D}} = BR$ where B is a constant. Therefore, the equation for the evolution of the baryon perturbations becomes

$$\ddot{\Delta}_{\mathrm{B}} + 2\left(\frac{\dot{R}}{R}\right)\dot{\Delta}_{\mathrm{B}} = 4\pi G \rho_{\mathrm{D}} BR$$

Since the background model is the critical model for which $R = (3H_0 t/2)^{2/3}$ and $3H_0^2 = 8\pi G\rho_{\mathrm{D}}$, this equation simplifies to

$$R^{3/2}\frac{\mathrm{d}}{\mathrm{d}R}\left(R^{-1/2}\frac{\mathrm{d}\Delta_{\mathrm{B}}}{\mathrm{d}R}\right) + 2\frac{\mathrm{d}\Delta_{\mathrm{B}}}{\mathrm{d}R} = \frac{3}{2}B \tag{8.4}$$

We find that the solution, $\Delta_{\mathrm{B}} = B(R - R_0)$, satisfies equation (8.4). This is a rather pleasant result because it shows that, even if the amplitude of the baryon perturbations is zero at the epoch corresponding to $R = R_0$, they grow to amplitude Δ by the epoch corresponding to R. In terms of redshift we can write

$$\Delta_{\mathrm{B}} = \Delta_{\mathrm{D}}\left(1 - \frac{z}{z_0}\right) \tag{8.5}$$

Thus, the amplitude of the perturbations in the baryons grows rapidly to the same amplitude as that of the dark matter perturbations, no matter how small they were to begin with. To put it crudely, the baryons fall into the pre-existing dark matter perturbations and, within a factor of two in redshift, have amplitude fluctuations half that of the dark matter perturbations.

A similar result is found in the early development of the perturbations when the dark matter and baryonic perturbations have scales greater than the horizon. Most of the inertial mass is in the radiation and so the development of the perturbation in the dark matter is closely tied to that of the radiation-dominated plasma.

Important differences occur when the perturbations enter the horizon during the radiation-dominated phases. As can be seen from Fig. 7.4, the baryonic perturbations are stabilised because the Jeans' length is of the same order as the horizon scale. Therefore, the baryonic perturbations become sound waves and oscillate with more or less exactly the same amplitude as when they entered the horizon right up to the epoch of recombination when the decoupling of the matter and radiation takes place. So long as the radiation-dominated plasma is the principal source of inertia, the dark matter perturbations are also stabilised and do not grow in amplitude. After the epoch of equality of the energy densities in the dark matter and the radiation, however, the dark matter perturbations grow independently of those in the radiation-dominated plasma. We see now why the above calculation is of considerable importance. The baryon perturbations are stabilised from the redshift at which they enter the horizon to the epoch of recombination but the amplitude of the perturbations in the dark matter grows from z_{eq} to the epoch of recombination. Therefore, the relative amplitudes of the fluctuations in the dark matter and the baryons is roughly $\Delta_{\rm B}/\Delta_{\rm D} \approx 1500/z_{\rm eq}$, that is, the baryon perturbations are of much smaller amplitude than those in the dark matter at the epoch of recombination. Perturbations on scales larger than those which come through the horizon at redshift $z_{\rm eq}$ have relatively smaller differences between $\Delta_{\rm D}$ and $\Delta_{\rm B}$. In the limit in which the perturbations come through the horizon at the epoch of recombination, the amplitudes of the fluctuations are of the same order of magnitude. As soon as the matter and radiation are decoupled, the amplitudes of the perturbations in the baryonic matter rapidly grow to values close to that in the dark matter by a redshift a few times smaller than the recombination redshift. Thus, even if the fluctuations in the matter were completely washed out by damping processes, the presence of fluctuations in the dark matter ensures that baryon fluctuations are regenerated after recombination.

8.4 The Evolution of Hot and Cold Dark Matter Perturbations

8.4.1 Hot Dark Matter

Let us consider first the case of hot dark matter. For the sake of definiteness, I will assume that the rest mass of the neutrino is 30 eV which means that they have $m_\nu c^2 = kT$ at a redshift $z = 1.26 \times 10^5$. This means that, during the processes of decoupling of the neutrinos and primordial nucleosynthesis which occurred when the Universe was between 1 and 1000 seconds old, the neutrinos were fully relativistic and none of the nucleosynthesis predictions of the standard Big Bang picture are affected.

The key process in the neutrino picture is *free streaming* which occurs as soon as the relativistic neutrinos enter the horizon. In all models, it is assumed that the perturbations are set up on scales much greater than the horizon. Therefore, although the particles are collisionless, they cannot escape from perturbations on scales larger than the horizon since that is as far as they can travel in the available cosmic time. As soon as they come through the horizon, however, if the neutrinos are relativistic, they can stream freely out of the perturbation. This process of free streaming means that the neutrino perturbations are damped as soon as they enter the horizon, provided the neutrinos are relativistic. In fact, it is only after the neutrinos become non-relativistic that they no longer escape freely from the perturbations. Thus, the only masses which can survive are those on very large scales. We can make a simple estimate of the range of masses which survive by working out the mass contained within the horizon when the neutrinos become non-relativistic. This mass corresponds to $M_\nu = (4\pi/3) r_{\rm H}^3 \rho_\nu$ where ρ_ν is the mass density in neutrinos at the epoch when they become non-relativistic. It is not quite clear in this simple presentation exactly what one means by the neutrinos becoming non-relativistic. Detailed calculations show that the particles may be considered non-relativistic by a redshift $z = 3 \times 10^4 (m_\nu/30\ {\rm eV})$ and the damping mass is

$$M_\nu = 4 \times 10^{15} \left(\frac{m_\nu}{30\ \mathrm{eV}}\right)^{-2} \mathrm{M}_\odot \tag{8.6}$$

This means that all smaller masses are damped out by the free streaming of the neutrinos.

The subsequent evolution of the fluctuations is straightforward. These perturbations begin to grow at the redshift z_{eq}. In the case of the relict neutrinos, this redshift is of the same order as the redshift at which the neutrinos become non-relativistic. The reason for this is that, according to the canonical hot Big Bang, the energy density in the neutrinos is more or less the same as the energy density in the photons. At the epoch when the neutrinos become non-relativistic, their inertial mass no longer decreases as the Universe expands in contrast to the case of the photons which continue to decrease in energy as R^{-1}.

In parallel with this development, adiabatic baryon fluctuations within the horizon may be damped by Silk damping. After the epoch of recombination, the baryons fall into the hot dark matter perturbations and attain the same amplitude as the neutrino perturbations. It will be noticed that this picture looks very like the standard adiabatic picture in which only the largest scale structures are formed. It is assumed that the subsequent behaviour of the perturbations is not dissimilar from the standard adiabatic picture in which smaller scale structures have to form by fragmentation of the large scale structures. The model has all the attendant advantages and disadvantages of the adiabatic baryonic model. Objects are formed late in the Universe with the added advantage that the dark haloes of clusters of galaxies are formed out of the neutrinos. It is difficult in this picture to account for the early formation of galaxies.

8.4.2 Cold Dark Matter

For definiteness, we consider the particles to have mass $m \gg 1$ GeV. The mass within the horizon when the particles became non-relativistic is therefore very small, $M \ll \mathrm{M}_\odot$ and consequently a very wide range of masses survives. Once again, the growth of the cold dark matter perturbations begins after z_{eq} and the baryonic matter falls into the growing perturbations after the epoch of recombination.

This model has a number of advantages. First of all, unlike the hot dark matter picture, perturbations on essentially all mass scales survive to the epoch of recombination. It is important that, in this model, globular cluster sized systems can begin to collapse immediately after recombination. Indeed, most of the baryonic mass in the Universe can now begin to collapse and begin the process of star formation soon after the epoch of recombination. Once discrete objects have formed, they then begin the process of hierarchical clustering under the gravitational influence of the initial fluctuation spectrum which extends to the largest masses. This process can be very conveniently described by the process of hierarchical clustering discussed in a classic paper by Press and Schechter (1974). In it, they show that the evolution of the mass spectrum of objects under a process of self-similar clustering can be described by the following expression

$$N(M,t)\mathrm{d}M \propto \left(\frac{M}{M^*}\right)^{(3+n)/6} \exp\left[-\left(\frac{M}{M^*}\right)^{(3+n)/3}\right]\frac{\mathrm{d}M}{M^2} \tag{8.7}$$

where n is the exponent of the power-spectrum of the initial fluctuations and the characteristic mass M^* varies with cosmic time as $M^* \propto t^{4/(n+3)}$. Computer simulations of the clustering of galaxies have shown that this remarkably simple formulation is a remarkably good representation of the evolution of the mass function with time (Efstathiou 1990).

The important difference as compared with the adiabatic model is that discrete objects can form early and therefore dissipative gas dynamical processes, which are crucial in forming thin pancakes on the scale of clusters and superclusters of galaxies, do not take place. Rather, the formation of large scale structures must take place through the process of hierarchical clustering under gravity.

8.5 How Well Do the Models Work?

I have given more details of the above arguments and how to proceed from there to the construction of quantitative models in my review *Galaxy Formation* (Longair 1989). I will summarise how the professionals proceed from here. For more details, the interested reader is referred to the papers by Bardeen et al (1986), Frenk (1986), Efstathiou (1990) and Davis et al (1992) which can all be very strongly recommended.

The procedures are as follows:

1. A spectrum of initial fluctuations is chosen. This is described in terms of the *power spectrum* of the density perturbations through the expression

$$\langle \Delta^2 \rangle = \frac{V}{(2\pi)^3}\int |\Delta_k|^2 \mathrm{d}^3k \tag{8.8}$$

The power spectrum is assumed to have no preferred scale and so is taken to be of power-law form

$$|\Delta_k|^2 \propto k^n \qquad \text{implying} \qquad \Delta = \frac{\delta\rho}{\rho} \propto M^{-(n+3)/6} \tag{8.9}$$

2. When these fluctuations enter the horizon, they become subject to various the damping and timing effects discussed above and which are described by the transfer functions derived by Bardeen et al (1986). The change in the input spectrum due to these effects is illustrated schematically in Fig. 8.1.

3. The fluctuations develop under gravity, in the case of the cold dark matter fluctuations by a process of hierarchical clustering. This process has been followed in computer simulations in which the particles interact gravitationally according to optimised computer codes which enable the mutual interactions of very large numbers of the particles to be followed.
4. *Biasing* is incorporated into the subsequent evolution of the perturbations so that the ultimate large scale structure resembles the observed large scale distribution of galaxies as closely as possible.

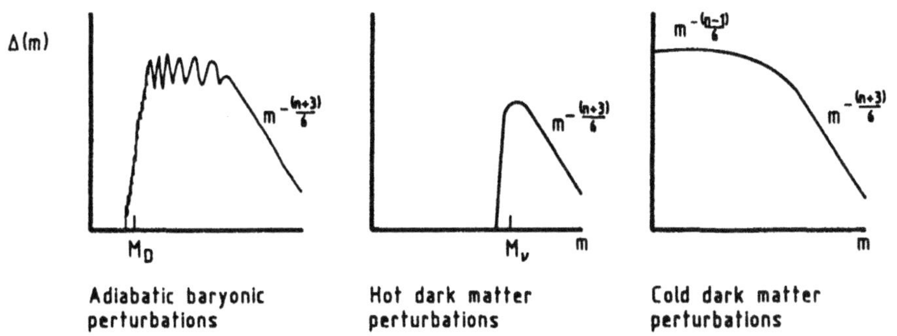

Fig. 8.1 A schematic diagram illustrating the form of the density contrast Δ as a function of mass at the epoch of recombination for an initial power spectrum of the form $|\Delta_k|^2 \propto k^n$. (i) Adiabatic baryonic fluctuations, (ii) hot dark matter fluctuations of neutrinos with rest mass 30 eV, (iii) the cold dark matter fluctuations.

Fig. 8.2 is a sample of the results of computer simulations of the hot and cold dark matter models carried out by Frenk (1986). In the case of the *hot dark matter* model it can be seen that the model is very effective in producing flattened structures like pancakes. The baryonic matter forms pancakes within the large neutrino haloes and their evolution is similar to the adiabatic picture from that point on. It can be seen that the model is too effective in producing flattened, stringy structures. Essentially everything collapses into the thin pancakes and filaments and the observed Universe is not as highly structured as this. In addition, galaxies must form rather late in this picture because it is only the most massive structures which survive to the recombination epoch. This means that it is difficult to produce stars and galaxies which are older than the structures which formed on the scale of $\sim 4 \times 10^{15} M_\odot$. Everything must have formed rather late in the Universe in this picture.

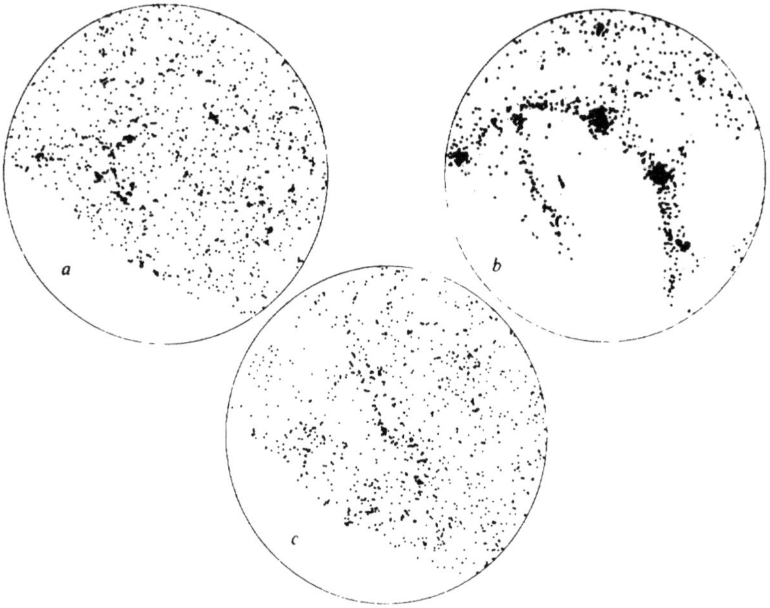

Fig. 8.2 Recent simulations of the expectations of (a) the cold dark matter and (b) the hot dark matter models of the origin of the large scale structure of the Universe compared with the observations (c) (Frenk 1986). The unbiased cold dark matter model does not produce sufficient large scale structure in the form of voids and filaments of galaxies whereas the unbiased hot dark matter model produces too much structure.

In the case of the *cold dark matter* picture, masses on all scales begin to collapse soon after recombination and star clusters and the first generations of stars can be old in this picture. Large scale systems like galaxies and clusters of galaxies are assembled from their component parts by dynamical processes which are simulated by computer modelling. Fig. 8.2 shows that structure indeed develops but is not as pronounced as in the observed Universe. This is because it is difficult to produce elongated structures by gravitational clustering which tends to make more symmetrical structures than the sheets and filaments of galaxies found in the Universe. One of the successes of the cold dark matter picture is, however, that it can account for the observed two-point correlation function of galaxies (expression 2.2), assuming $n = 1$ and that the phases of the waves which make up the power-spectrum $|\Delta_k|^2$ are random (Frenk 1986).

Evidently, neither model gives a particularly good fit to the observations and this has led to the idea that there may well be bias mechanisms which lead to the preferential formation of galaxies in certain regions of space rather than others. In the case of the cold dark matter picture, the biasing must be such that it results in galaxies being formed preferentially in regions of high density and not in the regions in between. In the case of the hot dark matter picture, some anti-biasing mechanism is needed so that all the formation of galaxies is not concentrated into the sheets and filaments which form too readily in this model.

Is there any evidence for biasing in the Universe? Many studies are underway to find out if there is a preference for galaxies forming in one region rather than another. A good example is the case of structures such as the Coma Cluster of galaxies. This cluster has been studied kinematically in greater detail than any other cluster and so the estimates of its mass and mass-to-light ratio are well known. There is a dark matter present which amounts to about a factor of 10 as compared with the visible matter but the mass-to-light ratio found when all the dark matter is included corresponds to a factor of only about one third of the value necessary to close the Universe. If the Universe really has the critical density, $\Omega_0 = 1$, this means that there must be biasing by a factor of three in favour of the formation of galaxies in the region of the Coma cluster as opposed to the general field.

Many possible biasing and anti-biasing mechanisms have been discussed by Dekel and Rees (1987) and by Dekel (1987). As they emphasise, perhaps the most important requirements are reliable observational estimates of how much biasing (or anti-biasing) really occurs in the Universe. I refer readers to their papers for a survey of the many possible bias mechanisms and how these ideas can be tested by observations of the nearby Universe. I mention some of these ideas to give a flavour of the astrophysical questions to be addressed.

1. An interesting example discussed by Kaiser (1986) is the influence of large scale perturbations upon a region in which the fluctuations from which galaxies are forming are already well developed. In this situation, the large-scale density perturbation enhances the density perturbations associated with the collapsing galaxies and so can lead to enhanced galaxy formation in this region. The study of the density peaks in the random density fluctuation fields is an important aspect of the detailed evolution of the perturbation spectrum (see e.g. Peacock and Heavens 1985).
2. Galactic explosions may sweep the gas from the vicinity of a galaxy, possibly having a positive or negative effect upon biasing. A violent explosion can remove the gas from the vicinity of the galaxy and make it too hot for further galaxy formation to occur in its vicinity. On the other hand, the swept-up gas may be greatly increased in density at the interface between the hot expanding gas and the intergalactic gas. By analogy with the case of galactic supernova remnants in which star formation can be stimulated by the passage of a shock wave, the same process on a galactic scale may stimulate the formation of new galaxies (Ostriker and Cowie 1981). One possibility is that this process might thicken the pancakes which form in the hot dark matter picture.

3. The gas in the voids between superclusters may be so hot that galaxies cannot form in these regions.

It is evident from these examples that the questions are largely astrophysical. In my view, the most important discrepancy is between the total mass in the largest bound systems for which we can make reasonable estimates of the mass density and the critical model for which $\Omega_0 = 1$. I find it striking that the determinations of the amount of gravitating matter in rich clusters of galaxies suggest that the matter they contain, including the dark matter, is insufficient to close the Universe. Therefore, if $\Omega_0 = 1$, most of the dark matter must be located between the clusters of galaxies where it is most difficult to make good mass estimates.

References – Chapter 8

Alcock, C., Allsman, R.A., Axelrod, T.S., Bennett, D.P., Cook, K.H., Park, H.-S., Marshall, S.L., Stubbs, C.W., Griest, K., Perlmetter, S., Sutherland, W., Freeman, K.C., Peterson, B.A., Quinn, P.J. and Rodgers, A.W. (1993a). In *Sky Surveys : Protostars to Protogalaxies*, (ed. T. Soifer), 291. San Francisco: Astron. Soc. Pac. Conf. Ser.

Alcock, C., Akerlof, C.W., Allsman, R.A., Axelrod, T.S., Bennett, D.P., Chan, S., Cook, K.H., Freeman, K.C., Griest, K., Marshall, S.L., Park, H.-S., Perlmutter, S., Peterson, B.A., Pratt, M.R., Quinn, P.J., Rodgers, A.W., Stubbs, C.W. and Sutherland, W. (1993b). Nat, **365**, 621.

Bardeen, J.M., Bond, J.R., Kaiser, N. and Szalay, A.S. (1986). ApJ, **304**, 15.

Canizares, C. (1987). In *Observational Cosmology*, (eds. A. Hewitt, G.R. Burbidge and Fang LiZhi), 729. Dordrecht: D. Reidel and Co.

Davis, M., Efstathiou, G., Frenk, C.S. and White, S.D.M. (1992). Nat, **356**, 489.

Dekel, A. (1987). Comments Astrophys., **11**, 235.

Dekel, A. and Rees, M.J. (1987). Nat, **326**, 455.

Efstathiou, G. (1990). In *Physics of the Early Universe*, (eds. J.A. Peacock, A.F. Heavens and A.T. Davies), 361. Edinburgh: SUSSP Publications.

Frenk, C.S. (1986). Phil. Trans. R. Soc. Lond., **A 330**, 517.

Hawking, S.W. (1975) Comm. Math. Phys.,**43**, 199.

Hewitt,J.N., Turner, E.L., Burke, B.F., Lawrence, C.R., Bennett, C.L., Langston, G.I. and Gunn, J.E. (1987). In *Observational Cosmology*, (eds. A. Hewitt, G.R. Burbidge and Fang LiZhi), 747. Dordrecht: D. Reidel and Co.

Kaiser, N. (1984). ApJ, **284**, L9.

Lyubimov, V.A., Novikov, E.G., Nozik, V.Z., Tretyakov, E.F. and Kozik, V.S. (1980). *Phys. Lett*, **94B**, 266.

Ostriker, J.P. and Cowie, L.L. (1981). ApJ, **243**, L127.

Peacock, J.A. and Heavens, A.F. (1985). MNRAS, **217**, 805.

Perkins, D.H. (1987). *Introduction to High Energy Physics*. Menlo Park, California: Addison-Wesley Publishing Co.

Press, W.H. and Schechter, P. (1974). ApJ, **187**, 425.

Tremaine, S.D. and Gunn, J.E. (1979). Phys. Rev. Lett., **42**, 467.

9 Fluctuations in the Cosmic Microwave Background Radiation

9.1 The Ionisation of the Intergalactic Gas Through the Epoch of Recombination

We have mentioned on several occasions the critical importance for cosmology of the limits to the temperature fluctuations in the Cosmic Microwave Background Radiation. The problem addressed in this chapter is how to relate the fluctuations in the dark and baryonic matter at the epoch of recombination to the temperature fluctuations which they imprint upon the background radiation at the epoch of recombiation. Of crucial importance in this calculation is the ionisation state of the intergalactic gas through the epoch of recombination. The optical depth of the intergalactic gas due to Thomson scattering is given by the expression (6.4) which shows how rapidly it increases to very large values at redshifts greater the redshift of recombination. Thus, temperature fluctuations which originate at redshifts greater than the redshift of recombination are damped out by scattering and the fluctuations we observe originate in a rather narrow redshift range about that at which the optical depth of the intergalactic gas is unity. This is why the precise determination of the ionisation history of the intergalactic gas is so important.

The problem, first discussed by Zeldovich, Kurt and Sunyaev (1968) and Peebles (1968), is well-known. During the recombination process, the photons released in the recombination of hydrogen atoms are sufficiently energetic to ionise other hydrogen atoms and thus there is no direct way of destroying the photons liberated in the recombination process. The answer is that Lyman-α photons are destroyed by the *two-photon process* in which two photons are liberated from the 2s state of hydrogen in a rare quadrupole transition. The spontaneous transition probability for this process is very small, $A = 8.23$ $\sec^{-1}$ but it turns out to be the dominant process which determines the rate of recombination of the intergalactic gas.

Detailed calculations of the degree of ionisation through the critical redshift range have been carried out by Jones and Wyse (1985) who find a very strong dependence of the fractional ionisation x upon redshift at recombination. They provide a convenient analytic expression for the degree of ionisation through the critical range of redshift:

$$x = 2.4 \times 10^{-3} \frac{(\Omega_0 h^2)^{\frac{1}{2}}}{\Omega_{\mathrm{B}} h^2} \left(\frac{z}{1,000}\right)^{12.75} \tag{9.1}$$

In this formula, Ω_0 is the density parameter for the Universe as a whole at the present epoch and Ω_{B} is the present density parameter of baryons. Using the same formalism which led to equation (6.4), we can find a simple expression for the optical depth of the intergalactic gas at redshifts $z \sim 1,000$

$$\tau = 0.37\left(\frac{z}{1,000}\right)^{14.25} \tag{9.2}$$

Because of the enormously strong dependence upon redshift, the redshift at which the optical depth of the intergalactic gas is unity is always very close to 1,070, independent of the exact values of Ω_0 and h. We can now work out the range of redshifts at which the photons of the background radiation were last scattered. This probability distribution is given by

$$\mathrm{d}p/\mathrm{d}\tau = e^{-\tau}\mathrm{d}\tau/\mathrm{d}z \tag{9.3}$$

This probability distribution can be closely approximated by a gaussian distribution with mean redshift 1,070 and standard deviation $\sigma = 80$ in redshift. This result formalises the statement that the last scattering of the photons did not take place at a single redshift but that half of the photons of the Microwave Background Radiation were last scattered between redshifts 1,010 and 1,130.

9.2 Fluctuations in the Background Radiation due to Large-Scale Density Perturbations

9.2.1 The Physical and Angular Scales of the Fluctuations

First of all, let us work out the physical scale at the present epoch corresponding to the thickness of the last scattering layer. The formula for the element of radial comoving distance at redshift z can be derived from equation (3.18)

$$\mathrm{d}r = \frac{c\mathrm{d}z}{H_0(1+z)(\Omega_0 z+1)^{1/2}} \tag{9.4}$$

Notice that because of the use of comoving coordinates, this is the distance element projected to the present epoch. Taking the approximation for large redshifts, we find $\mathrm{d}r = c\mathrm{d}z/(H_0 z^{3/2}\Omega_0^{1/2})$. Thus, the redshift interval of 120 at a redshift of 1,070 corresponds to a scale of $10(\Omega_0 h^2)^{-1/2}$ Mpc at the present day. On angular scales smaller than this value, we expect a number of independent fluctuations to be present along the line of sight through the redshift interval $1,010 < z < 1,130$. Consequently, the random superposition of these perturbations leads to a statistical reduction in the amplitude of the observed temperature fluctuations by a factor of roughly $N^{-1/2}$ where N is the number of fluctuations along the line of sight. The mass contained within this scale is $M \approx 3 \times 10^{14}(\Omega_0 h^2)^{1/2} \mathrm{M}_\odot$, corresponding roughly to the mass of a cluster of galaxies. The angular scale of these fluctuations as observed at the present epoch is $6\Omega_0^{1/2}$ arcmin. Notice, incidentally, that the tools for evaluating the fluctuations in the Cosmic Microwave Background Radiation are simpler than the $P(D)$ approach described in Section 4.5 because all the fluctuations arise at essentially the same redshift, $z \sim 1,070$.

The problem is now to convert the density fluctuations and their associated velocities into temperature fluctuations. The situation is depicted schematically in Fig. 9.1.

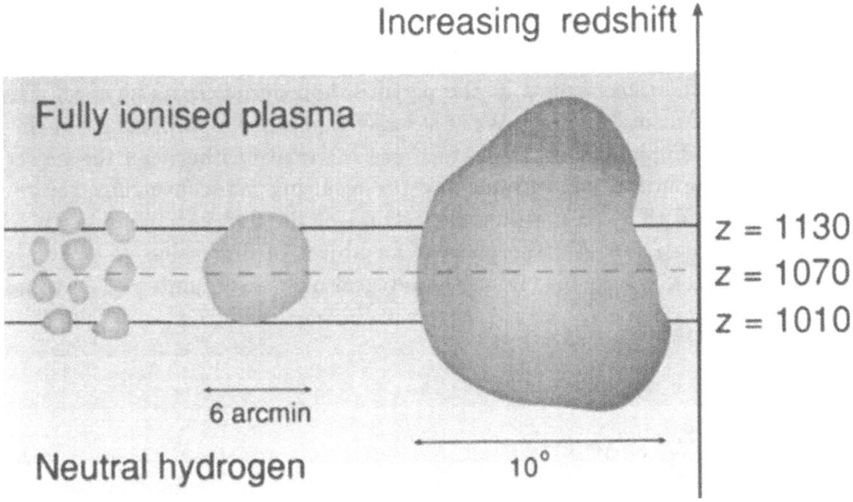

Fig. 9.1 A schematic diagram illustrating perturbations of different physical scales at the epoch of recombination. The largest scale perturbations are much larger than the width of the region within which most of the photons of the Cosmic Microwave Background Radiation were last scattered. On small scales, a number of perturbations is observed along the line of sight through the last scattering layer.

9.2.2 Large Angular Scales

Perturbations on angular scales very much greater than $6\Omega_0^{1/2}$ arcmin correspond to physical scales much larger than the thickness of the recombination layer, as depicted by the large fluctuation to the right in Fig. 9.1. On the very largest scales, the dominant source of temperature fluctuations results from the fact that the photons have to climb out of the gravitational potential well associated with the perturbation. During their subsequent propagation to the Earth, although the photons pass through gravitational potential fluctuations, what they gain by falling into them is exactly compensated by the gravitational redshift coming out and so it is the escape from the perturbations at the epoch of recombination which provides a first-order effect in the gravitational redshift. This phenomenon was first analysed by Sachs and Wolfe (1967) and is known as the *Sachs-Wolfe* effect. The fluctuations in the Cosmic Microwave Background Radiation observed by COBE are on an angular scale of 10° (Section 1.5) and so the Sachs-Wolfe effect is the predominant source of fluctuations.

We can illustrate the origin of the fluctuations by a simple calculation. First, we note that, to order of magnitude, the temperature perturbations $\Delta T/T$ are related to the gravitational redshift z_{grav} of the radiation by the approximate relations

$$\frac{\Delta T}{T} \sim z_{\mathrm{grav}} = \frac{\Delta\phi}{c^2} \approx \frac{G\Delta M}{Rc^2} \tag{9.5}$$

where ΔM is the mass excess in the perturbation and R is its physical size at the epoch of recombination. We can derive a number of important conclusions by some simple calculations which use the relations between the power spectrum of the initial fluctuations and the resulting mass spectrum, the relations (8.8) and (8.9). If the power-spectrum of the fluctuations is taken to have the form $|\Delta_k|^2 \propto k^n$, the mass of an object of dimension $R \sim k^{-1}$ is proportional to k^{-3} and so, taking the integral over wavenumber (8.8) to be

$$\langle \Delta^2 \rangle \propto |\Delta_k^2| k^3$$

we find that

$$\Delta = \frac{\delta\rho}{\rho} = \frac{\delta M}{M} \propto M^{-(n+3)/6} \tag{9.6}$$

Let us work out the fluctuation in the gravitational potential $\Delta\phi$ at the epoch of recombination in terms of the properties of the perturbations observed at the present epoch. Throughout this calculation, we assume that we are in the linear regime of the development of the perturbations and so the perturbations are still small to the present epoch. The physical size of the perturbation R at redshift z corresponds to a physical size R_0 at the present epoch $R(1+z) = R_0$. Assuming that $\Omega_0 = 1$, the density perturbation develops as $\Delta\rho/\rho \propto (1+z)^{-1}$ and so the density fluctuation $\Delta\rho$ at redshift z was

$$\Delta\rho = \frac{\Delta\rho_0}{(1+z)} \frac{\rho}{\rho_0} = \Delta\rho_0 (1+z)^2$$

Since $\Delta M \approx \Delta\rho R^3$ and $R = R_0/(1+z)$, it follows that

$$\Delta\phi \approx \frac{G\Delta M}{R} \approx G\Delta\rho_0 R_0^2 \tag{9.7}$$

This is the first remarkable result of this analysis — namely, that the perturbation to the gravitational potential is independent of cosmic epoch since all dependence upon redshift z has disappeared from the expression (9.7), that is, $\Delta\phi$ at any redshift is directly related to the properties of the same perturbations at the present epoch.

We can now incorporate the mass spectrum into the calculation. Since $\delta\rho_0 \propto \rho_0 M^{-(n+3)/6}$, and $M \approx \rho_0 R_0^3$, we find $\delta\rho_0 \propto R_0^{-(n+3)/2}$ and hence

$$\Delta\phi \approx G\Delta\rho_0 R_0^2 \propto R_0^{(1-n)/2} \tag{9.8}$$

It follows immediately from the expressions (9.5) and the expression relating the physical size R_0 of the perturbation to the angular size it would subtend at a large redshift, $R_0 = \theta D$, where $D = 2c/H_0\Omega_0$, that

$$\frac{\Delta T}{T} \approx \frac{\Delta\phi}{c^2} \propto \theta^{(1-n)/2} \tag{9.9}$$

This is an important result. The dependence of the amplitude of the temperature fluctuations upon angular scale depends upon the spectral index n of the initial power spectrum of the fluctuations. A spectral index $n = 0$ corresponds to a white noise spectrum. A spectral index $n = 1$ corresponds to the scale-invariant spectrum advocated by Harrison (1970) and by Zeldovich (1972). From the simple rules we developed for the evolution of the perturbations with time, it is straightforward to show that the spectrum with $n = 1$ corresponds to perturbations on different scales coming through the horizon with the same amplitude. This spectrum has the advantage that the amplitudes of the perturbations never become so great on small scales that large numbers of black holes were created in the early Universe. Furthermore, numerical simulations of the formation of structure in the cold dark matter picture show that the two-point correlation function for galaxies can be accounted for if $n = 1$.

It can be seen from the expression (9.9) that, if $n = 1$, the temperature fluctuations are expected to be of the same amplitude on all large angular scales. The results of detailed computations of the expected amplitude of the temperature fluctuations on a wide range of angular scales is shown in Fig. 9.2 (Bond and Efstathiou 1987, Efstathiou 1990). It can be seen that, on large angular scales, $\theta \geq 1°$, the temperature fluctuations have constant amplitude, in particular, on the scales which have been observed by COBE.

The physical scale of the perturbations on an angular scale of 10° at the present epoch is roughly $1,000(\Omega_0 h)^{-1}$ Mpc. Thus, the scale of the flucutations in the Cosmic Microwave Background Radiation correspond to scales at least an order of magnitude greater than the large holes seen in the distribution of galaxies in Fig. 2.2. We therefore have to extrapolate the mass spectrum of galaxies to large scales to estimate the amplitude of the temperature fluctuations. The problem is discussed in detail by Efstathiou (1990) and Peebles (1993). For the purposes of making a simple estimate, let us use Peebles' estimate for the amplitude of the density fluctuations on large scales from the second moment of two-point correlation function, the quantity J_3. He finds $\delta\rho/\rho = 0.3$ on scales $l = 50h^{-1}$ Mpc. This is similar to the value found from studies of the distribution of IRAS galaxies. Adopting the critical density, $\rho_0 = 2 \times 10^{-26} h^2$ kg m^{-2}, we find from the expression

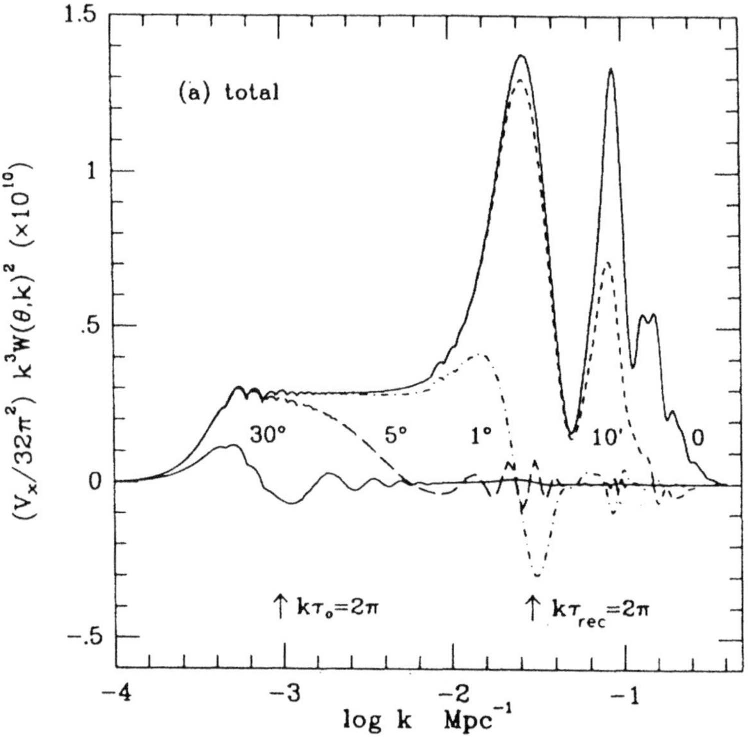

Fig. 9.2 The amplitude of the radiation autocorrelation function as a function of physical scale for a scale-invariant adiabatic cold dark matter model for which $\Omega_0 = 1$, $\Omega_B = 0.03$ and $h = 0.75$ (Efstathiou 1990).

(9.7) that $\Delta T/T \sim 10^{-5}$. This temperature fluctuation is roughly the same level of fluctuations as observed by the COBE DRM experiment. Obviously, a much more sophisticated analysis is needed to make a proper comparison with these magnificent observations.

The detailed analysis of Efstathiou (1990) shows that it is necessary to incorporate some degree of biasing into the simulations in order to obtain good agreement between the two-point correlation function for galaxies and the observed amplitude of the fluctuations in the Cosmic Microwave Background Radiation. This biasing is in the sense that the galaxies should show more structure than the underlying dark matter distribution and this would be in accord with the fact that the dark matter in clusters of galaxies accounts for only about 20-30% of the critical density. This biasing factor is also necessary to reconcile the predicted random velocities of field galaxies with their observed peculiar velocities.

9.2.3 Intermediate Mass Scales

Inspection of Fig. 9.2 shows that the predicted fluctuations oscillate on smaller angular scales, $1^\circ \geq \theta \geq 5$ arcmin. These oscillations are associated with the phasing of the perturbations as they come through the horizon. The fluctuations which grow to large amplitude are those which have the maximum amplitude when they come through the horizon in the radiation-dominated phases. As soon as they come through the horizon, they are stabilised and oscillate as sound waves until the fluctuations in the dark matter are dynamically dominant. They then grow in amplitude but it is apparent that the phase of the perturbation is crucial in determining its amplitude when it begins to grow. This is the origin of the oscillatory behaviour see in Fig. 9.2.

The other important difference is that the source of the temperature fluctuations is no longer the Sachs-Wolfe effect but rather the effects of first order *Doppler scattering* of the photons by the perturbations. The magnitudes of the velocities associated with the perturbations are given by

$$\frac{d\Delta}{dt} = -\nabla \cdot \delta \mathbf{v} \tag{9.10}$$

It can be seen from the considerations of section 9.2.1 that the fluctuations observed on scales greater than 10 arcmin are larger than the thickness of the layer from which the photons were last scattered and so the photons are only observed from a part of the cloud. As a result, first order Doppler scatterings from that part of the cloud cause fluctuations in the Cosmic Microwave Background Radiation. On scales smaller than the thickness of the last scattering layer, the fluctuations are damped because of the oppositely directed velocities across the wave contribute to the observed intensity.

9.3 Comparison with Observations

The considerations of the last section are summarised in Fig. 9.3 which shows a comparison of the predictions of the standard cold dark matter theory with a wide range of observations (For a discussion of many of these observations, see Readhead and Lawrence 1992). The predictions of the theory are indicated as crosses and the various upper limits and the COBE measurement are also shown. In this presentation, there is a difference of about a factor of two between the expectations of the theory and the COBE observations.

Since Fig. 9.3 was prepared, the MIT and Berkeley/UCSD Balloon experiments have reported positive detections at roughly the level of the COBE fluctuations and the ACME/HEMT UCSB experiments at the South Pole have made a possible detection at a level of about $\Delta T/T \approx 8 \times 10^{-6}$. The Tenerife experiment has reported the detection of fluctuations at roughly the same intensity level as the COBE observations. It is of particular importance that the Tenerife experiment has measured the fluctuations at two different

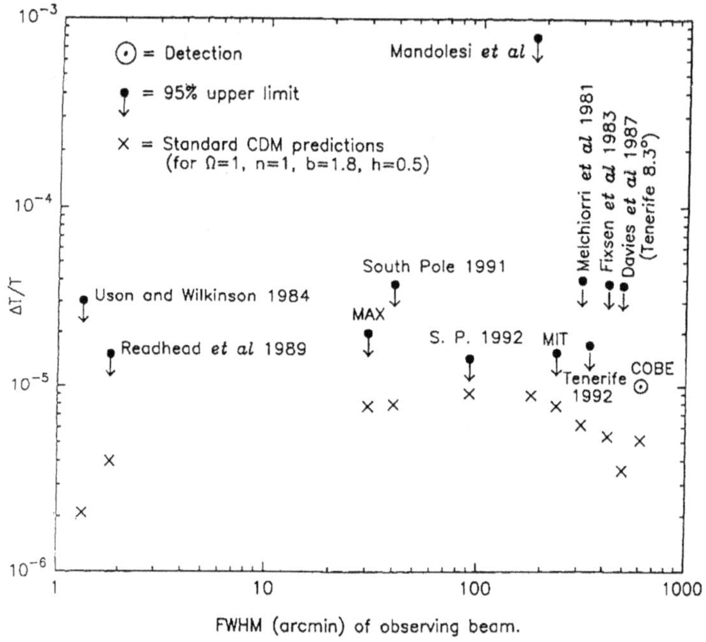

Fig. 9.3 Upper limits and observations of temperature fluctuations in the Cosmic Microwave Background Radiation. The expectations of a standard cold dark matter model for the formation of structure in the Universe is shown as crosses (Courtesy of Dr. Anthony Lasenby 1993).

frequencies and that the same pattern of fluctuations is seen at both of them at the same level of intensity fluctuation (Fig. 9.4). This result shows that the fluctuations have the same intensity spectrum as the Cosmic Microwave Background Radiation, thus demonstrating that the fluctuations are not associated with Galactic synchrotron radiation or with the thermal emission of the high latitude cirrus detected by the IRAS satellite, both of which have quite different spectral signatures.

This is a very rapidly developing area and a large number of experiments is being planned to determine the properties of the fluctuations much more precisely. There is now much better understanding of how these experiments can be carried out sucessfully and we can be confident that the study of background fluctuations will become a major growth area in observational cosmology.

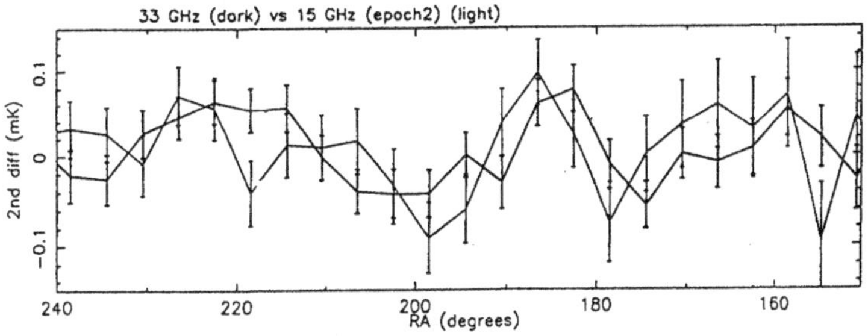

Fig. 9.4 Fluctuations in the Cosmic Microwave Background Radiation observed at frequencies of 15 and 33 GHz from the Tenerife experiment. The fluctuations have a similar intensity distribution at the two frequencies and correspond to the same amplitude of temperature fluctuation. (Courtesy of Dr. Anthony Lasenby)

9.4 Other Sources of Fluctuations

Whilst there is no question but that the discovery of fluctuations in the Cosmic Microwave Background Radiation is wonderful discovery (although not, with respect, the 'discovery of the century' claimed by some enthusiasts!), fluctuations had to be discovered at about this level of intensity or something would have been profoundly wrong with the cosmological models. As we have shown, the real problem is suppressing the temperature fluctuations to a level of $\Delta T/T \sim 10^{-5}$. In addition, there are a number of other effects which can cause fluctuations in the background at about this level and I will mention some of them just to indicate how rich these studies are likely to be for astrophysics and cosmology.

1. *The Reheating of the Intergalactic Gas* At some stage the intergalactic gas has to be heated and ionised. It is still not clear how this came about, although the consensus of opinion is that the most likely mechanism is photoionisation at redshifts less than 10 and quite possibly at redshifts about 4. We recall that the optical depth of the intergalactic gas to Thomson scattering is $\tau_{\mathrm{T}} = 0.04(\Omega_0 h^2)^{1/2} z^{3/2}$. Thus, if the gas were ionised back to redshifts of 10 with $\Omega_{\mathrm{B}} = 0.1$, $\tau_{\mathrm{T}} \sim 1$. This would lead to further damping of the fluctuations.
2. *The Sunyaev-Zeldovich Effect in Clusters of Galaxies* Wherever there is very hot gas, the Sunyaev-Zeldovich effect results in small decrements in the background radiation. As we have shown in section 5.3, the intensity fluctuation due to the passage of the Cosmic Microwave Background Radiation through a region of hot gas such as that contained within a cluster of galaxies is

$$\frac{\Delta I}{I} = -2y = -2\int\left(\frac{kT_{\mathrm{e}}}{m_{\mathrm{e}}c^2}\right)\sigma_{\mathrm{T}} N_{\mathrm{e}}\,\mathrm{d}l$$

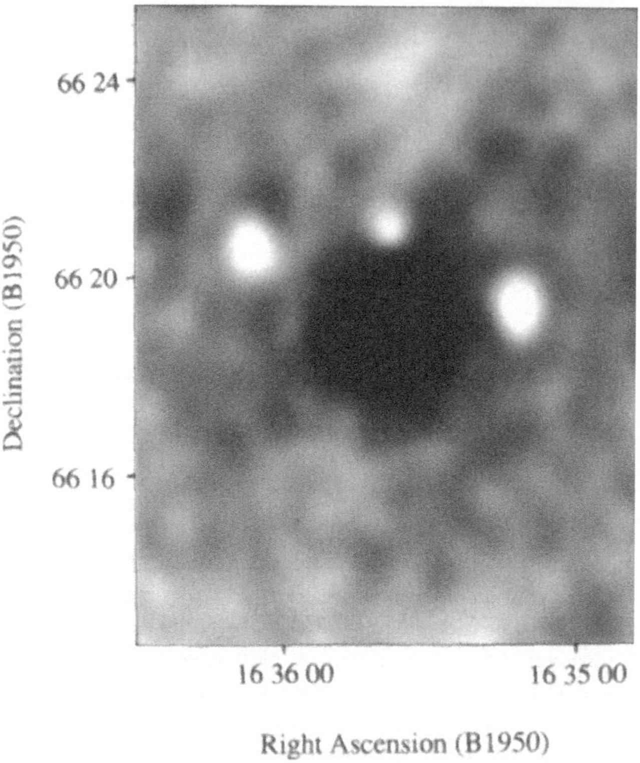

Fig. 9.5 A map of the Sunyaev-Zeldovich decrement in the Cosmic Microwave Background Radiation in the direction of the Abell cluster of galaxies Abell 2218 (Courtesy of Dr. Michael Jones).

in the Rayleigh-Jeans region of the spectrum. This effect has now been observed with certainty in a number of cluster of galaxies which are strong X-ray sources. Birkinshaw (1990) presented convincing scans showing a decrement in the cluster Abell 2218. Just this year, Michael Jones and his colleagues at the MRAO in Cambridge have upgraded the receiver systems of the Ryle 5-km Telescope and they have made the first synthesis map of the decrement in the background in the direction of this cluster (Fig. 9.5). This is an exciting result because, as is well known, this observation can be used in conjunction with the X-ray observations to estimate the physical size of the cloud of hot gas responsible for the X-ray emission, independent of the redshift of the cluster. Consequently Hubble's constant can be estimated by physical means directly. Although they are not quoting a number, Michael Jones tells me that the result is likely to favour low values of Hubble's constant. From the point of view of the observation of fluctuations in the background radiation, the important point is that the decrement has amplitude $\Delta I/I \sim 10^{-4}$, that is, significantly greater than the fluctuations in the Microwave Background Radiation. These fluctuations are, however, on an angular scale of a few arcminutes and suitable clusters for this experiment are quite difficult to find. The ROSAT survey should produce many more suitable candidates and consequently provide directly information about how significantly hot cluster gas could influence the observation of genuinely cosmological fluctuations.

References – Chapter 9

Birkinshaw, M. (1990). In *The Cosmic Microwave Background: 25 Years Later*, (eds. N. Mandolesi and N. Vittorio), 77. Dordrecht: Kluwer Academic Publishers.

Bond, and Efstathiou, G. (1987). MNRAS, **226**, 655.

Efstathiou, G. (1990). In *Physics of the Early Universe*, (eds. J.A. Peacock, A.F. Heavens and A.T. Davies), 361. Edinburgh: SUSSP Publications.

Harrison. E.R. (1970). Phys. Rev., **D1**, 2726.

Jones, B.J.T. and Wyse, R.F.G. (1985). A&A, **149**, 144.

Peebles, P.J.E. (1968). ApJ, **153**, 1.

Peebles, P.J.E. (1993). *Principles of Physical Cosmology*. Princeton: Princeton University Press.

Readhead, A.C.S. and Lawrence, C.R. (1992). ARA&A, **30**, 653.

Sachs, R.K. and Wolfe, A.M. (1967). ApJ, **147**, 73.

Zeldovich, Ya.B. (1972). MNRAS, **160**, 1P.

Zeldovich, Ya.B., Kurt, V.G. and Sunyaev, R.A. (1968). Zh. Exsp. Teor. Fiz., **55**, 278. [English translation: Sov. Phys. - JETP, **28**, 146.]

10 The Intergalactic Gas

The nature of the intergalactic gas is one of the more tantalising problems of modern cosmology. It is certain that there is intergalactic gas in rich clusters of galaxies as is indicated by the observation of intense diffuse bremsstrahlung at a temperature of about 10^8 K from clusters such as the Coma cluster. In such clusters, the mass of intergalactic gas is similar to the mass in the visible parts of the cluster galaxies. Furthermore, observations of the FeXXV and FeXXVI lines of highly ionised iron show that the heavy elements are underabundant relative to the cosmic abundances of the elements by a factor of about 4 or 5. It is inferred that the primordial gas has been enriched by the processes of nucleosynthesis taking place in galaxies.

It is much more difficult to study the diffuse intergalactic gas between clusters of galaxies. These is no definite evidence that there is any diffuse gas at all between the clusters of galaxies but there are reasons to suppose that there must be some gas present. For example, there must be some means of confining the hydrogen clouds which make up the Lyman-α forest. It seems unlikely that the process of galaxy formation was so efficient that all the matter was condensed into galaxies. In addition, the evidence of the enrichment of the intergalactic gas in clusters of galaxies suggests that processed gas can be recycled through the intergalactic medium.

Let us begin this brief survey by deriving expressions for the background emission from the intergalactic gas and its optical depth for absorption of the emission from background sources.

10.1 The Background Emission of and Absorption by the Intergalactic Gas

In section 4.2, we showed that the background intensity due to a uniform distribution of sources can be written in the form

$$I(\nu_0) = \frac{1}{4\pi}\int_0^\infty \frac{L(z,\nu_1)N_0}{(1+z)}\mathrm{d}r \tag{10.1}$$

where it is assumed that the comoving number density of sources is conserved but we have now allowed the luminosity of the sources to change with cosmic epoch. To adapt this expression for the case of the emission from the intergalactic gas, it is convenient to work in terms of the *proper* number density of objects at redshift z, $N(z) = N_0(1+z)^3$. Then, the luminosity per unit proper volume, that is the *emissivity* of the intergalactic medium is $\epsilon(\nu_1) = L(z,\nu_1)N(z)$ and so the expression for the background intensity from diffuse intergalactic emission processes is

$$I(\nu_0) = \frac{1}{4\pi}\int_0^\infty \frac{\epsilon(\nu_1)}{(1+z)^4}\mathrm{d}r \tag{10.2}$$

If we adopt the standard Friedman models, we find

$$I(\nu_0) = \frac{c}{4\pi H_0} \int_0^\infty \frac{\epsilon(\nu_1)}{(\Omega_0 z + 1)^{1/2}(1+z)^5} \mathrm{d}z \tag{10.3}$$

where, as usual, $\nu_0 = \nu_1/(1+z)$.

In exactly the same way, we can work out the optical depth of the gas at frequency ν_0 due to absorption by intergalactic matter. If $\alpha(\nu_1)$ is the absorption coefficient for radiation at frequency ν_1, then the increment of optical depth for the photons which will eventually be redshifted to frequency ν_0 by the time they reach the Earth is $\mathrm{d}\tau(\nu_0) = \alpha(\nu_1)\mathrm{d}l = \alpha(\nu_1)c\mathrm{d}t$ where $c\mathrm{d}t = \mathrm{d}l$ is the element of proper (or geodesic) distance at redshift z. Hence, integrating along the path of the photon, we find

$$\tau(\nu_0) = \frac{c}{H_0} \int_0^z \frac{\alpha(\nu_1)\mathrm{d}z}{(\Omega_0 z + 1)^{1/2}(1+z)^2} \tag{10.4}$$

Notice that, in the case of an absorption line, the absorption coefficient $\alpha(\nu_1)$ includes the profile of the absorption line.

10.2 The Gunn-Peterson Test

One of the most important tests for the presence of intergalactic neutral hydrogen was described independently by Gunn and Peterson (1965) and by Scheuer (1965), soon after the first quasar with redshift greater than 2, 3C9, was discovered. I am particularly fond of 3C9 since I made the optical identification of the quasar in 1963 as almost my first task for Martin Ryle (see Ryle and Sandage 1964, Longair 1965). The particular aspect of quasars which is used in the test is the fact that their continuum spectra are non-thermal and extend into the far ultraviolet and X-ray wavebands.

The test makes use of the fact that the Lyman-α absorption cross-section is very large and so, when the ultraviolet continuum of distant quasars is shifted to that redshift at which the continuum has wavelength 121.6 nm, the radiation is absorbed and re-emitted in some random direction many times so that, if there is sufficient neutral hydrogen at these redshifts, an absorption trough will be observed to the short wavelength side of the redshifted Lyman-α line. Only when quasars with redshifts $z > 2$ were discovered was the Lyman-α line redshifted into the observable visible waveband.

Let us carry out the simple calculation to show how the expression (10.4) can be used to work out the optical depth for Lyman-α scattering.

The expression for the photo-excitation cross-section for a Lyman-α transition is

$$\sigma(\nu) = \frac{e^2 f}{4\epsilon_0 m_e c} g(\nu - \nu_{\mathrm{Ly}}) \tag{10.5}$$

where $\nu_{\rm Ly}$ is the frequency of the Lyman-α transition, f is the oscillator strength, which is 0.416 for this transition, and the function $g(\nu - \nu_{\rm Ly})$ describes the profile of the Lyman-α absorption line. In this form, the function g has been normalised so that $\int g(x)\mathrm{d}x = 1$. Therefore, inserting $\alpha(\nu_1) = \sigma(\nu_1)N_{\rm H}(z)$ into the expression (10.4), we find the optical depth due to Lyman-α scattering

$$\begin{aligned}\tau(\nu_0) &= \frac{c}{H_0}\int_0^z \frac{\sigma(\nu_1)N_{\rm H}(z)\mathrm{d}z}{(\Omega_0 z+1)^{1/2}(1+z)^2} \\ &= \frac{e^2 f}{4\epsilon_0 m_e H_0}\int_1^{1+z_{\max}} \frac{N_{\rm H}(z)g[\nu_0(1+z)-\nu_{\rm Ly}]}{\nu_0(\Omega_0 z+1)^{1/2}(1+z)^2}\mathrm{d}[\nu_0(1+z)]\end{aligned} \tag{10.6}$$

Since $g(\nu)$ is very sharply peaked at the wavelength of Lyman-α, we can approximate it by a delta function and then

$$\tau(\nu_0) = \frac{e^2 f}{4\epsilon_0 m_e H_0 \nu_{\rm Ly}} \frac{N_{\rm H}(z)}{(\Omega_0 z+1)^{1/2}(1+z)} \tag{10.7}$$

Inserting the values of the constants, we find

$$\tau(\nu_0) = 4\times 10^4\, h^{-1}\frac{N_{\rm H}(z)}{(\Omega_0 z+1)^{1/2}(1+z)} \tag{10.8}$$

This continuum absorption feature has been searched for in those quasars which have such large redshifts that the Lyman-α line is redshifted into the observable optical waveband, that is, quasars with redshifts $1+z \geq (330\mathrm{nm})/\lambda_{\rm Ly}$, $z \geq 2$. There is no such restriction for observations with the IUE or with the Hubble Space Telescope. A typical spectrum of a large redshift quasar is shown in Fig. 10.1 and it can be observed that there is no evidence for a depression to the short wavelength side of either Lyman-α at 121.6 nm or HeI which is expected to be observed at a rest wavelength of 58.4 nm. Typically, an upper limit to the optical depth to the short wavelength side of Lyman-α is $\tau(\nu_0) \leq 0.1$. Substituting this value into the expression (10.8), we find that, for a quasar at a redshift of 3, the upper limit to the number density of neutral hydrogen atoms at that redshift is $N_{\rm H} \leq 10^{-5}$ m^{-3}. This is very small indeed compared with typical cosmological densities at that epoch which correspond to about $10(1+z)^3$ m^{-3}. Thus, if there is any hydrogen at all in the intergalactic medium, it must be very highly ionised.

Searches have been made for the Lyman-α trough in low redshift quasars which can be observed beyond the redshifted Lyman-α line from space by Davidsen (1993) and from the HST but there is little evidence for any absorption at all. It is inferred that, even at the present epoch, the intergalactic gas must be highly ionised.

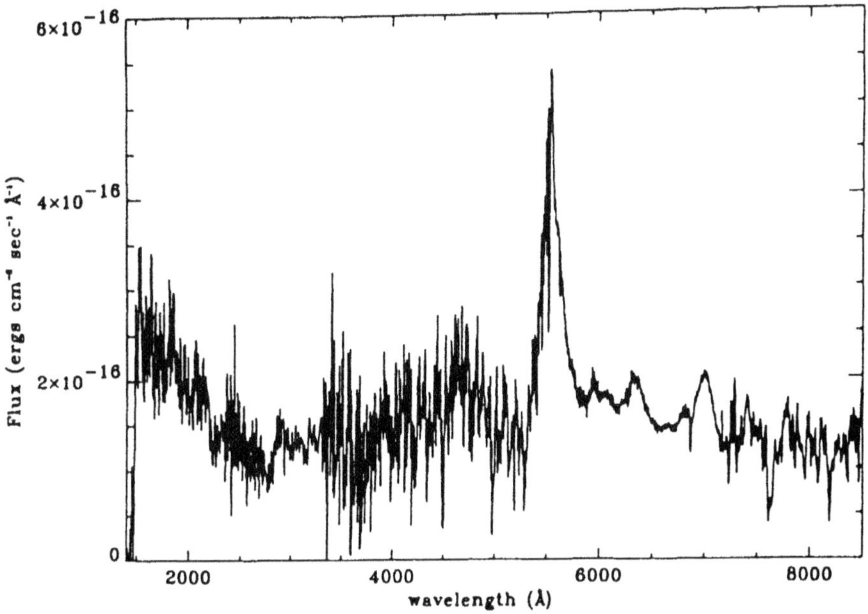

Fig. 10.1 The ultraviolet spectrum of the quasar OQ 172 observed by the Faint Object Spectrograph of the Hubble Space Telescope and at the Lick Observatory. Although there are lines present associated with the Lyman-α forest, there is no depression of the continuum intensity to the short wavelength side of either the Lyman-α line at 121.6 nm or the HeI line at 58.4 nm (from Beaver et al 1993).

10.3 The Collisional Excitation of the Intergalactic Gas

The severe problems of accounting for the X-ray background radiation as the bremsstrahlung of very hot intergalactic gas were described in Section 5.2.1. These arguments can be quantified using the expression (10.3) for the background intensity of the intergalactic medium and the standard expression for the bremsstrahling emissivity of a hot plasma. For ions of charge Z and number density N_Z, this emissivity can be written

$$\epsilon(\nu) = 6.8 \times 10^{-51} N_e N_Z Z^2 T^{-1/2} g(\nu, T) \exp\left(-\frac{h\nu}{kT}\right) \ \mathrm{W\ m^{-2}\ Hz^{-1}}$$

where $g(\nu, T)$ is a Gaunt factor which can be approximated by the form $g(\nu) = (3^{1/2}/\pi)\ln(kT/h\nu)$ for the case of the X-ray emission of a hot plasma. N_e is the electron number density. These arguments rule out the possibility that the intergalactic gas is of high density and at a very high temperature. It is, however, entirely possible that collisional excitation of the intergalactic gas by shocks was responsible for its heating and ionisation. In consequence, if the temperature of the gas lies in the appropriate temperature, it would be expected to be an emitter of collisionally excited HI and HeII Ly-α lines at $\lambda = 121.6$ nm and $\lambda = 30.4$ nm respectively. The observed intensity of the lines is given by the expression (10.3) in which the emissivity of the Lyman-α line per unit frequency interval can be written

$$\epsilon(\nu_1) = \epsilon_{\rm Ly}(z) g(\nu_1 - \nu_{\rm Ly})$$

where $\epsilon_{\rm Ly}(z)$ is the emissivity of the Lyman-α line in units of W m^{-3} and $g(\nu_1 - \nu_{\rm Ly})$ describes the line profile of the Lyman-α line such that $\int g(\nu_1 - \nu_{\rm Ly}) {\rm d}\nu_1 = 1$. Therefore, the intensity of the line per unit bandwidth is $\epsilon_{\rm Ly}(z) g(\nu_1 - \nu_{\rm Ly})$. The intensity of background radiation associated with the Lyman-α emission of the intergalactic gas is therefore

$$I(\nu_0) = \frac{c}{4\pi H_0} \int_0^\infty \frac{\epsilon_{\rm Ly}(z) g(\nu_1 - \nu_{\rm Ly})}{(\Omega_0 z + 1)^{1/2}(1+z)^5} {\rm d}z \tag{10.9}$$

As in section 10.2, we can take $g(\nu_1 - \nu_{\rm Ly})$ to be a δ-function in which case

$$\begin{aligned} I(\nu_0) &= \frac{c}{4\pi H_0} \int_0^\infty \frac{\epsilon_{\rm Ly}(z) \delta[\nu_0(1+z)]}{(\Omega_0 z + 1)^{1/2}(1+z)^5} {\rm d}z \\ &= \frac{c}{4\pi H_0} \frac{\epsilon_{\rm Ly}(z)}{\nu_{\rm Ly}(\Omega_0 z + 1)^{1/2}(1+z)^4} \end{aligned} \tag{10.10}$$

where $\nu_1 = \nu_{\rm Ly} = \nu_0(1+z)$. Notice that, in this development, the intensities are in units of W m^{-2} Hz^{-1} sr^{-1}. In ultraviolet astronomy, it is common to quote the intensites in terms of the numbers of photons s^{-1} m^{-2} Hz^{-1} sr^{-1} or of photons s^{-1} m^{-2} nm^{-1} sr^{-1}. To transform the expression (10.10) into these units, we note that $I(\nu_0) = h\nu_0 I_{\rm ph}(\nu_0)$ and $\epsilon_{\rm Ly}(z) = h\nu_{\rm Ly} N_{\rm Ly}(z)$ where $I(\nu_0)$ is the observed intensity in photons per unit frequency interval and $N_{\rm Ly}(z)$ is the emissivity of the intergalactic gas at redshift z in photons m^{-3}. It follows that

$$I_{\rm ph}(\nu_0) = \frac{c}{4\pi H_0} \frac{N_{\rm Ly}(z)}{\nu_{\rm Ly}(\Omega_0 z + 1)^{1/2}(1+z)^3} \tag{10.11}$$

Alternatively, in terms of photons per unit wavelength interval, we can write $I_{\rm ph}(\lambda_0) = I_{\rm ph}(\nu_0)({\rm d}\nu_0/{\rm d}\lambda_0)$ and so, recalling that $\lambda_0 = \lambda_{\rm Ly}(1+z)$, we find

$$I_{\rm ph}(\lambda_0) = \frac{c}{4\pi H_0} \frac{N_{\rm Ly}(z)}{\nu_{\rm Ly}(\Omega_0 z + 1)^{1/2}(1+z)^5} \tag{10.12}$$

As described by Jakobsen (1993), the emissivity of the intergalactic gas can be written in the form

$$N_{\rm Ly}(z) = N_0^2 (1+z)^6 \gamma_{\rm Ly}(T) \tag{10.13}$$

where $N_0 = 7.8\Omega_{\rm B} h^2$ is the number density of neutral hydrogen at the present epoch and $\Omega_{\rm B}$ is the density parameter corresponding to the density of the intergalactic gas. $\gamma_{\rm Ly}$ is a suitably normalised emission coefficient. Fig. 10.2

shows the variation of γ_{Ly} as a function of temperature for a standard mixture of hydrogen and helium in collisional and thermal equilibrium (Jakobsen 1993). The figure shows that collisionally excited HI and HeII Lyman-α lines are particularly intense at what are referred to as the 'thermostat' temperatures of $T \approx 2 \times 10^4$ K and $T \approx 8 \times 10^4$ K, at which the ionisation states of the intergalactic gas change from HI to HII and from HeII to HeIII respectively. These changes are illustrated in the lower half of Fig. 10.2. It is immediately apparent from Fig. 10.2(b) that, if the intergalactic gas were collisionally ionised, the temperature of the gas would have to be at least 10^5 K in order to result in $[N_{\rm HI}/N_{\rm HII}] \ll 10^{-5}$. These tools were used by a number of authors including Kurt and Sunyaev (1967), Weymann (1967) and others to show how the ultraviolet background radiation could be used to detect a 'luke-warm' intergalactic gas.

10.4 Quasar Absorption Lines and the Diffuse Intergalactic Far Ultraviolet Background Radiation

Important clues to the intensity of the far-ultraviolet background radiation have been derived from the absorption lines observed in the ultraviolet spectra of quasars. The nature and properties of the different types of absorption line system are vast subjects and for many more details reference should be made to the reviews in the volume *QSO Absorption Lines: Probing the Universe* (Blades, Turnshek and Norman 1988). In summary, there are two types of absorption line system. By far the most common are the 'Lyman-α forest' systems which dominate the spectra of large redshift quasars such as those seen in Fig. 10.1. These systems have neutral hydrogen column depths in the range $N_{\rm HI} \sim 10^{17} - 10^{21}$ m^{-2}. They show little evidence for heavy elements and are interpreted as intergalactic clouds, quite possibly consisting of unprocessed primordial material. In contrast, the rarer 'Lyman-limit' and 'damped Lyman-α' systems have column densities $N_{\rm HI} \sim 10^{21} - 10^{26}$ m^{-2} and have large optical depths for Lyman-α absorption. In the Lyman-limit systems, a corresponding Lyman continuum break is observed at 91.2 nm and these systems also display absorption lines of the common elements. The damped Lyman-α systems which have the largest column depths, $N_{\rm HI} \geq 10^{25}$ m^{-2} can be convincingly associated with the extended discs of spiral galaxies. The lower density Lyman-limit systems can be associated with the extended gaseous haloes of galaxies, similar to these observed about some nearby galaxies. The abundances of the heavy elements in the latter systems are less than 10% of the cosmic abundances of the elements but this is consistent with the requirement that the haloes be very extensive, $\sim 50 - 100$ kpc, and so the gas in these regions is not expected to be nearly as enriched as the gas in the disc of a galaxy.

Both types of system show variations of the number density of absorbers with redshift. It is convenient to parameterise the variation of the number density of absorbers with redshift by a power-law relation of the form

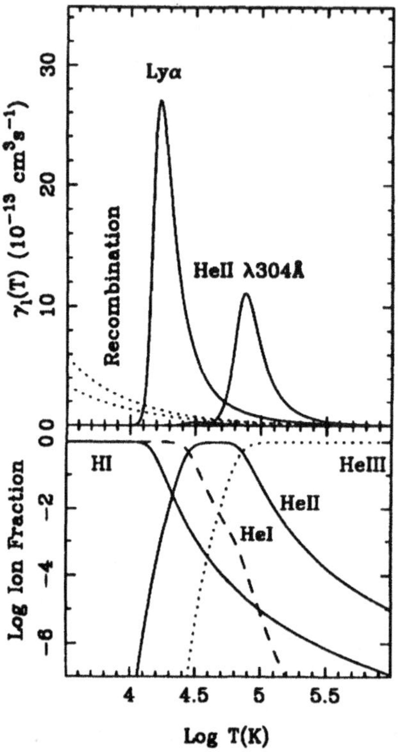

Fig. 10.2 (a) The emissivity per hydrogen atom of collisionally excited HI and HeII Ly-α emission. The dotted lines show the equivalent line emission due to recombination in a fully photoionised gas. (b) The ionisation structure of a cosmological mixture of hydrogen and helium in collisional equilibrium as a function of temperature. Emission from HeI is negligible compared with that of HI and HeII (from Jakobsen 1993).

$$N(z)\mathrm{d}z = A(1+z)^{\gamma}\mathrm{d}z \tag{10.14}$$

It is found that, for the Lyman-α forest systems, $A = 10$ and $\gamma = 2 - 3$ whereas, for the Lyman-limit systems $A \approx 1$ and $\gamma \sim 1$. It is interesting to compare these variations with redshift with what would be expected if the properties of the absorption systems were unchanging with cosmic epoch, that is, the absorbers have the same cross-sections and they have a constant *comoving* number density. The probability of absorption in the interval of *proper* length $\mathrm{d}r_{\mathrm{prop}}$ is

$$p(z)\mathrm{d}z = \sigma_{\mathrm{A}} N_{\mathrm{A}}(z)\mathrm{d}r_{\mathrm{prop}} \tag{10.15}$$

where σ_{A} is the cross-section of the absorbers and $N_{\mathrm{A}}(z)$ is their proper number density at redshift z. Now, $N_{\mathrm{A}}(z) = N_0(1+z)^3$, where N_0 is the number density of absorbers at the present epoch, and

$$\mathrm{d}r_{\mathrm{prop}} = c\mathrm{d}t = \frac{c}{H_0}\frac{\mathrm{d}z}{(\Omega_0 z+1)^{1/2}(1+z)^2}$$

Therefore,

$$p(z)\mathrm{d}z = \sigma_{\mathrm{A}} N_0 \frac{c}{H_0}\frac{(1+z)\mathrm{d}z}{(\Omega_0 z+1)^{1/2}} \tag{10.16}$$

Thus, if $\Omega_0 = 1$, $p(z) \propto (1+z)^{1/2}$ and, if $\Omega_0 = 0$, $p(z) \propto (1+z)$. Thus, the number density of Lyman-α systems changes more rapidly with cosmic epoch than is expected. The sense of the evolution is that there were more Lyman-α absorption systems at large redshifts as compared with lower redshifts. On the other hand, the Lyman-limit systems seem to show little variation with redshift other than what would be expected if their cross-sections and comoving number densities remained unchanged with cosmic epoch.

The number density of absorbers as a function of column depth is more or less the same for both classes, $\mathrm{d}p(N_{\mathrm{H}}) \propto N_{\mathrm{H}}^{-s}\mathrm{d}N_{\mathrm{H}}$ where $s \approx 1.2-1.6$. These are important results because they enable the optical depth of the Universe to absorption by neutral hydrogen in both types of clouds to be determined.

Observations of these absorption line systems impinge directly upon studies of the background radiation in a number of ways. Of particular interest is the 'proximity effect' (see Murdoch et al 1986). It is found that there is a deficit of Lyman-α absorption line systems close to the emission line redshift of the quasar relative to the expectations of expression (10.14). The standard interpretation of the Lyman-α absorption systems is that they are clouds which are almost fully ionised by the background of far ultraviolet Lyman continuum radiation. Close to the quasar, in addition to ionisation by intergalactic Lyman continuum radiation, the clouds are also ionised by the Lyman continuum radiation from the quasar itself. Since the intensity at any distance from the quasar can be estimated once its luminosity is known, the distance at which the cross-over takes place between ionisation predominantly due to the quasar and to the intergalactic continuum radiation can be found from the proximity effect and hence the intensity of the background ultraviolet radiation can be found. If it is assumed that the background emission spectrum has the form $I(\nu) \propto \nu^{-0.5}$, the background intensity at the Lyman limit is roughly 10^{-24} W m^{-2} Hz^{-1} sr^{-1} at redshift $z \sim 2-3$, corresponding to $\nu I_\nu \approx 3 \times 10^{-9}$ W m^{-2} sr^{-1}.

The same technique can be used to work out the flux of intergalactic ionising radiation at small redshifts thanks to observations made by the Hubble Space Telescope. Kulkarni and Fall (1993) have tentatively found evidence for the proximity effect in a sample of 13 low redshift quasars which have been observed by Bahcall and his colleagues as part of the quasar absorption line key project. They find a deficit of Lyman-α absorption clouds close to these quasars and their estimate of the intergalactic flux of ionising radiation at a typical redshift $z \sim 0.5$ lies in the range 4×10^{-26} to 2×10^{-27} W m^{-2} Hz^{-1} sr^{-1} with a best estimate of 6×10^{-27} W m^{-2} Hz^{-1} sr^{-1}. It is interesting to compare this estimate with other methods of estimating $I(\nu)$. $H\alpha$ emission has probably been detected from two high velocity clouds in the halo of our Galaxy and this provides an upper limit to the local ionising flux of radiation of $I_\nu \leq 2 \times 10^{-25}$ W m^{-2} Hz^{-1} sr^{-1} (Kutyrev and Reynolds 1989, Songaila, Bryant and Cowie 1989). Sunyaev (1968) first proposed using the existence of neutral hydrogen in the peripheries of galaxies to set limits to the flux of intergalactic Lyman continuum radiation. The results depend somewhat upon the assumptions made about the spectrum of the ionising radiation and the thickness of the layer of neutral hydrogen in the galaxies. Bochkarev and Sunyaev (1977), Corbelli and Salpeter (1993) and Maloney (1993) have found values in the range $(1 - 10) \times 10^{-26}$ W m^{-2} Hz^{-1} sr^{-1}.

Thus, the flux of ionising radiation at redshifts $z \sim 0.5$ may well be about two orders of magnitude less than the intensity at redshifts $z \sim 2 - 3$.

One of the thornier aspects of this subject has been the origin of these background intensities. One obvious source of ionising photons is the integrated emission of quasars which have non-thermal spectra which extend into the far ultraviolet and X-ray wavebands. The general conclusion of a number of studies is that, to account for the inferred background intensity of about 10^{-24} W m^{-2} Hz^{-1} sr^{-1} at $z \sim 2 - 3$, there must be very strong evolution of the quasar population with cosmic epoch. By pushing all the evolutionary parameters of the models to their limits, Bajtlik, Duncan and Ostriker (1988) were just able to account for this background intensity. Their computations assumed, however, that Lyman-continuum absorption in both types of absorption system could be neglected and, when this was incorporated into their computations, the predicted background intensity fell short by about a factor of 6 (Fig. 10.3). It is not at all clear what the sources of the ionising background is at large redshifts. Maybe a population of faint quasars at large redshifts can supply the deficit. Another possibility is that the ionising radiation from young galaxies or protogalaxies may make a contribution to the background. Steidel and Sargent (1989) have pointed out, however, that the predicted ionising spectrum from models of young galaxies is such that, if it were the principal contributor to the radiation which ionises the gas in Lyman-continuum systems, the line ratios would be quite different from those observed.

At low redshifts, the background flux density may well be two orders of magnitude less than that at large redshifts and so the problem of accounting for the background is somewhat less severe. The models due to Miralda-Esudé and Ostriker (1990), Madau (1992) and Zuo and Phinney (1993) could readily account for this background intensity.

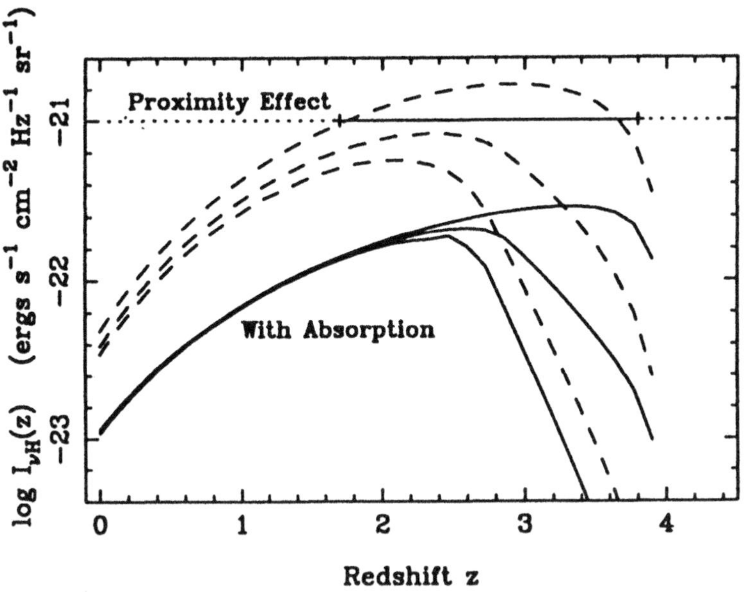

Fig. 10.3 The predicted background radiation at the Lyman limit due to quasars for several different models of their evolution with cosmic epoch (Bajtlik, Duncan and Ostriker 1988). The lower solid curves show the same results but now attenuated to take account of the absorption of the radiation in all types of Lyman forest and Lyman limit systems. The background intensity derived from the proximity effect is also shown (from Jakobsen 1993).

10.5 The Luke-Warm Intergalactic Gas

The absence of metals in the Lyman-α clouds strongly suggests that they consist of primordial material. Their column densities $\int N_{\rm H} dl$ can be determined from the strength of the absorption lines but it is much more difficult to determine the number density of the neutral hydrogen atoms. Only in exceptional cases can the number density of neutral hydrogen atoms can be found. A good example is the case of the Lyman-α clouds observed in front of the double quasar 2345+007 in which the same clouds are observed along the lines of sight to both quasars. Knowing the extent of the cloud, a number density of neutral hydrogen atoms of about 3×10^{-3} m^{-3} has been found (Carswell 1988). Knowing the intergalactic flux of ionising radiation, the number density of the ionised hydrogen atoms can be found by balancing the numbers of photoionisations by the number of the recombinations per second. Typically, the number density of ionised hydrogen amounts to about 10^2 m^{-3} so that the neutral hydrogen fraction is of the order 3×10^{-5}. The total mass of the Lyman-α cloud would be about 10^7 M$_\odot$. The typical masses of the Lyman-α clouds are thought to lie in the range 10^6 to 10^9 M$_\odot$.

The connection with the properties of the diffuse intergalactic gas arises from considerations of the means by which the clouds are confined. It should be emphasised that it is not at all certain how the clouds are held together. The binding might be gravitational or due to the pressure of the intergalactic gas. A popular picture is one in which the Lyman-α clouds are in rough pressure balance with a 'luke-warm' intergalactic gas at a temperature of about 3×10^4 K. The intergalactic flux of ionising radiation at a redshift of $2 - 3$ is so intense that the clouds are almost completely ionised. The temperature of the cloud is expected to be about 3×10^4 K and so, to attain rough pressure balance, the intergalactic gas at redshifts $z \sim 2 - 3$ would have to be about 10^2 m^{-3}. At low redshifts, this would correspond to a density parameter in the intergalactic gas $\Omega_{\rm IGG} \sim 0.1$ which is probably uncomfortably large.

The full treatment of this problem is quite complicated. A good example of the issues involved in these studies is provided by the papers by Ikeuchi and Ostriker (1983, 1987). A complete study involves consideration of the stability of the clouds, the role of evaporation, photoionisation balance within the clouds and so on. Ostriker (1988) favoured a picture in which the intergalactic gas was shock heated to a high temperature at redshifts $z \geq 5$ and the density parameter of the intergalactic gas corresponds to $\Omega \sim 0.02$.

The interesting question addressed by Jakobsen (1993) is whether or not the redshifted Lyman-α emission of such a luke-warm intergalactic gas could be detected. The easiest case to deal with is that in which it is supposed that the intergalactic gas is kept more or less fully ionised by the diffuse intergalactic ionising background radiation. The question is whether or not the redshifted diffuse Lyman-α emission of the recombining gas would be detectable in the ultraviolet waveband. Jakobsen considers all the possible cases involving the photoionisation and shock heating of the intergalactic plasma. In the case of a pure photoionisation model, it is straightforward to show that the diffuse redshifted Lyman-α background radiation observed at the present epoch must always be significantly less intense that the ionising background itself and so, even if the observed ionising background were responsible for the ionisation of the gas, the redshifted emission can make only a very small contribution to the background radiation. The same conclusion can be drawn for the case in which the intergalactic gas is heated by shocks. The basic reason that these backgrounds are so faint is twofold. First, the number density of neutral hydrogen must be very low because of the null result of the Gunn-Peterson test. Second, the constraints of primordial nucleosynthesis indicate the intergalactic gas is at least one and probably two orders of magnitude less than the critical cosmological density.

One important test of the ionisation state of the intergalactic gas is to search for He^+ using the equivalent of the Gunn-Peterson test, but now searching for the decrement associated with the redshifted He^+ Lyman-α, which occurs at a rest wavelength of 30.4 nm. If the intergalactic gas was photoionised by a diffuse intergalactic flux of ionising radiation with spectrum $I_\nu \propto \nu^{-0.5}$, it is expected that the HeII ions would be an order of magnitude more abundant than HI and so a Gunn-Peterson decrement might be observable in large redshift quasars in which the 30.4 nm absorption edge is redshifted into the observable ultraviolet waveband, that is, quasars with redshifts greater than 3. The problem with this test is that the 30.4 nm line falls below the absorption edge of neutral hydrogen and so the entire far ultraviolet spectrum is likely to be strongly attenuated by absorption in Lyman-α clouds and in Lyman limit systems (see Section 10.6).

Evidence for HeII absorption due to diffuse intergalactic gas has been found by Jakobsen and his colleagues (1994) in a remarkable set of observations made with the Faint Object Camera of the Hubble Space Telescope. The camera can be operated in a low resolution spectroscopic mode and, prior to the refurbishment of the HST, a search was made among 25 large redshift candidate quasars for the one or two examples in which the far ultraviolet continuum was not be absorbed by intervening Lyman-α clouds. The best candidate found was the quasar Q0302-003, which has redshift $z = 3.286$. The low-resolution spectrum of the quasar is shown in Fig. 10.4, in which is can be seen that there is break in the continuum at precisely the wavelength of the redshifted HeII line at 30.4 nm. At shorter wavelengths, the continuum intensity falls to zero. At the 90% confidence limit, the optical depth for HeII absorption is $\tau_{\rm HeII} \geq 1.7$. This limit corresponds to a lower bound to the diffuse intergalactic number density of HeII ions of $N_{\rm HeII} \geq 1.5h \times 10^{-3}$ m^{-3} at redshift $z = 3.29$. It is reassuring that this observation is direct evidence for primordial helium in the intergalactic gas at large redshifts.

Jakobsen and his colleagues argue, convincingly, that the absorption is likely to be associated with the diffuse intergalactic gas and so provides evidence that the gas is indeed lukewarm. What is remarkable is the strength of the HeII absorption compared with the hydrogen Lyman-α absorption. In most of the photoionisation models involving power-law continuum ionisation, it is expected that most of the helium is in the form of He^{++} and that only about 1% of the helium is in the form of He^{+}. Adopting the standard cosmic abundances of helium to hydrogen, the relative opacities of the gas for hydrogen and helium absorption can be reconciled if it is assumed that most of the helium is in the form of He^{+}. This has important implications for the spectrum of the radiation responsible for ionising the intergalactic gas. The spectrum would have to cut off rather abruptly at about 54 eV if too much He^{++} is not to be produced. It might be that such a cut-off is present in the continuum spectra of quasars. Perhaps a more natural explanation is that the ionising spectrum is associated with young galaxies which are expected to have steep far ultraviolet continua.

These pioneering observations are very important for undertanding the processes involved in heating and ionising the intergalactic gas but, as the authors point out, it is a very challenging programme to extend these obser-

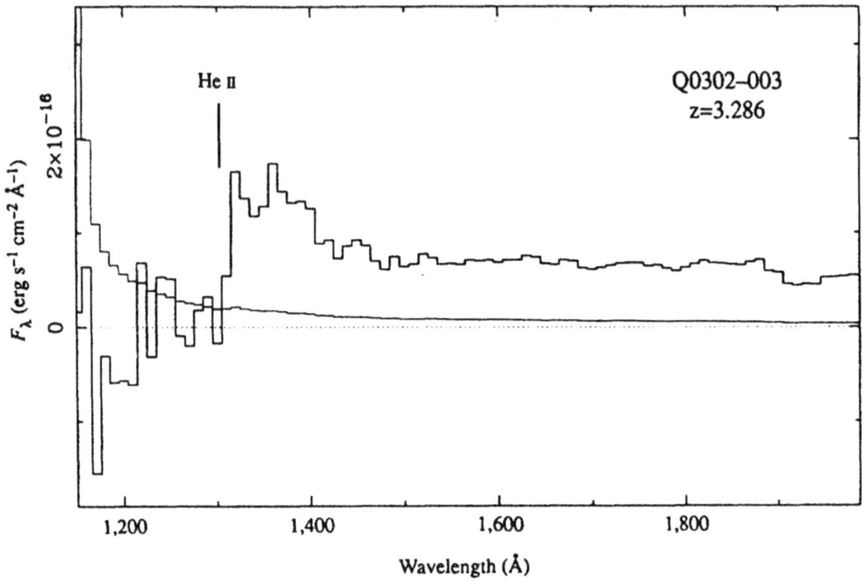

Fig. 10.4 The ultraviolet spectrum of the quasar Q0302-003 observed by the Faint Object Camera of the Hubble Space Telescope. The solid thin line gives the 1σ uncertainty per $\Delta\lambda = 10$ Å wavelength bin due to photon statistics. The position of the HeII line at 30.4 nm in the quasar rest frame is indicated. (Jakobsen et al 1994)

vations to larger samples of large redshift quasars. It does appear, however, that at last we have obtained a glimpse of the intergalactic gas at redshifts greater than 3.

10.6 The Lyman Continuum Opacity of the Intergalactic Gas

The null result of the Gunn-Peterson test shows that the diffuse intergalactic gas is transparent to far ultraviolet radiation. It has been appreciated over the last few years, however, that the cumulative effect of the Lyman-α forest and the Lyman-limit systems is to result in significant Lyman continuum absorption of the far ultraviolet radiation originating from large redshifts (Bechtold, Weymann, Lin and Malkan 1987, Møller and Jakobsen 1990). The fact that the number densities of both classes of absorption line systems, the range of column densities and their evolution with cosmic epoch are now quite well defined enable estimates of the opacity of the Universe to far ultraviolet photons to be made. Møller and Jakobsen have evaluated the opacity of the intergalactic medium in terms of the number density of absorption line systems as a function of redshift $N(z)$ and knowing the average absorption per absorber from the distribution function $N(N_{\rm H})$. They first work out the average optical depth of the cloud responsible for the absorption $\langle\tau\rangle = \langle\sigma_{\rm H}(\nu)N(N_{\rm H})\rangle$ and then the fraction of the background intensity on passing through a typical cloud is $1 - \exp(-\langle\tau\rangle)$. The absorption coefficent $\sigma_{\rm H}$ has the form $\sigma_0(\nu_{\rm H}/\nu)^3$ for all frequencies greater than $\nu_{\rm H}$ where $\nu_{\rm H}$ is the Lyman limit corresponding to 91.2 nm; for smaller frequencies, the absorption is taken to be zero. Then, the total optical depth on passing through a distribution of clouds $N(z)\mathrm{d}z$ is

$$\tau_{\rm tot} = \int_0^{z_e} N(z)[1 - \exp(-\langle\tau\rangle)]\mathrm{d}z$$

and the fractional transmitted intensity is $E = \exp(-\tau_{\rm tot})$.

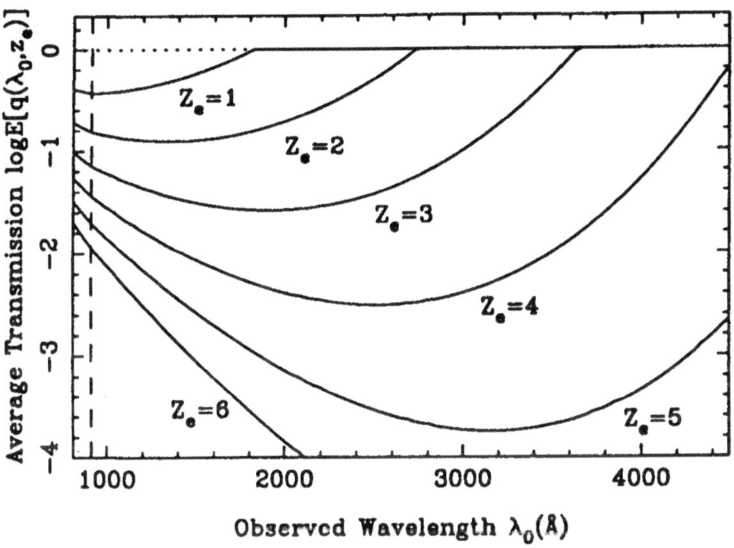

Fig. 10.5 The average transmission of the ultraviolet universe out to large redshifts as a function of wavelength. The 'Lyman valley' shown in the figure includes the Lyman continuum opacity due to both the Lyman forest and to Lyman-limit systems (from Møller and Jakobsen 1990).

This expression has been evaluated by Møller and Jakobsen (1990) and their results are displayed in Fig. 10.5. In the figure, $z_{\rm e}$ is the emission redshift and transmitted fraction of the radiation as observed at different ultraviolet wavelengths. The diagram can be interpreted as follows. Suppose we observe at a wavelength of 300 nm. Then, it is apparent that the sources at redshifts of 2 and less do not suffer any Lyman continuum absorption because at a redshift of 2, the emission wavelength is 100 nm which is a longer wavelength than the Lyman limit. However, by a redshift of 3, the emission wavelength is 75 nm and there is strong absorption by the clouds in the redshift interval 3 to $z = (300/91.2) - 1 = 2.29$. It can be seen that the average transmission decreases very rapidly with increasing redshift and this has important consequences for the observability of sources at the very largest redshifts. For example, suppose we wished to observe the Gunn-Peterson decrement for He^+ at rest wavelength 30.4 nm or the HeII Lyman-α emission at the same wavelength at an observing wavelength of 160 nm. The radiation would then originate from redshift $z = (160/30.4) - 1 = 4.3$. Inspection of Fig. 10.5 shows that the intensity of such radiation would be attenuated by a factor of about 100, rendering its observation very difficult indeed. It is apparent that the far-ultraviolet Universe at large redshifts is likely to be heavily obscured. As Jakobsen (1993) expresses, 'Even if the intergalactic medium did go through a phase of intense HeII emission during reheating, the resulting far-ultraviolet radiation will in all likelihood remain forever hidden form our view.'

There is some slight relief from this conclusion in that the absorption is due to discrete clouds rather than to a diffuse medium and the strongest absorbers are the optically thick Lyman-limit systems. Therefore, along some lines of sight, there may be little absorption. It is therefore expected that any large redshift, far-ultraviolet diffuse emission should be extremely patchy.

References – Chapter 10

Bajtlik, S., Duncan, R.C. and Ostriker, J.P. (1988). ApJ, **327**, 570.

Beaver, E., Burbidge, E.M., Cohen, R., Junkkarinen, V., Lyons, R. and Rosenblatt, E. (1992). In *Science with the Hubble Space Telescope*, (eds. P. Benvenuti and E. Schreier), 53. Garching-bei-München: ESO Publications.

Bechtold, J., Weymann, R.J., Lin, Z. and Malkan, M.A. (1987). ApJ, **315**, 180.

Blades, J.C., Turnshek, D. and Norman, C.A. (eds) (1988). *QSO Absorption Lines: Probing the Universe*. Cambridge: Cambridge University Press.

Bochkarev, N.G. and Sunyaev, R.A. (1977). Sov. Astr., **21**, 542.

Carswell, R.F. (1988). In *QSO Absorption Lines: Probing the Universe*, (eds. J.C. Blades, D. Turnshek and C.A. Norman), 91. Cambridge: Cambridge University Press.

Corbelli, E. and Salpeter, E.E. (1993). ApJ. (in press)

Davidsen, A. (1993). Science, **259**, 327.

Gunn, J.E. and Peterson, B.A. (1965). ApJ, **142**, 1633.

Ikeuchi, S. and Ostriker, J.P. (1987). ApJ,**301**, 552.

Jakobsen, P. (1993). In *Extragalactic Background Radiation - 1993 Moriond Conference*, (eds. B. Rocca-Volmerange and J.M. Deharveng), (in press).

Jakobsen, P. (1994). In *The Extragalactic Background Radiation*, (eds. M. Livio, M. Fall, D. Calzetti and P. Madau), (in press). Cambridge: Cambridge University Press.

Jakobsen, P., Boksenberg, A., Deharveng, J.M., Greenfield, P, Jedrzejewski, R. and Paresce, F. (1994). Nature, **370**, 35.

Møller, P and Jakobsen, P. (1990). A&A, **228**, 299.

Kulkarni, V.P. and Fall, S.M. (1993). ApJ, **413**, 62.

Kurt, V.G. and Sunyaev, R.A. (1967). *Cosm. Res.*, **5**, 496.

Kutyrev, A.S. and Reynolds, R.J. (1989). ApJ, **344**, L9.

Longair, M.S. (1965). MNRAS, **129**, 419.
Madau, P. (1992). ApJ, **389**, L1.
Maloney, P. (1993). ApJ, **414**, 41.
Miralda-Escudé, J. and Ostriker, J.P. (1990). ApJ, **350**, 1.
Murdoch, H.S., Hunstead, R.W., Pettini, M. and Blades, J.C. (1986). ApJ, **309**, 19.
Ostriker, J.P. (1988). In *QSO Absorption Lines: Probing the Universe*, (eds. J.C. Blades, D. Turnshek and C.A. Norman), 319. Cambridge: Cambridge University Press.
Ostriker, J.P. and Ikeuchi, S. (1983). ApJ, **268**, L63.
Ryle, M. and Sandage, A.R. (1964). ApJ, **139**, 419.
Scheuer, P.A.G. (1965). Nat, **207**, 963.
Songaila, A., Bryant, W. and Cowie, L.L. (1988). ApJ, **345**, L71.
Steidel, C.C. and Sargent, W.L.W. (1989). ApJ, **382**, 433.
Sunyaev, R.A. (1969). Astrophys. Lett., **3**, 33.
Weymann, R. (1967). ApJ, **147**, 887.
Zuo, L. and Phinney, S. (1993). ApJ, (in press).

11 Galaxy Formation and the Background Radiation

11.1 When Were the Heavy Elements Formed?

To conclude these lectures, I want to look at the problem of galaxy formation from a completely different point of view. This material will be presented much more as work in progress than as a formal pedagogical lecture. The primary question I want to address is 'When were the heavy elements created?' We know that heavy elements are being synthesised today, as is beautifully demonstrated by the observations of the synthesis of cobalt, nickel and iron in the supernova SN 1987A. However, what we want to ask is when were the bulk of the heavy elements synthesised? There are a number of suggestive pieces of evidence which indicate that there must have been some element synthesis in galaxies at large redshifts. I would emphasise that none of these arguments are 100% watertight but taken together they are suggestive.

1. Even in the most distant galaxies and quasars we can observe, the abundances of the heavy elements are not so different from the cosmic abundances of the elements today.
2. The quasar absorption line systems which are believed to be associated with the discs of galaxies at large redshifts have somewhat lower heavy element abundances as compared with cosmic abundances of the elements but they are certainly not zero and there must have been a substantial amount of element formation by a redshift of, say, 3.
3. The abundances of the elements in the G-K dwarfs are remarkably uniform despite the fact that some of these stars are as old as the Galaxy. This observation has been interpreted as evidence for prompt initial element formation in our Galaxy, perhaps associated with an initial burst of star formation.
4. Even in the very oldest globular clusters, the abundances of the elements are only about 1% of the present cosmic abundances.
5. It may be that the ultraviolet emission from young galaxies is needed to ionise the intergalactic gas (see Chapter10).

The burden of this evidence is that there must have been at least some element formation at large redshifts and the question is how to find more direct evidence for this and how it can be related to observations of the background radiation. First of all, I will present a very pleasant argument given by Lilly and Cowie (1987).

11.2 The Lilly and Cowie Argument

This argument was first presented by Lilly and Cowie (1987) and in a more general context by Cowie (1988). The analysis begins with the observation that a prolonged burst of star formation in a galaxy has a remarkably flat intensity spectrum out to the Lyman limit at 91.2 nm. This is illustrated by the model star-bursts synthesised by Bruzual and presented by White (1989) (Fig. 11.1). The figure shows the spectrum of a starburst galaxy at different ages, assuming that the star formation rate is constant and that the stars are formed with the same Salpeter mass function. The flatness of the spectrum is due to the fact that, although the most luminous blue stars have short lifetimes, these are constantly being replaced by new stars. Furthermore, the intensity of the flat part of the spectrum is directly proportional to the rate of formation of heavy elements since their energy is primarily derived from the conversion of hydrogen into helium which is the first stage in the synthesis of the heavy elements — these are only formed in stars with mass greater than about $8M_\odot$. From simple physical arguments, we can work out the intensity of the flat region of the spectrum of the star-forming galaxy and hence the rate at which metals are being created.

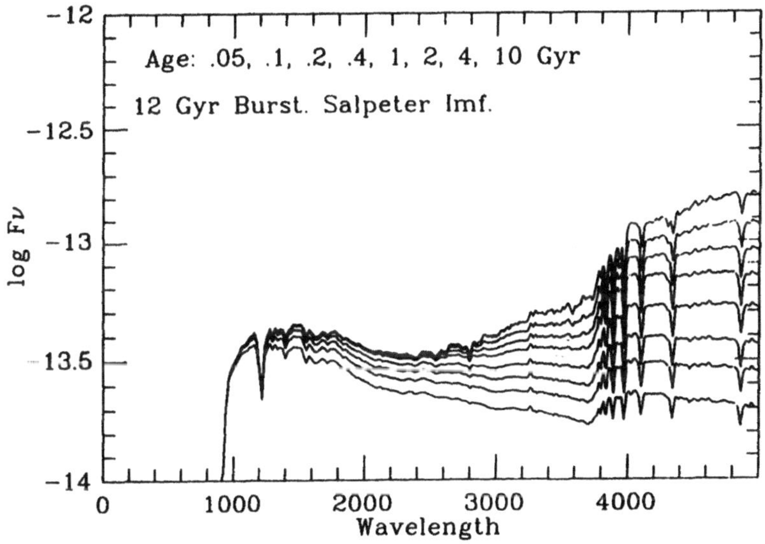

Fig. 11.1 Synthetic spectra for a region with constant star formation rate at the ages indicated. A Salpeter initial mass function has been assumed with cut-offs at 75 and 0.08 $M_\odot$. The spectra were generated by Gustavo Bruzual from a recent version of his evolutionary synthesis programmes (from White 1989).

The energy liberated when hydrogen is converted into helium, which is the essential first stage in the synthesis of the heavy elements, is 0.007 times the rest mass energy of the hydrogen atoms consumed. Inspection of models of stars which burn hydrogen all the way through to synthesise elements such as carbon and oxygen show that this is a good approximation for the energy release during the main part of the lifetime of the star. Therefore, if a mass of metals $(Z\dot{M})$ is created per unit time, the total luminosity of the system is $0.007c^2(Z\dot{M})$. But this is just the energy which is emitted with a flat spectrum I_ν = constant up to the Lyman limit $\nu_{\rm Ly}$ as shown in Fig. 11.1. Therefore,

$$\int L_\nu \mathrm{d}\nu = L_\nu \nu_{\rm Ly} = 0.007c^2(Z\dot{M})$$

Expressing this result in terms of a metal formation rate, we find

$$L_\nu = 2 \times 10^{22} \left(\frac{\dot{M}Z}{1M_\odot \ \mathrm{year}^{-1}}\right) \quad \mathrm{W\ Hz^{-1}} \tag{11.1}$$

at all wavelengths longer than the Lyman continuum limit at 91.2 nm. Similar results are found from detailed modelling of the metals produced by star-forming galaxies (see, for example, Meier 1976).

It is now a simple calculation to work out the background intensity due to such sources. We can use the integral (4.13) in the form

$$I(\nu_0) = \frac{1}{4\pi}\int_0^\infty \frac{L(\nu_0, z)N_0}{(1+z)^{1+\alpha}}\mathrm{d}r \tag{11.2}$$

where we have allowed the luminosity to be an arbitrary function of redshift. Now $L(\nu_0, z)N_0$ is just the comoving luminosity density per unit bandwidth due to the the formation of metals. From the above calculations, this is just $0.007c^2\dot{\rho}_{\rm m}/\nu_{Ly}$ where $\dot{\rho}_{\rm m}$ is the *comoving* rate of formation of heavy elements and the spectral index $\alpha = 0$. Therefore,

$$I(\nu_0) = \frac{0.007c^2}{4\pi\nu_{\rm Ly}}\int_0^\infty \dot{\rho}_{\rm m}\frac{\mathrm{d}r}{(1+z)}$$

But $\mathrm{d}r/(1+z) = c\mathrm{d}t$ and so the intensity of the background is

$$I(\nu_0) = \frac{0.007c^3}{4\pi\nu_{\rm Ly}}\int_0^\infty \dot{\rho}_{\rm m}\mathrm{d}t \tag{11.3}$$

always provided the Lyman limit is not redshifted beyond the observing waveband. This is the remarkable result which Lilly and Cowie found — the background intensity due to star-forming galaxies is directly related to the rate at which elements are formed and is *completely independent* of the cosmological model. Cowie (1988) recommends a correction factor of 1.5 to take account of the fraction of metals which is not returned to the general interstellar medium but is locked up in stellar remnants, resulting in better agreement with Meier's simulations. Inserting the values of the constants,

$$I_\nu = 7 \times 10^{-25} \left(\frac{\int_0^\infty \dot{\rho}_\mathrm{m} \mathrm{d}t}{10^{-31}\ \mathrm{kg\ m^{-3}}} \right) \quad \mathrm{W\ m^{-2}\ Hz^{-1}\ sr^{-1}} \tag{11.4}$$

Notice that the density used in this relation is the density of heavy elements observed at the present epoch and that a density of 10^{-31} kg m^{-3} of heavy elements would correspond roughly to $Z = 0.01$ in a Universe in which the density parameter in baryons is $\Omega_\mathrm{B} = 0.01$. The beautiful thing about this relation is that, by inserting the intensity of the background radiation originating in a particular redshift interval Δz due to flat spectrum star-forming galaxies, we can immediately read off the density of metals synthesised in that interval.

Lilly and Cowie have observed a class of flat spectrum objects which appear to be similar to the objects illustrated in Fig. 11.1 in their very deep optical surveys. Originally, it was though that these objects lay at large redshifts but it is now believed that they have redshifts roughly one. The background intensity due to such objects amounted to about 10^{-24} W m^{-2} Hz^{-1} sr^{-1}. Lilly and Cowie interpret this result as meaning that a significant fraction of the heavy elements, about 1.5×10^{-31} kg m^{-3}, must have been produced at about a redshift of one. In fact, this heavy element abundance is significantly less than the maximum permissible. If we were to assume that $H_0 = 50$ km s^{-1} Mpc^{-1}, the upper limit to the baryon density in the Universe as determined by the need to produce at least the observed abundance of deuterium is about 0.1. If a metal abundance of $Z = 0.01$ is adopted, the corresponding mass density in metals would be 5×10^{-30} kg m^{-3} so that the background intensity would be about 30 times this value, $I \approx 3.5 \times 10^{-23}$ W m^{-2} Hz^{-1} sr^{-1} or $\nu I_\nu \approx 2.4 \times 10^{-8}$ W m^{-2} sr^{-1} at 440 nm. Thus, the total background intensity due to metal formation could be up to about 30 or 40 times the intensity already detected by Lilly and Cowie. In fact, such an intensity would exceed the upper limit to the background intensity reported by Toller (1990).

Now, it is well known that galaxies undergoing bursts of star formation are not only sources of ultraviolet continuum radiation but also are strong emitters in the far infrared waveband because of the presence of dust in the star forming regions. According to Weedman (1994), in a sample of star forming galaxies studied by the IUE, most of the galaxies emit much more of their luminosities in the far infrared rather than in the ultraviolet region of the spectrum. As a result, it is quite possible that most of the radiation associated with the formation of the heavy elements is not radiated in the ultraviolet-optical region of the spectrum but in the far infrared region and this would permit a higher abundance of the elements as compared with the existing optical and ultraviolet limits.

This was one of the motivations for undertaking a study of the feasibility of detecting the far-infrared emission from star-forming galaxies at large redshifts in the submillimetre waveband.

11.3 Submillimetre Cosmology

Andrew Blain and I have been carrying out some computations of the expected source counts and background emission expected from star-forming galaxies at large redshifts in the submillimetre and millimetre wavebands (Blain and Longair 1993a). Until recently, the prospects for making surveys of sources in the submillimetre waveband have not been very encouraging because of the lack of array detectors which would allow a significant region of sky to be surveyed. The situation will change dramatically in the near future with the introduction of submillimetre bolometer array detectors on telescopes such as the James Clerk Maxwell Telescope. Specifically, the Submillimetre Common User Bolometer Array (SCUBA) currently being completed for that telescope will enable the mapping of regions of the sky in these wavebands to be carried out about 10,000 times faster than is possible with the current generation of single element detectors.

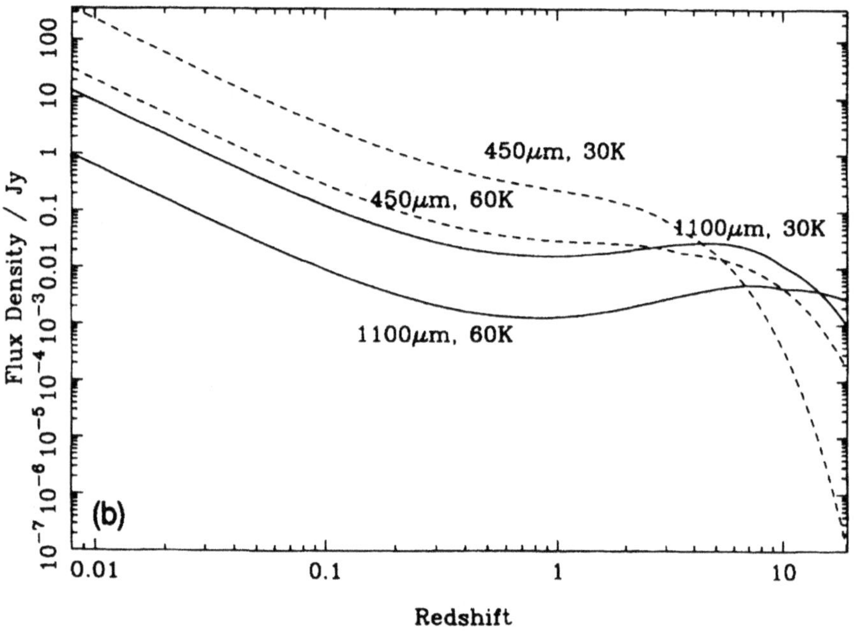

Fig. 11.2 The flux density-redshift relations for a standard dust emission spectrum from a source of far infrared luminosity $10^{13} L_{\odot}$ evaluated for dust temperatures of 30 and 60 K and for wavelengths of 450 and 1100 μm (Blain and Longair 1993a).

It might be thought that the detection of star-forming galaxies at cosmologically interesting distances in the submillietre waveband would be very

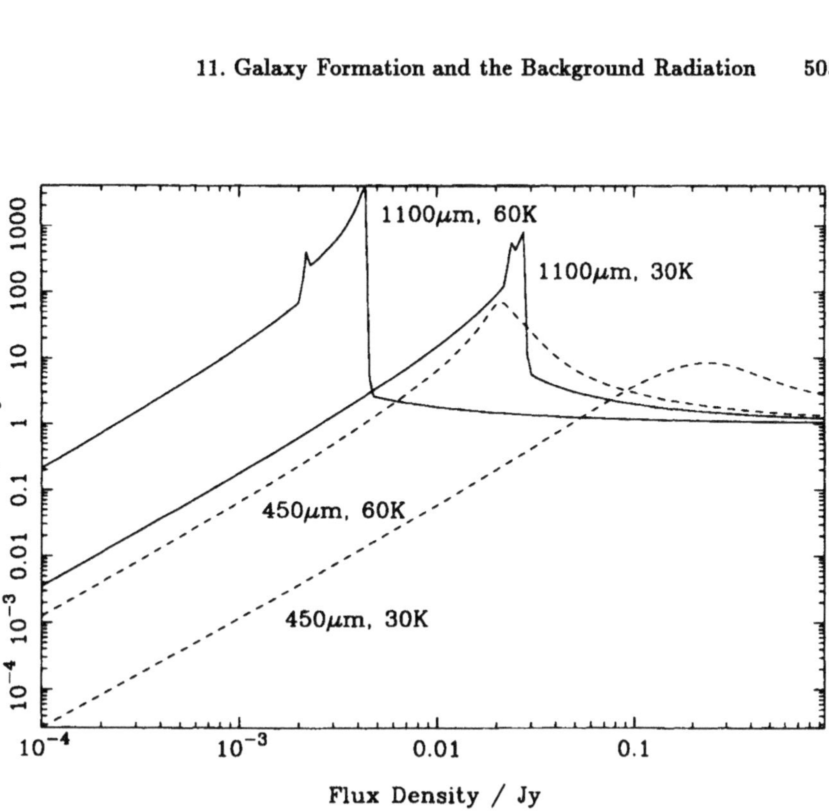

Fig. 11.3 Differential source counts normalised to the expectations of a Euclidean world model for a uniform distribution of standard dust sources at temperatures of 30 and 60 K as observed at wavelengths of 450 and 1100 μm. The bolometric luminosity of the dust source is assumed to be 10^{13} $L_{\odot}$ (Blain and Longair 1993a).

difficult because nearby examples of these types of galaxy are only weak submillimetre emitters. This problem is, however, more than offset by the enormous far infrared luminosities of these galaxies which are redshifted into the submillimetre waveband at redshifts greater than about 1. Specifically, the far infrared spectra of IRAS galaxies peak about 100 μm and have very steep spectra, $I_{\nu} \propto \nu^{\alpha}$ where α is about 3-4 at longer wavelengths. As a consequence, the 'K-corrections' are very large and negative at submillimetre wavelengths. The result is that, at redshifts greater than 1, the flux density of a standard IRAS galaxy is more or less independent of redshift until the far infrared maximum is redshifted through the submillimetre wavebands. This is illustrated in Fig. 11.2 which shows the expected flux density-redshift relations for a galaxy emitting 10^{13} $L_{\odot}$ with a standard dust emission spectrum at temperatures of 30 and 60 K as observed at 450 and 1100 μm.

Correspondingly, the counts of submillimetre sources show a remarkable behaviour at those flux densities at which the 'coasting phase' in the flux density-redshift relation is reached. The predicted differential number counts for a single luminosity class of source at different wavelengths and for differ-

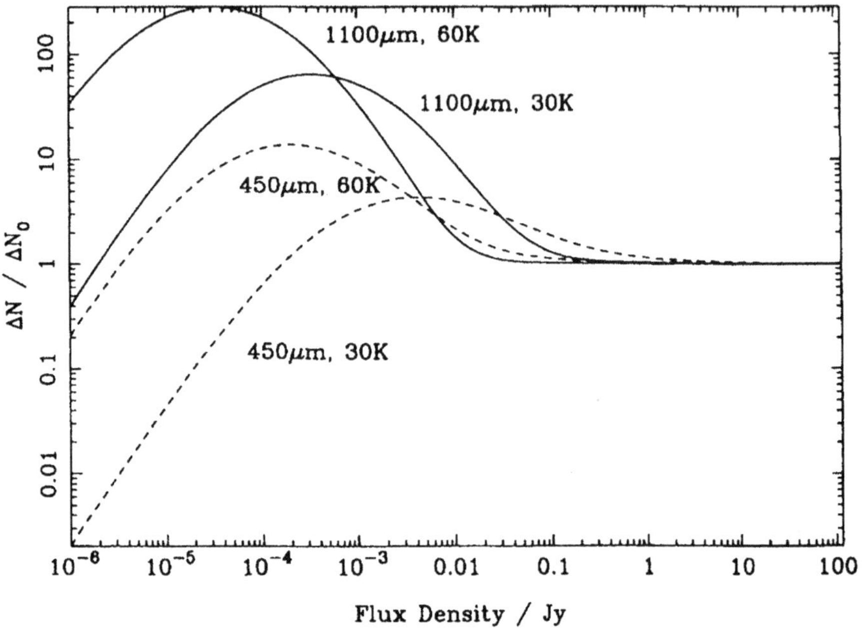

Fig. 11.4 The normalised differential source counts of all IRAS galaxies at wavelengths of 450 and 1100 μm for assumed dust temperatures of 30 and 60 K. It is assumed that the comoving number densities and luminosities of the sources are unchanged with cosmic epoch. (Blain and Longair 1993a).

ent assumed temperatures of the dust grains are shown in Fig. 11.3. These differential counts have to be convolved with the luminosity function of the sources and this can be found from the IRAS luminosity function derived by Saunders et al (1990). The differential source counts for a uniform population of sources is shown in Fig. 11.4 in which it can be seen that there is an enormous excess over the expectations of a 'Euclidean' model. It must be emphasised that these computations are carried out for a *uniform* world model and that the apparent 'excess' is entirely due to the large and negative K-corrections. If the effects of cosmological evolution are included, an even more remarkable excess of faint sources and extraordinarily steep source counts are predicted. Fig. 11.5 shows the results of incorporating the effects of luminosity evolution of the form $L \propto (1+z)^3$ in the redshift interval $0 \leq z \leq 2$ and a constant value at larger redshifts, $L = 27L_0$ where L_0 is the luminosity of sources at zero redshift; according to Peacock (1993), this form of evolution can account not only for the radio and optical counts of quasars and radio sources but also for the counts of IRAS galaxies. In this case, there would be very large surface densities of submillimetre sources at flux densities which will be accessible to instruments such as SCUBA.

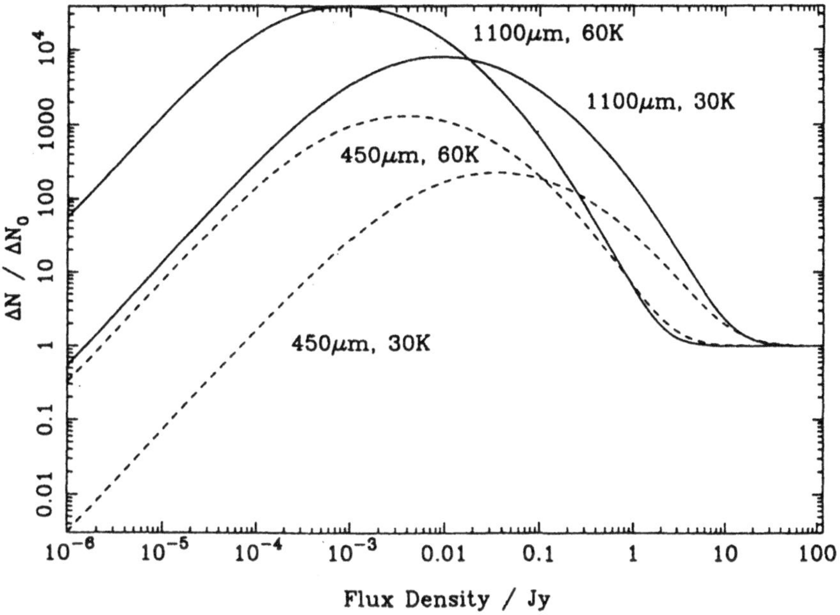

Fig. 11.5 The normalised differential source counts for IRAS galaxies at wavelengths of 450 and 1100 μm for assumed dust temperatures of 30 and 60 K. It is assumed that the comoving number densities are unchanged with cosmic epoch but that the luminosities of the sources evolve as $(1+z)^3$ in the redshift interval $0 < z < 2$ and remain constant at 27 times the local luminosity at all redshifts greater than 2 (Blain and Longair 1993a).

11.4 The Millimetre Background Radiation and Galaxy Formation

As part of our analysis, we have investigated the feasibility of distinguishing different models of galaxy formation by submillimetre observations. The expectations of hot dark matter or 'pancake' models in which large scale structures form first at relatively late epochs can be well approximated by the evolution models described at the end of the last section and result in very steep number counts of submillimetre sources. In fact, it is a quite general result that any model in which the bulk of the star and galaxy formation takes place at redshifts of the order 2-5 results in large number densities of sources at accessible submillimetre flux densities. The background radiation from such populations are strongly constrained by the fact that the spectrum of the Cosmic Microwave Background Radiation is known to be very precisely of black-body form in the wavelength interval 2500 to 500 μm (Fig. 11.6). The problem can be alleviated if it is assumed that the dust grains in the large redshift star-forming galaxies are at a higher temperature, say, 60 - 80 K.

The more intriguing case concerns the currently favoured picture of galaxy formation involving cold dark matter (Blain and Longair 1993b). We assume that galaxies form by hierarchical clustering according to the analytic prescription given by Press and Schechter (1974). This formalism for the evolution of the mass spectrum of objects as a function of time is known to be a remarkably good representation of the results of more detailed computer simulations (Efstathiou 1990, Kauffmann and White 1993) and is the favoured means by which structure forms in the cold dark matter picture of galaxy formation. From this, we can derive a rate of formation of objects of all masses as a function of epoch. We assume that in each coalescence, a fixed fraction of the masses of the colliding galaxies is converted into stars resulting in the formation of a fixed fraction of heavy elements. The binding energy of these elements is liberated in the optical and ultraviolet wavebands but we assume that it is absorbed by dust in the young galaxy and reradiated in the far-infrared region of the spectrum at the temperature to which the dust is heated.

In the full calculations, the far-infrared radiation has a standard dust emission spectrum (Hildebrand 1983) which is strongly peaked at a frequency ν_0 correponding to the temperature of the dust grains — if emitted at redshift z, this spectrum is observed at frequencies $\nu \sim \nu_0/(1+z)$. Because we have to integrate over a wide range of redshifts in the hierarchical models, we can obtain a good approximation to the more complete results by assuming that all the radiation is emitted at a single frequency and therefore the dust emission from a star-forming galaxy at redshift z can be approximated by a delta-function. Let us suppose that the total power liberated in the far-infrared waveband per unit comoving volume is ρ_L at frequency ν_0. Then, the background spectrum for the Einstein-de Sitter critical model ($\Omega_0 = 1$) is

$$I_\nu = \frac{c}{4\pi H_0} \int \rho_L(z)(1+z)^{-5/2} \delta[\nu(1+z) - \nu_0] \mathrm{d}z. \tag{11.5}$$

We make the standard assumption that, if a fraction x of the mass of objects with comoving mass density ρ is converted into metals, then the binding energy of the heavy elements is released producing an energy density which is roughly, $\rho_E = 0.007 x \rho c^2$ (see, for example, Cowie 1988). Therefore, the density of metals formed by the redshift z_0 is

$$\Omega_m(z_0) = \frac{1}{0.007 c^2 H_0} \int_{z_{max}}^{z_0} \rho_L(z)(1+z)^{-5/2} \mathrm{d}z, \tag{11.6}$$

where z_{max} is the redshift at which star formation in the coalescing galaxies is assumed to begin.

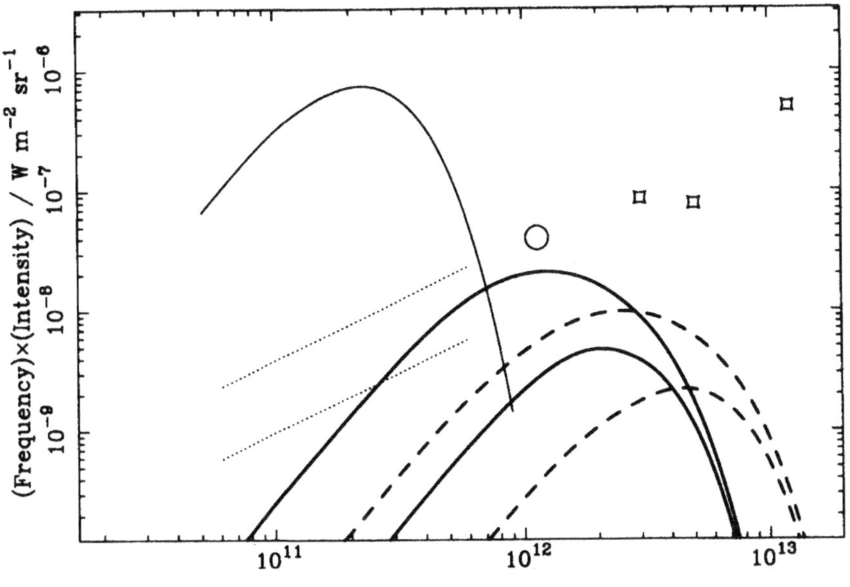

Fig. 11.6 The millimetre and submillimetre background radiation due to IRAS galaxies assuming that the dust temperatures are 30 and 60 K and that the sources are either uniformly ditributed in space or exhibit the form of strong cosmological evolution described in the text. (Blain and Longair 1993a).

The Press-Schechter mass spectrum $N_{\rm PS}$ evolves with time according to the expression

$$N_{\rm PS}(M,t)\,{\rm d}M = \frac{\bar{\rho}}{\sqrt{\pi}}\frac{\gamma}{M^2}\left(\frac{M}{M^*}\right)^{\gamma/2}\exp\left[-\left(\frac{M}{M^*}\right)^{\gamma}\right]{\rm d}M, \tag{11.7}$$

where the mass M^* evolves with cosmic time as $M^* \propto t^{4/3\gamma}$ and $\bar{\rho}$ is the mean mass density per unit comoving volume. The exponent $\gamma = (n+3)/3$ where n is the spectral index of the power spectrum of the initial fluctuations as defined in the expression (8.9). Taking the derivative of the Press-Schechter function with respect to time gives the net rate of change of the number of objects with masses in the range M to $M + {\rm d}M$ which is the difference between the rate of formation of these objects, $\dot{N}_{\rm form}$, and the rate at which they are destroyed by coalescence to form more massive objects, $\dot{N}_{\rm dest}$,

$$\dot{N}_{\rm PS} = \dot{N}_{\rm form} - \dot{N}_{\rm dest}. \tag{11.8}$$

For masses greater than M^*, the rate of formation of objects is well represented by the derivative of the Press-Schechter function but at smaller masses $\dot{N}_{\rm PS}$ becomes negative reflecting the fact that the rate of loss of objects is greater than their rate of formation. $\dot{N}_{\rm form}$ must, however, be positive for all masses. The self-similar nature of hierarchical clustering means that there must be no preferred mass-scale for mergers. Consequently, the formation and destruction rates have the same power-law slope as $\dot{N}_{\rm PS}$ at low masses. There is no unique solution to the problem of deriving $\dot{N}_{\rm form}$ from the Press-Schechter function because it depends upon assumptions about which low mass objects are likely to coalesce to form more massive objects.

We have therefore made numerical and analytic simulations of the formation rate $\dot{N}_{\rm form}$ in the Press-Schechter formalism. These simulations require additional information about the mass spectrum of objects undergoing mergers and are thus model dependent. It turns out, however, that the results are remarkably similar for reasonable models. We find that our simulated merger rate can be very well represented by a function of the form

$$\dot{N}_{\rm form}(M,t) = \dot{N}_{\rm PS} + \frac{\phi N_{\rm PS}}{t} \exp\left[(1-\alpha)\left(\frac{M}{M^*}\right)^{\gamma}\right], \tag{11.9}$$

where α and ϕ are constants. The first term on the right-hand side is the time derivative of the Press-Schechter function. The second term ensures that the formation rate of objects is correct at masses $M \leq M^*$. It has the same power-law slope as $\dot{N}_{\rm PS}$ and $N_{\rm PS}$ at low masses and the constant ϕ defines the formation rate. The exponential term containing the parameter α is included to steepen the cutoff of the Press-Schechter function near M^* to ensure that $\dot{N}_{\rm form}$ tends to $\dot{N}_{\rm PS}$ at higher masses, while leaving the low-mass power law slope unchanged. The parameter α is always greater than 1 but cannot be very much greater than 1 or else the cut-off of the formation rate would occur at too low masses. In other words, the second term on the right-hand side of expression (11.9) is constrained to be of power-law form with a cut-off roughly at M^*. In practice, the parameters α and ϕ are chosen to set $\dot{N}_{\rm form}$ equal to the values found in our simulations at $M = M^*$ and $M \ll M^*$ respectively. Our approximation to $\dot{N}_{\rm form}$ is accurate to within 5 per cent and is generally very much better than this value. Typical values for these parameters are $1.3 \leq \phi \leq 1.7$ and $1.31 \leq \alpha \leq 1.37$.

Assuming that a fraction x of the mass in a merger is converted into heavy elements and that an energy $0.007xMc^2$ is liberated in the far-infrared waveband when a mass xM is converted into heavy elements, we find that the luminosity density at redshift z, $\rho_{\rm L}(z)$, is

$$\rho_{\rm L}(z) = 0.007xc^2 \int M\dot{N}_{\rm form}\,{\rm d}M. \tag{11.10}$$

Inserting expression (11.9) for $\dot{N}_{\rm form}$, we obtain

$$\rho_{\mathrm{L}}(z) = 0.007xc^2 \left\{ \int M\dot{N}_{\mathrm{PS}}\mathrm{d}M + \int \frac{\phi N_{\mathrm{PS}} M}{t} \exp\left[(1-\alpha)\left(\frac{M}{M^*}\right)^{\gamma}\right] \mathrm{d}M \right\}. \tag{11.11}$$

Because of mass conservation, the first integral is zero and the expression becomes

$$\begin{aligned} \rho_{\mathrm{L}}(z) &= 0.007\bar{\rho}c^2\beta t^{-1} \\ &= 0.007\bar{\rho}c^2\beta\frac{3H_0}{2}(1+z)^{3/2}. \end{aligned} \tag{11.12}$$

where $\beta = \phi x/\sqrt{\alpha}$. Expression (11.12) is the result we have been seeking. The comoving luminosity density changes with redshift as $(1+z)^{3/2}$; it is independent of γ and depends only upon ϕ, x and α. Now substituting into expression (11.5) for the background emission, we find

$$\begin{aligned} I_\nu &= 0.007\bar{\rho}c^3\frac{3}{8\pi}\beta\int_0^{z_{\max}} \frac{\delta[\nu(1+z)-\nu_0]}{(1+z)}\,\mathrm{d}z \\ &= 0.007\bar{\rho}c^3\frac{3}{8\pi}\frac{\beta}{\nu_0}, \end{aligned} \tag{11.13}$$

that is, the observed background I_ν is independent of frequency in the frequency range $\nu_0 > \nu > \nu_0/(1+z_{\max})$ where ν_0 is the frequency at which the far-infrared emission is liberated in its rest frame and $z_{\max}$ is the redshift at which star formation begins.

Expression (11.6) for the density of heavy elements created by redshift z becomes

$$\Omega_{\mathrm{m}}(z) = \frac{3}{2}\beta\ln\left(\frac{1+z_{\max}}{1+z}\right). \tag{11.14}$$

The same mass of metals is produced in each logarithmic redshift interval. We assume that sufficient metal formation occurs near $z_{\max}$ to allow large scale dust absorption of starlight in coalescing galaxies.

The background intensity can be directly related to the total amount of heavy elements created

$$I_\nu = \frac{1}{4\pi}0.007\bar{\rho}c^2\frac{1}{\nu_0}\frac{\Omega_{\mathrm{m}}(0)}{\ln(1+z_{\max})}. \tag{11.15}$$

11.5 Results and Conclusions

As an example of how these results can be used to constrain models of galaxy and element formation in the hierarchical clustering scenario, we give the following specific examples for comparison with the COBE observations and the results of more detailed calculations. In Fig. 11.7, we show the predicted

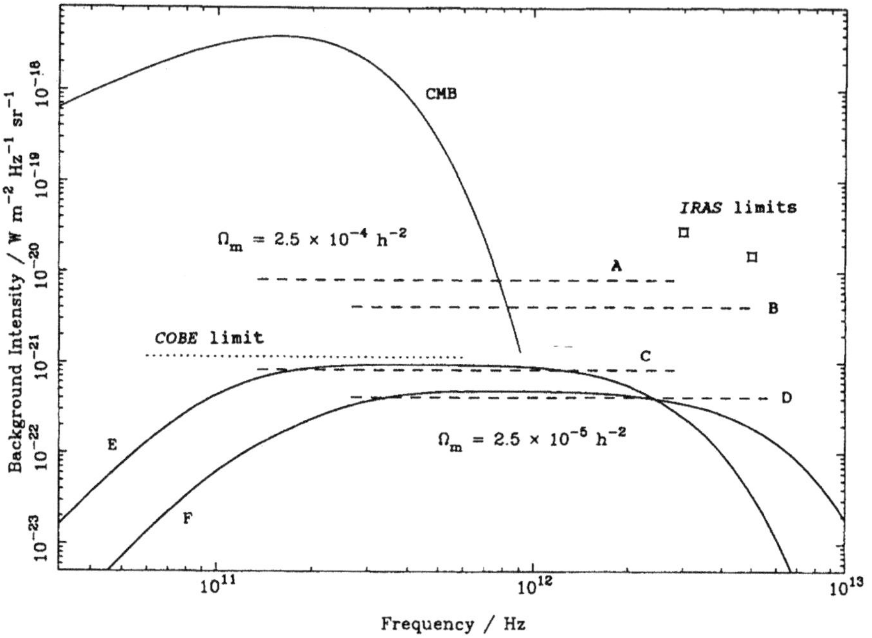

Fig. 11.7 The background radiation intensity predicted by our models. The cosmic microwave background and upper limits to the extragalactic background light from the IRAS satellite are also shown. Horizontal dashed lines show the flat spectrum predicted by expression (11.15). Results are presented for different dust temperatures, 30 and 60 K, and metal densities $\Omega_m(0) = 2.5 \times 10^{-4} h^{-2}$ and $2.5 \times 10^{-5} h^{-2}$. Lower backgrounds extending over a higher frequency range are found for the higher temperatures. The background intensity is proportional to the density of heavy elements produced. For $\Omega_m(0) = 2.5 \times 10^{-5} h^{-2}$, we show the results of our detailed calculations including the full dust spectra at temperatures of 30 and 60 K (Blain and Longair 1993b)

spectrum of the millimetre background for values of the metal density parameter $\Omega_m = 2.5 \times 10^{-4} h^{-2}$ and $2.5 \times 10^{-5} h^{-2}$ and dust temperatures of 30 and 60 K (dashed lines). z_{max} has been taken to be 20. Also shown are the spectrum of the cosmic microwave background radiation as well as the COBE limits for any deviation from a perfect black-body spectrum (dotted line - from Mather 1993). The predictions of our more detailed models for the case $\Omega_m = 2.5 \times 10^{-5} h^{-2}$ and dust temperatures of 30 and 60 K are shown as heavy solid lines. It can be seen that the remarkable precision with which the background spectrum is known to be of black-body form sets useful constaints upon the parameters of the models. Already, models in which $\Omega_m \sim 2.5 \times 10^{-5} h^{-2}$ are close to the limit of the *COBE* observations. We note that the temperature of the dust comes into thermal equilibrium with the cosmic background radiation at large redshifts and thus sets a lower limit to the dust temperature at these epochs.

According to our models, the density of metals today corresponds to $\Omega_{\rm m} \leq 2.5 \times 10^{-5}h^{-2}$. If this value is combined with the assumption that the metals comprise 1% of the baryonic mass of the Universe today, it is apparent that $\Omega_{\rm B} \leq 2.5 \times 10^{-3}h^{-2}$. Even if $h = 0.5$, this is a very low value for the baryonic mass density, $\Omega_{\rm B} \leq 0.01$ and consequently essentially all the dark matter would have to be in non-baryonic form.

The model can be simply adjusted for different assumptions — for example, Fig. 11.7 shows that, as the temperature of the dust changes from 30 K to 60 K, the flat spectrum is shifted to higher frequencies by a factor of two and reduced in intensity by the same factor, if the same amount of metals is created. From these simple rules, it is apparent how changing the star-formation rate or the temperature of the dust can produce features in the background spectrum. To a good approximation, a constant fractional star-formation rate at a fixed temperature produces a 'top-hat' spectrum and the predicted spectrum can be synthesised from such 'top-hats' for each redshift interval.

We are not claiming that this is how galaxies form but our formalism provides a simple baseline with which more realistic variants can be compared. The exciting prospect is that high precision measurements of the spectrum of the cosmic microwave background spectrum can provide direct evidence for the hierarchical clustering of structure back to large redshifts.

References – Chapter 11

Blain, A.W. and Longair, M.S. (1993a). MNRAS, **264**, 509.

Blain, A.W. and Longair, M.S. (1993b). MNRAS, **265**, L21.

Cowie, L.L. (1988). In *The Post-Recombination Universe*, (eds. N. Kaiser and A. Lasenby), 1. Dordrecht: Kluwer Academic Publishers.

Efstathiou, G. (1990). In *The Physics of the Early Universe* (eds. J.A. Peacock, A.F. Heavens and A.T. Davies), 361. Edinburgh: SUSSP Publications.

Hildebrand, R.H., (1983). QJRAS, **24**, 267.

Kauffmann, G. and White, S.D.M. (1993). MNRAS, **261**, 921.

Lilly, S.J. and Cowie, L.L. (1987). In *Infrared Astronomy with Arrays*, (eds C.G. Wynn-Williams and E.E. Becklin), 473. Honolulu: University of Hawaii Publications.

Mather, J.C. (1993). In *The Frontiers of Space and Ground-based Astronomy*, (eds. W. Wamsteker, M.S. Longair and Y. Kondo), (in press). Dordrecht: Kluwer Academic Publishers.

Meier, D.L. (1975). ApJ, **207**, 343.

Peacock, J.A. (1993). In *The Nature of Compact Objects in Active Galactic Nuclei*, (eds. A. Robinson and R.J. Terlevich R. J.), (in press). Cambridge: Cambridge University Press.

Press, W.H. and Schechter P. (1974). ApJ, **187**, 425.

Saunders, W., Rowan-Robinson, M., Lawrence, A., Efstathiou, G., Kaiser, N., Ellis R.S. and Frenk, C.S. (1990). MNRAS, **242**, 318.

Toller, G. In *The Galactic and Extragalactic Background Radiation*, (eds. S. Bowyer and C. Leinert), 21. Dordrecht: Kluwer Academic Publishers.
Weedman, D.W. (1994). In *The Physics of Active Galaxies - First Mount Stromlo Symposium*, (ed. G. Bicknell), (in press). Cambridge: Cambridge University Press.
White, S.D.M. (1989). In *The Epoch of Galaxy Formation*, (eds. C.S. Frenk, R.S. Ellis, T. Shanks, A.F. Heavens and J.A. Peacock), 15. Dordrecht: Kluwer Academic Publishers.

Printing: Mercedesdruck, Berlin
Binding: Buchbinderei Lüderitz & Bauer, Berlin

Index

GPSR Compliance

The European Union's (EU) General Product Safety Regulation (GPSR) is a set of rules that requires consumer products to be safe and our obligations to ensure this.

If you have any concerns about our products, you can contact us on ProductSafety@springernature.com

In case Publisher is established outside the EU, the EU authorized representative is:

Springer Nature Customer Service Center GmbH
Europaplatz 3
69115 Heidelberg, Germany

Batch number: 10844669

Printed by Printforce, the Netherlands